Semismooth Newton Methods for Variational Inequalities and Constrained Optimization Problems in Function Spaces

MOS-SIAM Series on Optimization

This series is published jointly by the Mathematical Optimization Society and the Society for Industrial and Applied Mathematics. It includes research monographs, books on applications, textbooks at all levels, and tutorials. Besides being of high scientific quality, books in the series must advance the understanding and practice of optimization. They must also be written clearly and at an appropriate level for the intended audience.

Series Volumes

Ulbrich, Michael, *Semismooth Newton Methods for Variational Inequalities and Constrained Optimization Problems in Function Spaces*

Biegler, Lorenz T., *Nonlinear Programming: Concepts, Algorithms, and Applications to Chemical Processes*

Shapiro, Alexander, Dentcheva, Darinka, and Ruszczyński, Andrzej, *Lectures on Stochastic Programming: Modeling and Theory*

Conn, Andrew R., Scheinberg, Katya, and Vicente, Luis N., *Introduction to Derivative-Free Optimization*

Ferris, Michael C., Mangasarian, Olvi L., and Wright, Stephen J., *Linear Programming with MATLAB*

Attouch, Hedy, Buttazzo, Giuseppe, and Michaille, Gérard, *Variational Analysis in Sobolev and BV Spaces: Applications to PDEs and Optimization*

Wallace, Stein W. and Ziemba, William T., editors, *Applications of Stochastic Programming*

Grötschel, Martin, editor, *The Sharpest Cut: The Impact of Manfred Padberg and His Work*

Renegar, James, *A Mathematical View of Interior-Point Methods in Convex Optimization*

Ben-Tal, Aharon and Nemirovski, Arkadi, *Lectures on Modern Convex Optimization: Analysis, Algorithms, and Engineering Applications*

Conn, Andrew R., Gould, Nicholas I. M., and Toint, Phillippe L., *Trust-Region Methods*

Semismooth Newton Methods for Variational Inequalities and Constrained Optimization Problems in Function Spaces

Michael Ulbrich
Technische Universität München
München, Germany

Society for Industrial and Applied Mathematics
Philadelphia

Mathematical Optimization Society
Philadelphia

10 9 8 7 6 5 4 3 2 1

Partial support of this work by the Deutsche Forschungsgemeinschaft is gratefully acknowledged.

Library of Congress Cataloging-in-Publication Data

Ulbrich, Michael.
Semismooth Newton methods for variational inequalities and constrained optimization problems in function spaces / Michael Ulbrich.
p. cm. – (MOS-SIAM series on optimization)
Includes bibliographical references and index.
ISBN 978-1-611970-68-5
1. Function spaces. 2. Constrained optimization. 3. Variational inequalities (Mathematics) 4. Newton-Raphson method. I. Title.
QA323.U43 2011
515'.64–dc22 2011005282

To my wife Jessica

and my brother Stefan.

Contents

Notation

General Notation

$\|\cdot\|_Y$	Norm of the Banach space Y.
$(\cdot,\cdot)_Y$	Inner product of the Hilbert space Y.
Y^*	Dual space of the Banach space Y.
$\langle\cdot,\cdot\rangle_{Y^*,Y}$	Dual pairing of the Banach space Y and its dual space Y^*.
$\langle\cdot,\cdot\rangle$	Dual pairing $\langle u,v\rangle = \int_\Omega u(\omega)v(\omega)d\omega$.
$\mathcal{L}(X,Y)$	Space of bounded linear operators $M : X \to Y$ from the Banach space X to the Banach space Y, equipped with the norm $\|\cdot\|_{X,Y}$.
$\|\cdot\|_{X,Y}$	Strong operator norm on $\mathcal{L}(X,Y)$, i.e., $\|M\|_{X,Y} = \sup\{\|Mx\|_Y : x \in X,\ \|x\|_X = 1\}$.
M^*	Adjoint operator of $M \in \mathcal{L}(X,Y)$, i.e., $M^* \in \mathcal{L}(Y^*,X^*)$ and $\langle Mx, y'\rangle_{Y,Y^*} = \langle x, M^*y'\rangle_{X,X^*}$ for all $x \in X$, $y' \in Y^*$.
B_Y	Open unit ball about 0 in the Banach space Y.
$\bar{B}_Y$	Closed unit ball about 0 in the Banach space Y.
B_p^n	Open unit ball about 0 in $(\mathbb{R}^n, \|\cdot\|_p)$.
$\bar{B}_p^n$	Closed unit ball about 0 in $(\mathbb{R}^n, \|\cdot\|_p)$.
$\partial\Omega$	Boundary of the domain Ω.
$\mathrm{cl}\ M$	Topological closure of the set M.
$\mathrm{co}\ M$	Convex hull of the set M.
$\overline{\mathrm{co}}\ M$	Closed convex hull of the set M.
$\mathrm{meas}(M)$	Lebesgue measure of the set M.
$\mathbf{1}_{\Omega'}$	Characteristic function of a measurable set $\Omega' \subset \Omega$, taking the value 1 on Ω' and 0 on its complement $\Omega \setminus \Omega'$.

Derivatives

F' Fréchet derivative of the operator $F : X \to Y$, i.e., $F'(x) \in \mathcal{L}(X,Y)$ and $\|F(x+s) - F(x) - F'(x)s\|_Y = o(\|s\|_X)$ as $\|s\|_X \to 0$.

F_x Partial Fréchet-derivative of the operator $F : X \times Y \to Z$ with respect to $x \in X$.

F'' Second Fréchet derivative.

F_{xy} Second partial Fréchet derivative.

$\partial_B f$ B-differential of the locally Lipschitz function $f : \mathbb{R}^n \to \mathbb{R}^m$.

∂f Clarke's generalized Jacobian of the locally Lipschitz continuous function $f : \mathbb{R}^n \to \mathbb{R}^m$.

$\partial_C f$ Qi's C-subdifferential of the locally Lipschitz function $f : \mathbb{R}^n \to \mathbb{R}^m$.

$\partial^* f$ Generalized differential of an operator $f : X \to Y$; see section 3.2.

$\partial^\circ \Psi$ Generalized differential of a superposition operator $\Psi(u) = \psi(G(u))$; see section 3.3.

Function Spaces

$L^p(\Omega)$ $p \in [1,\infty)$; Banach space of equivalence classes of Lebesgue measurable functions $u : \Omega \to \mathbb{R}$ such that $\|u\|_{L^p} \stackrel{\text{def}}{=} \left(\int_\Omega |u(x)|^p dx\right)^{1/p} < \infty$. $L^2(\Omega)$ is a Hilbert space with inner product $(u,v)_{L^2} = \int_\Omega u(x)v(x)dx$.

$L^\infty(\Omega)$ Banach space of equivalence classes of Lebesgue measurable functions $u : \Omega \to \mathbb{R}$ that are essentially bounded on Ω; i.e., $\|u\|_{L^\infty} \stackrel{\text{def}}{=} \operatorname{ess\,sup}_{x\in\Omega} |u(x)| < \infty$.

$C_c^\infty(\Omega)$ Space of infinitely differentiable functions $u : \Omega \to \mathbb{R}$, $\Omega \subset \mathbb{R}^n$ open, with compact support $\mathrm{cl}\{x : u(x) \neq 0\} \subset \Omega$.

$H^{k,p}(\Omega)$ $k \geq 0$, $p \in [1,\infty]$; Sobolev space of functions $u \in L^p(\Omega)$, $\Omega \subset \mathbb{R}^n$ open, such that $D^\alpha u \in L^p(\Omega)$ for all weak derivatives up to order k, i.e., for all $|\alpha| \leq k$. Here $D^\alpha = \frac{\partial^{\alpha_1}}{\partial x_1^{\alpha_1}} \cdots \frac{\partial^{\alpha_n}}{\partial x_n^{\alpha_n}}$ and $|\alpha| = \alpha_1 + \cdots + \alpha_n$. $H^{k,p}(\Omega)$ is a Banach space with norm $\|u\|_{H^{k,p}} = \left(\sum_{|\alpha|\leq k} \|D^\alpha u\|_{L^p}^p\right)^{1/p}$ and similarly for $p = \infty$.

$H^k(\Omega)$ $k \geq 0$; short notation for the Hilbert space $H^{k,2}(\Omega)$.

$H_0^k(\Omega)$ $k \geq 1$; closure of $C_c^\infty(\Omega)$ in $H^k(\Omega)$.

$H^{-k}(\Omega)$ $k \geq 1$; dual space of $H_0^k(\Omega)$ with respect to the distributional dual pairing.

Several vector-valued function spaces are introduced in section 10.2.

Preface

This book provides a comprehensive treatment of a very successful class of methods for solving optimization problems with PDE and inequality constraints as well as variational inequalities in function spaces. The approach combines the idea of nonsmooth pointwise reformulations of systems of inequalities with the concept of semismooth Newton methods. The book originates from the author's Habilitation thesis, in which the by then intensively investigated semismooth approach for finite-dimensional complementarity problems and variational inequalities was extended to and investigated in a function space setting. It was not predictable in 2000 that ten years later semismooth Newton methods would be one of the most important approaches for solving inequality constrained optimization problems in function spaces. The book develops this theory in detail; discusses recent progress, such as results on mesh independence, state constraints, and L^1-optimization; and shows applications ranging from obstacle problems to flow control. It is the author's hope that this book will be helpful for the future development of this exciting field.

The success of the semismooth approach in PDE constrained optimization and related fields was preceded by exciting research on semismooth Newton methods in finite dimensions and their application to complementarity problems. Mifflin's (1977) notion of semismoothness and the first papers on semismooth Newton methods, authored by Qi (1993), Qi and Sun (1993), and Pang and Qi (1993) formed important pillars for these developments. On the infinite-dimensional side, several abstract concepts for nonsmooth Newton methods in Banach spaces had been developed, e.g., by Kummer (1988, 1992), Robinson (1994), and Chen, Nashed, and Qi (2000). For transferring the full power of nonsmooth Newton methods to the function space setting, it was, however, crucial to investigate superposition operators with nonsmooth outer function, which occur when pointwise complementarity systems are reformulated as nonsmooth equations. This step was first done by the author (2001, 2002) and by Hintermüller, Ito, and Kunisch (2002). The latter paper also contains the important observation that the primal dual active set strategy can be interpreted as a semismooth Newton method. Since then, many contributions have been made to the field and the research is ongoing. Due to space limitations, I had to make a selection of topics that are presented in this book. I think that this choice is attractive and well suited for enabling the reader to follow the ongoing research in the field. Particular features of this book are

- rigorous development of the theory of semismooth Newton methods in a function space setting.
- mesh-independence results for semismooth Newton methods.
- regularizations and their rate of convergence for problems with state constraints.

- a globalization strategy based on a trust region framework.
- applications to elliptic optimal control, obstacle, and instationary flow control problems.

Acknowledgments

From the very beginnings of this book to its completion I received support from all sides. In the following, I would like to express my gratitude by mentioning several of those who, in one way or another, supported this book project. My Habilitation that formed the basis of this book was made possible by Klaus Ritter who offered me a PostDoc position and a scientific perspective at his chair. My visits to Rice University, especially those in 1999/2000 and in 2006, were very fruitful in developing essential material of this book. In particular, the results of Chapter 11 are strongly based on joint work with Matthias Heinkenschloss, Scott Collis, Kaveh Ghayour, and Stefan Ulbrich during my visit to Rice University in 1999/2000. I am thankful to John Dennis and to Matthias Heinkenschloss for their hospitality. John Dennis, Stephen Robinson, and Klaus Ritter were the reviewers of my Habilitation thesis and encouraged me with their positive feedback. My work on complementarity and nonsmooth Newton methods profited from fruitful discussions with Michael Ferris, Matthias Heinkenschloss, Michael Hintermüller, Christian Kanzow, Danny Ralph, Philippe Toint, Stefan Ulbrich, and others. Especially, I would like to thank my brother Stefan Ulbrich for a long and enjoyable collaboration. John Dennis and Danny Ralph helped me to get the opportunity to publish this book in the MOS-SIAM Series on Optimization, and Danny Ralph accompanied this long-lasting project with his continuous support and patience. I am thankful to Linda Thiel and Sara Murphy from SIAM for their help and kind cooperation. I also would like to acknowledge the consistently good working conditions at Technische Universität München. Furthermore, the work presented in this book was partially funded by the DFG.

My deepest thanks are reserved for my dear wife, Jessica, for her love and support. She always gave me encouragement and she generously accepted that I had to spend (too) many weekend days on writing the book rather than spending this time with her.

München, January 2011 Michael Ulbrich

Chapter 1

Introduction

A central theme of applied mathematics is the design of accurate mathematical models for a variety of technical, financial, medical, and many other applications, and the development of efficient numerical algorithms for their solution. Often, these models contain parameters that should be adjusted in an optimal way, either to maximize the accuracy of the model (parameter identification), or to control the simulated system in a desired way (optimal control). Since optimization with simulation constraints is more challenging than simulation alone (which already can be very involved), the development and analysis of efficient optimization methods is crucial for the viability of this approach. Besides the optimization of systems, minimization problems and variational inequalities have often arisen in the process of building mathematical models; this, e.g., applies to contact problems, free boundary problems, and elastoplastic problems [63, 78, 79, 137, 138, 165].

Most of the variational problems mentioned so far share the property that they are continuous in time and/or space, so that infinite-dimensional function spaces provide the appropriate setting for their analysis. Since essential information on the problem to solve is carried by the properties of the underlying infinite-dimensional spaces, the successful design of robust and mesh-independent optimization methods requires a thorough convergence analysis in this infinite-dimensional function space setting. The purpose of this work is to develop and analyze a class of Newton-type methods for the solution of optimization problems and variational inequalities that are posed in function spaces and contain *pointwise inequality constraints*.

1.1 An Engineering Example: Vortex Control

We illustrate the class of problems we are interested in by a concrete example, vortex control, that is relevant in engineering applications.

The effect of blade-vortex interaction (BVI) is a serious source of noise and of vibrations in turbines and in machines with rotors. The rotating blades generate vortices that interact with consecutive blades. The collision of blades and vortices generates, e.g., the typical helicopter noise.

The following optimal control problem, considered in more detail in Chapter 11, is a simplified model problem that was used in [44, 46, 45] to investigate the potential of control mechanisms for reducing the strength of vortices and thus also the effects of BVI (such as aeroacoustic noise). For ease of presentation and for computational tractability, the problem of controlling blade-vortex interaction is idealized as follows: We consider the motion of a pair of two counter-rotating viscous vortices in air above an infinite plane wall in two dimensions located at $x_2 = 0$. The fluid flow is modeled by the compressible Navier–Stokes equations, where the state of the system is represented by the density $\rho(t,x)$, the velocity field $(v_1(t,x), v_2(t,x))$, and the temperature $\theta(t,x)$ of the fluid (air). All these quantities are time (t) and space (x) dependent. The setup, described by the initial state $(\rho_0(x), v_{01}(x), v_{02}(x), \theta_0(x))$ of the fluid at time $t = 0$, is such that, due to the interaction of the two rotations, the vertices move towards the wall. Without a control mechanism, the vertices hit the wall and bounce back. Our control mechanism consists suction and blowing on a part $\Gamma_c \times \{0\}$ of the wall $\mathbb{R} \times \{0\}$, i.e., control of the normal velocity $u(t,x_1)$, $x_1 \in \Gamma_c$, of the fluid on $\Gamma_c \times \{0\}$, which is part of the wall. This is a control mechanism that usually is not directly implementable in practice, but there exist micro devices that can generate comparable effects. Our control objective is to reduce the final time kinetic energy of the fluid (in particular, the energy in the vortices) as much as possible. This is a simplified objective function compared to measuring noise, but in principle a more complicated objective function tailored to quantifying noise would be possible as well. It turns out that a regularization term that penalizes oscillations of the control needs to be added to the objective function in order to keep the control reasonably smooth. In addition, due to physical restrictions, it is not possible to generate arbitrarily strong suction and blowing. Hence, we pose pointwise lower and upper bounds on the control, i.e., $a(x_1) \le u(t,x_1) \le b(x_1)$ for all $x_1 \in \Gamma_c$.

The objective function is chosen as follows:

$$J(\rho, v_1, v_2, \theta, u) = \int_\Omega \left[\frac{\rho}{2}(v_1^2 + v_2^2)\right]_{t=T} dx + \frac{\alpha}{2}\|u\|_U^2.$$

The first term is the kinetic energy at the final time $t = T$, whereas the second term serves as a regularization for the control. The control is (t,x_1)-dependent and the control space U is equipped with a norm that penalizes oscillatory behavior, e.g., the H^1 Sobolev norm, which is built from the L^2-norm and the L^2-norm of the first derivative. Furthermore, $\alpha > 0$ denotes the regularization parameter and is usually chosen small.

The resulting problem has the following structure:

$$\underset{\rho, v_1. v_2, \theta, u}{\text{minimize}} \quad J(\rho, v_1. v_2, \theta, u)$$

subject to

- State Equation:
  The state (ρ, v_1, v_2, θ) satisfies
  - the compressible Navier–Stokes equations,
  - the wall boundary conditions corresponding to the boundary control u,
  - the initial conditions $(\rho(0,x), v_1(0,x), v_2(0,x), \theta(0,x)) = (\rho_0(x), v_1(x), v_2(x), \theta(x)) \ \ \forall\, x.$
- Control bounds: $a \le u \le b$.

We have not written down the detailed compressible Navier–Stokes equations nor the boundary conditions, since these would be quite lengthy.

1.2 Important Aspects of PDE-Constrained Optimization Problems

The above optimal flow control problem shows several important features of the class of optimization problems that we will consider in this book:

- The unknowns appearing in the optimization problem are functions, not finite-dimensional vectors. Therefore, the optimization problem is posed in infinite-dimensional function spaces. From an abstract point of view, we thus have to deal with an optimization problem in Banach spaces.
- The compressible Navier–Stokes equations, including initial and boundary conditions, appear as an equality constraint in the problem. This constraint can be interpreted as an operator equation in appropriately chosen function spaces.
- The problem contains pointwise inequality constraints.

Written in short notation, the above problem has the form

$$\min_{y,u} \ J(y,u) \quad \text{subject to} \quad E(y,u)=0, \ \ g(y,u)\in C, \tag{1.1}$$

where the state space Y and the control space U are function spaces, the state equation $E(y,u)=0$ is a (system of) PDE(s), and $g(y,u)\in C$ is an abstract inequality constraint. Here, g maps to a space of $\mathbb{R}^m$-valued functions that are defined on a measurable (i.e., open) set $\Omega\subset\mathbb{R}^n$. Furthermore, $C\subset\mathbb{R}^m$ is a closed convex set in which $g(y,u)$ has to lie pointwise almost everywhere (a.e.). In the constraint $g(y,u)\in C$, and throughout this work, relations between measurable functions are meant to hold pointwise almost everywhere on Ω in the Lebesgue sense.

The main aim of this work is to develop fast convergent Newton-type methods for solving problems of the form (1.1) and to provide a rigorous convergence analysis. Formally, the method that we will consider is very universally applicable. For rigorous convergence results, we need to assume additional structure. In particular, we require Fréchet-differentiability assumptions on J, E, and g. In addition, the inequality constraint has to be posed in a Lebesgue space; i.e.,

$$g: Y\times U \to Z = L^q(\Omega)^m, \quad q\in[1,\infty), \quad \Omega\subset\mathbb{R}^n.$$

Here, $L^q(\Omega)$ is the space of (equivalence classes of) Lebesgue integrable functions $v:\Omega\to\mathbb{R}$ such that

$$\|v\|_{L^q(\Omega)} \overset{\text{def}}{=} \left(\int_\Omega |v(\omega)|^q\, d\omega\right)^{1/q} < \infty.$$

Extensions of these requirements, in particular to the case of state constraints, are topics of ongoing research, which in part will also be sketched in this book.

A crucial role is played by the first-order necessary optimality conditions. Denote by W the image space of the operator E. To focus on the essentials, let us consider the case $m=1$ and $C=(-\infty,0]$. We then can write the inequality constraint in the form

$$g(y,u)\le 0 \quad \text{a.e. on } \Omega.$$

Under a suitable constraint qualification (which is a condition that the constraints have to satisfy at the solution) the following first-order optimality conditions (Karush–Kuhn–Tucker (KKT) conditions) hold:

There exist a Lagrange multiplier $w \in W^*$ for the state equation $E(y,u) = 0$ and a Lagrange multiplier $z \in Z^* = L^{q'}(\Omega)$, $1/q + 1/q' = 1$, for the inequality constraint $g(y,u) \le 0$ such that

$$J_y(y,u) + E_y(y,u)^* w + g_y(y,u)^* z = 0, \tag{1.2}$$

$$J_u(y,u) + E_u(y,u)^* w + g_u(y,u)^* z = 0, \tag{1.3}$$

$$E(y,u) = 0, \tag{1.4}$$

$$g(y,u) \le 0, \quad z \ge 0, \quad g(y,u)z = 0 \quad \text{a.e. on } \Omega. \tag{1.5}$$

For readers with a finite-dimensional optimization background, we briefly compare this with the corresponding situation in finite-dimensional optimization as it would arise, e.g., by finite element discretization of the optimal control problem. To distinguish between the function space setting and the finite-dimensional setting, we use boldface notation for the latter:

$$\min_{\mathbf{y},\mathbf{u}} \mathbf{J}(\mathbf{y},\mathbf{u}) \quad \text{subject to} \quad \mathbf{E}(\mathbf{y},\mathbf{u}) = 0, \quad \mathbf{g}(\mathbf{y},\mathbf{u}) \le 0.$$

Here, $\mathbf{y} \in \mathbb{R}^m$ and $\mathbf{u} \in \mathbb{R}^l$ are the discrete state and control, respectively, and $\mathbf{J} : \mathbb{R}^m \times \mathbb{R}^l \to \mathbb{R}$, $\mathbf{E} : \mathbb{R}^m \times \mathbb{R}^l \to \mathbb{R}^m$, $\mathbf{g} : \mathbb{R}^m \times \mathbb{R}^l \to \mathbb{R}^r$.

For this problem, under a constraint qualification, the first-order optimality conditions read

$$\nabla_{\mathbf{y}} \mathbf{J}(\mathbf{y},\mathbf{u}) + \mathbf{E}_{\mathbf{y}}(\mathbf{y},\mathbf{u})^T \mathbf{w} + \mathbf{g}_{\mathbf{y}}(\mathbf{y},\mathbf{u})^T \mathbf{z} = 0, \tag{1.6}$$

$$\nabla_{\mathbf{u}} \mathbf{J}(\mathbf{y},\mathbf{u}) + \mathbf{E}_{\mathbf{u}}(\mathbf{y},\mathbf{u})^T \mathbf{w} + \mathbf{g}_{\mathbf{u}}(\mathbf{y},\mathbf{u})^T \mathbf{z} = 0, \tag{1.7}$$

$$\mathbf{E}(\mathbf{y},\mathbf{u}) = 0, \tag{1.8}$$

$$\mathbf{g}_i(\mathbf{y},\mathbf{u}) \le 0, \quad \mathbf{z}_i \ge 0, \quad \mathbf{g}_i(\mathbf{y},\mathbf{u})\mathbf{z}_i = 0, \quad i = 1,\dots,r. \tag{1.9}$$

We make a couple of observations and remarks:

- As usual, we have identified the finite-dimensional spaces with $\mathbb{R}^d$, d as appropriate.
- Pointwise (a.e.) assertions in function space correspond to componentwise assertions in $\mathbb{R}^d$.
- Pointwise inequalities in the function space setting correspond to componentwise inequalities in the finite-dimensional framework.
- A very common discretization technique is to use continuous piecewise linear functions or discontinuous piecewise constant functions on a finite element triangulation. This means that the vector $\mathbf{w}$ corresponding to the discretization of the function w can be interpreted as a function itself. In fact, if the discretization is continuous piecewise linear then $\mathbf{w}$ carries the nodal values of the finite element function. If the discretization is piecewise constant, then $\mathbf{w}$ carries the values of the finite element function on the elements (e.g., triangles). It is then obvious in both cases that the condition $\mathbf{w} \ge 0$ is equivalent to the finite element function being nonnegative.
- In the finite-dimensional problem, the distinction between a space and its dual space is usually not clearly made. This is because the dual space of $\mathbb{R}^d$ (i.e., the space of all continuous linear forms on $\mathbb{R}^d$) can be conveniently represented by $\mathbb{R}^d$ itself. In fact, every continuous linear form $\ell : \mathbb{R}^d \to \mathbb{R}$ can be uniquely written as $\ell(x) = \sum_{i=1}^d y_i x_i = y^T x$ with appropriate $y \in \mathbb{R}^d$.

- The Jacobian matrix is the matrix representation of the Fréchet derivative w.r.t. the standard basis $e_j = (\delta_{ij})_{1 \le i \le d}$ of $\mathbb{R}^d$.
- A further word on dual spaces and dual pairings might be helpful. Let W_1, W_2 be Banach spaces with dual spaces W_1^*, W_2^*, and let $\langle \cdot,\cdot \rangle_{W_i^*,W_i}$ be the corresponding dual pairings. Then the dual operator $A^* : W_2^* \to W_1^*$ of the linear continuous operator $A : W_1 \to W_2$ is defined by

$$\langle A^* w_2^*, w_1 \rangle_{W_1^*,W_1} = \langle w_2^*, A w_1 \rangle_{W_2^*,W_2} \quad \forall\, w_1 \in W_1,\ w_2^* \in W_2^*.$$

In the case $W_1 = \mathbb{R}^{d_1}$, $W_2 = \mathbb{R}^{d_2}$, the usual choice is $W_i^* = W_i$, $\langle w_i^*, w_i \rangle_{W_i^*,W_i} = \sum_{j=1}^{d_i} w_{ij}^* w_{ij} = (w_i^*)^T w_i$, where $w_i = (w_{i1},\dots,w_{id_i})^T$ and $w_i^* = (w_{i1}^*,\dots,w_{id_i}^*)^T$. If in this situation we denote by $\mathbf{A}$ the matrix representation of the continuous linear operator $A : W_1 \to W_2$ w.r.t. the canonical bases of W_1 and W_2, then $\mathbf{A}^T$ is the matrix representation of A^*. In fact,

$$\begin{aligned}\langle A^* w_2^*, w_1 \rangle_{W_1^*,W_1} &= \langle w_2^*, A w_1 \rangle_{W_2^*,W_2} = (w_2^*)^T (\mathbf{A} w_1) = (\mathbf{A}^T w_2^*)^T w_1 \\ &= \langle \mathbf{A}^T w_2^*, w_1 \rangle_{W_1^*,W_1}.\end{aligned}$$

If we would choose other dual pairings, e.g., $\langle w_i^*, w_i \rangle_{W_i^*,W_i} = (w_i^*)^T \mathbf{M}_i w_i$ with invertible matrices $\mathbf{M}_i \in \mathbb{R}^{d_i \times d_i}$, then A^* would be represented by the matrix $\mathbf{M}_1^{-T} \mathbf{A}^T \mathbf{M}_2^T$. This follows from

$$\begin{aligned}\langle A^* w_2^*, w_1 \rangle_{W_1^*,W_1} &= \langle w_2^*, A w_1 \rangle_{W_2^*,W_2} = (w_2^*)^T \mathbf{M}_2 (A w_1) = (w_2^*)^T \mathbf{M}_2 (\mathbf{A} w_1) \\ &= (\mathbf{M}_1^{-T} \mathbf{A}^T \mathbf{M}_2^T w_2^*)^T \mathbf{M}_1 w_1 = \langle \mathbf{M}_1^{-T} \mathbf{A}^T \mathbf{M}_2^T w_2^*, w_1 \rangle_{W_1^*,W_1}.\end{aligned}$$

Based on these comments on the connections between the Banach space setting and the finite-dimensional case, the similarities between (1.2)–(1.5) and (1.6)–(1.9) should be apparent.

1.3 Nonsmooth Reformulation of Complementarity Conditions

In this book, the following approach will be systematically investigated and applied: The complementarity condition

$$x_1 \ge 0, \quad x_2 \ge 0, \quad x_1 x_2 = 0 \tag{1.10}$$

with $x = (x_1, x_2) \in \mathbb{R}^2$ is equivalently reformulated as

$$\phi(x_1, x_2) = 0, \tag{1.11}$$

where the function $\phi : \mathbb{R}^2 \to \mathbb{R}$ is an *NCP-function*, i.e., a function for which (1.10) and (1.11) are equivalent. One possible choice for ϕ is $\phi(x_1,x_2) = \min\{x_1,x_2\}$. This function has the typical properties of state-of-the-art NCP-functions: it is Lipschitz continuous, but not everywhere differentiable.

By using this reformulation pointwise in (1.5), it is then obvious that the complementarity condition (1.5) can be written as

$$\phi(-g(y,u),z) = 0 \quad \text{a.e. on } \Omega, \tag{1.12}$$

where the left-hand side is meant to be a function $\Omega \to \mathbb{R}$,

$$\phi(-g(y,u),z)(\omega) = \phi(-g(y,u)(\omega), z(\omega)), \quad \omega \in \Omega.$$

Similarly, in the discretized problem, we can use this reformulation componentwise instead of pointwise to rewrite (1.9) equivalently as

$$\phi(-\mathbf{g}_i(\mathbf{y},\mathbf{u}),\mathbf{z}_i) = 0, \quad 1 \le i \le r. \tag{1.13}$$

For illustration, we further specialize and consider $U = L^p(\Omega)$ and the control constraint

$$u \le b$$

with $b \in L^p(\Omega)$. In the above terminology, this corresponds to the choice

$$g(y,u) = u - b.$$

The complementarity condition (1.5) then reads

$$u \le b, \quad z \ge 0, \quad (u-b)z = 0 \quad \text{a.e. on } \Omega,$$

and the reformulated complementarity condition (1.12) has the form

$$\phi(b-u,z) = 0 \quad \text{a.e. on } \Omega.$$

Returning to the more general setting $g(y,u) \le 0$, the optimality system (1.2)–(1.5) can therefore be rewritten as the system of operator equations (1.2)–(1.4), (1.12). The algorithms developed and investigated in this book are Newton-type methods that are targeted at this kind of equation. The analytically difficult part is the reformulated complementarity conditions, since, as already said, ϕ is in general nonsmooth and thus also the operator equation (1.12) is nonsmooth. Therefore, it will be essential to analyze superposition operators of the form $u \mapsto \phi(u)$ with nonsmooth outer functions. We will carry out such an analysis in the framework of L^p-spaces. The main concept that will be developed for the analysis of a nonsmooth version of the Newton method is the notion of *semismoothness*.

1.4 Nonsmooth Reformulation of Variational Inequalities

There is a very close relationship between complementarity conditions and variational inequalities which will be used throughout this book. In fact, the reformulation approach introduced before just used the fact that variational inequalities can be reformulated as nonsmooth equations. We demonstrate this for the following case.

Bound-Constrained Variational Inequality Problem (VIP)

Find $u \in L^2(\Omega)$ such that

$$\begin{aligned} &u \in \mathcal{B} \overset{\text{def}}{=} \{v \in L^2(\Omega) : a \le v|_{\Omega_a},\ v|_{\Omega_b} \le b\}, \\ &(F(u), v-u)_{L^2} \ge 0 \quad \forall\, v \in \mathcal{B}. \end{aligned} \tag{1.14}$$

Here, $(u,v)_{L^2} = \int_\Omega u(\omega)v(\omega)d\omega$, and $F : L^2(\Omega) \to L^2(\Omega)$ is a continuous linear or nonlinear operator, where $L^2(\Omega)$ is the usual Lebesgue space of square integrable functions on the bounded Lebesgue measurable set $\Omega \subset \mathbb{R}^n$. We assume that Ω has positive Lebesgue measure such that $0 < \text{meas}(\Omega) < \infty$. The lower and upper bounds satisfy $a \in L^p(\Omega_a)$ and $b \in L^p(\Omega_b)$ with $p \geq 2$. Furthermore, $\Omega_a, \Omega_b \subset \Omega$ are measurable with $a \leq b$ on $\Omega_a \cap \Omega_b$. The case of unilateral bounds corresponds to choosing $\Omega_a = \emptyset$ or $\Omega_b = \emptyset$.

In many situations, the VIP (1.14) describes the first-order necessary optimality conditions of a bound-constrained minimization problem of the form

$$\underset{u \in L^2(\Omega)}{\text{minimize}} \quad j(u) \quad \text{subject to} \quad u \in \mathcal{B}. \tag{1.15}$$

In this case, F is the Fréchet derivative $j' : L^2(\Omega) \to L^2(\Omega)$ of the objective functional $j : L^2(\Omega) \to \mathbb{R}$.

The connection to complementarity conditions is most apparent for the unilateral case $\Omega_a = \Omega$, $\Omega_b = \emptyset$ with lower bound $a \equiv 0$. The resulting problem is an NCP:

$$u \in L^2(\Omega), \quad u \geq 0, \qquad (F(u), v - u)_{L^2} \geq 0 \quad \forall\, v \in L^2(\Omega),\ v \geq 0. \tag{1.16}$$

Then, as we will see, and as might already be obvious to the reader, (1.16) is equivalent to the pointwise complementarity system

$$u \geq 0, \quad F(u) \geq 0, \quad uF(u) = 0 \quad \text{on } \Omega. \tag{1.17}$$

This is a pointwise complementarity condition as in (1.5) and thus, using an NCP-function, can be reformulated equivalently as a nonsmooth operator equation. In fact, if $\phi : \mathbb{R}^2 \to \mathbb{R}$ is an NCP-function, i.e., satisfies

$$\phi(x) = 0 \quad \iff \quad x_1, x_2 \geq 0,\ x_1 x_2 = 0,$$

then, using the same trick as in (1.12), the NCP (1.16) is equivalent to

$$\Phi(u) = 0, \quad \text{where} \quad \Phi(u) = \phi\big(u(\omega), F(u)(\omega)\big), \quad \omega \in \Omega. \tag{1.18}$$

We now address the more general problem class (1.14), and start by showing that (1.14) can be reformulated equivalently as the following system of pointwise inequalities: $u \in L^2(\Omega)$ satisfies

$$\text{(i) } a \leq u \leq b, \quad \text{(ii) } (u-a)F(u) \leq 0, \quad \text{(iii) } (u-b)F(u) \leq 0 \quad \text{on } \Omega. \tag{1.19}$$

Here, for compact notation, we have set $a|_{\Omega \setminus \Omega_a} \equiv -\infty$, and $b|_{\Omega\setminus\Omega_b} \equiv +\infty$. Furthermore, on $\Omega \setminus \Omega_a$, condition (ii) has to be interpreted as $F(u) \leq 0$, and on $\Omega \setminus \Omega_b$ condition (iii) means $F(u) \geq 0$. Note that in the case of an NCP, i.e., $a \equiv 0$ and $b \equiv +\infty$, $\Omega_a = \Omega$, $\Omega_b = \emptyset$, the conditions (1.19) reduce to $u \geq 0$, $uF(u) \leq 0$, $F(u) \geq 0$, which is equivalent to the NCP (1.17).

We now verify the equivalence of (1.14) and (1.19). In fact, if u is a solution of (1.14) then (i) holds. Now assume that (ii) is violated on a set Ω' of positive measure. Then $(u-a)F(u) > 0$ on $\Omega' \cap \Omega_a$ and $F(u) > 0$ on $\Omega' \setminus \Omega_a$. We define $v \in \mathcal{B}$ by $v = a$ on $\Omega' \cap \Omega_a$, $v = u - 1$ on $\Omega' \setminus \Omega_a$, and $v = u$ on $\Omega \setminus \Omega'$, and obtain the contradiction

$$(F(u), v-u)_{L^2} = -\int_{\Omega' \cap \Omega_a} F(u)(u-a)d\omega - \int_{\Omega' \setminus \Omega_a} F(u)d\omega < 0.$$

In the same way, (iii) can be shown to hold.

Conversely, if $u \in L^2(\Omega)$ solves (1.19), then (i)–(iii) imply that Ω is the union of the disjoint sets

$$\begin{aligned}
\Omega_= &= \{\omega : a(\omega) < u(\omega) < b(\omega),\ F(u)(\omega) = 0\},\\
\Omega_\geq &= \{\omega : u(\omega) = a(\omega) \neq b(\omega),\ F(u)(\omega) \geq 0\},\\
\Omega_\leq &= \{\omega : u(\omega) = b(\omega) \neq a(\omega),\ F(u)(\omega) \leq 0\},\\
\Omega_f &= \{\omega : u(\omega) = a(\omega) = b(\omega)\}.
\end{aligned}$$

Now, for arbitrary $v \in \mathcal{B}$, we have $v|_{\Omega_f} = u|_{\Omega_f}$ and thus

$$(F(u), v-u)_{L^2} = \int_{\Omega_\geq} F(u)(v-a)d\omega + \int_{\Omega_\leq} F(u)(v-b)d\omega \geq 0,$$

so that u solves (1.14).

1.4.1 Finite-Dimensional Variational Inequalities

In finite dimensions, the NCP and, more generally, the box-constrained variational inequality problem (which is also called the mixed complementarity problem (MCP)) have been extensively investigated and there exists a significant, still growing body of literature on numerical algorithms for their solution; see below. Although we consider finite-dimensional problems throughout this subsection, we will work with the same notations as in the function space setting (a, b, u, F, etc.), since there is no danger of ambiguity. In analogy to (1.17), the finite-dimensional MCP consists in finding $u \in \mathbb{R}^m$ such that

$$a_i \leq u_i \leq b_i, \quad (u_i - a_i)F_i(u) \leq 0, \quad (u_i - b_i)F_i(u) \leq 0, \quad i = 1,\dots,m, \tag{1.20}$$

where $a \in (\mathbb{R} \cup \{-\infty\})^m$, $b \in (\mathbb{R} \cup \{+\infty\})^m$, and $F : \mathbb{R}^m \to \mathbb{R}^m$ are given. In the case $a_i = -\infty$ the second condition is to be understood in the sense $F_i(u) \leq 0$. Similarly, in the case $b_i = +\infty$ the third condition means $F_i(u) \geq 0$.

We begin with an early approach by Eaves [64] who observed (in the more general framework of VIPs on closed convex sets) that (1.20) can be equivalently written in the form

$$u - P_{[a,b]}(u - F(u)) = 0, \tag{1.21}$$

where $P_{[a,b]}(u) = \max\{a, \min\{u, b\}\}$ (componentwise) is the Euclidean projection onto $[a,b] = \prod_{i=1}^m [a_i, b_i]$. Here, in the case $a_i = -\infty$, $[a_i, b_i]$ stands for $(-\infty, b_i]$, etc. Note that if the function F is C^k, then the left-hand side of (1.21) is piecewise C^k and thus, as we will see, *semismooth*. Semismoothness is a central concept in this book that will be introduced and analyzed in detail.

The reformulation (1.21) can be embedded in a more general framework. To this end, we interpret (1.20) as a system of m conditions of the form

$$\alpha \leq x_1 \leq \beta, \quad (x_1 - \alpha)x_2 \leq 0, \quad (x_1 - \beta)x_2 \leq 0, \tag{1.22}$$

which have to be fulfilled by $x = (u_i, F_i(u))$ for $[\alpha, \beta] = [a_i, b_i]$, $i = 1,\dots,m$. Given any function $\phi_{[\alpha,\beta]} : \mathbb{R}^2 \to \mathbb{R}$ with the property

$$\phi_{[\alpha,\beta]}(x) = 0 \iff (1.22)\ \text{holds}, \tag{1.23}$$

we can write (1.20) equivalently as

$$\phi_{[a_i,b_i]}(u_i, F_i(u)) = 0, \quad i = 1,\dots,m. \tag{1.24}$$

A function with the property (1.23) is called an *MCP-function* for the interval $[\alpha,\beta]$ (also the name BVIP-function is used, where BVIP stands for box-constrained variational inequality problem). The link between (1.21) and (1.24) consists in the fact that the function $\phi_{[\alpha,\beta]}:\mathbb{R}^2\to\mathbb{R}^2$,

$$\phi^E_{[\alpha,\beta]}(x) = x_1 - P_{[\alpha,\beta]}(x_1 - x_2) \quad \text{with} \quad P_{[\alpha,\beta]}(t) = \max\{\alpha, \min\{t,\beta\}\} \tag{1.25}$$

defines an MCP-function for the interval $[\alpha,\beta]$. Also, since, with arbitrary $\sigma > 0$, the condition (1.22) is equivalent to (1.22) with x_2 replaced by σx_2, we can make the following conclusion. If $\phi_{[\alpha,\beta]}$ is an MCP-function for the interval $[\alpha,\beta]$, then also $x\mapsto\phi_{[\alpha,\beta]}(x_1,\sigma x_2)$ is an MCP-function for the interval $[\alpha,\beta]$. Furthermore, if $\phi_{[0,1]}$ is an MCP-function for the interval $[0,1]$, then for arbitrary finite bounds $-\infty<\alpha<\beta<+\infty$, the function $x\mapsto\phi_{[0,1]}(\frac{x_1-\alpha}{\beta-\alpha}, x_2)$ is an MCP-function for the interval $[\alpha,\beta]$.

The canonical MCP-function for the infinite intervals $[\alpha,\infty)$, $(-\infty,\beta]$, and $\mathbb{R}$ with $\alpha,\beta\in\mathbb{R}$, are

$$\phi_{[\alpha,\infty)}(x) = \phi(x_1-\alpha, x_2), \quad \phi_{(-\infty,\beta]}(x) = -\phi(\beta - x_1, -x_2), \quad \phi_{\mathbb{R}}(x) = x_2,$$

where $\phi:\mathbb{R}^2\to\mathbb{R}$ is an *NCP-function*, i.e., an MCP-function for the interval $[0,\infty)$. According to (1.23), $\phi:\mathbb{R}^2\to\mathbb{R}$ is an NCP-function if and only if

$$\phi(x) = 0 \quad\iff\quad x_1, x_2 \ge 0, \quad x_1x_2 = 0. \tag{1.26}$$

The corresponding reformulation of the NCP then is

$$\Phi(u) \stackrel{\text{def}}{=} \begin{pmatrix} \phi(u_1, F_1(u)) \\ \vdots \\ \phi(u_m, F_m(u)) \end{pmatrix} = 0, \tag{1.27}$$

and the NCP-function $\phi^E_{[0,\infty)}$ can be written in the form

$$\phi^E(x) = \phi^E_{[0,\infty)}(x) = \min\{x_1, x_2\}.$$

A further important reformulation, which is due to Robinson [175], uses the normal map

$$F_{[a,b]}(z) := F(P_{[a,b]}(z)) + z - P_{[a,b]}(z).$$

It is not difficult to see that every solution z of the normal map equation

$$F_{[a,b]}(z) = 0 \tag{1.28}$$

gives rise to a solution $u = P_{[a,b]}(z)$ of (1.20), and, conversely, that for any solution u of (1.28), the vector $z = u - F(u)$ solves (1.28). Therefore, the MCP (1.20) and the normal equation (1.28) are equivalent. Again, the normal map is piecewise C^k if F is C^k. In contrast to the reformulation based on NCP- and MCP-functions, the normal map approach evaluates F only at feasible points, which can be advantageous in certain situations.

Many modern algorithms for finite-dimensional NCPs and MCPs are based on reformulations by means of the Fischer–Burmeister NCP-function

$$\phi^{FB}(x) = x_1 + x_2 - \sqrt{x_1^2 + x_2^2}, \tag{1.29}$$

which was introduced by Fischer [71]. This function is Lipschitz continuous and 1-order semismooth on $\mathbb{R}^2$ (the definition of semismoothness is given below, and in more detail in Chapter 2). Further, ϕ^{FB} is C^∞ on $\mathbb{R}^2 \setminus \{0\}$, and $(\phi^{FB})^2$ is continuously differentiable on $\mathbb{R}^2$. The latter property implies that if F is continuously differentiable, the function $\frac{1}{2}\Phi^{FB}(u)^T\Phi^{FB}(u)$ can serve as a continuously differentiable merit function for (1.27). It is also possible to obtain 1-order semismooth MCP-functions from the Fischer–Burmeister function; see [24, 70] and section 5.1.1.

The described reformulations were successfully used as the basis for the development of locally superlinearly convergent Newton-type methods for the solution of (mixed) NCPs [24, 52, 53, 60, 66, 68, 69, 70, 126, 127, 133, 163, 172, 190]. This is remarkable, since all these reformulations are nonsmooth systems of equations. However, the underlying functions are *semismooth*, a concept introduced by Mifflin [160] for real-valued functions on $\mathbb{R}^n$, and extended to mappings between finite-dimensional spaces by Qi [168] and Qi and Sun [170]. Here a function $f : \mathbb{R}^l \to \mathbb{R}^m$ is called semismooth at $x \in \mathbb{R}^l$ if it is Lipschitz continuous near x, directionally differentiable at x, and

$$\sup_{M \in \partial f(x+h)} \|f(x+h) - f(x) - Mh\| = o(\|h\|) \quad \text{as } h \to 0,$$

where the set-valued function

$$\partial f : \mathbb{R}^l \rightrightarrows \mathbb{R}^{m\times l},\ \partial f(x) = \mathrm{co}\{M \in \mathbb{R}^{m\times l} : x_k \to x,\ f \text{ is differentiable at } x_k, \text{ and } f'(x_k) \to M\}$$

denotes Clarke's generalized Jacobian ("co" is the convex hull). Details are given in Chapter 2. It can be shown that piecewise C^1-functions are semismooth; see section 2.5.3. Further, it is easy to prove that the Newton method (where, in the Newton equation, the Jacobian is replaced by an arbitrary element of ∂f) converges superlinearly in a neighborhood of a CD-regular ("CD" for Clarke-differential) solution x^*, i.e., a solution where all elements of $\partial f(x^*)$ are invertible. More details on semismoothness in finite dimensions can be found in Chapter 2.

It should be mentioned that continuously differentiable NCP-functions can also be constructed. In fact, already in the 1970s Mangasarian [154] had proved the equivalence of the NCP to a system of equations, which, in our terminology, he obtained by choosing the NCP-function

$$\phi^M(x) = \theta(|x_2 - x_1|) - \theta(x_2) - \theta(x_1),$$

where $\theta : \mathbb{R} \to \mathbb{R}$ is any strictly increasing function with $\theta(0) = 0$. The most straightforward choice perhaps is $\theta(t) = t$, which gives $\phi^M = -2\phi^E$. If, in addition, θ is C^1 with $\theta'(0) = 0$, then ϕ^M is C^1. This is, e.g., satisfied by $\theta(t) = t|t|$. Nevertheless, most modern approaches prefer nondifferentiable, semismooth reformulations. This has a good reason. In fact, consider (1.27) with a differentiable NCP-function. Then the Jacobian of Φ is given by

$$\Phi'(u) = \mathrm{diag}\big(\phi_{x_1}(u_i, F(u_i))\big) + \mathrm{diag}\big(\phi_{x_2}(u_i, F(u_i))\big)F'(u).$$

Now, since $\phi(t,0) = 0 = \phi(0,t)$ for all $t \geq 0$, we see that $\phi'(0,0) = 0$. Thus, if strict complementarity is violated for the ith component, i.e., if $u_i = 0 = F_i(u)$, then the ith row of $\Phi'(u)$ is zero, and thus the Newton method is not applicable if strict complementarity is violated at the solution. This can be avoided by using nonsmooth NCP-functions, because they can be constructed in such a way that every element of the generalized gradient $\partial\phi(x)$ is bounded away from zero at every point $x \in \mathbb{R}^2$. For the Fischer–Burmeister function, e.g., there holds $\phi^{FB'}(x) = (1,1) - x/\|x\|_2$ for all $x \neq 0$, and thus $\|g\|_2 \geq \sqrt{2} - 1$ for all $g \in \partial\phi^{FB}(x)$ and all $x \in \mathbb{R}^2$.

The development of nonsmooth Newton methods [143, 144, 176, 168, 170, 166], especially the unifying notion of semismoothness [168, 170], has led to considerable research on numerical methods for the solution of finite-dimensional VIPs that are based on semismooth reformulations [24, 52, 53, 66, 68, 69, 70, 126, 127, 133, 163, 190]. These investigations confirm that this approach admits an elegant and general theory (in particular, no strict complementarity assumption is required) and leads to very efficient numerical algorithms [70, 162, 163].

Related Approaches

The research on semismoothness-based methods is still in progress. Closely connected to semismooth approaches are Jacobian smoothing methods and continuation methods [39, 37, 132]. Here, a family of functions $(\phi_\mu)_{\mu \geq 0}$ is introduced such that ϕ_0 is a semismooth NCP- or MCP-function, ϕ_μ, $\mu > 0$, is smooth, and $\phi_\mu \to \phi_0$ in a suitable sense as $\mu \to 0$. These functions are used to derive a family of equations $\Phi_\mu(u) = 0$ in analogy to (1.27). In the continuation approach [37], a sequence (u_k) of approximate solutions corresponding to parameter values $\mu = \mu_k$ with $\mu_k \to 0$ is generated such that u_k converges to a solution of the equation $\Phi_0(u) = 0$. Steps are usually obtained by solving the smoothed Newton equation $\Phi'_{\mu_k}(u_k)s_k^c = -\Phi_{\mu_k}(u_k)$, yielding centering steps towards the central path $\{x : \Phi_\mu(x) = 0 \text{ for some } \mu > 0\}$, or by solving the Jacobian smoothing Newton equation $\Phi'_{\mu_k}(u_k)s_k = -\Phi_0(u_k)$, yielding fast steps towards the solution set of $\Phi_0(u) = 0$. The latter steps are also used as trial steps in the Jacobian smoothing methods [39, 132]. Since the limit operator Φ_0 is semismooth, the analysis of these methods relies heavily on the properties of $\partial\Phi_0$ and the semismoothness of Φ_0.

The smoothing approach is also used in the development of algorithms for mathematical programs with equilibrium constraints (MPECs) [67, 73, 128, 153, 184, 199]. In this difficult class of problems, an objective function $f(u,v)$ has to be minimized under the constraint $u \in \mathcal{S}(v)$, where $\mathcal{S}(v)$ is the solution set of a VIP that is parameterized by v. Under suitable conditions on this inner problem, $\mathcal{S}(v)$ can be characterized equivalently by its KKT conditions. These, however, when taken as constraints for the outer problem, violate any standard constraint qualification. Alternatively, the KKT conditions can be rewritten as a system of semismooth equations by means of an NCP-function. This, however, introduces the (mainly numerical) difficulty of nonsmooth constraints, which can be circumvented by replacing the NCP-function with a smoothing NCP-function and considering a sequence of solutions of the smoothed MPEC corresponding to $\mu = \mu_k$, $\mu_k \to 0$.

In conclusion, semismooth Newton methods are at the heart of many modern algorithms in finite-dimensional optimization, and hence should also be investigated in the framework of optimal control and infinite-dimensional VIPs. This is the goal of this book.

1.4.2 Infinite-Dimensional Variational Inequalities

A main concern of this work is to present important progress, which has been made since the end of the 1990s, in extending the concept of semismooth Newton methods to a class of nonsmooth operator equations that is sufficiently rich to cover appropriate reformulations of the infinite-dimensional VIP (1.14). This book is based on the author's Habilitation, in which such an extension was systematically developed for the first time [191] and which resulted in the papers [192, 193, 194]. Further important contributions in this direction were made, on an abstract level, by Kummer [143, 144] and by Chen, Nashed, and Qi [38]. Reformulations based on the min-NCP-function were considered in Hintermüller, Ito, and Kunisch [102], where it was also observed that the primal dual active set strategy, developed by Bergounioux, Ito, Kunisch, et al. [21, 20, 119, 120, 145] and closely related to a method proposed by Hoppe [113], can be interpreted as a special case of the semismooth Newton method in function space.

For extending the semismooth approach to variational inequalities in function spaces, in a first step we derive analogues of the reformulations in section 1.4.1, but now in the function space setting. We begin with the NCP (1.17). Replacing componentwise operations by pointwise (a.e.) operations, we can apply an NCP-function ϕ pointwise to the pair of functions $(u, F(u))$ to obtain the superposition operator

$$\Phi(u)(\omega) = \phi\big(u(\omega), F(u)(\omega)\big). \tag{1.30}$$

Under appropriate assumptions, this defines an operator $\Phi : L^2(\Omega) \to L^2(\Omega)$ that is semismooth as a mapping $L^p(\Omega) \to L^2(\Omega)$ for suitably chosen $p \in [2, \infty]$; see section 3.3.1. Obviously, (1.17) is equivalent to the nonsmooth operator equation

$$\Phi(u) = 0. \tag{1.31}$$

In the same way, the more general problem (1.14) can be converted into an equivalent nonsmooth equation. To this end, we use a Lipschitz continuous, semismooth NCP-function ϕ and a Lipschitz continuous, semismooth MCP-function $\phi_{[\alpha,\beta]}$, $-\infty < \alpha < \beta < +\infty$. Now, we define the operator $\Phi : L^2(\Omega) \to L^2(\Omega)$,

$$\Phi(u)(\omega) = \begin{cases} F(u)(\omega), & \omega \in \Omega \setminus (\Omega_a \cup \Omega_b), \\ \phi\big(u(\omega) - a(\omega), F(u)(\omega)\big), & \omega \in \Omega_a \setminus \Omega_b, \\ -\phi\big(b(\omega) - u(\omega), -F(u)(\omega)\big), & \omega \in \Omega_b \setminus \Omega_a, \\ \phi_{[a(\omega),b(\omega)]}\,(u(\omega), F(u)(\omega)), & \omega \in \Omega_a \cap \Omega_b. \end{cases} \tag{1.32}$$

Again, Φ is a superposition operator on the four different subsets of Ω distinguished in (1.32). Along the same line, the normal map approach can be generalized to the function space setting. We will concentrate on NCP-function-based reformulations and their generalizations.

This approach is applicable whenever it is possible to write the problem under consideration as an operator equation in which the underlying operator is obtained by the superposition $\Psi = \psi \circ G$ of a Lipschitz continuous and semismooth function ψ and a continuously Fréchet differentiable operator G with reasonable properties, which maps into a direct product of Lebesgue spaces. We will show that the results for finite-dimensional semismooth equations can be extended to superposition operators in function spaces. To this end, we first develop a general semismoothness concept for operators in Banach spaces and then use these results to analyze superlinearly convergent Newton methods for semismooth operator equations. Then we apply this theory to superposition operators in function spaces of the form $\Psi = \psi \circ G$. We work with a set-valued generalized differential $\partial^\circ \Psi$ that is motivated by Qi's finite-dimensional C-subdifferential [169]. The semismoothness result we establish is an estimate of the form

$$\sup_{M \in \partial^\circ \Psi(y+s)} \|\Psi(y+s) - \Psi(y) - Ms\|_{L^r} = o(\|s\|_Y) \quad \text{as} \quad \|s\|_Y \to 0.$$

We also prove semismoothness of order $\alpha > 0$, which means that the above estimate holds with $o(\|s\|_Y)$ replaced by $O(\|s\|_Y^{1+\alpha})$. This semismoothness result enables us to apply the class of semismooth Newton methods that we analyzed in the abstract setting. If applied to nonsmooth reformulations of VIPs, these methods can be regarded as infinite-dimensional analogues of finite-dimensional semismooth Newton methods for this class of problems. As a consequence, we can adjust to the function space setting many of the ideas that were developed for finite-dimensional VIPs in recent years.

This conceptually simple idea, which was developed in the 1990s for the numerical solution of finite-dimensional NCPs, led to very successful Newton-based algorithms for NCPs. We will develop and investigate a semismoothness concept that is applicable to the operators arising in (1.18) and that allows us to develop a class of Newton-type methods for the solution of (1.18).

1.5 Properties of Semismooth Newton Methods

The nonsmooth Newton methods that we will systematically investigate in this book have, like their finite-dimensional counterparts—the semismooth Newton methods—several remarkable properties:

(a) The methods are locally superlinearly convergent, and they converge with q-rate > 1 under slightly stronger assumptions.

(b) Although an inequality-constrained problem is solved, only one linear operator equation has to be solved per iteration. Thus, the cost per iteration is comparable to that of the Newton method for smooth operator equations. We remark that sequential quadratic programming (SQP) algorithms, which are very efficient in practice, require the solution of an inequality constrained quadratic program per iteration, which can be significantly more expensive. Thus, it is also attractive to combine SQP methods with the class of Newton methods we describe here, either by using the Newton method for solving subproblems, or by rewriting the complementarity conditions in the Kuhn–Tucker system as an operator equation.

(c) The convergence analysis does not require a strict complementarity condition to hold. Thus, we can prove fast convergence also for the case when the set $\{\omega : \bar{u}(\omega) = 0, F(\bar{u})(\omega) = 0\}$ has positive measure at the solution $\bar{u}$.

(d) The systems that have to be solved in each iteration are of the form

$$[d_1 \cdot I + d_2 \cdot F'(u)]s = -\Phi(u), \tag{1.33}$$

where $I : u \mapsto u$ is the identity and F' denotes the Fréchet derivative of F. Further, d_1, d_2 are suitably chosen nonnegative L^∞-functions and $d_i \cdot A$ stands for the operator $s \in L^2(\Omega) \mapsto d_i \cdot (As) \in L^2(\Omega)$. The functions d_i are chosen depending on u and satisfy $0 < \gamma_1 < d_1 + d_2 < \gamma_2$ on Ω uniformly in u. More precisely (with all required concepts thoroughly introduced later on), the pair of L^∞-functions (d_1, d_2) is a measurable selection of the measurable multifunction

$$\omega \in \Omega \mapsto \partial\phi\big(u(\omega), F(u)(\omega)\big),$$

where $\partial\phi$ is Clarke's generalized gradient of ϕ. As we will see, in typical applications the system (1.33) can be symmetrized and is not much harder to solve than a system involving only the operator $F'(u)$, which would arise for the unconstrained problem $F(u) = 0$. In particular, fast solvers like multigrid methods, preconditioned iterative solvers, etc., can be applied to solve (1.33).

(e) The method is not restricted to the problem class (1.14). Among the possible extensions we also investigate VIPs of the form (1.14), but with the feasible set $\mathcal{B}$ replaced by

$$\mathcal{C} = \{u \in L^p(\Omega)^m : u(\omega) \in C \text{ on } \Omega\}, \quad C \subset \mathbb{R}^m \text{ closed and convex.}$$

Furthermore, we will consider mixed problems, where $F(u)$ is replaced by $F(y,u)$ and where we have the additional operator equation $E(y,u) = 0$. In particular, such problems arise as the first-order necessary optimality conditions (KKT conditions) of optimization problems with optimal control structure

$$\text{minimize} \quad J(y,u) \quad \text{subject to} \quad E(y,u) = 0, \quad u \in \mathcal{C}.$$

(f) Various other extensions are possible. For instance, certain quasi-variational inequalities [16, 18], i.e., variational inequalities for which the feasible set depends on u (e.g., $a = A(u)$, $b = B(u)$), can be solved by our class of semismooth Newton methods.

1.6 Examples

For illustration, we continue by giving examples of two problem classes that fit in the above framework.

1.6.1 Optimal Control Problem with Control Constraints

In section 1.1 we considered an optimal flow control problem, which is a particular instance of an optimal control problem. We now discuss this class of problems in a general setting, restricting ourselves to control-constrained problems.

Let the *state space* Y (a Banach space), the *control space* U (a Banach space), and the set $U_{\text{ad}} \subset U$ of *admissible* or *feasible controls* be given. The *state* $y \in Y$ of the system under consideration is governed by the *state equation*

$$E(y,u) = 0, \tag{1.34}$$

where $E : Y \times U \to W$, and W denotes a Banach space. In our context, the state equation is usually given by the weak formulation of a partial differential equation (PDE), including all boundary conditions that are not already contained in the definition of Y. The optimal control problem consists of finding a control $\bar{u} \in U_{\text{ad}}$ and a corresponding state $\bar{y}$ such that the state equation $E(\bar{y}, \bar{u}) = 0$ is satisfied and $J(\bar{y}, \bar{u})$ is minimized under all pairs $(y, u) \in Y \times U_{\text{ad}}$ with $E(y, u) = 0$.

Thus, the control problem is given by

$$\underset{y \in Y, u \in U}{\text{minimize}} \quad J(y,u) \quad \text{subject to} \quad (1.34) \text{ and } u \in U_{\text{ad}}. \tag{1.35}$$

There are now two possibilities. Either we address (1.35) directly, with unknowns (y, u) and the state equation considered as an equality constraint, or we use the state equation to eliminate the state from the problem. In the present section, we decide to follow the second approach. To this end, we assume that for every control $u \in U_{\text{ad}}$, the state equation (1.34) possesses a unique solution $y = y(u) \in Y$. Then, the state equation can be used to express the state in terms of the control, $y = y(u)$, and to write the control problem in the equivalent reduced form

$$\text{minimize} \quad j(u) \quad \text{subject to} \quad u \in U_{\text{ad}}, \tag{1.36}$$

with the reduced objective function $j(u) \stackrel{\text{def}}{=} J(y(u), u)$. Since our approach is based on optimality conditions and Newton-type methods, efficient formulas for the derivatives of j are essential. Therefore, we discuss this issue in more detail.

By the implicit function theorem, the continuous differentiability of $y(u)$ in a neighborhood of $\hat{u}$ follows if E is continuously differentiable in a neighborhood of $(y(\hat{u}), \hat{u})$ and $E_y(y(\hat{u}), \hat{u})$ is continuously invertible. Further, if in addition J is continuously differentiable in a neighborhood of $(y(\hat{u}), \hat{u})$, then j is continuously differentiable in a neighborhood of $\hat{u}$. In the same way, differentiability of higher order can be ensured. For problem (1.36), the derivative $j'(u) \in U^*$ of j is given by

$$j'(u) = J_u(y,u) + y_u(u)^* J_y(y,u),$$

with $y = y(u)$. In fact,

$$\begin{aligned} \langle j'(u), v \rangle_{U^*,U} &= \langle J_u(y,u), v \rangle_{U^*,U} + \langle J_y(y,u), y_u(u)v \rangle_{Y^*,Y} \\ &= \langle J_u(y,u), v \rangle_{U^*,U} + \langle y_u(u)^* J_y(y,u), v \rangle_{U^*,U}. \end{aligned}$$

Alternatively, j' can be expressed via the following *adjoint representation*:

$$j'(u) = J_u(y,u) + E_u(y,u)^* w.$$

Here, the *adjoint state* $w = w(u) \in W^*$ is the solution of the *adjoint equation*

$$E_y(y,u)^* w = -J_y(y,u), \tag{1.37}$$

with $y = y(u)$. We give a brief derivation of this formula here, and refer to section A.1 in the appendix for more details. Adjoint-based expressions for the second derivative j'' are also available; see section A.1. To derive the adjoint representation, we start with differentiating the equation $E(y(u), u) = 0$ with respect to u. This gives

$$E_y(y,u) y_u(u) + E_u(y,u) = 0,$$

where $y = y(u)$. Therefore,

$$y_u(u) = -E_y(y,u)^{-1}E_u(y,u)$$

and thus $y_u(u)^* = -E_u(y,u)^*(E_y(y,u)^*)^{-1}$, where we have used $(AB)^* = B^*A^*$ and $(A^{-1})^* = (A^*)^{-1}$. Hence,

$$\begin{aligned} j'(u) &= J_u(y,u) + y_u(u)^* J_y(y,u) = J_u(y,u) - E_u(y,u)^*(E_y(y,u)^*)^{-1} J_y(y,u) \\ &= J_u(y,u) + E_u(y,u)^* w, \end{aligned}$$

where $w = w(u)$ solves the adjoint equation (1.37). The adjoint representation of j' is remarkable since only one state equation solve (which is needed for computing $j(u)$ anyway) and one adjoint equation solve are required to obtain $j'(u)$. If, as most of the time in this book, $E(y,u) = 0$ is a PDE, then the adjoint equation is a linear PDE of the same (or related) type.

Example 1.1 (elliptic optimal control problem). We now make the example more concrete and consider as the state equation a Poisson problem with distributed control on the right-hand side,

$$-\Delta y = u \quad \text{on } \Omega, \quad y = 0 \quad \text{on } \partial\Omega, \tag{1.38}$$

and an objective function of tracking type

$$J(y,u) = \frac{1}{2}\int_\Omega (y - y_d)^2 dx + \frac{\lambda}{2}\int_\Omega u^2 dx.$$

Here, $\Omega \subset \mathbb{R}^n$ is a nonempty and bounded open set with boundary $\partial\Omega$, $y_d \in L^2(\Omega)$ is a target state that we would like to achieve as well as possible by controlling u, and the second term is for the purpose of regularization (the parameter $\lambda > 0$ is typically very small, e.g., $\lambda = 10^{-3}$).

As usual, we will work with weak solutions and a weak (variational) form of the Poisson equation, which is given by

$$y \in H_0^1(\Omega), \quad \int_\Omega [\nabla y^T \nabla v - uv]\,dx = 0 \quad \forall\, v \in H_0^1(\Omega). \tag{1.39}$$

Here, the Sobolev space $H_0^1(\Omega)$ is the space of all L^2-functions v that vanish on $\partial\Omega$ and satisfy $\nabla v \in L^2(\Omega)^n$. More precisely, $H_0^1(\Omega)$ is the completion of $C_c^\infty(\Omega)$, the C^∞-functions with compact support $\operatorname{supp} v \subset \Omega$, with respect to the norm $\|v\|_{H^1(\Omega)} = (\|v\|_{L^2(\Omega)}^2 + \|\nabla v\|_{L^2(\Omega)^n}^2)^{1/2}$. The space $H_0^1(\Omega)$ is a closed subspace of the Sobolev space $H^1(\Omega)$, which is the set of all functions $v \in L^2(\Omega)$ such that $\|v\|_{H^1(\Omega)} < \infty$, where ∇v is the weak derivative of v.

An appropriate choice of spaces for (1.39) is then $Y = H_0^1(\Omega)$, $W = H^{-1}(\Omega) = H_0^1(\Omega)^* = Y^*$. For the control space we choose $U = L^2(\Omega)$ and for the feasible set we make the choice $U_{\text{ad}} = \mathcal{B}$ with $\mathcal{B}$ as defined in (1.14). For convenience, we extend the bounds a and b to Ω by setting $a|_{\Omega\setminus\Omega_a} \equiv -\infty$, $b|_{\Omega\setminus\Omega_b} \equiv +\infty$. The case of unilateral bounds corresponds to choosing $\Omega_a = \emptyset$, $a \equiv -\infty$ or $\Omega_b = \emptyset$, $b \equiv +\infty$.

The state equation is given by $E(y,u) = 0$ with $E : Y \times U \mapsto W$,

$$E(y,u) = Ay - Bu.$$

Here, $A \in \mathcal{L}(H_0^1(\Omega), H^{-1}(\Omega))$ and $B \in \mathcal{L}(L^2(\Omega), H^{-1}(\Omega))$ are defined by

$$\begin{aligned}
\langle Ay, v\rangle_{H^{-1},H_0^1} &= \int_\Omega \nabla y^T \nabla v \, dx && \forall\, y, v \in H_0^1(\Omega),\\
\langle Bu, v\rangle_{H^{-1},H_0^1} &= \int_\Omega uv \, dx && \forall\, u \in L^2(\Omega),\ v \in H_0^1(\Omega).
\end{aligned}$$

The control problem thus reads

$$\begin{aligned}
&\underset{y\in H_0^1(\Omega), u\in L^2(\Omega)}{\text{minimize}} \quad \frac{1}{2}\int_\Omega (y - y_d)^2 dx + \frac{\lambda}{2}\int_\Omega u^2 dx \\
&\text{subject to} \quad Ay - Bu = 0, \quad u \in \mathcal{B}.
\end{aligned} \tag{1.40}$$

The state equation has a unique solution operator $U \ni u \mapsto y(u) \in Y$ and the reduced problem has the form (1.15). We apply the adjoint calculus to derive a formula for $j'(u) \in U^* = L^2(\Omega) = U$.

There holds, for all $y, s \in Y = H_0^1(\Omega)$, $u, d \in U = L^2(\Omega)$, $v \in W^* = Y = H_0^1(\Omega)$,

$$\begin{aligned}
\langle J_y(y,u), s\rangle_{Y^*,Y} &= \langle y - y_d, s\rangle_{H^{-1},H_0^1},\\
(J_u(y,u), d)_U &= (\lambda u, d)_{L^2},\\
\langle E_y(y,u)s, v\rangle_{W,W^*} &= \langle As, v\rangle_{H^{-1},H_0^1} = \int_\Omega \nabla s^T \nabla v \, dx,\\
\langle E_u(y,u)d, v\rangle_{W,W^*} &= \langle Bd, v\rangle_{H^{-1},H_0^1} = -\int_\Omega dv \, dx.
\end{aligned}$$

Therefore, the adjoint state $w \in W^* = H_0^1(\Omega)$ is given by

$$\int_\Omega \nabla z^T \nabla w \, dx = -\int_\Omega (y - y_d) z \, dx \quad \forall\, z \in Y = H_0^1(\Omega), \tag{1.41}$$

where y solves (1.39). This is the variational form of the following elliptic PDE:

$$-\Delta w = -(y - y_d) \quad \text{on } \Omega, \quad w = 0 \quad \text{on } \partial\Omega. \tag{1.42}$$

The adjoint representation of the derivative of the reduced objective function $j(u) = J(y(u), u)$ is given by

$$\begin{aligned}
(j'(u), d)_U &= (J_u(y,u), d)_U + (E_u(y,u)^* w, d)_U\\
&= (J_u(y,u), d)_{L^2} + (E_u(y,u)d, w)_{H^{-1},H_0^1}\\
&= \int_\Omega \lambda u d \, dx - \int_\Omega dw \, dx,
\end{aligned}$$

where y solves (1.39) and w solves (1.41). Hence, the derivative of the reduced objective function j is

$$j'(u) = \lambda u - w. \qquad \blacksquare$$

The elliptic optimal control problem in Example 1.1 has the following properties that are common to many control problems and will be of use later on:

- The mapping $u \mapsto w(u)$ possesses a *smoothing property*. In fact, it is a smooth (in this simple example even affine linear and bounded) mapping from $U = L^2(\Omega)$ to $W = H_0^1(\Omega)$, which is continuously embedded in $L^q(\Omega)$ for appropriate $q > 2$. If the boundary of Ω is sufficiently smooth, elliptic regularity results even imply that the mapping $u \mapsto w(u)$ maps smoothly into $H_0^1(\Omega) \cap H^2(\Omega)$.

- The solution $\bar{u}$ enjoys the *additional regularity* property that it is contained in $L^p(\Omega) \hookrightarrow U$ (note that Ω is bounded) for appropriate $p \in (2, \infty]$ if the bounds satisfy $a|_{\Omega_a} \in L^p(\Omega_a)$, $b|_{\Omega_b} \in L^p(\Omega_b)$. In fact, let $p \in (2, \infty]$ be such that $H_0^1(\Omega) \hookrightarrow L^p(\Omega)$. Due to the convexity of the problem, the reduced optimal control problem $\min_{u \in U_{ad}} j(u)$ is equivalent to the VIP

$$\bar{u} \in U_{ad}, \quad (j'(\bar{u}), u - \bar{u})_{L^2(\Omega)} \geq 0 \quad \forall\, u \in U_{ad}.$$

  Since $U_{ad} = \mathcal{B}$, we have available our earlier observation that the VIP is equivalent to (1.19) with $F \equiv j'$, u replaced by $\bar{u}$, and the interpretation on $\{a = -\infty\}$ and $\{b = +\infty\}$ given after (1.19). From this, we see that $j'(\bar{u}) = \lambda \bar{u} - \bar{w}$ vanishes on $\Omega_0 := \{x \in \Omega : a(x) < \bar{u}(x) < b(x)\}$. Hence, using $\bar{w} \in H_0^1(\Omega) \hookrightarrow L^p(\Omega)$, we conclude $\bar{u}|_{\Omega_0} = \lambda^{-1} \bar{w}|_{\Omega_0} \in L^p(\Omega_0)$. On $\Omega_a \setminus \Omega_0$ we have $\bar{u} = a$, and on $\Omega_b \setminus \Omega_0$ there holds $\bar{u} = b$. This shows $\bar{u} \in L^p(\Omega)$.

As mentioned, due to convexity, the reduced optimal control problem can be written in the form (1.14) with $F = j'$, and it enjoys the following properties.

There exist $p, q \in (2, \infty]$ such that

- $F : L^2(\Omega) \to L^2(\Omega)$ is continuously differentiable (here even continuous affine linear).

- F has the form $F(u) = \lambda u + G(u)$, where $G : L^2(\Omega) \to L^q(\Omega)$ is locally Lipschitz continuous (here even continuous affine linear).

- The solution is contained in $L^p(\Omega)$.

This problem arises as special case in the class of semilinear elliptic control problems that we discuss in detail in section 9.1.

Example 1.2. The distributed control of the right-hand side in Example 1.1 can be replaced by a variety of other control mechanisms. One alternative is Neumann boundary control. To describe this briefly, let us assume that the boundary $\partial\Omega$ is sufficiently smooth with positive and finite Hausdorff measure. We consider the problem

$$\begin{aligned} &\underset{y \in H^1(\Omega), u \in L^2(\partial\Omega)}{\text{minimize}} && \frac{1}{2}\int_\Omega (y - y_d)^2 dx + \frac{\lambda}{2}\int_{\partial\Omega} u^2 dS \\ &\text{subject to} && -\Delta y + y = f \quad \text{on } \Omega, \quad \frac{\partial y}{\partial n} = u \quad \text{on } \partial\Omega, \quad u \in U_{\text{ad}}, \end{aligned} \tag{1.43}$$

where $U_{\text{ad}} \subset U = L^2(\partial\Omega)$, $f \in L^2(\Omega)$, and $\partial/\partial n$ denotes the outward normal derivative. Setting $Y = H^1(\Omega)$ and $W = Y^* = H^1(\Omega)^*$, the state equation in weak form reads

$$\begin{aligned} &\forall\, v \in W^* = H^1(\Omega): \\ &(\nabla y, \nabla v)_{L^2(\Omega)^2} + (y, v)_{L^2(\Omega)} = (f, v)_{L^2(\Omega)} + (u, v|_{\partial\Omega})_{L^2(\partial\Omega)}, \end{aligned}$$

where $y \in Y = H^1(\Omega)$. This can be written in the form $E(y,u) = 0$ with $E : H^1(\Omega) \times L^2(\partial\Omega) \to H^1(\Omega)^*$. A calculation similar to the above yields for the reduced objective function

$$j'(u) = \lambda u - w|_{\partial\Omega},$$

where the adjoint state $w = w(u) \in W^* = H^1(\Omega)$ is the solution of

$$\begin{aligned}&\forall\, z \in Y = H^1(\Omega):\\ &(\nabla z, \nabla w)_{L^2(\Omega)^2} + (z,w)_{L^2(\Omega)} = -(y - y_d, z)_{L^2(\Omega)}.\end{aligned}$$

This is the variational formulation of the following elliptic PDE:

$$-\Delta w + w = -(y - y_d) \quad \text{on } \Omega, \quad \frac{\partial w}{\partial n} = 0 \quad \text{on } \partial\Omega.$$

Using standard results on Neumann problems, we see that the mappings

$$u \in L^2(\partial\Omega) \mapsto y(u) \in H^1(\Omega) \mapsto w(u) \in H^1(\Omega)$$

are continuous affine linear, and thus

$$u \in L^2(\partial\Omega) \mapsto w(u)|_{\partial\Omega} \in H^{1/2}(\partial\Omega) \hookrightarrow L^q(\partial\Omega)$$

for appropriate $q > 2$. Therefore, we have a scenario comparable to the distributed control problem, but now posed on the boundary of Ω. ∎

1.6.2 Variational Inequalities

As further application, we discuss a variational inequality arising from obstacle problems. For $q \in [2,\infty)$, let $g \in H^{2,q}(\Omega)$ represent a (lower) obstacle located over the nonempty bounded open set $\Omega \subset \mathbb{R}^2$ with sufficiently smooth boundary; denote by $y \in H_0^1(\Omega)$ the position of a membrane, and by $f \in L^q(\Omega)$ external forces. For compatibility we assume $g \le 0$ on $\partial\Omega$. Then y solves the problem

$$\underset{y \in H_0^1(\Omega)}{\text{minimize}} \quad \frac{1}{2} a(y,y) - (f,y)_{L^2} \quad \text{subject to} \quad y \ge g, \tag{1.44}$$

where

$$a(y,z) = \sum_{i,j} a_{ij} \frac{\partial y}{\partial x_i} \frac{\partial z}{\partial x_j},$$

$a_{ij} = a_{ji} \in C^1(\bar{\Omega})$, and a is H_0^1-elliptic. Let $A \in \mathcal{L}(H_0^1, H^{-1})$ be the operator induced by a; i.e., $a(y,z) = \langle y, Az\rangle_{H_0^1, H^{-1}}$.

It can be shown, see section 9.2 and [29], that (1.44) possesses a unique solution $\bar{y} \in H_0^1(\Omega)$ and that, in addition, $\bar{y} \in H^{2,q}(\Omega)$. Using Fenchel–Rockafellar duality [65], an equivalent dual problem can be derived, which assumes the form

$$\underset{u \in L^2(\Omega)}{\text{maximize}} \quad -\frac{1}{2}(f + u, A^{-1}(f+u))_{L^2} + (g,u)_{L^2} \quad \text{subject to} \quad u \ge 0. \tag{1.45}$$

The dual problem admits a unique solution $\bar{u} \in L^2(\Omega)$, which in addition satisfies $\bar{u} \in L^q(\Omega)$. From the dual solution $\bar{u}$ we can recover the primal solution $\bar{y}$ via $\bar{y} = A^{-1}(f + \bar{u})$.

Obviously, the concave quadratic objective function in (1.45) is not L^2-coercive, which we compensate for by adding a regularization. This yields the objective function $-j_\lambda(u)$, where

$$j_\lambda(u) = \frac{1}{2}(f + u, A^{-1}(f + u))_{L^2} - (g, u)_{L^2} + \frac{\lambda}{2}\|u - u_d\|_{L^2}^2,$$

$\lambda > 0$ is a (small) parameter, and $u_d \in L^q(\Omega)$, $q \in [2, \infty)$, is chosen appropriately. We will show in section 9.2 that the solution $\bar{u}_\lambda$ of the regularized problem

$$\underset{u \in L^2(\Omega)}{\text{maximize}} \quad -j_\lambda(u) \quad \text{subject to} \quad u \geq 0 \tag{1.46}$$

lies in $L^q(\Omega)$ and satisfies $\|\bar{u}_\lambda - \bar{u}\|_{H^{-1}} = o(\lambda^{1/2})$, which implies $\|\bar{y}_\lambda - \bar{y}\|_{H_0^1} = o(\lambda^{1/2})$, where $\bar{y}_\lambda = A^{-1}(f + \bar{u}_\lambda)$.

Since j_λ is strictly convex, problem (1.46) can be written in the form (1.14) with $F = j_\lambda'$. We have

$$F(u) = \lambda u + A^{-1}(f + u) - g - \lambda u_d \stackrel{\text{def}}{=} \lambda u + G(u).$$

Using that $A \in \mathcal{L}(H_0^1, H^{-1})$ is a homeomorphism, and that $H_0^1(\Omega) \hookrightarrow L^p(\Omega)$ for all $p \in [1, \infty)$, we conclude that the operator G maps $L^2(\Omega)$ continuously affine linearly into $L^q(\Omega)$. Therefore, we see the following:

- $F : L^2(\Omega) \to L^2(\Omega)$ is continuously differentiable (here even continuous affine linear).
- F has the form $F(u) = \lambda u + G(u)$, where $G : L^2(\Omega) \to L^q(\Omega)$ is locally Lipschitz continuous (here even continuous affine linear).
- The solution is contained in $L^q(\Omega)$.

A detailed discussion of this problem including numerical results is given in section 9.2. In a similar way, obstacle problems on the boundary can be treated. Furthermore, time-dependent parabolic variational inequality problems can be reduced, by semidiscretization in time, to a sequence of elliptic variational inequality problems.

1.7 Organization

We now give an overview on the organization of this book.

In Chapter 2 we collect important results of finite-dimensional nonsmooth analysis. Several generalized differentials known from the literature (Clarke's generalized Jacobian, B-differential, and Qi's C-subdifferential) and their properties are considered. Furthermore, finite-dimensional semismoothness is discussed and semismooth Newton methods are introduced. Finally, we give important examples for semismooth functions, i.e., piecewise smooth functions, and discuss finite-dimensional generalizations of the semismoothness concept.

In the first part of Chapter 3 we establish semismoothness results for operator equations in Banach spaces. The definition is based on a set-valued generalized differential and requires an approximation condition to hold. Furthermore, semismoothness of higher order is introduced. It is shown that continuously differentiable operators are semismooth with respect to their Fréchet derivative, and that the sum, composition, and direct product of semismoothness operators is again semismooth. The semismoothness concept is used to develop a Newton method for semismooth operator equations that is superlinearly convergent (with q-order $1+\alpha$ in the case of α-order semismoothness). Several variants of this method are considered, including an inexact version that allows us to work with approximate generalized differentials in the Newton system, and a version that includes a projection in order to stay feasible with respect to a given closed convex set containing the solution.

In the second part of Chapter 3 this abstract semismoothness concept is applied to the concrete situation of operators obtained by superposition of a Lipschitz continuous semismooth function and a smooth operator mapping into a product of Lebesgue spaces. This class of operators is of significant practical importance as it contains reformulations of variational inequalities by means of semismooth NCP-, MCP-, and related functions. We first develop a suitable generalized differential that has simple structure and is closely related to the finite-dimensional C-subdifferential. Then we show that the considered superposition operators are semismooth with respect to this differential. We also develop results to establish semismoothness of higher order. The theory is illustrated by applications to the NCP. The semismoothness of superposition operators enables us, via nonsmooth reformulations, to develop superlinearly convergent Newton methods for the solution of the NCP (1.17), and, as we show in Chapter 5, for the solution of the VIP (1.14) and even more general problems. Finally, further properties of the generalized differential are considered.

In Chapter 4 we investigate two ingredients that are needed in the analysis of Chapter 3. In Chapter 3 it becomes apparent that in general a smoothing step is required to close a gap between two different L^p-norms. This necessity was already observed in similar contexts before semismooth Newton methods were systematically investigated; see, e.g., [135, 195]. In section 4.1 we describe a way in which smoothing steps can be constructed. The approach is based on an idea by Kelley and Sachs [135]. Furthermore, in section 4.2 we investigate a particular choice of the MCP-function that leads to reformulations for which no smoothing step is required. For this choice, semismooth Newton methods are identical to the primal-dual active set strategy, as was observed by Hintermüller, Ito, and Kunisch in [102]. The analysis of semismooth Newton methods in Chapter 3 relies on a regularity condition that ensures the uniform invertibility (between appropriate spaces) of the generalized differentials in a neighborhood of the solution. In section 4.3 we develop sufficient conditions for this regularity assumption.

In Chapter 5 we show how the developed concepts can be applied to solve more general problems than NCPs. In particular, we propose semismooth reformulations for bound-constrained VIPs and, more generally, for VIPs with pointwise convex constraints. These reformulations allow us to apply semismooth Newton methods for their solution. Furthermore, we discuss how semismooth Newton methods can be applied to solve mixed problems, i.e., systems of VIPs and smooth operator equations. We concentrate on mixed problems arising as the KKT conditions of constrained optimization problems with optimal control structure. A close relationship between reformulations based on the black-box approach, in which the reduced problem is considered, and reformulations based on the all-at-once approach, where the full KKT-system is considered, is established. We observe that the generalized differentials of the black-box reformulation appear as Schur complements in the generalized differentials of the all-at-once reformulation. This can be used to relate regularity conditions of both approaches. We also describe how smoothing steps can be computed.

Chapter 6 is devoted to the study of mesh-independence results of semismooth Newton methods for complementarity problems in L^p spaces. The mesh-independence theory for the classical Newton method cannot be directly extended to the semismooth case since the order of semismoothness is *not* stable with respect to perturbations of the evaluation point. Therefore, new techniques are needed to develop mesh-independence results for semismooth Newton methods. The first such result was proved by Hintermüller and Ulbrich [106]. The investigations in Chapter 6 develop mesh-independent order of semismoothness results and a corresponding mesh-independence theory for semismooth Newton methods that extends the available results significantly. In fact, while [106] proved mesh independence of any desired linear rate of convergence, we develop, in addition, a mesh-independent q-order of superlinear convergence. The results are illustrated by a semilinear elliptic optimal control problem.

In Chapter 7 we describe a way to make the developed class of semismooth Newton methods globally convergent by embedding them in a trust-region method. To this end, we propose three variants of minimization problems such that solutions of the semismooth operator equation are critical points of the minimization problem. Then we develop and analyze a class of nonmonotone trust-region methods for the resulting optimization problems in a general Hilbert space setting. The trial steps have to fulfill a model decrease condition which, as we show, can be implemented by means of a generalized fraction of Cauchy decrease condition. For this algorithm, global convergence results are established. Further, it is shown how semismooth Newton steps can be used to compute trial steps, and it is proved that, under appropriate conditions, eventually Newton steps are always taken. Therefore, the rate of local convergence to regular solutions is at least q-superlinear.

Chapter 8 is devoted to state-constrained optimal control and related problems. It investigates a class of penalization methods that includes the Moreau–Yosida regularization. The significant difficulty of state constraints and related problems is that pointwise inequality constraints are posed in a function space that is more regular than L^p, $p < \infty$; for instance, in a Sobolev space or in the space of continuous functions. The Lagrange multiplier corresponding to this constraint then lives in the dual space and thus is not a measurable function, but rather a measure. Therefore, the complementarity condition is not posed in a pointwise a.e. sense and thus cannot be rewritten by means of an NCP-function. In Chapter 8, an approach of regularizing the problem is considered such that smooth or semismooth reformulations of the optimality system are possible. Error estimates in terms of the regularization parameter are also derived and an interpretation of the approach in terms of dual regularization is given.

In Chapter 9 the developed algorithms are applied to concrete problems. Section 9.1 discusses in detail the applicability of semismooth Newton methods to a semilinear elliptic control problem with bounds on the control. Furthermore, a finite element discretization is discussed and it is shown that the application of finite-dimensional semismooth Newton methods to the discretized problem can be viewed as a discretization of the infinite-dimensional semismooth Newton method. Furthermore, it is discussed how multigrid methods can be used to solve the semismooth Newton system efficiently. The efficiency of the method is documented by various numerical tests. Both black-box and all-at-once approaches are tested. Furthermore, a nested iteration is proposed that first solves the problem approximately on a coarse grid to obtain a good initial point on the next finer grid and proceeds in this way until the finest grid is reached. As a second application we investigate the obstacle problem of section 1.6.2 in detail. An equivalent dual problem is derived, which is augmented by a regularization term to make it coercive. An error estimate for the regularized solution is established in terms of the regularization parameter. We then show that our class of semismooth Newton methods is applicable to the regularized dual problem. Numerical results for a finite element discretization are presented. In the implementation we again use multigrid methods to solve the semismooth Newton system. The chapter is concluded by a short section on the recently intensively investigated field of L^1-optimization.

In Chapter 10 we show that our class of semismooth Newton methods can be applied to solve control-constrained distributed optimal control problems governed by the incompressible Navier–Stokes equations. To this end, differentiability and local Lipschitz continuity properties of the control-to-state mapping are investigated. Furthermore, results for the adjoint equation are established that allow us to prove a smoothing property of the reduced gradient mapping. These results show that semismooth Newton methods can be applied to the flow control problem and that these methods converge superlinearly in a neighborhood of regular critical points. Numerical results are presented for the case of control of the right-hand side. As control constraints, pointwise bounds as well as pointwise ball constraints are considered. The discrete problem has about 74,000,000 state unknowns and about 3,300,000 control unknowns (500 time steps with about 148,700 state unknowns and about 66,000 control unknowns per time level).

In Chapter 11 we present applications of our method to the boundary control of the time-dependent compressible Navier–Stokes equations. As already described in section 1.1, we control the normal velocity of the fluid on part of the boundary (suction and blowing), subject to pointwise lower and upper bounds. As a control objective, the terminal kinetic energy is minimized. In the algorithm, the Hessian is approximated by BFGS matrices. This problem is quite large scale, with over 75,000 unknown controls and over 29,000,000 state variables (distributed over 600 time levels). The numerical results show that our approach is viable and efficient also for quite large scale, state-of-the-art control problems.

The appendix contains some useful supplementary material. In section A.1 we describe the adjoint-based gradient and Hessian representation for the reduced objective function of optimal control problems. Section A.2 collects several frequently used inequalities. In section A.3 we state elementary properties of multifunctions. Finally, in section A.4, the differentiability properties of Nemytskii operators are considered.

Chapter 2

Elements of Finite-Dimensional Nonsmooth Analysis

In this chapter we collect several results of finite-dimensional nonsmooth analysis that are required for our investigations. In particular, finite-dimensional semismoothness and semismooth Newton methods are considered. The concepts introduced in this section will serve as a motivation and guideline for the developments in subsequent sections.

All generalized differentials considered here are set-valued functions (or multifunctions). Basic properties of multifunctions, like upper semicontinuity, can be found in section A.3 of the appendix.

Throughout, we denote by $\|\cdot\|$ both arbitrary but fixed norms on the respective $\mathbb{R}^n$-spaces and induced matrix norms. The open unit ball $\{x \in \mathbb{R}^n : \|x\| < 1\}$ is denoted by B^n.

2.1 Generalized Differentials

On the nonempty open set $V \subset \mathbb{R}^n$, we consider the function

$$f : V \to \mathbb{R}^m$$

and denote by $D_f \subset V$ the set of all $x \in V$ at which f admits a (Fréchet-) derivative $f'(x) \in \mathbb{R}^{m\times n}$. Now suppose that f is Lipschitz continuous near $x \in V$, i.e., that there exists an open neighborhood $V(x) \subset V$ of x on which f is Lipschitz continuous. Then, according to Rademacher's theorem [207], $V(x) \setminus D_f$ has Lebesgue measure zero. Hence, the following constructions make sense.

Definition 2.1. [40, 166, 170] *Let $V \subset \mathbb{R}^n$ be open and $f : V \to \mathbb{R}^m$ be Lipschitz continuous near $x \in V$. The set*

$$\partial_B f(x) \stackrel{\text{def}}{=} \{M \in \mathbb{R}^{m\times n} : \exists (x_k) \subset D_f : \ x_k \to x, \ f'(x_k) \to M\}$$

is called B-subdifferential *("B" for Bouligand) of f at x. Moreover,* Clarke's generalized Jacobian *of f at x is the convex hull $\partial f(x) \stackrel{\text{def}}{=} \mathrm{co}(\partial_B f(x))$, and*

$$\partial_C f(x) \stackrel{\text{def}}{=} \partial f_1(x) \times \cdots \times \partial f_m(x)$$

denotes Qi's C-subdifferential.

The differentials $\partial_B f$, ∂f, and $\partial_C f$ have the following properties.

Proposition 2.2. *Let $V \subset \mathbb{R}^n$ be open and $f : V \to \mathbb{R}^m$ be locally Lipschitz continuous. Then for $x \in V$ the following hold:*

(a) *$\partial_B f(x)$ is nonempty and compact.*

(b) *$\partial f(x)$ and $\partial_C f(x)$ are nonempty, compact, and convex.*

(c) *The set-valued mappings $\partial_B f$, ∂f, and $\partial_C f$, respectively, are locally bounded and upper semicontinuous.*

(d) $\partial_B f(x) \subset \partial f(x) \subset \partial_C f(x)$.

(e) *If f is continuously differentiable in a neighborhood of x, then*

$$\partial_C f(x) = \partial f(x) = \partial_B f(x) = \{f'(x)\}.$$

Proof. The results for $\partial_B f(x)$ and $\partial f(x)$ as well as (d) are established in [40, Prop. 2.6.2]. Part (e) immediately follows from the definition of the respective differentials. The remaining assertions on $\partial_C f$ are immediate consequences of the properties of $\partial f_i(x)$. □

The following chain rule holds.

Proposition 2.3. [40, Cor. 2.6.6] *Let $V \subset \mathbb{R}^n$ and $W \subset \mathbb{R}^l$ be nonempty open sets, $g : V \to W$ be Lipschitz continuous near $x \in V$, and $h : W \to \mathbb{R}^m$ be Lipschitz continuous near $g(x)$. Then, $f = h \circ g$ is Lipschitz continuous near x and for all $v \in \mathbb{R}^n$, it holds that*

$$\partial f(x)v \subset \mathrm{co}(\partial h(g(x))\partial g(x)v) = \mathrm{co}\{M_h M_g v : M_h \in \partial h(g(x)),\ M_g \in \partial g(x)\}.$$

If, in addition, h is continuously differentiable near $g(x)$, then, for all $v \in \mathbb{R}^n$,

$$\partial f(x)v = h'(g(x))\partial g(x)v.$$

If f is real-valued (i.e., if $m = 1$), then in both chain rules the vector v can be omitted.

In particular, choosing $h(y) = e_i^T y = y_i$ and $g = f$, where e_i is the ith unit vector, we see the following.

Corollary 2.4. *Let $V \subset \mathbb{R}^n$ be open and $f : V \to \mathbb{R}^m$ be Lipschitz continuous near $x \in V$. Then*

$$\partial f_i(x) = e_i^T \partial f(x) = \{M_i : M_i \text{ is the } i\text{th row of some } M \in \partial f(x)\}.$$

2.2 Semismoothness

The notion of semismoothness was introduced by Mifflin [160] for real-valued functions defined on finite-dimensional spaces, and extended to mappings between finite-dimensional spaces by Qi [168] and Qi and Sun [170]. The importance of semismooth equations results from the fact that, although the underlying mapping is in general nonsmooth, the Newton method is still applicable and converges locally with q-superlinear rate to a regular solution.

Definition 2.5. [160, 166, 170] *Let $V \subset \mathbb{R}^n$ be nonempty and open. The function $f : V \to \mathbb{R}^m$ is* semismooth *at $x \in V$ if it is Lipschitz continuous near x and if the following limit exists for all $s \in \mathbb{R}^n$:*

$$\lim_{\substack{M \in \partial f(x+\tau d) \\ d \to s,\ \tau \to 0^+}} Md.$$

If f is semismooth at all $x \in V$, we call f semismooth (on V).

Note that we include the local Lipschitz condition in the definition of semismoothness. Hence, if f is semismooth at x, it is also Lipschitz continuous near x. Semismoothness admits different, yet equivalent, characterizations. To formulate them, we first recall directional and Bouligand- (or B-) differentiability.

Definition 2.6. *Let the function $f : V \to \mathbb{R}^m$ be defined on the open set V.*

(a) *f is* directionally differentiable *at $x \in V$ if the* directional derivative

$$f'(x,s) \stackrel{\text{def}}{=} \lim_{\tau \to 0^+} \frac{f(x+\tau s) - f(x)}{\tau}$$

exists for all $s \in \mathbb{R}^n$.

(b) *f is* B-differentiable *at $x \in V$ if f is directionally differentiable at x and*

$$\|f(x+s) - f(x) - f'(x,s)\| = o(\|s\|) \quad \textit{as } s \to 0.$$

(c) *f is α-order* B-differentiable *at $x \in V$, $0 < \alpha \leq 1$, if f is directionally differentiable at x and*

$$\|f(x+s) - f(x) - f'(x,s)\| = O(\|s\|^{1+\alpha}) \quad \textit{as } s \to 0.$$

Note that $f'(x,\cdot)$ is positive homogeneous. Furthermore, it is known that directional differentiability and B-differentiability are equivalent for locally Lipschitz continuous mappings between finite-dimensional spaces [182]. The following Proposition gives alternative definitions of semismoothness.

Proposition 2.7. *Let $f : V \to \mathbb{R}^m$ be defined on the open set $V \subset \mathbb{R}^n$. Then for $x \in V$ the following statements are equivalent:*

(a) *f is semismooth at x.*

(b) *f is Lipschitz continuous near x, $f'(x,\cdot)$ exists, and*

$$\sup_{M \in \partial f(x+s)} \|Ms - f'(x,s)\| = o(\|s\|) \quad \textit{as } s \to 0.$$

(c) *f is Lipschitz continuous near x, $f'(x,\cdot)$ exists, and*

$$\sup_{M\in\partial f(x+s)} \|f(x+s)-f(x)-Ms\| = o(\|s\|) \quad \textit{as } s\to 0. \tag{2.1}$$

Proof. Concerning the equivalence of (a) and (b), see [170, Thm. 2.3]. If f is Lipschitz continuous near x and directionally differentiable at x, then, as noted above, f is also B-differentiable at x. Hence, it is now easily seen that (b) and (c) are equivalent, since for all $M\in\partial f(x+s)$

$$\begin{aligned}&\big|\|f(x+s)-f(x)-Ms\| - \|Ms-f'(x,s)\|\big|\\ &\qquad\le \|f(x+s)-f(x)-f'(x,s)\| = o(\|s\|) \quad \text{as } s\to 0. \qquad \square\end{aligned}$$

The version (c) is especially well suited for the analysis of Newton-type methods. To give a first example of semismooth functions, we note the following immediate consequence of Proposition 2.7.

Proposition 2.8. *Let $V\subset\mathbb{R}^n$ be open. If $f:V\to\mathbb{R}^n$ is continuously differentiable in a neighborhood of $x\in V$, then f is semismooth at x and $\partial f(x)=\partial_B f(x)=\{f'(x)\}$.*

Further, the class of semismooth functions is closed under composition.

Proposition 2.9. [72, Lem. 18] *Let $V\subset\mathbb{R}^n$ and $W\subset\mathbb{R}^l$ be open sets. Let $g:V\to W$ be semismooth at $x\in V$ and $h:W\to\mathbb{R}^m$ be semismooth at $g(x)$ with $g(V)\subset W$. Then the composite map $f\stackrel{\text{def}}{=}h\circ g:V\to\mathbb{R}^m$ is semismooth at x. Moreover,*

$$f'(x,\cdot)=h'(g(x),g'(x,\cdot)).$$

It is natural to ask whether f is semismooth when its component functions are semismooth and vice versa. This is in fact true.

Proposition 2.10. *The function $f:V\to\mathbb{R}^m$, $V\subset\mathbb{R}^n$ open, is semismooth at $x\in V$ if and only if its component functions are semismooth at x.*

Proof. We use the characterization of semismoothness given in Proposition 2.7. If f is semismooth at x, then the functions f_i are Lipschitz continuous near x and directionally differentiable at x. Furthermore, by Corollary 2.4,

$$\begin{aligned}&\sup_{v\in\partial f_i(x+s)} |f_i(x+s)-f_i(x)-vs|\\ &\qquad= \sup_{M\in\partial f(x+s)} |e_i^T(f(x+s)-f(x)-Ms)| = o(\|s\|) \quad \text{as } s\to 0,\end{aligned}$$

which proves the semismoothness of f_i at x. The reverse direction is an immediate consequence of the inclusion $\partial f(x)\subset\partial_C f(x)$. $\square$

2.3 Semismooth Newton Method

We now analyze the following Newton-like method for the solution of the equation

$$f(x) = 0, \tag{2.2}$$

where $f : V \to \mathbb{R}^n$, $V \subset \mathbb{R}^n$ open, is semismooth at the solution $\bar{x} \in V$.

Algorithm 2.11 (semismooth Newton method).

0. Choose an initial point x_0 and set $k = 0$.
1. If $f(x_k) = 0$, then STOP.
2. Choose $M_k \in \partial f(x_k)$ and compute s_k from
$$M_k s_k = -f(x_k).$$
3. Set $x_{k+1} = x_k + s_k$, increment k by one, and go to step 1.

Under a regularity assumption on the matrices M_k, this iteration converges locally q-superlinearly.

Proposition 2.12. *Let $f : V \to \mathbb{R}^n$ be defined on the open set $V \subset \mathbb{R}^n$ and denote by $\bar{x} \in \mathbb{R}^n$ an isolated solution of* (2.2)*. Assume the following:*

(a) *Estimate* (2.1) *holds at $x = \bar{x}$ (which, in particular, is satisfied if f is semismooth at $\bar{x}$).*

(b) *One of the following conditions holds:*

  (i) *There exists a constant $C > 0$ such that, for all k, the matrices M_k are nonsingular with $\|M_k^{-1}\| \le C$.*

  (ii) *There exist constants $\eta > 0$ and $C > 0$ such that, for all $x \in \bar{x} + \eta B^n$, every $M \in \partial f(x)$ is nonsingular with $\|M^{-1}\| \le C$.*

  (iii) *The solution $\bar{x}$ is CD-regular ("CD" for Clarke-differential); i.e., every $M \in \partial f(\bar{x})$ is nonsingular with $\|M^{-1}\| \le C$.*

Then there exists $\delta > 0$ such that, for all $x_0 \in \bar{x} + \delta B^n$, (i) *holds and Algorithm* 2.11 *either terminates with $x_k = \bar{x}$ or generates a sequence (x_k) that converges q-superlinearly to $\bar{x}$.*

Various results of this type can be found in the literature [143, 144, 166, 168, 170]. In particular, Kummer [144] develops a general abstract framework of essentially two requirements (CA) and (CI), under which the Newton method is well defined and converges superlinearly. The condition (2.1) is a special case of the approximation condition (CA), whereas (CI) is a uniform injectivity condition, which, in our context, corresponds to assumption (b) (ii).

Since the proof of Proposition 2.12 is not difficult and quite helpful in getting familiar with the notion of semismoothness, we sketch it here.

Proof. First, we prove (iii) $\Longrightarrow$ (ii). Assume that (ii) does not hold. Then there exist sequences $x^i \to \bar{x}$ and $A_i \in \partial f(x^i)$ such that, for any i, either A_i is singular or $\|(A_i)^{-1}\| \ge i$.

Since ∂f is upper semicontinuous and compact-valued, we can select a subsequence such that $A_i \to A \in \partial f(\bar{x})$. Due to the properties of the matrices A_i, A cannot be invertible, and thus (iii) does not hold.

Further, observe that (ii) implies (i) whenever $x_k \in \bar{x} + \eta B^n$ for all k. Therefore, if one of the conditions in (b) holds, we have (i) at hand as long as $x_k \in \bar{x} + \delta B^n$ and $\delta > 0$ is sufficiently small. Denoting the error by $v_k = x_k - \bar{x}$ and using $M_k s_k = -f(x_k)$, $f(\bar{x}) = 0$, we obtain for such x_k

$$\begin{aligned} M_k v_{k+1} &= M_k(s_k + v_k) = -f(x_k) + M_k v_k \\ &= -[f(\bar{x} + v_k) - f(\bar{x}) - M_k v_k]. \end{aligned} \tag{2.3}$$

Invoking (2.1) yields

$$\|M_k v_{k+1}\| = o(\|v_k\|) \quad \text{as } \|v_k\| \to 0. \tag{2.4}$$

Hence, for sufficiently small $\delta > 0$, we have

$$\|M_k v_{k+1}\| \le \frac{1}{2C}\|v_k\|,$$

and thus by (i)

$$\|v_{k+1}\| \le \|M_k^{-1}\|\,\|M_k v_{k+1}\| \le \frac{1}{2}\|v_k\|.$$

This shows $x_{k+1} \in \bar{x} + (\delta/2)B^n$ and inductively $x_k \to \bar{x}$ (in the nontrivial case $x_k \ne \bar{x}$ for all k). Now we conclude from (2.4) that the rate of convergence is q-superlinear. □

2.4 Higher-Order Semismoothness

The rate of convergence of the semismooth Newton method can be improved if instead of (2.1) an estimate of higher order is available. This leads to the following definition of higher-order semismoothness, which can be interpreted as a semismooth relaxation of Hölder-continuous differentiability.

Definition 2.13. [170] *Let the function $f : V \to \mathbb{R}^m$ be defined on the open set $V \subset \mathbb{R}^n$. Then, for $0 < \alpha \le 1$, f is called* α-order semismooth *at $x \in V$ if f is locally Lipschitz continuous near x, $f'(x,\cdot)$ exists, and*

$$\sup_{M \in \partial f(x+s)} \|Ms - f'(x,s)\| = O(\|s\|^{1+\alpha}) \quad \text{as } s \to 0.$$

If f is α-order semismooth at all $x \in V$, we call f α-order semismooth (on V).

For α-order semismooth functions, a counterpart of Proposition 2.7 can be established.

Proposition 2.14. *Let $f : V \to \mathbb{R}^m$ be defined on the open set $V \subset \mathbb{R}^n$. Then for $x \in V$ and $0 < \alpha \le 1$ the following statements are equivalent:*

(a) *f is α-order semismooth at x.*

(b) *f is Lipschitz continuous near x, α-order B-differentiable at x, and*

$$\sup_{M\in\partial f(x+s)} \|f(x+s)-f(x)-Ms\| = O(\|s\|^{1+\alpha}) \quad \text{as } s\to 0. \tag{2.5}$$

Proof. According to results in [170], α-order semismoothness at x implies α-order B-differentiability at x. Now we can proceed as in the proof of Proposition 2.7. □

Of course, α-Hölder continuously differentiable functions are α-order semismooth. More precisely, we have the following.

Proposition 2.15. *Let $V\subset\mathbb{R}^n$ be open. If $f:V\to\mathbb{R}^m$ is differentiable in a neighborhood of $x\in V$ with α-Hölder continuous derivative, $0<\alpha\le 1$, then f is α-order semismooth at x and $\partial f(x)=\partial_B f(x)=\{f'(x)\}$.*

The class of α-order semismooth functions is closed under composition.

Proposition 2.16. [72, Thm. 21] *Let $V\subset\mathbb{R}^n$ and $W\subset\mathbb{R}^l$ be open sets and $0<\alpha\le 1$. Let $g:V\to W$ be α-order semismooth at $x\in V$ and $h:W\to\mathbb{R}^m$ be α-order semismooth at $g(x)$ with $g(V)\subset W$. Then the composite map $f\stackrel{\text{def}}{=}h\circ g:V\to\mathbb{R}^m$ is α-order semismooth at x. Moreover,*

$$f'(x,\cdot)=h'(g(x),g'(x,\cdot)).$$

Further, we obtain the following by a straightforward modification of the proof of Proposition 2.10.

Proposition 2.17. *Let $V\subset\mathbb{R}^n$ be open. The function $f:V\to\mathbb{R}^m$ is α-order semismooth at $x\in V$, $0<\alpha\le 1$, if and only if its component functions are α-order semismooth at x.*

Concerning the rate of convergence of Algorithm 2.11, the following holds.

Proposition 2.18. *Let the assumptions in Proposition 2.12 hold, but assume that instead of (2.1) the stronger condition (2.5), with $0<\alpha\le 1$, holds at the solution $\bar{x}$. Then there exists $\delta>0$ such that, for all $x_0\in\bar{x}+\delta B^n$, Algorithm 2.11 either terminates with $x_k=\bar{x}$ or generates a sequence (x_k) that converges to $\bar{x}$ with rate $1+\alpha$.*

Proof. In light of Proposition 2.12, we only have to establish the improved rate of convergence. But from $v_k\to 0$, (2.3), and (2.5) it follows immediately that

$$\|v_{k+1}\| = O(\|v_k\|^{1+\alpha}). \qquad \square$$

2.5 Examples of Semismooth Functions

2.5.1 The Euclidean Norm

The Euclidean norm $e: x\in\mathbb{R}^n\mapsto\|x\|_2=(x^Tx)^{1/2}$ is an important example of a 1-order semismooth function that arises, e.g., as the nonsmooth part of the Fischer–Burmeister function. Obviously, e is Lipschitz continuous on $\mathbb{R}^n$, and C^∞ on $\mathbb{R}^n \setminus \{0\}$ with

$$e'(x) = \frac{x^T}{\|x\|_2}.$$

Therefore,

$$\partial e(x) = \partial_B e(x) = \left\{ \frac{x^T}{\|x\|_2} \right\} \quad \text{for } x \neq 0,$$
$$\partial_B e(0) = \{v^T : v \in \mathbb{R}^n,\ \|v\|_2 = 1\}, \quad \text{and} \quad \partial e(0) = \{v^T : v \in \mathbb{R}^n,\ \|v\|_2 \le 1\}.$$

By Proposition 2.15, e is 1-order semismooth on $\mathbb{R}^n \setminus \{0\}$, since it is smooth there. On the other hand, for all $s \in \mathbb{R}^n \setminus \{0\}$ and $v \in \partial e(s)$ there holds $v = s^T/\|s\|_2$ and

$$e(s) - e(0) - vs = \|s\|_2 - \|s\|_2 = 0.$$

Hence, e is also 1-order semismooth at 0.

2.5.2 The Fischer–Burmeister Function

The Fischer–Burmeister function was already defined in (1.29):

$$\phi^{FB} : \mathbb{R}^2 \to \mathbb{R}, \quad \phi^{FB}(x) = x_1 + x_2 - \sqrt{x_1^2 + x_2^2}.$$

$\phi = \phi^{FB}$ is the difference of the linear function $f(x) = x_1 + x_2$ and the 1-order semismooth and Lipschitz continuous function $\|x\|_2$; see section 2.5.1. Therefore, ϕ is Lipschitz continuous and 1-order semismooth by Propositions 2.15 and 2.16. Further, from the definition of $\partial_B \phi$ and $\partial \phi$, it is immediately clear that

$$\partial_B \phi(x) = f'(x) - \partial_B \|x\|_2, \quad \partial \phi(x) = f'(x) - \partial \|x\|_2.$$

Hence, for $x \neq 0$,

$$\partial \phi(x) = \partial_B \phi(x) = \left\{ (1,1) - \frac{x^T}{\|x\|_2} \right\},$$

and

$$\partial_B \phi(0) = \{(1,1) - y^T : \|y\|_2 = 1\}, \quad \partial \phi(0) = \{(1,1) - y^T : \|y\|_2 \le 1\}.$$

From this one can see that for all $x \in \mathbb{R}^2$ and all $v \in \partial \phi^{FB}(x)$ there holds $v_1, v_2 \ge 0$, $2 - \sqrt{2} \le v_1 + v_2 \le 2 + \sqrt{2}$, showing that all generalized gradients are bounded above (a consequence of the global Lipschitz continuity) and are bounded away from zero.

2.5.3 Piecewise Differentiable Functions

Piecewise continuously differentiable functions are an important subclass of semismooth functions. We refer to Scholtes [181] for a thorough treatment of the topic, where the results of this section can be found. For the reader's convenience, we include selected proofs.

Definition 2.19. [181] *A function $f : V \to \mathbb{R}^m$ defined on the open set $V \subset \mathbb{R}^n$ is called* PC^k-function *("P" for piecewise), $1 \leq k \leq \infty$, if f is continuous and if at every point $x_0 \in V$ there exist a neighborhood $W \subset V$ of x_0 and a finite collection of C^k-functions $f^i : W \to \mathbb{R}^m$, $i = 1, \ldots, N$, such that*

$$f(x) \in \{f^1(x), \ldots, f^N(x)\} \quad \forall\, x \in W.$$

We say that f is a continuous selection *of $\{f^1, \ldots, f^N\}$ on W. The set*

$$I(x) = \{i : f(x) = f^i(x)\}$$

is the active index set *at $x \in W$, and*

$$I^e(x) = \{i \in I(x) : x \in cl(int\{y \in W : f(y) = f^i(y)\})\}$$

is the essentially active index set *at x.*

The following is obvious.

Proposition 2.20. *The class of PC^k-functions is closed under composition, finite summation, and multiplication (in case the respective operations make sense).*

Example 2.21. The functions $t \in \mathbb{R} \mapsto |t|$, $x \in \mathbb{R}^2 \mapsto \max\{x_1, x_2\}$, and $x \in \mathbb{R}^2 \mapsto \min\{x_1, x_2\}$ are PC$^\infty$-functions. As a consequence, the projection onto the interval $[\alpha, \beta]$, $P_{[\alpha,\beta]}(t) = \max\{\alpha, \min\{t, \beta\}\}$ is PC$^\infty$, and thus also the MCP-function $\phi^E_{[\alpha,\beta]}$ defined in (1.25). ■

Proposition 2.22. *Let the* PC^k*-function $f : V \to \mathbb{R}^m$ be a continuous selection of the C^k-functions $\{f^1, \ldots, f^N\}$ on the open set $V \subset \mathbb{R}^n$. Then, for $x \in V$, there exists a neighborhood W of x on which f is also a continuous selection of $\{f^i : i \in I^e(x)\}$.*

Proof. Assume the contrary. Then the open sets

$$V_r = \{y \in V : \|y - x\| < 1/r,\ f(y) \neq f^i(y)\ \forall\, i \in I^e(x)\}$$

are nonempty for all $r \in \mathbb{N}$. Let $\{i_1, \ldots, i_q\}$ enumerate the set $\{1, \ldots, N\} \setminus I^e(x)$. Set $V_r^0 = V_r$ and, for $l = 1, \ldots, q$, generate the open sets

$$V_r^l = V_r^{l-1} \cap \{y \in V : f(y) \neq f^{i_l}(y)\}.$$

Since for all $y \in V$ there exists $i \in I^e(x) \cup \{i_1, \ldots, i_q\}$ with $f(y) = f^i(y)$, we see that $V_r^q = \emptyset$. Hence, there exists a maximal l_r with $V_r^{l_r} \neq \emptyset$. With $j_r = i_{l_r+1}$ we have

$$\emptyset \neq V_r^{l_r} \subset \{y \in V : f(y) = f^{j_r}(y)\}.$$

We can select a constant subsequence $(j_r)_{r \in K}$; i.e., $j_r = j \notin I^e(x)$ for all $r \in K$. Now, there holds

$$\bigcup_{r \in K} V_r^{l_r} \subset \{y \in V : f(y) = f^j(y)\}.$$

The set on the left is open and has x as a limit point, since $\emptyset \neq V_r^{l_r} \subset x + \frac{1}{r}B^n$ for all $r \in K$. Therefore, $j \in I^e(x)$, which is a contradiction. □

Proposition 2.23. [181, Cor. 4.1.1] *Every* PC^1*-function* $f : V \to \mathbb{R}^m$, $V \subset \mathbb{R}^n$ *open, is locally Lipschitz continuous.*

Proposition 2.24. *Let the* PC^1*-function* $f : V \to \mathbb{R}^m$, $V \subset \mathbb{R}^n$ *open, be a continuous selection of the* C^1*-functions* $\{f^1, \ldots, f^N\}$ *in a neighborhood* W *of* $x \in V$. *Then* f *is B-differentiable at* x *and, for all* $y \in \mathbb{R}^n$,

$$f'(x, y) \in \{(f^i)'(x)y : i \in I^e(x)\}.$$

Further, if f *is differentiable at* x, *then*

$$f'(x) \in \{(f^i)'(x) : i \in I^e(x)\}.$$

Proof. The first part restates [181, Prop. 4.1.3.1]. Now assume that f is differentiable at x. Then, for all $y \in \mathbb{R}^n$, $f'(x)y \in \{(f^i)'(x)y : i \in I^e(x)\}$. Denote by $q \geq 1$ the cardinality of $I^e(x)$. Now choose $l = q(n-1)+1$ vectors $y_r \in \mathbb{R}^n$, $r = 1, \ldots, l$, such that every selection of n of these vectors is linearly independent (the vectors y_r can be obtained, e.g., by choosing l pairwise different numbers $t_r \in \mathbb{R}$, and setting $y_r = (1, t_r, t_r^2, \ldots, t_r^{n-1})^T$). For every r, choose $i_r \in I^e(x)$ such that $f'(x)y_r = (f^{i_r})'(x)y_r$. Since r ranges from 1 to $q(n-1)+1$ and i_r can assume only q different values, we can find n pairwise different indices $r_1, \ldots, r_n$ such that $i_{r_1} = \cdots = i_{r_n} = j$. Since the columns of $Y = (y_{r_1}, \ldots, y_{r_n})$ are linearly independent and $f'(x)Y = (f^j)'(x)Y$, we conclude that $f'(x) = (f^j)'(x)$. □

Proposition 2.25. *Let the* PC^1*-function* $f : V \to \mathbb{R}^m$, $V \subset \mathbb{R}^n$ *open, be a continuous selection of the* C^1*-functions* $\{f^1, \ldots, f^N\}$ *in a neighborhood of* $x \in V$. *Then*

$$\partial_B f(x) = \{(f^i)'(x) : i \in I^e(x)\}, \tag{2.6}$$

$$\partial f(x) = \mathrm{co}\{(f^i)'(x) : i \in I^e(x)\}. \tag{2.7}$$

Proof. We know from Proposition 2.23 that f is locally Lipschitz continuous, and thus the subdifferentials are well defined. By Proposition 2.22, f is a continuous selection of $\{f^i : i \in I^e(x)\}$ in a neighborhood W of x. Further, for $M \in \partial_B f(x)$, there exists $x_k \to x$ in W such that $f'(x_k) \to M$. Among the functions f^i, $i \in I^e(x)$, exactly those with indices $i \in I^e(x) \cap I^e(x_k)$ are essentially active at x_k. Hence, by Proposition 2.22, f is a continuous selection of $\{f^i : i \in I^e(x) \cap I^e(x_k)\}$ in a neighborhood of x_k. Proposition 2.24 now yields that $f'(x_k) = (f^{i_k})'(x_k)$ for some $i_k \in I^e(x) \cap I^e(x_k)$. Now we select a subsequence $k \in K$ on which i_k is constant with value $i \in I^e(x)$. Since $(f^i)'$ is continuous, this proves $M = (f^i)'(x)$, and thus "$\subset$" in (2.6). For every $i \in I^e(x)$ there exists, by definition, a sequence $x_k \to x$ such that $f \equiv f^i$ in an open neighborhood of every x_k. In particular, f is differentiable at x_k (since f^i is C^1), and $f'(x_k) = (f^i)'(x_k) \to (f^i)'(x)$. This completes the proof of (2.6). Assertion (2.7) is an immediate consequence of (2.6). □

We now establish the semismoothness of PC^1-functions.

Proposition 2.26. *Let $f : V \to \mathbb{R}^m$ be a* PC^1*-function on the open set $V \subset \mathbb{R}^n$. Then f is semismooth. If f is a* PC^2*-function, then f is* 1*-order semismooth.*

Proof. The local Lipschitz continuity and B-differentiability of f is guaranteed by Propositions 2.23 and 2.24. Now consider $x \in V$. In a neighborhood W of x, f is a continuous selection of C^1-functions $\{f^1, \dots, f^N\}$ and, without restriction, we may assume that all f^i are active at x. For all $x+s \in W$ and all $M \in \partial f(x+s)$ we have, by Proposition 2.25, $M = \sum_{i \in I^e(x+s)} \lambda_i (f^i)'(x+s)$, $\lambda_i \geq 0$, $\sum_i \lambda_i = 1$. Hence, by Taylor's theorem, using $f^i(x+s) = f(x+s)$ for all $i \in I^e(x+s)$,

$$\begin{aligned} \|f(x+s) - f(x) - Ms\| &= \sum_{i \in I^e(x+s)} \lambda_i \|f^i(x+s) - f^i(x) - (f^i)'(x+s)s\| \\ &\leq \max_{i \in I^e(x+s)} \int_0^1 \|(f^i)'(x+\tau s)s - (f^i)'(x+s)s\| d\tau = o(\|s\|), \end{aligned}$$

which establishes the semismoothness of f. If the f^i are C^2, we obtain

$$\|f(x+s) - f(x) - Ms\| \leq \max_{i \in I^e(x+s)} \int_0^1 \tau \|s^T (f^i)''(x+\tau s)s\| d\tau = O(\|s\|^2),$$

showing that f is 1-order semismooth in this case. □

2.6 Extensions

It is obvious that useful semismoothness concepts can also be obtained for other suitable generalized derivatives. This was investigated in a general, finite-dimensional framework by Jeyakumar [123, 124]. He introduced the concept of $\partial^* f$-semismoothness, where $\partial^* f$ is an *approximate Jacobian*. For the definition of approximate Jacobians we refer to [125]. In what follows, it is sufficient to know that an approximate Jacobian of $f : \mathbb{R}^n \mapsto \mathbb{R}^m$ is a closed-valued multifunction $\partial^* f : \mathbb{R}^n \rightrightarrows \mathbb{R}^{m \times n}$ and that $\partial_B f$, ∂f, and $\partial_C f$ are approximate Jacobians. To avoid confusion with the infinite-dimensional semismoothness concept introduced later (which essentially corresponds to weak J-semismoothness), we denote Jeyakumar's semismoothness concept by J-semismoothness ("J" for Jeyakumar).

Definition 2.27. *Let $f : \mathbb{R}^n \mapsto \mathbb{R}^m$ be a function with approximate Jacobian $\partial^* f$.*

(a) *The function f is called* weakly $\partial^* f$-J-semismooth *at x if it is continuous near x and*

$$\sup_{M \in \overline{co}\partial^* f(x+s)} \|f(x+s) - f(x) - Ms\| = o(\|s\|) \quad \text{as } s \to 0. \tag{2.8}$$

(b) *The function f is* $\partial^* f$-J-semismooth *at x if*

(i) *f is B-differentiable at x (e.g., locally Lipschitz continuous near x and directionally differentiable at x, see [182]), and*

(ii) *f is* weakly $\partial^* f$-J-semismooth *at x.*

Obviously, we can define weak $\partial^* f$-J-semismoothness of order α by requiring the order $O(\|s\|^{1+\alpha})$ in (2.8), and $\partial^* f$-J-semismoothness of order α by the additional requirement that f be α-order B-differentiable at x.

Note that for locally Lipschitz continuous functions $\partial_B f$-, ∂f-, and $\partial_C f$-J-semismoothness all coincide with the usual semismoothness; cf. Proposition 2.10 in the case of $\partial_C f$-J-semismoothness. The same holds true for α-order semismoothness.

Algorithm 2.11 can be extended to weakly $\partial^* f$-J-semismoothness equations by choosing $M_k \in \partial^* f(x_k)$ in step 2. The proof of Proposition 2.12 can be left unchanged, with the only difference that in assumption (b) (iii) we have to require that $\partial^* f$ is compact-valued and upper semicontinuous at $\bar{x}$. If f is weakly $\partial^* f$-J-semismoothness of order α at $\bar{x}$, then an analogue of Proposition 2.18 holds.

Finally, we point out that the concept of H-differentials for functions $f : \mathbb{R}^n \supset V \to \mathbb{R}^m$ introduced by Gowda and Ravindran [81] is closely related to approximate Jacobians, and that for H-differentiable functions a semismoothness concept can be developed; see [80].

Chapter 3

Newton Methods for Semismooth Operator Equations

3.1 Introduction

It was shown in Chapter 1 that semismooth NCP- and MCP-functions can be used to reformulate the VIP (1.14) as (one or more) nonsmooth operator equation(s) of the form

$$\Phi(u) = 0, \quad \text{where} \quad \Phi(u)(\omega) = \phi\big(G(u)(\omega)\big) \quad \text{on } \Omega, \tag{3.1}$$

with G mapping $u \in L^p(\Omega)$ to a vector of Lebesgue functions. In particular, for NCPs we have $G(u) = (u, F(u))$ with $F : L^p(\Omega) \to L^{p'}(\Omega)$, $p, p' \in (1, \infty]$. In finite dimensions this reformulation technique is well investigated and yields a semismooth system of equations, which can be solved by semismooth Newton methods. Naturally, the question arises if it is possible to develop a similar semismoothness theory for operators of the form (3.1). This question is of significant practical importance since the performance of numerical methods for infinite-dimensional problems is intimately related to the infinite-dimensional problem structure. In particular, it is desirable that the numerical method can be viewed as a discrete version of a well-behaved abstract algorithm for the infinite-dimensional problem. Then, for increasing accuracy of discretization, the convergence properties of the numerical algorithm can be expected to be (and usually are) predicted very well by the infinite-dimensional convergence analysis. Therefore, the investigation of algorithms in the original infinite-dimensional problem setting is very helpful for the development of robust, efficient, and mesh-independent numerical algorithms.

In the following, we carry out such an analysis for semismooth Newton methods that are applicable to operator equations of the form (3.1). We split our investigations in two parts. First, we develop

- a general semismoothness concept for operators $f : Y \supset V \to Z$ in Banach spaces, which is based on a set-valued generalized differential $\partial^* f$,
- a locally q-superlinearly convergent Newton-like method for the solution of $\partial^* f$-semismoothness operator equations,
- extensions of these methods that (a) allow inexact computations and (b) incorporate a projection to stay feasible with respect to a closed convex set containing the solution,

- α-order $\partial^* f$-semismoothness and, based on this, convergence with q-order $1+\alpha$ for the developed Newton methods,
- results on the (α-order) semismoothness of the sum, composition, and direct product of semismooth operators with respect to suitable generalized differentials.

The presentation of this chapter follows [191] and section 3.3 is closely related to [193]. In this second part, we fill the abstract concepts with life by considering the concrete case of superposition operators in function spaces. We investigate operators of the form $\Psi(y)(\omega) = \psi(G(y)(\omega))$, a class that includes the operators arising in reformulations (3.1) of VIPs. In particular,

- We introduce a suitable generalized differential $\partial^\circ\Psi$ that is easy to compute and has a natural finite-dimensional counterpart.
- We prove that, under suitable assumptions, the operators Ψ are $\partial^\circ\Psi$-semismooth; under additional assumptions, we establish α-order semismoothness.
- We apply the general semismoothness theory to develop locally fast convergent Newton-type methods for the operator equation $\Psi(y) = 0$.

The publications [102, 191, 193] provided a rigorous basis for the later intensively investigated and successfully applied semismooth Newton methods in function spaces. In [102] the important connection between the primal dual active set strategy and the semismooth Newton method for a reformulation of complementarity systems by means of the max-NCP-function was observed and investigated. Recently, the paper [180] introduced an alternative way of proving semismoothness of superposition operators.

In carrying out our program of investigating the semismoothness of superposition operators, we want to achieve a reasonable compromise between generality and applicability of the developed concepts. Concerning generality, it is possible to pose abstract conditions on an operator and its generalized differential such that superlinearly convergent Newton-type methods can be developed. We refer to Kummer [144], where a nice such framework is developed. Similarly, on the abstract level, we work with the following general concept: Given an operator $f : Y \supset V \to Z$ (V open) between Banach spaces and a set-valued mapping $\partial^* f : V \rightrightarrows \mathcal{L}(Y, Z)$ with nonempty images, i.e., $\partial^* f(y) \neq \emptyset$ for all $y \in V$, we say that f is $\partial^* f$-semismooth at $y \in V$ if f is continuous near y and

$$\sup_{M \in \partial^* f(y+s)} \|f(y+s) - f(y) - Ms\|_Z = o(\|s\|_Y) \quad \text{as } \|s\|_Y \to 0.$$

If the remainder term is of the order $O(\|s\|_Y^{1+\alpha})$, $0 < \alpha \leq 1$, we call f α-order $\partial^* f$-semismooth at y. The class of $\partial^* f$-semismooth operators allows a relatively straightforward development and analysis of Newton-type methods. The reader should be aware that in view of section 2.6 it would be more precise to use the term "weakly $\partial^* f$-semismooth" instead of "semismooth," since we do not require the B-differentiability of f at y. Nevertheless, we prefer the term "semismooth" for brevity, and this is in agreement with the common use of this notion as it is evolving recently. Therefore, our definition of semismoothness is slightly weaker than finite-dimensional semismoothness, but, as already said, still powerful enough to admit the design of superlinearly convergent Newton-type methods, which is our main objective. It is also weaker than the abstract semismoothness concept that was proposed by Chen, Nashed, and Qi [38]; to avoid ambiguity, we call this concept CNQ-semismoothness ("CNQ" for Chen, Nashed, and Qi). In [38], the notions of a slanting function f° and of slant differentiability of f are introduced and a generalized derivative $\partial_S f(y)$, the slant derivative, is obtained as the collection of all possible limits $\lim_{y_k \to y} f^\circ(y_k)$. CNQ-semismoothness is then defined by imposing appropriate conditions on the approximation properties of the slanting function and the slant derivative. These conditions are equivalent [38, Thm. 3.3] to the requirements that (i) f is slantly differentiable in a neighborhood of y, (ii) f is $\partial_S f$-semismoothness at y, and (iii) f is B-differentiable at y; i.e., the directional derivative $f'(y,s) = \lim_{t \to 0^+}(f(y+ts) - f(y))/t$ exists and satisfies $\|f(x+s) - f(x) - f'(x,s)\|_Z = o(\|s\|_Y)$ as $\|s\|_Y \to 0$.

For $\partial^* f$-semismooth equations we develop Newton-like methods and prove q-superlinear convergence. For this, we impose regularity assumptions that are similar to their finite-dimensional counterparts (e.g., those in Proposition 2.12). For α-order $\partial^* f$-semismooth equations, convergence of order $\geq 1+\alpha$ will be proved. In view of our applications to reformulations of the VIP, and, more generally, semismooth superposition operators, it is advantageous to formulate and analyze the Newton method in a two-norm framework, which requires us to augment the Newton iteration by a smoothing step. Further, we allow for inexactness in the computations and also analyze a projected version of the algorithm which generates iterates that stay within a prescribed closed convex set.

Unfortunately, from the viewpoint of applications, the abstract framework of $\partial^* f$-semismoothness (as well as other general approaches) leaves two important questions unanswered:

(a) Given a particular operator f, how should $\partial^* f$ be chosen?

(b) Is there an easy way to verify that f is $\partial^* f$-semismooth?

The same questions arise in the case of CNQ-semismoothness. Then part (a) consists of finding an appropriate slanting function, and part (b) becomes even more involved since CNQ-semismoothness is stronger than $\partial_S f$-semismoothness.

The major, second part of this chapter is intended to develop satisfactory answers to these two questions for a class of nonsmooth operators which includes the mappings Φ arising from reformulations of NCPs and MCPs; see (3.1). More precisely, we consider superposition operators of the form

$$\Psi : Y \to L^r(\Omega), \quad \Psi(y)(\omega) = \psi\big(G(y)(\omega)\big), \tag{3.2}$$

with mappings $\psi : \mathbb{R}^m \to \mathbb{R}$ and $G : Y \to \prod_{i=1}^m L^{r_i}(\Omega)$, where $1 \leq r \leq r_i < \infty$, Y is a real Banach space, and $\Omega \subset \mathbb{R}^n$ is a bounded measurable set with positive Lebesgue measure. Essentially, our working assumptions are that ψ is Lipschitz continuous and semismooth, and that G is continuously Fréchet-differentiable. The detailed assumptions are given below. As a generalized differential for Ψ we introduce an appropriate multifunction

$$\partial^\circ \Psi : Y \rightrightarrows \mathcal{L}(Y, L^r)$$

(the superscript "o" is used to indicate that ∂° is designed especially for superposition operators), which is easy to compute and is motivated by Qi's finite-dimensional C-subdifferential [169]; this addresses question (a) raised above. In our main result we prove the $\partial^\circ\Psi$-semismoothness of Ψ:

$$\sup_{M \in \partial^\circ \Psi(y+s)} \|\Psi(y+s) - \Psi(y) - Ms\|_{L^r} = o(\|s\|_Y) \quad \text{as } \|s\|_Y \to 0. \tag{3.3}$$

This answers question (b) for superposition operators of the form (3.2). We also give conditions under which Ψ is α-order $\partial^\circ\Psi$-semismooth, $0 < \alpha \le 1$.

Based on (3.3), we use the abstract results of the first part to develop a locally q-superlinearly convergent Newton method for the nonsmooth operator equation

$$\Psi(y) = 0. \tag{3.4}$$

Moreover, in the case where Ψ is α-order semismooth we prove convergence with q-order $1+\alpha$. As was observed in earlier work on related local convergence analyses in function space [135, 195], we have to incorporate a smoothing step (explicitly or implicitly) to overcome the nonequivalence of norms. We also give an example showing that this smoothing step can be indispensable.

Although the differentiability properties of superposition operators with smooth ψ are well investigated (see, e.g., the expositions [12, 13]), this was not the case for nonsmooth functions ψ until the publication of [102, 192, 193, 191] in 2001–2002.

As already said, an important application of our results, which motivates our investigations, are reformulations of VIPs (1.14) posed in function spaces. Throughout this chapter, our investigations of the operator Ψ will be accompanied by illustrations at the example of NCP-function-based reformulations of NCPs, which, briefly recalled, consists in finding $u \in L^p(\Omega)$ such that a.e. on Ω there holds

$$u \ge 0, \quad F(u) \ge 0, \quad uF(u) = 0, \tag{3.5}$$

where the operator $F : L^p(\Omega) \to L^{p'}(\Omega)$, $1 < p', p \le \infty$, is given. As always, $\Omega \subset \mathbb{R}^n$ is assumed to be bounded and measurable with positive Lebesgue measure. Using a Lipschitz continuous, semismooth NCP-function $\phi : \mathbb{R}^2 \to \mathbb{R}$, (3.5) is equivalent to the operator equation (3.1). Obviously, choosing $Y = L^p(\Omega)$, $r_2 = r \in [1, p') \cap [1, p)$, $r_1 \in [r, p)$, $\psi \equiv \phi$, and $G : L^p(\Omega) \ni u \mapsto (u, F(u))$, we have $\Psi \equiv \Phi$ with Ψ as in (3.2). The most frequent situation is that F is given as an operator $L^2(\Omega) \to L^2(\Omega)$ and that there exist $p, p' > 2$ such that the solution $\bar{u}$ of the NCP satisfies $\bar{u} \in L^p(\Omega)$ and F maps $L^p(\Omega)$ into $L^{p'}(\Omega)$. The resulting problem then can be viewed to have the form (3.5) with $p, p' > 2$ as specified and $r = 2$. Our focus on the NCP as the main example rather than reformulations of the more general VIP is just for notational convenience. In fact, as can be seen from (1.32), the general VIP requires us to use different reformulations on different parts of Ω, depending on the kind of bounds (none, only lower, only upper, lower and upper), a burden we want to avoid in this chapter.

To establish the semismoothness of Ψ we have to choose an appropriate vector-valued generalized differential. Although the available literature on generalized differentials and subdifferentials is mainly focused on real-valued functions (see, e.g., [26, 40, 41, 178] and the references therein), several authors have proposed and analyzed generalized differentials for nonlinear operators between infinite-dimensional spaces [48, 77, 118, 171, 186]. In our approach, we work with a generalized differential that exploits the structure of Ψ. Roughly speaking, our general guidance hereby is to transcribe, at least formally, componentwise operations in $\mathbb{R}^k$ to pointwise operations in function spaces. To sketch the idea, note that the finite-dimensional analogue of the operator Ψ is the mapping

$$\Psi^{\mathrm{f}} : \mathbb{R}^k \to \mathbb{R}^l, \quad \Psi^{\mathrm{f}}_j(x) = \psi\big(G^j(x)\big), \quad j = 1, \dots, l,$$

with ψ as above and C^1-mappings $G^j : \mathbb{R}^k \to \mathbb{R}^m$. We have the correspondences $\omega \in \Omega \leftrightarrow j \in \{1,\ldots,l\}$, $y \in Y \leftrightarrow x \in \mathbb{R}^k$, and $G(y)(\omega) \leftrightarrow G^j(x)$. Componentwise application of the chain rule for Clarke's generalized gradient [40, Thm. 2.3.10] shows that the C-subdifferential of Ψ^{f} consists of matrices $M \in \mathbb{R}^{l\times k}$ having rows of the form

$$M_j = \sum_{i=1}^{m} d_i^j (G_i^j)'(x) \quad \text{with} \quad d^j \in \partial\psi\big(G^j(x)\big).$$

For completeness, let us note that, conversely, every such matrix is an element of $\partial_C \Psi^{\mathrm{f}}$ if, e.g., ψ or $-\psi$ is regular in the sense of Clarke [40, Def. 2.3.4, Thm. 2.3.10]. Carrying out the same construction for Ψ in a purely formal manner suggests choosing a generalized differential for Ψ consisting of operators of the form

$$Y \ni v \mapsto \sum_{i=1}^{m} d_i \cdot \big(G_i'(x)v\big) \quad \text{with} \quad (d_1,\ldots,d_m)(\omega) \in \partial\psi\big(G(y)(\omega)\big) \text{ a.e. on } \Omega,$$

where the inclusion on the right is meant in the sense of measurable selections. One advantage of this approach, which motivates our choice of the generalized differential $\partial^\circ\Psi$, is that it consists of relatively "concrete" objects as compared to those investigated in, e.g., [48, 77, 118, 171, 186], which necessarily are more abstract since they are not restricted to a particular structure of the underlying operator. It is not the objective of this chapter to investigate the connections between the generalized differential $\partial^\circ\Psi$ and other generalized differentials. There are close relationships, but we leave it as a topic for future research. Here, we concentrate on the development of a semismoothness concept based on $\partial^\circ\Psi$, a related nonsmooth Newton method, and the relations to the respective finite-dimensional analogues.

As already mentioned, the aim is to develop and analyze Newton-like methods for the solution of NCP or, closely related, bound-constrained optimization problems posed in function spaces. Here, we call an iteration Newton-like if each iteration essentially requires the solution of a linear operator equation. We point out that in this sense sequential quadratic programming (SQP) methods for problems involving inequality constraints [4, 5, 6, 8, 9, 98, 189] are not Newton-like, since each iteration requires the solution of a quadratic programming problem (or, put differently, a linearized generalized equation) which is in general significantly more expensive than solving a linear operator equation. Therefore, instead of applying the methods considered in this chapter directly to the nonlinear problem, they also could be of interest as subproblem solvers for SQP methods.

Important earlier investigations that are closely related to the systematic investigation of semismoothness in function spaces started in [102, 191, 193] and presented in the following, include the analysis of Bertsekas' projected Newton method by Kelley and Sachs [135], and the investigation of affine-scaling interior-point Newton methods by Ulbrich and Ulbrich [195]. Both papers deal with bound-constrained minimization problems in function spaces and prove the local q-superlinear convergence of their respective Newton-like methods. In both approaches the convergence results are obtained by estimating directly the remainder terms appearing in the analysis of the Newton iteration. Here, specific properties of the solution are exploited, and a strict complementarity condition is assumed in both papers. We develop our results for the general problem class (3.4) and derive the applicability to NCPs as a simple, but important, special case. In the context of NCPs and optimization, we do not have to assume any strict complementarity condition.

Notation

In this chapter we equip product spaces $\prod_i Y_i$ with the norm $\|y\|_{\prod_i Y_i} = \sum_i \|y\|_{Y_i}$. In particular $\|v\|_{\prod_i L^{q_i}} = \sum_i \|v_i\|_{L^{q_i}}$. Further, for convenience, we write $\sum_i$ and $\prod_i$ instead of $\sum_{i=1}^m$ and $\prod_{i=1}^m$.

3.2 Abstract Semismooth Operators and the Newton Method

3.2.1 Semismooth Operators in Banach Spaces

In the previous section we outlined the following abstract semismoothness concept for general operators between Banach spaces.

Definition 3.1. *Let $f : Y \supset V \to Z$ be defined on an open subset V of the Banach space Y with images in the Banach space Z. Further, let a set-valued mapping $\partial^* f : V \rightrightarrows \mathcal{L}(Y,Z)$ be given with nonempty images; i.e., $\partial^* f(y) \neq \emptyset$ for all $y \in V$, and let $y \in V$.*

(a) *We say that f is $\partial^* f$-*semismooth *at y if f is continuous near y and*

$$\sup_{M \in \partial^* f(y+s)} \|f(y+s) - f(y) - Ms\|_Z = o(\|s\|_Y) \quad \text{as } \|s\|_Y \to 0.$$

(b) *We say that f is α-*order $\partial^* f$-semismooth *at y, $0 < \alpha \le 1$, if f is continuous near y and*

$$\sup_{M \in \partial^* f(y+s)} \|f(y+s) - f(y) - Ms\|_Z = O(\|s\|_Y^{1+\alpha}) \quad \text{as } \|s\|_Y \to 0.$$

(c) *The multifunction $\partial^* f$ will be called the* generalized differential *of f, and the nonemptiness of the images of $\partial^* f$ will always be assumed. In particular, the $\partial^* f$-semismoothness of f at a point $y \in V$ shall automatically imply that the images of $\partial^* f$ are nonempty on V.*

Remark 3.2. *The mapping $Y \ni y \mapsto \partial^* f(y) \subset \mathcal{L}(Y,Z)$ can be interpreted as a set-valued point-based approximation; see Robinson* [176], *Kummer* [144], *and Xu* [202].

3.2.2 Basic Properties

We begin by showing several fundamental properties of semismooth operators. First, it is important to know that continuously differentiable operators f are $\{f'\}$-semismooth. Here, we use the following notation.

Definition 3.3. *Let $F : A \to B$ be a mapping from the set A to the set B. Then we denote by $\{F\} : A \rightrightarrows B$ the single-valued set-valued mapping $A \ni a \mapsto \{F(a)\} \subset B$.*

Proposition 3.4. *Let $f : Y \supset V \to Z$ be differentiable on the neighborhood V of y with its derivative f' being continuous near y. Then f is $\{f'\}$-semismooth at y. If f' is α-Hölder continuous near y, $0 < \alpha \le 1$, then f is α-order $\{f'\}$-semismooth at y.*

Proof. We have by the fundamental theorem of calculus

$$\begin{aligned}\|f(y+s)-f(y)-f'(y+s)s\|_Z &\le \int_0^1 \|(f'(y+ts)-f'(y+s))s\|_Z dt \\ &\le \sup_{0\le t\le 1} \|f'(y+ts)-f'(y+s)\|_{Y,Z}\|s\|_Y = o(\|s\|_Y) \quad \text{as } \|s\|_Y \to 0.\end{aligned}$$

Thus f is $\{f'\}$-semismooth at y. If f' is α-Hölder continuous near y, we obtain

$$\begin{aligned}&\sup_{0\le t\le 1} \|f'(y+ts)-f'(y+s)\|_{Y,Z} \\ &\qquad \le \sup_{0\le t\le 1} O(\|(t-1)s\|_Y^\alpha) = O(\|s\|_Y^\alpha) \quad \text{as } \|s\|_Y \to 0,\end{aligned}$$

which shows the α-order $\{f'\}$-semismoothness of f at y. □

We proceed by proving the semismoothness of the sum of semismooth operators.

Proposition 3.5. *Let $V \subset Y$ be open and let $f_i : V \to Z$ be (α-order) $\partial^* f_i$-semismooth at $y \in V$, $i = 1,\dots,m$. Consider the operator*

$$f : Y \supset V \to Z, \quad f(y) = f_1(y) + \cdots + f_m(y)$$

and define $(\partial^* f_1 + \cdots + \partial^* f_m) : V \rightrightarrows \mathcal{L}(Y,Z)$ as follows:

$$(\partial^* f_1 + \cdots + \partial^* f_m)(y) = \{M_1 + \cdots + M_m : M_i \in \partial^* f_i(y),\ i = 1,\dots,m\}.$$

Let $\partial^* f : V \rightrightarrows \mathcal{L}(Y,Z)$ satisfy $\emptyset \ne \partial^* f(y) \subset (\partial^* f_1 + \cdots + \partial^* f_m)(y)$ for all $y \in V$. Then f is (α-order) $\partial^* f$-semismooth at y.

Proof. By the $\partial^* f_i$-semismoothness of f_i,

$$\begin{aligned}&\sup_M \|f(y+s)-f(y)-Ms\|_Z \\ &\qquad \le \sum_i \sup_{M_i} \|f_i(y+s)-f_i(y)-M_i s\|_Z = o(\|s\|_Y) \quad \text{as } \|s\|_Y \to 0,\end{aligned}$$

where the suprema are taken over $M \in \partial^* f(y+s)$ and $M_i \in \partial^* f_i(y+s)$, respectively. In the case of α-order semismoothness, we can replace $o(\|s\|_Y)$ by $O(\|s\|_Y^{1+\alpha})$. □

The next result shows that the direct product of semismooth operators is itself semismooth with respect to the direct product of the generalized differentials of the components.

Proposition 3.6. *Let $V \subset Y$ be open and assume that the operators $f_i : V \to Z_i$, $i = 1,\dots,m$, are (α-order) $\partial^* f_i$-semismooth at $y \in V$ with generalized differentials $\partial^* f_i : V \rightrightarrows \mathcal{L}(Y,Z_i)$. Consider the operator*

$$f = (f_1,\dots,f_m) : V \ni y \mapsto \big(f_1(y),\dots,f_m(y)\big) \in Z \stackrel{\text{def}}{=} Z_1 \times \cdots \times Z_m$$

and define $(\partial^* f_1 \times \cdots \times \partial^* f_m) : V \rightrightarrows \mathcal{L}(Y,Z)$, *where* $(\partial^* f_1 \times \cdots \times \partial^* f_m)(y)$ *is the set of all operators* $M \in \mathcal{L}(Y,Z)$ *of the form*

$$M : v \mapsto (M_1 v, \ldots, M_m v) \quad \textit{with} \quad M_i \in \partial^* f_i(y),\ i = 1,\ldots,m.$$

Let $\partial^* f : V \rightrightarrows \mathcal{L}(Y,Z)$ *satisfy* $\emptyset \neq \partial^* f(y) \subset (\partial^* f_1 \times \cdots \times \partial^* f_m)(y)$ *for all* $y \in V$. *Then the operator* f *is* (α-*order*) $\partial^* f$-*semismooth at* y.

Proof. By definition, for all $M \in \partial^* f(y+s)$ there exist $M_i \in \partial^* f_i(y+s)$ with $Mv = (M_1 v, \ldots, M_m v)$. Hence, using the norm $\|z\|_Z = \|z_1\|_{Z_1} + \cdots + \|z_m\|_{Z_m}$, and writing $\sup_M$ and $\sup_{M_i}$ for suprema taken over $M \in \partial^* f(y+s)$ and $M_i \in \partial^* f_i(y+s)$, respectively, we obtain

$$\begin{aligned}\sup_M \|f(y+s) - f(y) - Ms\|_Z &\leq \sum_{i=1}^{m} \sup_{M_i} \|f_i(y+s) - f_i(y) - M_i s\|_{Z_i} \\ &= o(\|s\|_Y) \quad \text{as } \|s\|_Y \to 0.\end{aligned}$$

In the case of α-order semismoothness, the above holds with $o(\|\cdot\|)$ replaced by $O(\|\cdot\|^{1+\alpha})$. □

Remark 3.7. *We stress that the construction of* $\partial^* f_1 \times \cdots \times \partial^* f_m$ *from* $\partial^* f_i$ *is analogous to that of the C-subdifferential* $\partial_C f$ *from* ∂f_i.

Next, we give conditions under which the composition of two semismooth operators is semismooth.

Proposition 3.8. *Let* $U \subset X$ *and* $V \subset Y$ *be open. Further, let* $f_1 : U \to Y$ *be Lipschitz continuous near* $x \in U$ *and* (α-*order*) $\partial^* f_1$-*semismooth at* x. *Further, let* $f_2 : V \to Z$ *be* (α-*order*) $\partial^* f_2$-*semismooth at* $y = f_1(x)$ *with* $\partial^* f_2$ *being bounded near* y. *Let* $f_1(U) \subset V$ *and consider the operator* $f \stackrel{\text{def}}{=} f_2 \circ f_1 : X \supset U \to Z$, $f(x) = f_2(f_1(x))$. *Further, define the set-valued mapping* $(\partial^* f_2 \circ \partial^* f_1) : U \rightrightarrows \mathcal{L}(X,Z)$ *as follows:*

$$(\partial^* f_2 \circ \partial^* f_1)(x) = \{M_2 M_1 : M_1 \in \partial^* f_1(x),\ M_2 \in \partial^* f_2\big(f_1(x)\big)\}.$$

Let $\partial^* f : U \rightrightarrows \mathcal{L}(X,Z)$ *satisfy* $\emptyset \neq \partial^* f(x) \subset (\partial^* f_2 \circ \partial^* f_1)(x)$ *for all* $x \in U$. *Then* f *is* (α-*order*) $\partial^* f$-*semismooth at* x.

Proof. We set $h = f_1(x+s) - f_1(x)$, $x+s \in U$. For all $x+s \in U$ and all $M \in \partial^* f(x+s)$ there exist $M_1 \in \partial^* f_1(x+s)$ and $M_2 \in \partial^* f_2\big(f_1(x+s)\big) = \partial^* f_2(y+h)$ with $M = M_2 M_1$. Due to the Lipschitz continuity of f_1 near x, we have

$$\|h\|_Y = \|f_1(x+s) - f_1(x)\|_Y = O(\|s\|_X) \quad \text{as } \|s\|_X \to 0. \tag{3.6}$$

Further, since $\partial^* f_2$ is bounded near y, we can use the semismoothness of f_1, f_2, and (3.6) to see that for all sufficiently small $s \in X$ there holds

$$\begin{aligned}
&\sup_M \|f(x+s) - f(x) - Ms\|_Z \\
&= \sup_{M_1,M_2} \|f_2(y+h) - f_2(y) - M_2 M_1 s\|_Z \\
&\le \sup_{M_1,M_2} \left(\|f_2(y+h) - f_2(y) - M_2 h\|_Z + \|M_2(h - M_1 s)\|_Z\right) \\
&\le o(\|h\|_Y) + \sup_{M_2} \|M_2\|_{Y,Z} \sup_{M_1} \|f_1(x+s) - f_1(x) - M_1 s\|_Y \\
&= o(\|h\|_Y) + o(\|s\|_X) = o(\|s\|_X) \quad \text{as } \|s\|_X \to 0,
\end{aligned}$$

where the suprema are taken over $M \in \partial^* f(x+s)$, $M_1 \in \partial^* f_1(x+s)$, and $M_2 \in \partial^* f_2(y+h)$, respectively. Therefore, f is $\partial^* f$-semismooth at x. In the case of α-order semismoothness, we can replace $o(\|\cdot\|)$ with $O(\|\cdot\|^{1+\alpha})$ in the above calculations, which yields the α-order $\partial^* f$-semismooth of f at x. $\square$

Remark 3.9. *The developed results provide a variety of ways to combine semismooth operators to construct new semismooth operators.*

3.2.3 Semismooth Newton Method in Banach Spaces

In analogy to Algorithm 2.11, we now consider a Newton-like method for the solution of the operator equation

$$f(y) = 0, \tag{3.7}$$

which uses the generalized differential $\partial^* f$. We will assume that $f : V \to Z$, $V \subset Y$ open, is $\partial^* f$-semismooth at the solution $\bar{y} \in V$ of (3.7). As we will see, it can be important for applications to incorporate an additional device, the "smoothing step," in the algorithm, which enables us to work with two-norm techniques. The following short discussion explains, for an important problem class, why this two-norm approach is in general required. To this end, consider the NCP

$$u \ge 0, \quad F(u) \ge 0, \quad uF(u) = 0$$

with $u \in L^2(\Omega)$ and an operator $F : L^2(\Omega) \to L^2(\Omega)$ having suitable structure. Then only a very particular choice of the NCP-function ϕ results in a reformulation

$$\Phi(u) = 0 \quad \text{with} \quad \Phi(u) = \phi(u, F(u))$$

such that $\Phi : L^2(\Omega) \to L^2(\Omega)$ is semismooth. In contrast, we will be able to prove in a quite general setting that $\Phi : L^p(\Omega) \to L^2(\Omega)$ is semismooth for $p > 2$. Then, however, to apply the semismooth Newton iteration, it would be necessary to assume that the operators $M_k \in \partial\Phi(u_k)$ are boundedly invertible in $\mathcal{L}(L^p, L^2)$, which is usually not satisfied. The smoothing step introduced below enables us to work in a framework where, given the availability of a suitable smoothing step, only the semismoothness of $\Phi : L^p(\Omega) \to L^2(\Omega)$

for some $p > 2$ and the bounded invertibility of M_k in $\mathcal{L}(L^2, L^2)$ are required. These turn out to be appropriate assumptions that are verifiable. Although this will be discussed in more detail later on, we briefly indicate a possibility of constructing smoothing steps. To this end, assume that F has the particular structure $F(u) = \lambda u + G(u)$, where $\lambda > 0$ and $G : L^2(\Omega) \to L^2(\Omega)$ has a smoothing property in the sense that G maps $L^2(\Omega)$ locally Lipschitz continuously to $L^p(\Omega)$. For the special NCP-function $\tilde{\phi}(x) = \min\{x_1, x_2/\lambda\}$ the NCP can be written as $\min\{u, \lambda^{-1}F(u)\} = 0$, and from

$$\min\{u, \lambda^{-1}F(u)\} = \min\{u, u + \lambda^{-1}G(u)\} = u + \min\{0, \lambda^{-1}G(u)\}$$

we see that the solution $\bar{u}$ of the NCP satisfies

$$\bar{u} = \min\{0, \lambda^{-1}G(\bar{u})\}.$$

This shows that $\bar{u}$ is a fixed point of $S(u) = \min\{0, \lambda^{-1}G(u)\}$. The operator S is locally Lipschitz continuous from $L^2(\Omega)$ to $L^p(\Omega)$ and satisfies all conditions that we will require from a smoothing step. In particular, we also see that $\bar{u} \in L^2(\Omega)$ enjoys the additional regularity $\bar{u} = S(\bar{u}) \in L^p(\Omega)$. This smoothing step can then also be used in combination with other NCP-functions. In addition, we will see that for the above NCP-function $\tilde{\phi}$, the operator $\tilde{\Phi}(u) = \tilde{\phi}(u, F(u)) = u - S(u)$ is semismooth from $L^2(\Omega)$ to $L^2(\Omega)$ if the above smoothing property holds and $G : L^2(\Omega) \to L^2(\Omega)$ is continuously Fréchet differentiable.

Returning to the general abstract setting, we introduce a further Banach space Y_0 (playing the role of $L^2(\Omega)$ in the above NCP example), in which Y (playing the role of $L^p(\Omega)$ in the NCP example) is continuously and densely embedded, and augment the semismooth Newton iteration by a smoothing step.

Algorithm 3.10 (semismooth Newton method).

0. Choose an initial point $y_0 \in V$ and set $k = 0$.
1. Choose $M_k \in \partial^* f(y_k)$, compute $s_k \in Y_0$ from
$$M_k s_k = -f(y_k),$$
and set $y_{k+1}^0 = y_k + s_k$.
2. Perform a smoothing step: $Y_0 \ni y_{k+1}^0 \mapsto y_{k+1} = S_k(y_{k+1}^0) \in Y$.
3. If $y_{k+1} = y_k$, then STOP with result $y^* = y_{k+1}$.
4. Increment k by one and go to step 1.

Remark 3.11. *The stopping test in step* 3 *is certainly not standard. In fact, we could remove step* 3 *and perform the following simpler test at the beginning of step* 1*: "If $f(y_k) = 0$, then STOP with result $y^* = y_k$." But then, under the hypotheses stated in Assumption* 3.12*, we only could prove that y^* is a solution of* (3.7)*, but we would not know if $y^* = \bar{y}$ or not. For Algorithm* 3.10*, however, we are able to prove that $y^* = \bar{y}$ holds in the case of finite termination.*

If we strengthen Assumption 3.12 (b) *slightly, we can show $y^* = \bar{y}$ even for the case when we terminate with $y^* = y_k$ if $f(y_k) = 0$. This will be discussed in Algorithm* 3.14 *and Theorem* 3.15.

Before we prove the fast local convergence of this algorithm, a comment on the smoothing step is in order. First, it is clear that the smoothing step can be eliminated from the algorithm by choosing $Y_0 = Y$ and $S_k(y^0_{k+1}) = y^0_{k+1}$. However, as we will see later, in many important situations the operators M_k are not continuously invertible in $\mathcal{L}(Y,Z)$. Fortunately, the following framework, which turns out to be widely applicable, provides an escape from this difficulty.

Assumption 3.12. *The space Y is continuously and densely embedded in a Banach space Y_0 such that*

(a) *(Regularity condition) The operators M_k map Y_0 continuously into Z with bounded inverses, and there exists a constant $C_{M^{-1}} > 0$ such that*

$$\|M_k^{-1}\|_{Z,Y_0} \le C_{M^{-1}}.$$

(b) *(Smoothing condition) The smoothing steps in step 1 satisfy*

$$\|S_k(y^0_{k+1}) - \bar{y}\|_Y \le C_S \|y^0_{k+1} - \bar{y}\|_{Y_0}$$

for all k, where $\bar{y} \in Y$ solves (3.7).

Theorem 3.13. *Let $f : Y \supset V \to Z$ be an operator between Banach spaces, defined on the open set V, with generalized differential $\partial^* f : V \rightrightarrows \mathcal{L}(Y,Z)$. Denote by $\bar{y} \in V$ a solution of* (3.7) *and let Assumption* 3.12 *hold. Then there holds*

(a) *If f is $\partial^* f$-semismooth at $\bar{y}$, then there exists $\delta > 0$ such that, for all $y_0 \in \bar{y} + \delta B_Y$, Algorithm* 3.10 *either terminates with $y^* = y_k = \bar{y}$ or generates a sequence $(y_k) \subset V$ that converges q-superlinearly to $\bar{y}$ in Y.*

(b) *If in* (a) *the mapping f is α-order $\partial^* f$-semismooth at $\bar{y}$, $0 < \alpha \le 1$, then the q-order of convergence is at least $1+\alpha$.*

The proof is similar as that of Proposition 2.12.

Proof. (a) Denote the errors before/after smoothing by $v^0_{k+1} = y^0_{k+1} - \bar{y}$ and $v_{k+1} = y_{k+1} - \bar{y}$, respectively. Now let $\delta > 0$ be so small that $\bar{y} + \delta B_Y \subset V$ and consider $y_k \in \bar{y} + \delta B_Y$. Using $M_k s_k = -f(y_k)$ and $f(\bar{y}) = 0$, we obtain

$$M_k v^0_{k+1} = M_k(s_k + v_k) = -f(y_k) + M_k v_k = -[f(\bar{y} + v_k) - f(\bar{y}) - M_k v_k]. \qquad (3.8)$$

This and the $\partial^* f$-semismoothness of f at $\bar{y}$ yield

$$\|M_k v^0_{k+1}\|_Z = o(\|v_k\|_Y) \quad \text{as } \|v_k\|_Y \to 0. \qquad (3.9)$$

Hence, for sufficiently small $\delta > 0$, we have

$$\|M_k v^0_{k+1}\|_Z \le \frac{1}{2C_{M^{-1}}C_S}\|v_k\|_Y, \qquad (3.10)$$

and thus by Assumption 3.12 (a)

$$\|v^0_{k+1}\|_{Y_0} \le \|M_k^{-1}\|_{Z,Y_0}\|M_k v^0_{k+1}\|_Z \le \frac{1}{2C_S}\|v_k\|_Y.$$

Therefore, using Assumption 3.12 (b),

$$\|v_{k+1}\|_Y \le C_S \|v_{k+1}^0\|_{Y_0} \le \frac{1}{2}\|v_k\|_Y. \tag{3.11}$$

This shows

$$y_{k+1} \in \bar{y} + \frac{\|v_k\|_Y}{2}\bar{B}_Y \subset \bar{y} + \frac{\delta}{2}B_Y \subset V. \tag{3.12}$$

If the algorithm terminates in step 3, then

$$\|v_k\|_Y = \|v_{k+1}\|_Y \le \frac{1}{2}\|v_k\|_Y,$$

hence $v_k = 0$, and thus $y^* = y_k = \bar{y}$.

On the other hand, if the algorithm runs infinitely, then (3.12) inductively yields $V \ni y_k \to \bar{y}$ in Y. Now we conclude from the derived estimates and (3.9) that

$$\begin{aligned} \|v_{k+1}\|_Y &\le C_S\|v_{k+1}^0\|_{Y_0} \le C_S\|M_k^{-1}\|_{Z,Y_0}\|M_k v_{k+1}^0\|_Z \\ &\le C_S C_{M^{-1}}\|M_k v_{k+1}^0\|_Z = o(\|v_k\|_Y), \end{aligned} \tag{3.13}$$

which completes the proof of (a).

(b) If, in addition, f is α-order $\partial^* f$-semismooth at $\bar{y}$, then we can write $O(\|v_k\|_Y^{1+\alpha})$ on the right-hand side of (3.9) and obtain, as in (3.13),

$$\|v_{k+1}\|_Y = O(\|v_k\|_Y^{1+\alpha}). \qquad \square$$

As discussed already in Remark 3.11, the standard stopping criterion for Newton-type methods is to terminate if $f(y_k) = 0$. We now analyze the semismooth Newton iteration with smoothing step for this standard termination condition.

Algorithm 3.14 (semismooth Newton method, second version).

0. Choose an initial point $y_0 \in V$ and set $k = 0$.
1. If $f(y_k) = 0$, then STOP with result $y^* = y_k$.
2. Choose $M_k \in \partial^* f(y_k)$, compute $s_k \in Y_0$ from

   $$M_k s_k = -f(y_k),$$

   and set $y_{k+1}^0 = y_k + s_k$.
3. Perform a smoothing step: $Y_0 \ni y_{k+1}^0 \mapsto y_{k+1} = S_k(y_{k+1}^0) \in Y$.
4. Increment k by one and go to step 1.

Theorem 3.15. *Let $f : Y \supset V \to Z$ be an operator between Banach spaces, defined on the open set V, with generalized differential $\partial^* f : V \rightrightarrows \mathcal{L}(Y, Z)$. Denote by $\bar{y} \in V$ a solution of* (3.7) *and let Assumption* 3.12 *hold for all iterations k in which steps 2–4 are executed. Then there holds*

(a) *If* f *is* $\partial^* f$*-semismooth at* $\bar{y}$*, then there exists* $\delta > 0$ *such that, for all* $y_0 \in \bar{y} + \delta B_Y$*, Algorithm* 3.14 *either terminates with a solution* $y^* = y_k \in \bar{y} + \delta B_Y$ *or generates a sequence* $(y_k) \subset V$ *that converges q-superlinearly to* $\bar{y}$ *in* Y*. If the algorithm terminates in step* 1 *with* $y^* = y_k \in \bar{y} + \delta B_Y$ *such that* $f(y^*) = 0$, $\|y^* - \bar{y}\|_Y \leq C_S \|y^* - \bar{y}\|_{Y_0}$*, and there exists* $M_* \in \partial^* f(y^*) \cap \mathcal{L}(Y_0, Z)$ *satisfying the regularity condition* $\|M_*^{-1}\|_{Z,Y_0} \leq C_{M^{-1}}$*, then there holds* $y^* = \bar{y}$*.*

(b) *If in* (a) *the operator* f *is* α*-order* $\partial^* f$*-semismooth at* $\bar{y}$, $0 < \alpha \leq 1$*, then the q-order of convergence is at least* $1 + \alpha$*.*

Proof. The proof of the first part of (a) is identical to the proof of the corresponding assertion of Theorem 3.13 (a). Now consider the case of termination in iteration k and let the assumptions in the second part of assertion (a) hold. Then $y^* = y_k$ and $f(y_k) = 0$. Setting $S_k(y) = y$ and $M_k = M_*$, Assumption 3.12 is satisfied also for iteration k. Hence, with $y_{k+1}^0 := y_k - M_k^{-1} f(y_k) = y_k = y^*$, $y_{k+1} := S_k(y_{k+1}^0) = S_k(y_k) = y_k = y^*$, $v_{k+1}^0 := y_{k+1}^0 - \bar{y} = y^* - \bar{y}$, and $v_{k+1} := y_{k+1} - \bar{y} = y^* - \bar{y}$, the estimate (3.11) holds; i.e.,

$$\begin{aligned}\|y^* - \bar{y}\|_Y = \|y_{k+1} - \bar{y}\|_Y &= \|v_{k+1}\|_Y \leq C_S \|v_{k+1}^0\|_{Y_0} \leq \frac{1}{2}\|v_k\|_Y \\ &= \frac{1}{2}\|y_k - \bar{y}\|_Y = \frac{1}{2}\|y^* - \bar{y}\|_Y.\end{aligned}$$

From this, we obtain $y^* = \bar{y}$.

The proof of (b) is exactly as for Theorem 3.13 (b). □

We conclude this subsection by considering Algorithm 3.14 for the case without a smoothing step. This corresponds to $Y_0 = Y$ and $S_k(y) = y$ for all k, and thus $y_{k+1} = y_{k+1}^0 = y_k - M_k^{-1} f(y_k)$ for all iterations in which steps 2–4 are executed. Assumption 3.12 (a) then reduces to

There exists a constant $C_{M^{-1}} > 0$ such that

$$\|M_k^{-1}\|_{Z,Y} \leq C_{M^{-1}}.$$

Assumption 3.12 (b) can be removed since it trivially holds with $C_S = 1$.

The assumption in the second part of Theorem 3.15 (a) reduces to the requirement that there exists $M_* \in \partial^* f(y^*)$ with $\|M_*^{-1}\|_{Z,Y} \leq C_{M^{-1}}$.

3.2.4 Inexact Semismooth Newton Method in Banach Spaces

From a computational point of view, due to discretization and finite precision arithmetic, we only can compute approximate elements of $\partial^* f$ in general. We address this issue by allowing a certain amount of inexactness in the operators M_k.[1] We incorporate the possibility of inexact computations in our algorithm by modifying step 1 of Algorithm 3.10 as follows.

[1] We stress that an inexact solution of a linear operator equation $Ms = b$, $M \in \mathcal{L}(Y_0, Z)$, can always be interpreted as an exact solution of a system with perturbed operator: If $Md = b + e$, then there holds $(M + \delta M)d = b$ with, e.g., $\delta M v = \langle w, v\rangle_{Y_0^*, Y_0} e$ for all $v \in Y_0$, where $w \in Y_0^*$ is chosen such that $\langle w, d\rangle_{Y_0^*, Y_0} = -1$.

Algorithm 3.16 (inexact semismooth Newton method).
As Algorithm 3.10, but with step 1 replaced by

1. Choose a boundedly invertible operator $B_k \in \mathcal{L}(Y_0, Z)$, compute $s_k \in Y_0$ from

$$B_k s_k = -f(y_k),$$

and set $y_{k+1}^0 = y_k + s_k$.

On the operators B_k we pose a Dennis–Moré–type condition [54, 56, 157, 173], which we formulate in two versions, a weaker one required for superlinear convergence and a stronger variant to prove convergence with q-order $1+\alpha$.

Assumption 3.17.

(a) *There exist operators $M_k \in \partial^* f(y_k + s_k)$ such that*

$$\|(B_k - M_k)s_k\|_Z = o(\|s_k\|_{Y_0}) \quad \text{as } \|s_k\|_Y \to 0, \tag{3.14}$$

where $s_k \in Y_0$ is the step computed in step 1.

(b) *Condition* (a) *holds with* (3.14) *replaced by*

$$\|(B_k - M_k)s_k\|_Z = O(\|s_k\|_{Y_0}^{1+\alpha}) \quad \text{as } \|s_k\|_Y \to 0.$$

Theorem 3.18. *Let $f : Y \supset V \to Z$ be an operator between Banach spaces, defined on the open set V, with generalized differential $\partial^* f : V \rightrightarrows \mathcal{L}(Y, Z)$. Let $\bar{y} \in V$ be a solution of* (3.7) *and let f be Lipschitz continuous near $\bar{y}$. Further, let Assumptions* 3.12 *and* 3.17 (a) *hold (with the same operators M_k in both assumptions). Then,*

(a) *If f is $\partial^* f$-semismooth at $\bar{y}$, then there exists $\delta > 0$ such that, for all $y_0 \in \bar{y} + \delta B_Y$, Algorithm* 3.16 *either terminates with $y^* = y_k = \bar{y}$ or generates a sequence $(y_k) \subset V$ that converges q-superlinearly to $\bar{y}$ in Y.*

(b) *If in* (a) *the mapping f is α-order $\partial^* f$-semismooth at $\bar{y}$, $0 < \alpha \le 1$, and if Assumption* 3.17 (b) *is satisfied, then the q-order of convergence is at least $1+\alpha$.*

Proof. We use the same notations as in the proof of Theorem 3.13 and set $\mu_k = \|(B_k - M_k)s_k\|_Z$. Throughout, consider $y_k \in \bar{y} + \delta B_Y$ and let $\delta > 0$ be so small that f is Lipschitz continuous on $\bar{y} + \delta B_Y \subset V$ with modulus $L > 0$. Then there holds

$$\|f(y_k)\|_Z \le L\|v_k\|_Y.$$

We estimate the Y_0-norm of s_k:

$$\begin{aligned} \|s_k\|_{Y_0} &\le \|M_k^{-1}\|_{Z,Y_0}(\|B_k s_k\|_Z + \|(M_k - B_k)s_k\|_Z) \\ &\le C_{M^{-1}}(\|f(y_k)\|_Z + \mu_k) \le C_{M^{-1}}(L\|v_k\|_Y + \mu_k). \end{aligned} \tag{3.15}$$

By reducing δ, we achieve that $C_{M^{-1}}\mu_k \le \|s_k\|_{Y_0}/2$. Hence,

$$\|s_k\|_{Y_0} \le 2C_{M^{-1}}L\|v_k\|_Y. \tag{3.16}$$

Next, using $f(\bar{y}) = 0$ and $B_k s_k = -f(y_k) = -f(\bar{y} + v_k)$, we derive

$$\begin{aligned} M_k v_{k+1}^0 &= M_k(s_k + v_k) = (M_k - B_k)s_k + B_k s_k + M_k v_k \\ &= (M_k - B_k)s_k - [f(\bar{y} + v_k) - f(\bar{y}) - M_k v_k]. \end{aligned} \tag{3.17}$$

This, Assumption 3.17 (a), the $\partial^* f$-semismoothness of f at $\bar{y}$, and (3.16) yield

$$\|M_k v_{k+1}^0\|_Z = o(\|s_k\|_{Y_0}) + o(\|v_k\|_Y) = o(\|v_k\|_Y) \quad \text{as } \|v_k\|_Y \to 0. \tag{3.18}$$

Now we can proceed as in the part of the proof of Theorem 3.13 (a) starting after (3.9) to show assertion (a).

(b) If, in addition, f is α-order $\partial^* f$-semismooth at $\bar{y}$ and Assumption 3.17 (b) holds, then we can improve (3.18) to

$$\|M_k v_{k+1}^0\|_Z = O(\|s_k\|_{Y_0}^{1+\alpha}) + O(\|v_k\|_Y^{1+\alpha}) = o(\|v_k\|_Y^{1+\alpha}) \quad \text{as } \|v_k\|_Y \to 0.$$

Now we can proceed as in the proof of Theorem 3.13 (b). □

In the same way, we can formulate an inexact version of Algorithm 3.14.

Algorithm 3.19 (inexact semismooth Newton method, second version).
As Algorithm 3.14, but with step 2 replaced by

2. Choose a boundedly invertible operator $B_k \in \mathcal{L}(Y_0, Z)$, compute $s_k \in Y_0$ from

$$B_k s_k = -f(y_k),$$

and set $y_{k+1}^0 = y_k + s_k$.

Theorem 3.20. *Let $f : Y \supset V \to Z$ be an operator between Banach spaces, defined on the open set V, with generalized differential $\partial^* f : V \rightrightarrows \mathcal{L}(Y, Z)$. Let $\bar{y} \in V$ be a solution of* (3.7) *and let f be Lipschitz continuous near $\bar{y}$. Further, let Assumptions* 3.12 *and* 3.17 (a) *hold for all iterations k in which steps 2–4 are executed (with the same operators M_k in both assumptions). Then there holds*

(a) *If f is $\partial^* f$-semismooth at $\bar{y}$, then there exists $\delta > 0$ such that, for all $y_0 \in \bar{y} + \delta B_Y$, Algorithm* 3.19 *either terminates with a solution $y^* = y_k \in \bar{y} + \delta B_Y$ or generates a sequence $(y_k) \subset V$ that converges q-superlinearly to $\bar{y}$ in Y. If the algorithm terminates in step* 1 *with $y^* = y_k \in \bar{y} + \delta B_Y$ such that $f(y^*) = 0$, $\|y^* - \bar{y}\|_Y \le C_S \|y^* - \bar{y}\|_{Y_0}$, and there exists $M_* \in \partial^* f(y^*) \cap \mathcal{L}(Y_0, Z)$ satisfying the regularity condition $\|M_*^{-1}\|_{Z,Y_0} \le C_{M^{-1}}$, then there holds $y^* = \bar{y}$.*

(b) *If in* (a) *the mapping f is α-order $\partial^* f$-semismooth at $\bar{y}$, $0 < \alpha \le 1$, and if Assumption* 3.17 (b) *is satisfied, then the q-order of convergence is at least $1 + \alpha$.*

Proof. The proof of the first part of (a) is identical to the proof of the corresponding assertion of Theorem 3.18 (a). Now consider the case of termination in iteration k and let the assumptions in the second part of assertion (a) hold. Then $y^* = y_k$ and $f(y_k) = 0$. Setting $S_k(y) = y$ and $B_k = M_k = M_*$, Assumptions 3.12 and 3.17 (a) are satisfied also for iteration k. Hence, with $y_{k+1}^0 := y_k - B_k^{-1} f(y_k) = y_k = y^*$, $y_{k+1} := S_k(y_{k+1}^0) = S_k(y_k) = y_k = y^*$, $v_{k+1}^0 := y_{k+1}^0 - \bar{y} = y^* - \bar{y}$, and $v_{k+1} := y_{k+1} - \bar{y} = y^* - \bar{y}$, we can proceed exactly as in the proof of Theorem 3.15 (a).

The proof of (b) is exactly as for Theorem 3.18 (b). □

3.2.5 Projected Inexact Semismooth Newton Method in Banach Spaces

As a last variant of semismooth Newton methods, we develop a projected version of Algorithm 3.16 that is applicable to the constrained semismooth operator equation

$$f(y) = 0 \quad \text{subject to} \quad y \in \mathcal{K}, \tag{3.19}$$

where $\mathcal{K} \subset Y$ is a closed convex set. Here, let $f : Y \supset V \to Z$ be defined on the open set V and assume that (3.19) possesses a solution $\bar{y} \in V \cap \mathcal{K}$. Sometimes it is desirable to have an algorithm for (3.19) that stays feasible with respect to $\mathcal{K}$. To achieve this, we augment Algorithm 3.19 by a projection onto $\mathcal{K}$. We assume that an operator $P_{\mathcal{K}} : Y \to \mathcal{K} \subset Y$ is available with the following properties.

Assumption 3.21.

(a) *$P_{\mathcal{K}}$ is a projection onto $\mathcal{K}$; i.e., for all $y \in Y$ there holds*

$$\|P_{\mathcal{K}}(y) - y\|_Y = \min_{v \in \mathcal{K}} \|v - y\|_Y.$$

(b) *For all y in a Y-neighborhood of $\bar{y}$ there holds*

$$\|P_{\mathcal{K}}(y) - \bar{y}\|_Y \le L_P \|y - \bar{y}\|_Y$$

with a constant $L_P > 0$.

These two requirements are easily seen to be satisfied in all situations we encounter in this work. In particular, it holds with $L_P = 1$ if Y is a Hilbert space or if $\mathcal{K} = \mathcal{B}$ and $Y = L^p(\Omega)$, $p \in [1, \infty]$. In the latter case, we use

$$P_{\mathcal{B}}(u)(\omega) = P_{[a(\omega), b(\omega)]}(u(\omega)) = \max\{a(\omega), \min\{u(\omega), b(\omega)\}\} \quad \text{on } \Omega,$$

which satisfies the assumptions (for $p \in [1, \infty)$, and $P_{\mathcal{B}}$ is the unique metric projection onto $\mathcal{B}$). We are now in a position to formulate the algorithm.

Algorithm 3.22 (projected inexact semismooth Newton method).

0. Choose an initial point $y_0 \in V \cap \mathcal{K}$ and set $k = 0$.
1. Choose an invertible operator $B_k \in \mathcal{L}(Y_0, Z)$, compute $s_k \in Y_0$ from

$$B_k s_k = -f(y_k),$$

and set $y_{k+1}^0 = y_k + s_k$.

2. Perform a smoothing step: $Y_0 \ni y_{k+1}^0 \mapsto y_{k+1}^1 = S_k(y_{k+1}^0) \in Y$.
3. Project onto $\mathcal{K}$: $y_{k+1} = P_{\mathcal{K}}(y_{k+1}^1)$.
4. If $y_{k+1} = y_k$, then STOP with result $y^* = y_{k+1}$.
5. Increment k by one and go to step 1.

Remark 3.23.

(a) *Since $y_0 \in \mathcal{K}$ and all iterates y_k, $k \geq 1$, are obtained by projection onto $\mathcal{K}$, we have $y_k \in \mathcal{K}$ for all k.*

(b) *It is interesting to observe that by composing the smoothing step and the projection step, we obtain a step*

$$S_k^P(y_{k+1}^0) = P_{\mathcal{K}}(S_k(y_{k+1}^0))$$

that has the smoothing property in a Y_0-neighborhood of $\bar{y}$. In fact, for y_{k+1}^0 near $\bar{y}$ (in Y_0), there holds by Assumptions 3.12 *and* 3.21

$$\|S_k^P(y_{k+1}^0) - \bar{y}\|_Y \leq L_P\|S_k(y_{k+1}^0) - \bar{y}\|_Y \leq C_S L_P \|y_{k+1}^0 - \bar{y}\|_{Y_0}.$$

Theorem 3.24. *Let $f : Y \supset V \to Z$ be an operator between Banach spaces, defined on the open set V, with generalized differential $\partial^* f : V \rightrightarrows \mathcal{L}(Y,Z)$. Let $\mathcal{K} \subset Y$ be closed and convex with corresponding projection operator $P_{\mathcal{K}}$ and let $\bar{y} \in V \cap \mathcal{K}$ be a solution of* (3.19). *Further, assume that f is Lipschitz continuous on $\mathcal{K}$ near $\bar{y}$ and let Assumptions* 3.12, 3.17 (a), *and* 3.21 *hold. Then,*

(a) *If f is $\partial^* f$-semismooth at $\bar{y}$, then there exists $\delta > 0$ such that, for all $y_0 \in (\bar{y} + \delta B_Y) \cap \mathcal{K}$, Algorithm* 3.22 *either terminates with $y_k = \bar{y}$ or generates a sequence $(y_k) \subset V \cap \mathcal{K}$ that converges q-superlinearly to $\bar{y}$ in Y.*

(b) *If in* (a) *the mapping f is α-order $\partial^* f$-semismooth at $\bar{y}$, $0 < \alpha \leq 1$, and if Assumption* 3.17 (b) *is satisfied, then the q-order of convergence is at least $1+\alpha$.*

Proof. We only sketch the modifications required to adjust the proof of Theorem 3.18 to the present situation. We choose $\delta > 0$ sufficiently small to ensure that f is Lipschitz on $\mathcal{K}_\delta = (\bar{y} + \delta B_Y) \cap \mathcal{K}$. Then, for all $y_k \in \mathcal{K}_\delta$ we can establish (3.15), (3.16), and, by reducing δ, (3.17) and (3.18). A further reduction of δ yields, instead of (3.10),

$$\|M_k v_{k+1}^0\|_{Y_0} \leq (2C_{M^{-1}} C_S L_P)^{-1} \|v_k\|_Y$$

and thus, analogous to (3.11),

$$\|v_{k+1}^1\|_Y \leq C_S \|v_{k+1}^0\|_Y \leq C_{M^{-1}} C_S \|M_k v_{k+1}^0\|_Y \leq (2L_P)^{-1} \|v_k\|_Y,$$

where $v_{k+1}^1 = y_{k+1}^1 - \bar{y}$. Hence, for δ small enough, Assumption 3.21 (b) can be used to derive

$$\|v_{k+1}\|_Y \leq L_P \|v_{k+1}^1\|_Y \leq \|v_k\|_Y / 2.$$

The rest of the proof, including for part (b), can be transcribed directly from Theorem 3.18. $\square$

A projected version of the inexact semismooth Newton method with a stopping criterion as stated in step 1 of Algorithm 3.19 can also be formulated, and a convergence result of the form of Theorem 3.20, with adjustments along the lines of Theorem 3.24, can be proved.

3.2.6 Alternative Regularity Conditions

In the convergence theorems we used the regularity condition of Assumption 3.12 (a), which requires uniform invertibility in $\mathcal{L}(Y_0, Z)$ of all operators M_k. Since $M_k \in \partial^* f(y_k)$, we also could require the uniform invertibility of all $M \in \partial^* f(y)$ on a neighborhood of $\bar{y}$ more precisely.

Assumption 3.25. *There exist $\eta > 0$ and $C_{M^{-1}} > 0$ such that, for all $y \in \bar{y} + \eta B_Y$, every $M \in \partial^* f(y)$ is an invertible element of $\mathcal{L}(Y_0, Z)$ with $\|M^{-1}\|_{Z,Y_0} \leq C_{M^{-1}}$.*

Then obviously the following holds.

Theorem 3.26. *Let the operator $f : Y \to Z$ and a corresponding generalized differential $\partial^* f : Y \rightrightarrows \mathcal{L}(Y, Z)$ be given. Denote by $\bar{y} \in Y$ a solution of* (3.7) *and let Assumption* 3.25 *hold. Further assume that $y_k \in \bar{y} + \eta B_Y$ for all k. Then Assumption* 3.12 (a) *holds. In particular, Theorems* 3.13, 3.15, 3.18, 3.20, *and* 3.24 *remain true if Assumption* 3.12 (a) *is replaced by Assumption* 3.25.

Proof. The first part follows directly from the fact that $M_k \in \partial^* f(y_k)$. The proofs of the Theorems 3.13, 3.15, 3.18, 3.20, and 3.24 can be applied without change as long as $y_k \in \bar{y} + \eta B_Y$. In particular it follows for $y_k \in \bar{y} + \delta B_Y$ and $\delta \in (0, \eta]$ small enough that $y_{k+1} \in \bar{y} + (\delta/2) B_Y \subset \bar{y} + \eta B_Y$; see, e.g., (3.12). Therefore, all iterates remain in $\bar{y} + \eta B_Y$, and the proofs are applicable without change. $\square$

Remark 3.27.

(a) *The requirement on M_* in Theorems* 3.15 *and* 3.20 *is automatically satisfied under Assumption* 3.25 *since there holds $y^* = y_k \in \bar{y} + \delta B_Y \subset \bar{y} + \eta B_Y$ and $M_* \in \partial^* f(y^*)$.*

(b) *For the projected Newton method, the requirement of Assumption* 3.25 *can be restricted to all $y \in (\bar{y} + \eta B_Y) \cap \mathcal{K}$.*

A further variant, which corresponds to the finite-dimensional CD-regularity is obtained by restricting the bounded invertibility to all $M \in \partial^* f(\bar{y})$.

Assumption 3.28. *The multifunction $Y \ni y \mapsto \partial^* f(y) \subset \mathcal{L}(Y_0, Z)$ is upper semicontinuous at $\bar{y}$, and there exists $C_{M^{-1}} > 0$ such that every $M \in \partial^* f(\bar{y})$ is invertible in $\mathcal{L}(Y_0, Z)$ with $\|M^{-1}\|_{Z,Y_0} \leq C_{M^{-1}}$.*

Theorem 3.29. *Assumption* 3.28 *implies Assumption* 3.25. *In particular, Theorems* 3.13, 3.15, 3.18, 3.20, *and* 3.24 *remain true if Assumption* 3.12 (a) *is replaced by Assumption* 3.28.

Proof. Let Assumption 3.28 hold and choose $\varepsilon = 1/(2C_{M^{-1}})$. By upper semicontinuity there exists $\eta > 0$ such that $\partial^* f(y) \subset \partial^* f(\bar{y}) + \varepsilon B_{\mathcal{L}(Y_0,Z)}$ for all $y \in \bar{y} + \eta B_Y$. Now consider any $y \in \bar{y} + \eta B_Y$ and any $M \in \partial^* f(y)$. Then there exists $\bar{M} \in \partial^* f(\bar{y})$ with

$$\|M - \bar{M}\|_{Y_0,Z} < \varepsilon = \frac{1}{2C_{M^{-1}}} \leq \frac{1}{2\|\bar{M}^{-1}\|_{Z,Y_0}}.$$

Therefore, by Banach's theorem [129, p. 155], M is invertible in $\mathcal{L}(Y_0, Z)$ with

$$\begin{aligned}\|M^{-1}\|_{Z,Y_0} &\leq \frac{\|\bar{M}^{-1}\|_{Z,Y_0}}{1 - \|\bar{M}^{-1}\|_{Z,Y_0}\|M - \bar{M}\|_{Y_0,Z}} \\ &\leq \frac{C_{M^{-1}}}{1 - C_{M^{-1}}/(2C_{M^{-1}})} = 2C_{M^{-1}}.\end{aligned}$$

Thus, Assumption 3.25 holds with $C_{M^{-1}}$ replaced by $2C_{M^{-1}}$. ☐

Remark 3.30. *Theorem* 3.29 *is conveniently applicable in finite dimensions. In the general Banach space setting, however, upper semicontinuity of $\partial^* f$ with respect to the operator norm topology is a quite strong requirement. More realistic is usually upper semicontinuity with respect to the weak operator topology on the image space, which is generated by the seminorms $M \mapsto |\langle w, My\rangle_{Z^*,Z}|$, $w \in Z^*$, $y \in Y_0$. However, this weak form of upper semicontinuity is (except for the finite-dimensional case) not strong enough to prove results as in Theorem* 3.29. *In conclusion, we observe that in the infinite-dimensional setting the regularity conditions stated in Assumption* 3.12 (a) *and in Assumption* 3.25 *are much more widely applicable than Assumption* 3.28.

3.3 Semismooth Superposition Operators and the Newton Method

We now concentrate on nonsmooth superposition operators of the form

$$\Psi : Y \to L^r(\Omega), \quad \Psi(y)(\omega) = \psi\big(G(y)(\omega)\big), \tag{3.20}$$

with mappings $\psi : \mathbb{R}^m \to \mathbb{R}$ and $G : Y \to \prod_{i=1}^m L^{r_i}(\Omega)$. Throughout we assume that $1 \leq r \leq r_i < \infty$, Y is a real Banach space, and $\Omega \subset \mathbb{R}^n$ is a bounded measurable set with positive Lebesgue measure.

Remark 3.31. *Since all our investigations are of local nature, it would be sufficient if G is only defined on a nonempty open subset of Y. Having this in mind, we prefer to work on Y to avoid notational inconveniences.*

Throughout, our investigations are illustrated by the example

$$\Phi(u) = 0, \quad \text{where} \quad \Phi(u)(\omega) = \phi\big(u(\omega), F(u)(\omega)\big) \quad \text{on } \Omega \tag{3.21}$$

with $F : L^p(\Omega) \to L^{p'}(\Omega)$, $p, p' \in (1,\infty]$. Here, $\phi : \mathbb{R}^2 \to \mathbb{R}$ is an NCP-function and the above operator thus occurs in the reformulated NCP. As already observed, Φ can be cast in the form Ψ.

3.3.1 Assumptions

In the rest of the chapter, we will impose the following assumptions on G and ψ.

Assumption 3.32. *There are $1 \le r \le r_i < q_i \le \infty$, $1 \le i \le m$ such that*

(a) *The operator $G : Y \to \prod_i L^{r_i}(\Omega)$ is continuously Fréchet differentiable.*

(b) *The mapping $Y \ni y \mapsto G(y) \in \prod_i L^{q_i}(\Omega)$ is locally Lipschitz continuous; i.e., for all $y \in Y$ there exists an open neighborhood $U = U(y)$ and a constant $L_G = L_G(U)$ such that*

$$\sum_i \|G_i(y_1) - G_i(y_2)\|_{L^{q_i}} \le L_G \|y_1 - y_2\|_Y \quad \forall\, y_1, y_2 \in U.$$

(c) *The function $\psi : \mathbb{R}^m \to \mathbb{R}$ is Lipschitz continuous of rank $L_\psi > 0$; i.e.,*

$$|\psi(x_1) - \psi(x_2)| \le L_\psi \|x_1 - x_2\|_1 \quad \forall\, x_1, x_2 \in \mathbb{R}^m.$$

(d) *ψ is semismooth.*

Remark 3.33. *Since by assumption the set Ω is bounded, we have the continuous embedding $L^q(\Omega) \hookrightarrow L^p(\Omega)$ whenever $1 \le p \le q \le \infty$.*

Remark 3.34. *It is important to note that the norm of the image space in* (b) *is stronger than in* (a).

For semismoothness of order > 0 we will strengthen Assumptions 3.32 as follows.

Assumption 3.35. *As Assumption* 3.32, *but with* (a) *and* (d) *replaced by the following: There exists $\alpha \in (0,1]$ such that*

(a) *The operator $G : Y \to \prod_i L^{r_i}(\Omega)$ is Fréchet differentiable with locally α-Hölder continuous derivative.*

(d) *ψ is α-order semismooth.*

Note that for the special case $Y = \prod_i L^{q_i}(\Omega)$ and $G = I$ we have

$$\Psi : Y \ni y \mapsto \psi(y),$$

and it is easily seen that Assumptions 3.32 and 3.35, respectively, reduce to parts (c) and (d).

Under Assumptions 3.32, the operator Ψ defined in (3.20) is well defined and locally Lipschitz continuous.

Proposition 3.36. *Let Assumptions* 3.32 *hold. Then for all* $1 \le q \le q_i$, $1 \le i \le m$, *and thus in particular for* $q = r$, *the operator* Ψ *defined in* (3.20) *maps* Y *locally Lipschitz continuously into* $L^q(\Omega)$.

Proof. Using Lemma A.5, we first prove $\Psi(Y) \subset L^q(\Omega)$, which follows from

$$\begin{aligned}\|\Psi(y)\|_{L^q} &= \|\psi\big(G(y)\big)\|_{L^q} \le \|\psi(0)\|_{L^q} + \|\psi\big(G(y)\big) - \psi(0)\|_{L^q} \\ &\le c_{q,\infty}(\Omega)|\psi(0)| + L_\psi \sum\nolimits_i \|G_i(y)\|_{L^q} \\ &\le c_{q,\infty}(\Omega)|\psi(0)| + L_\psi \sum\nolimits_i c_{q,q_i}(\Omega)\|G_i(y)\|_{L^{q_i}}.\end{aligned}$$

To prove the local Lipschitz continuity, denote by L_G the local Lipschitz constant in Assumption 3.32 (b) on the set U and let $y_1, y_2 \in U$ be arbitrary. Then, again by Lemma A.5,

$$\begin{aligned}\|\Psi(y_1) - \Psi(y_2)\|_{L^q} &\le L_\psi \sum\nolimits_i \|G_i(y_1) - G_i(y_2)\|_{L^q} \\ &\le L_\psi \sum\nolimits_i c_{q,q_i}(\Omega)\|G_i(y_1) - G_i(y_2)\|_{L^{q_i}} \\ &\le L_\psi L_G \left(\max_{1\le i\le m} c_{q,q_i}(\Omega) \right) \|y_1 - y_2\|_Y. \qquad \square\end{aligned}$$

For the special case Φ in (3.21), the nonsmooth NCP-reformulation, and the choices

$$\begin{gathered} Y = L^p(\Omega), \quad q_1 = p, \quad q_2 = p', \quad r_2 = r \in [1, p') \cap [1, p), \quad r_1 \in [r, p), \\ \psi \equiv \phi, \quad G(u) = \big(u, F(u)\big), \end{gathered} \tag{3.22}$$

we have $\Psi \equiv \Phi$, and Assumption 3.32 can be expressed in the following simpler form.

Assumption 3.37. *There exists* $r \in [1, p) \cap [1, p')$ *such that*

(a) *The mapping* $L^p(\Omega) \ni u \mapsto F(u) \in L^r(\Omega)$ *is continuously Fréchet differentiable.*

(b) *The operator* $F : L^p(\Omega) \to L^{p'}(\Omega)$ *is locally Lipschitz continuous.*

(c) *The function* $\phi : \mathbb{R}^2 \to \mathbb{R}$ *is Lipschitz continuous.*

(d) ϕ *is semismooth.*

In fact, (a) and the continuous embedding $L^p(\Omega) \hookrightarrow L^{r_1}(\Omega)$ imply 3.32 (a). Further, (b) and the Lipschitz continuity of the identity $L^p(\Omega) \ni u \mapsto u \in L^p(\Omega)$ yield Assumption 3.32 (b). Finally, (c),(d) imply Assumption 3.32 (c),(d).

In the same way, Assumption 3.35 for Φ becomes the following.

Assumption 3.38. *As Assumption* 3.37, *but with* (a) *and* (d) *replaced by the following:*

There exist $r \in [1, p) \cap [1, p')$ *and* $\alpha \in (0, 1]$ *such that*

(a) *The operator* $F : L^p(\Omega) \to L^r(\Omega)$ *is Fréchet differentiable with locally* α*-Hölder continuous derivative.*

(d) ϕ *is* α*-order semismooth.*

Remark 3.39. *The three different L^p-spaces deserve an explanation. Usually, we have the following scenario: $F : L^2(\Omega) \to L^2(\Omega)$ is (often even twice) continuously differentiable and has the property that there exist $p, p' > 2$ such that the mapping $L^p(\Omega) \ni u \mapsto F(u) \in L^{p'}(\Omega)$ is locally Lipschitz continuous. A typical example arises from optimal control problems as the problem* (1.40) *that we discussed in section* 1.6.1. *In this problem, which in view of many applications can be considered to be typical, $F = j'$ is the reduced gradient of the optimal control problem, which, in adjoint representation, is given by*

$$F(u) = \lambda u - w(u),$$

where $w(u)$ is the adjoint state. The mapping $u \mapsto w(u)$ is locally Lipschitz continuous (for the problem under consideration even continuous affine linear) from $L^2(\Omega)$ to the Sobolev space $H_0^1(\Omega)$ and thus, via continuous embedding, also to $L^{p'}(\Omega)$ for suitable $p' > 2$. Hence, for arbitrary $p \geq p'$, F maps $L^p(\Omega)$ locally Lipschitz continuously to $L^{p'}(\Omega)$. Often, we can invoke regularity results for the adjoint equation to prove the local Lipschitz continuity of the mapping $L^2(\Omega) \ni u \mapsto w(u) \in H_0^1(\Omega) \cap H^2(\Omega)$ which allows us to choose p' even larger, if desired.

Therefore, as a rule of thumb, usually we are dealing with the case where F is smooth as a mapping $L^2(\Omega) \to L^2(\Omega)$ and locally Lipschitz continuous as a mapping $L^p(\Omega) \to L^{p'}(\Omega)$, $p, p' > 2$. Obviously, these conditions imply the weaker Assumption 3.37 *for $1 \leq r \leq 2$ and $p, p' > 2$ as specified.*

3.3.2 A Generalized Differential for Superposition Operators

For the development of a semismoothness concept for the operator Ψ defined in (3.20) we have to choose an appropriate generalized differential. As we already mentioned in the introduction, our aim is to work with a differential that is as closely connected to finite-dimensional generalized Jacobians as possible. Hence, we will propose a generalized differential $\partial^\circ\Psi$ in such a way that its natural finite-dimensional discretization contains Qi's C-subdifferential.

The construction is motivated by a formal pointwise application of the chain rule. In fact, suppose for the moment that the operator $Y \ni y \mapsto G(y) \in C(\overline{\Omega})^m$ is continuously differentiable, where $C(\overline{\Omega})$ denotes the space of continuous functions equipped with the max-norm. Then for fixed $\omega \in \Omega$ the function $f : y \mapsto G(y)(\omega)$ is continuously differentiable with derivative $f'(y) \in \mathcal{L}(Y, \mathbb{R}^m)$,

$$f'(y) : v \mapsto \big(G'(y)v\big)(\omega).$$

The chain rule for generalized gradients [40, Thm. 2.3.10] applied to the real-valued mapping $y \mapsto \Psi(y)(\omega) = \psi(f(y))$ yields

$$\begin{aligned} &\partial\big(\Psi(y)(\omega)\big) \subset \partial\psi\big(f(y)\big) \circ f'(y) \\ &= \big\{g \in Y^* : \langle g, v\rangle_{Y^*,Y} = \textstyle\sum_i d_i(\omega)\big(G_i'(y)v\big)(\omega),\ d(\omega) \in \partial\psi\big(G(y)(\omega)\big)\big\}. \end{aligned} \tag{3.23}$$

Furthermore, we can replace $\subset$ by $=$ if ψ is regular (e.g., convex or concave) or if the linear operator $f'(y)$ is onto, see [40, Thm. 2.3.10]. Inspired by the idea of the finite-dimensional C-subdifferential, and following the above motivation, we return to the general setting of Assumption 3.32, and define the generalized differential $\partial^\circ\Psi(y)$ in such a way that for all $M \in \partial^\circ\Psi(y)$, the linear form $v \mapsto (Mv)(\omega)$ is an element of the right-hand side in (3.23).

Definition 3.40. *Let Assumptions* 3.32 *hold. For* Ψ *as defined in* (3.20) *we define the generalized differential* $\partial^\circ\Psi : Y \rightrightarrows \mathcal{L}(Y, L^r)$,

$$\partial^\circ\Psi(y) \stackrel{\text{def}}{=} \left\{ M \in \mathcal{L}(Y, L^r) : \begin{array}{l} M : v \mapsto \sum_i d_i \cdot \big(G_i'(y)v\big), \\ d \text{ measurable selection of } \partial\psi\big(G(y)\big) \end{array} \right\}. \tag{3.24}$$

Remark 3.41. *The superscript "$\circ$" is chosen to indicate that this generalized differential is designed for superposition operators.*

The generalized differential $\partial^\circ\Psi(y)$ is nonempty. To show this, we first prove the following.

Lemma 3.42. *Let Assumption* 3.32 (a) *hold and let* $d \in L^\infty(\Omega)^m$ *be arbitrary. Then the operator*

$$M : Y \ni v \mapsto \sum_i d_i \cdot \big(G_i'(y)v\big)$$

is an element of $\mathcal{L}(Y, L^r)$ *and*

$$\|M\|_{Y,L^r} \le \sum_i c_{r,r_i}(\Omega)\|d_i\|_{L^\infty}\|G_i'(y)\|_{Y,L^{r_i}}. \tag{3.25}$$

Proof. By Assumption 3.32 (a) and Lemma A.5,

$$\begin{aligned}
\|Mv\|_{L^r} &= \Big\|\sum_i d_i \cdot \big(G_i'(y)v\big)\Big\|_{L^r} \le \sum_i \|d_i\|_{L^\infty}\|G_i'(y)v\|_{L^r} \\
&\le \Big(\sum_i c_{r,r_i}(\Omega)\|d_i\|_{L^\infty}\|G_i'(y)\|_{Y,L^{r_i}}\Big)\|v\|_Y \quad \forall\, v \in Y,
\end{aligned}$$

which shows that (3.25) holds and $M \in \mathcal{L}(Y, L^r)$. ◻

In the next step, we show that the multifunction

$$\partial\psi\big(G(y)\big) : \Omega \ni \omega \mapsto \partial\psi\big(G(y)(\omega)\big) \subset \mathbb{R}^m$$

is measurable (see Definition A.8 or [177, p. 160]).

Lemma 3.43. *Any closed-valued, upper semicontinuous multifunction* $\Gamma : \mathbb{R}^k \rightrightarrows \mathbb{R}^l$ *is Borel measurable.*

Proof. Let $C \subset \mathbb{R}^l$ be compact. We show that $\Gamma^{-1}(C)$ is closed. To this end, let $(x_k) \subset \Gamma^{-1}(C)$ be arbitrary with $x_k \to x^*$. Then there exist $z_k \in \Gamma(x_k) \cap C$, and, due to the compactness of C, we achieve by transition to a subsequence that $z_k \to z^* \in C$. Since $x_k \to x^*$, upper semicontinuity yields that there exist $\hat{z}_k \in \Gamma(x^*)$ with $(z_k - \hat{z}_k) \to 0$ and thus $\hat{z}_k \to z^*$. Therefore, since $\Gamma(x^*)$ is closed, we obtain $z^* \in \Gamma(x^*) \cap C$. Hence, $x^* \in \Gamma^{-1}(C)$, which proves that $\Gamma^{-1}(C)$ is closed and therefore a Borel set. ◻

Corollary 3.44. *The multifunction* $\partial\psi(G(y)) : \Omega \rightrightarrows \mathbb{R}$ *is measurable.*

Proof. By Lemma 3.43, the compact-valued and upper semicontinuous multifunction $\partial\psi$ is Borel measurable. Now, for all closed sets $C \subset \mathbb{R}^m$, we have, setting $w = G(y) \in \prod_i L^{r_i}(\Omega)$,

$$\partial\psi\big(G(y)\big)^{-1}(C) = \{\omega \in \Omega : w(\omega) \in \partial\psi^{-1}(C)\}.$$

This set is measurable, since $\partial\psi^{-1}(C)$ is a Borel set and w is a (class of equivalent) measurable function(s). □

The next result is a direct consequence of Lipschitz continuity; see [40, Proposition 2.1.2].

Lemma 3.45. *Under Assumption* 3.32 (c) *there holds* $\partial\psi(x) \subset [-L_\psi, L_\psi]^m$ *for all* $x \in \mathbb{R}^m$.

Combining this with Corollary 3.44 yields the following.

Lemma 3.46. *Let Assumptions* 3.32 *hold. Then for all* $y \in Y$, *the set*

$$K(y) = \big\{d : \Omega \to \mathbb{R}^m : d \text{ measurable selection of } \partial\psi\big(G(y)\big)\big\} \tag{3.26}$$

is a nonempty subset of $L_\psi \bar{B}^m_{L^\infty} \subset L^\infty(\Omega)^m$.

Proof. By the Theorem on Measurable Selections [177, Cor. 1C] and Corollary 3.44, $\partial\psi(G(y))$ admits at least one measurable selection $d : \Omega \to \mathbb{R}^m$; i.e.,

$$d(\omega) \in \partial\psi\big(G(y)(\omega)\big) \quad \text{a.e. on } \Omega.$$

From Lemma 3.45 it follows $d \in L_\psi \bar{B}^m_{L^\infty}$. □

We now can prove the following.

Proposition 3.47. *Under Assumptions* 3.32, *for all* $y \in Y$ *the generalized differential* $\partial^\circ\Psi(y)$ *is nonempty and bounded in* $\mathcal{L}(Y, L^r)$.

Proof. Lemma 3.46 ensures that there exist measurable selections d of $\partial\psi(G(y))$ and that all these d are contained in $L_\psi \bar{B}^m_{L^\infty}$. Hence, Lemma 3.42 shows that

$$M : v \mapsto \sum_i d_i \cdot \big(G_i'(y)v\big)$$

is in $\mathcal{L}(Y, L^r)$. The boundedness of $\partial^\circ\Psi(y)$ follows from (3.25). □

We now have everything at hand to introduce a semismoothness concept that is based on the generalized differential $\partial^\circ\Psi$. We postpone the investigation of further properties of $\partial^\circ\Psi$ to sections 3.3.7 and 3.3.8. There, we will establish chain rules, and the convex-valuedness, weak compact-valuedness, and weak graph closedness of $\partial^\circ\Psi$.

3.3.3 Semismoothness of Superposition Operators

In this section, we prove the main result of this chapter, which asserts that under Assumption 3.32 the operator Ψ is $\partial^\circ\Psi$-semismooth. Under Assumption 3.35 and a further condition we prove $\partial^\circ\Psi$-semismooth of order >0. For convenience, we will use the term semismoothness instead of $\partial^\circ\Psi$-semismoothness in what follows. Therefore, applying the general Definition 3.1 to the current situation, we have the following.

Definition 3.48. *The operator Ψ is called ($\partial^\circ\Psi$-)*semismooth *at $y \in Y$ if it is continuous near y and*

$$\sup_{M \in \partial^\circ\Psi(y+s)} \|\Psi(y+s)-\Psi(y)-Ms\|_{L^r} = o(\|s\|_Y) \quad \text{as } s \to 0 \text{ in } Y. \tag{3.27}$$

*Ψ is α-*order *($\partial^\circ\Psi$-)* semismooth *at $y \in Y$, $0<\alpha\le 1$, if it is continuous near y and*

$$\sup_{M \in \partial^\circ\Psi(y+s)} \|\Psi(y+s)-\Psi(y)-Ms\|_{L^r} = O\left(\|s\|_Y^{1+\alpha}\right) \quad \text{as } s \to 0 \text{ in } Y. \tag{3.28}$$

In the following main theorems we prove the semismoothness and the β-order semismoothness, respectively, of the operator Ψ.

Theorem 3.49. *Under Assumptions* 3.32, *the operator Ψ is semismooth on Y.*

Under slightly stronger assumptions, we can also show β-order semismoothness of Ψ.

Theorem 3.50. *Let Assumptions* 3.35 *hold and let $y \in Y$. Assume that there exists $\gamma > 0$ such that the set*

$$\Omega_\varepsilon = \left\{\omega : \max_{\|h\|_1 \le \varepsilon} \left(\rho\big(G(y)(\omega),h\big) - \varepsilon^{-\alpha}\|h\|_1^{1+\alpha}\right) > 0\right\}, \quad \varepsilon > 0,$$

with the residual function $\rho : \mathbb{R}^m \times \mathbb{R}^m \to \mathbb{R}$ given by

$$\rho(x,h) = \max_{z^T \in \partial\psi(x+h)} |\psi(x+h)-\psi(x)-z^T h|,$$

has the following decrease property:

$$meas(\Omega_\varepsilon) = O(\varepsilon^\gamma) \quad \text{as } \varepsilon \to 0^+. \tag{3.29}$$

Then the operator Ψ is β-order semismooth at y with

$$\begin{aligned} &\beta = \min\left\{\frac{\gamma\nu}{1+\gamma/q_0}, \frac{\alpha\gamma\nu}{\alpha+\gamma\nu}\right\}, \quad \text{where} \\ &q_0 = \min_{1\le i\le m} q_i, \qquad \nu = \frac{q_0 - r}{q_0 r} \quad \text{if } q_0 < \infty, \qquad \nu = \frac{1}{r} \quad \text{if } q_0 = \infty. \end{aligned} \tag{3.30}$$

The proofs of both theorems will be presented in section 3.3.5. We will follow our original proof given in [191, 193]. In his recent paper [180], Schiela takes a different approach and uses distribution functions to obtain very similar results.

Remark 3.51. *Condition* 3.29 *requires the measurability of the set* Ω_ε, *which will be verified in the proof.*

Remark 3.52. *As we will see in Lemma* 3.59, *it would be sufficient to require only the local* β*-order Hölder continuity of* G' *in Assumption* 3.35 (a) *with* $\beta \le \alpha$ *as defined in* (3.30).

It might be helpful to give an explanation of the abstract condition (3.29). For convenient notation, let $x = G(y)(\omega)$. Due to the α-order semismoothness of ψ provided by Assumption 3.35, we have $\rho(x,h) = O(\|h\|_1^{1+\alpha})$ as $h \to 0$. In essence, Ω_ε is the set of all $\omega \in \Omega$ where there exists $h \in \varepsilon \bar{B}_1^m$ for which this asymptotic behavior is not yet observed, because the remainder term $\rho(x,h)$ exceeds $\|h\|_1^{1+\alpha}$ by a factor of at least $\varepsilon^{-\alpha}$, which grows infinitely as $\varepsilon \to 0$. From the continuity of the Lebesgue measure it is clear that $\text{meas}(\Omega_\varepsilon) \to 0$ as $\varepsilon \to 0$. The decrease condition (3.29) essentially states that the measure of the set Ω_ε where $G(y)$ takes "bad values," i.e., values at which the radius of small residual is very small, decreases with the rate ε^γ.

The following subsection applies Theorems 3.49 and 3.50 to reformulated nonlinear complementarity problems. Furthermore, it provides a very concrete interpretation of condition (3.29).

Application to NCPs

We apply the semismoothness result to the operator Φ that arises in the reformulation (3.21) of nonlinear complementarity problems (3.5). In this situation, Assumption 3.32 can be expressed in the form of Assumption 3.37. Hence, Theorem 3.49 becomes the following.

Theorem 3.53. *Under Assumption* 3.32, *the operator* $\Phi : L^p(\Omega) \to L^r(\Omega)$ *defined in* (3.21) *is semismooth on* $L^p(\Omega)$.

Remark 3.54. *Due to the structure of* Φ, *we have for all* $M \in \partial^\circ \Phi(u)$ *and* $v \in L^p(\Omega)$

$$Mv = d_1 v + d_2 \cdot \big(F'(y)v\big), \tag{3.31}$$

where $d \in L^\infty(\Omega)^2$ *is a measurable selection of* $\partial\phi(u, F(u))$.

Theorem 3.50 is applicable as well. Once we have chosen a particular NCP-function, condition (3.29) can be made very concrete, so that we can write Theorem 3.50 in a more elegant form. We discuss this for the Fischer–Burmeister function $\phi = \phi^{FB}$, which is Lipschitz continuous and 1-order semismooth, and thus satisfies Assumptions 3.35 (c) and (d) with $\alpha = 1$. Then the following theorem holds.

Theorem 3.55. *Let Assumptions* 3.38 (a) *and* (b) *hold and consider the operator* Φ *with* $\phi = \phi^{FB}$. *Assume that for* $u \in L^p(\Omega)$ *there exists* $\gamma > 0$ *such that*

$$\text{meas}(\{0 < |u| + |F(u)| < \varepsilon\}) = O(\varepsilon^\gamma) \quad \text{as } \varepsilon \to 0. \tag{3.32}$$

Then Φ is β-order semismooth at u with

$$\begin{aligned} &\beta = \min\left\{\frac{\gamma\nu}{1+\gamma/q}, \frac{\alpha\gamma\nu}{\alpha+\gamma\nu}\right\}, \quad \textit{where} \\ &q = \min\{p, p'\}, \qquad \nu = \frac{q-r}{qr} \quad \textit{if } q < \infty, \qquad \nu = \frac{1}{r} \quad \textit{if } q = \infty. \end{aligned} \tag{3.33}$$

Proof. We only have to prove the equivalence of (3.29) and (3.32). Obviously, this follows easily when we have established the following relation:

$$\{0 < \|G(u)\|_1 < \varepsilon\} \subset \Omega_\varepsilon \subset \left\{0 < \|G(u)\|_1 < \left(1+2^{-1/2}\right)\varepsilon\right\} \tag{3.34}$$

with $G(u) = (u, F(u))$. The function $\phi = \phi^{FB}$ is C^∞ on $\mathbb{R}^2 \setminus \{0\}$, see section 2.5.2, with derivative

$$\phi'(x) = (1,1) - \frac{x^T}{\|x\|_2}.$$

To show the first inclusion in (3.34), let ω be such that $x = G(u)(\omega)$ satisfies $0 < \|x\|_1 < \varepsilon$. We observe that, for all $\lambda \in \mathbb{R}$, there holds

$$\phi(\lambda x) = \lambda(x_1 + x_2) - |\lambda|\|x\|_2,$$

and thus, for all $\sigma > 0$,

$$\rho(x, -(1+\sigma)x) = -\sigma\|x\|_2 + \|x\|_2 + (1+\sigma)\frac{x^T x}{\|x\|} = 2\|x\|_2.$$

Hence, for the choice $h = -tx$ with $t \in (1, \sqrt{2})$ such that $\|h\|_1 \leq \varepsilon$, we obtain

$$\rho(x,h) = 2\|x\|_2 \geq \sqrt{2}\|x\|_1 = \frac{\sqrt{2}}{t}\|h\|_1 > \|h\|_1 \geq \varepsilon^{-\alpha}\|h\|_1^{1+\alpha}.$$

This implies $\omega \in \Omega_\varepsilon$ and thus proves the first inclusion.

Next, we prove the second inclusion in (3.34). On $\mathbb{R}^2 \setminus \{0\}$ there holds

$$\phi''(x) = \frac{1}{\|x\|_2^3}\begin{pmatrix} -x_2^2 & x_1 x_2 \\ x_1 x_2 & -x_1^2 \end{pmatrix}.$$

The eigenvalues of $\phi''(x)$ are 0 and $-\|x\|_2^{-1}$. In particular, we see that $\|\phi''(x)\|_2 = \|x\|_2^{-1}$ explodes as $x \to 0$. If $0 \notin [x, x+h]$, then Taylor expansion of $\phi(x)$ about $x+h$ yields with appropriate $\tau \in [0,1]$

$$\rho(x,h) = |\phi(x+h) - \phi(x) - \phi'(x+h)h| = \frac{1}{2}|h^T\phi''(x+\tau h)h| \leq \frac{\|h\|_2^2}{2\|x+\tau h\|_2}.$$

Further, $\rho(0,h) = 0$ and $\rho(x,0) = 0$.

Now consider any $\omega \in \Omega$ that is not contained in the right-hand side of (3.34) and set $x = G(u)(\omega)$. If $x = 0$, then certainly $\omega \notin \Omega_\varepsilon$, since then $\rho(x,\cdot) \equiv 0$. If on the other hand

$\|x\|_1 \geq (1+2^{-1/2})\varepsilon$, then we have for all $h \in \varepsilon \bar{B}_1^2$

$$\rho(x,h) \leq \frac{\|h\|_2^2}{2\|x+\tau h\|_2} \leq \frac{\|h\|_1^2}{\sqrt{2}\|x+\tau h\|_1} \leq \varepsilon^{-1}\|h\|_1^2 \leq \varepsilon^{-\alpha}\|h\|_1^{1+\alpha},$$

and thus $\omega \notin \Omega_\varepsilon$. $\square$

Remark 3.56. *The meaning of* (3.29), *which was shown to be equivalent to* (3.32), *can be interpreted in the following way: The set* $\{0 < |u|+|F(u)| < \varepsilon\}$ *on which the decrease rate in measure is assumed is the set of all* ω *where strict complementarity holds, but is less than* ε. *In a neighborhood of these points the curvature of* ϕ *is very large since* $\|\phi''(G(u)(\omega))\|_2 = \|G(u)(\omega)\|_2^{-1}$ *is big. This requires that* $|G(u+s)(\omega) - G(u)(\omega)|$ *must be very small in order to have a sufficiently small residual* $\rho(G(u)(\omega), G(u+s)(\omega) - G(u)(\omega))$.

We stress that a violation of strict complementarity, i.e., $u(\omega) = F(u)(\omega) = 0$, *does not cause any problems since then* $\rho(G(u)(\omega),\cdot) \equiv \rho(0,\cdot) \equiv 0$.

3.3.4 Illustrations

In this section we give two examples to illustrate the above analysis by pointing out the necessity of the main assumptions and by showing that the derived results cannot be improved in several respects:

- Example 3.57 shows the necessity of the norm gap between the L^{q_i}- and L^r-norms.
- Example 3.58 discusses the sharpness of our order of semismoothness β in Theorem 3.49 for varying values of γ.

In order to prevent our examples from being too academic, we will not work with the simplest choices possible. Rather, we will throughout use reformulations of NCPs based on the Fischer–Burmeister function.

In the proofs of Theorems 3.49 and 3.50, more precisely in the derivation of (3.41) and (3.42), we need the gap between the L^{q_i}- and L^r-norms in order to apply Hölder's inequality. The following example illustrates that both theorems do not in general hold if we drop the condition $r_i < q_i$ in Assumptions 3.32 and 3.35.

Example 3.57 (necessity of the L^{q_i}–L^r-norm gap). We consider the operator Φ arising in semismooth reformulations of the NCP by means of the Fischer–Burmeister function. Theorem 3.53 ensures that, under Assumption 3.37, Φ is semismooth. Our aim here is to show that the requirement $r < q = \min\{p, p'\}$ is indispensable in the sense that in general (3.27) (with $\Psi \equiv \Phi$) is violated for $r \geq q$.

In section 3.2 we developed and analyzed semismooth Newton methods. A central requirement for superlinear convergence is the semismoothness of the underlying operator at the solution. Hence, we will construct a simple NCP with a unique solution for which (3.27) fails to hold whenever $r \geq q$.

Let $1 < p \leq \infty$ be arbitrary, choose $\Omega = (0,1)$, and set

$$F(u)(\omega) = u(\omega) + \omega.$$

Obviously, $\bar{u} \equiv 0$ is the unique solution of the NCP. Choosing $p' = p$, $\phi = \phi^{FB}$, and $\alpha = 1$, Assumptions 3.32 and 3.35 are satisfied for all $r \in [1, p)$. To show that the requirement $r < p$ is really necessary to obtain the semismoothness of Φ, we will investigate the residual

$$R(s) \stackrel{\text{def}}{=} \Phi(\bar{u}+s) - \Phi(\bar{u}) - Ms, \quad M \in \partial^\circ \Phi(\bar{u}+s), \tag{3.35}$$

at $\bar{u} \equiv 0$ with $s \in L^\infty(\Omega)$, $s \geq 0$, $s \neq 0$. Our aim is to show that, for all $r \in [1, \infty]$, there holds

$$\|R(s)\|_{L^r} = o(\|s\|_{L^p}) \quad \text{as } s \to 0 \text{ in } L^\infty \quad \Longrightarrow \quad r < p. \tag{3.36}$$

Setting $\sigma = s(\omega)$, we have for all $\omega \in (0, 1)$

$$(Ms)(\omega) = d_1(\omega)s(\omega) + d_2(\omega)(F'(0)s)(\omega) = d_1(\omega)\sigma + d_2(\omega)\sigma$$

with

$$d(\omega) \in \partial\phi\big(s(\omega), F(s)(\omega)\big) = \partial\phi(\sigma, \sigma+\omega) = \{\phi'(\sigma, \sigma+\omega)\},$$

where we have used $\sigma + \omega > 0$ and ϕ is smooth at $x \neq 0$. Hence, with $e = (1, 1)^T$, noting that the linear part of ϕ cancels in $R(s)(\omega)$, we derive

$$\begin{aligned}
R(s)(\omega) &= \phi(\sigma, \sigma+\omega) - \phi(0, \omega) - \phi'(\sigma, \sigma+\omega)\sigma e \\
&= -\|(\sigma, \sigma+\omega)\|_2 + \|(0, \omega)\|_2 + \frac{\sigma(\sigma, \sigma+\omega)e}{\|(\sigma, \sigma+\omega)\|_2} \\
&= \omega - \frac{\sigma^2 + (\sigma+\omega)^2 - \sigma(2\sigma+\omega)}{\|(\sigma, \sigma+\omega)\|_2} = \omega - \frac{\omega(\sigma+\omega)}{(2\sigma^2 + 2\sigma\omega + \omega^2)^{1/2}}.
\end{aligned}$$

Now let $0 < \varepsilon < 1$. For the special choice $s_\varepsilon \stackrel{\text{def}}{=} \varepsilon \mathbf{1}_{(0,\varepsilon)}$, i.e., $s_\varepsilon(\omega) = \varepsilon$ for $\omega \in (0, \varepsilon)$ and $s_\varepsilon(\omega) = 0$ otherwise, we obtain

$$\|s_\varepsilon\|_{L^p} = \varepsilon^{\frac{p+1}{p}} \quad (1 < p < \infty), \quad \|s_\varepsilon\|_{L^\infty} = \varepsilon.$$

In particular, $s_\varepsilon \to 0$ in L^∞ as $\varepsilon \to 0$. For $0 < \omega < \varepsilon$, there holds

$$|R(s_\varepsilon)(\omega)| \geq \omega\left(1 - \sup_{0<t<1} \frac{1+t}{\sqrt{2+2t+t^2}}\right) = \frac{5-2\sqrt{5}}{5}\omega \geq \frac{\omega}{10}.$$

Hence, $\|R(s_\varepsilon)\|_{L^\infty} \geq \dfrac{\varepsilon}{10} \geq \dfrac{\|s_\varepsilon\|_{L^p}}{10}$, and for all $r \in [p, \infty)$

$$\|R(s_\varepsilon)\|_{L^r} \geq \frac{1}{10}\left(\int_0^\varepsilon \omega^r \, d\omega\right)^{\frac{1}{r}} = \frac{\varepsilon^{\frac{r+1}{r}}}{10(r+1)^{\frac{1}{r}}} \geq \frac{\|s_\varepsilon\|_{L^p}}{10(r+1)^{\frac{1}{r}}}.$$

Therefore, (3.36) is proved. This shows that in (3.27) the norm on the left must be stronger than on the right. ∎

Next, we show that, at least in the case $q_0 \leq (1+\alpha)r$, the order of our semismoothness result is sharp. By showing this for varying values of γ, we also observe that decreasing values of γ reduce the maximum order of semismoothness exactly as stated in Theorem 3.49. Hence, our result does not overestimate the role of γ.

Example 3.58 (order of semismoothness and its dependence on γ). We consider the following NCP, which generalizes the one in Example 3.57: Let $1 < p \le \infty$ be arbitrary, set $\Omega = (0,1)$, and choose

$$F(y)(\omega) = u(\omega) + \omega^\theta, \quad \theta > 0.$$

Obviously, $\bar{u} \equiv 0$ is the unique solution of the NCP. Choosing $p' = p$, $\phi = \phi^{FB}$, and $\alpha = 1$, Assumption 3.35 is satisfied for all $r \in [1, p)$.

From $F(\bar{u})(\omega) = (0, \omega^\theta)$ it follows that $\gamma = 1/\theta$ is the maximum value for which condition (3.32), and thus the equivalent condition (3.29), is satisfied.

With the residual $R(s)$ as defined in (3.35) we obtain

$$|R(s)(\omega)| = \omega^\theta - \frac{\omega^\theta(s(\omega) + \omega^\theta)}{\sqrt{2s(\omega)^2 + 2s(\omega)\omega^\theta + \omega^{2\theta}}}.$$

For $\varepsilon \in (0,1)$ and $s_\varepsilon \stackrel{\text{def}}{=} \varepsilon^\theta \mathbf{1}_{(0,\varepsilon)}$ we have

$$\|s_\varepsilon\|_{L^p} = \varepsilon^{\frac{p\theta+1}{p}} \quad (1 < p < \infty), \quad \|s_\varepsilon\|_{L^\infty} = \varepsilon^\theta.$$

Further, for $0 < \omega < \varepsilon$, there holds

$$|R(s_\varepsilon)(\omega)| \ge \omega^\theta \left(1 - \sup_{0<t<1} \frac{1+t}{\sqrt{2+2t+t^2}}\right) = \frac{5-2\sqrt{5}}{5}\omega^\theta \ge \frac{\omega^\theta}{10}.$$

Hence, for all $r \in [1, p)$

$$\begin{aligned}\|R(s_\varepsilon)\|_{L^r} &\ge \frac{1}{10}\left(\int_0^\varepsilon \omega^{r\theta} d\omega\right)^{\frac{1}{r}} = \frac{\varepsilon^{\frac{r\theta+1}{r}}}{10(r\theta+1)^{\frac{1}{r}}} \\ &\ge \frac{\|s_\varepsilon\|_{L^p}^{\frac{pr\theta+p}{pr\theta+r}}}{10(r\theta+1)^{\frac{1}{r}}} = \frac{\|s_\varepsilon\|_{L^p}^{1+\frac{\gamma\nu}{1+\gamma/q_0}}}{10(r\theta+1)^{\frac{1}{r}}}\end{aligned}$$

with $q_0 = p' = p$, $\gamma = 1/\theta$, and ν as in (3.30). This shows that the value of β given in Theorem 3.49 is sharp for all values of θ (and thus γ) at least as long as $q_0 \le (1+\alpha)r$, which in the current setting can be written as $p \le (1+\alpha)r$. ∎

In the case $q_0 > (1+\alpha)r$ our value of β can still be improved slightly, see [180], by splitting Ω in more than the two parts $\Omega_{\beta\varepsilon}$ and $\Omega_{\beta\varepsilon}^c$ by choosing different values ε_k for ε that correspond to different powers of $\|v\|_{\prod_i L^{q_i}}$. In order to keep the analysis as clear as possible, we do not pursue this idea any further in the current work.

3.3.5 Proof of the Main Theorems

We can simplify the analysis by exploiting the following fact.

Lemma 3.59. *Let Assumptions* 3.32 *hold and suppose that the operator*

$$\Lambda : \prod_i L^{q_i}(\Omega) \ni u \mapsto \psi(u) \in L^r(\Omega)$$

is semismooth. Then the operator $\Psi : Y \to L^r(\Omega)$ *defined in* (3.20) *is also semismooth. Further, if Assumptions* 3.35 *hold and* Λ *is* α*-order semismooth, then* Ψ *is* α*-order semismooth.*

Proof. We first observe that, given any $M \in \partial^\circ \Psi(y+s)$, there is $M_\Lambda \in \partial^\circ \Lambda(G(y+s))$ such that $M = M_\Lambda G'(y+s)$. In fact, there exists a measurable selection $d \in L^\infty(\Omega)^m$ of $\partial\psi(\omega)$ such that $M = \sum_i d_i \cdot G_i'(y+s)$, and obviously $M_\Lambda : v \mapsto \sum_i d_i v_i$ yields an element of $\partial^\circ \Lambda(G(y+s))$ with the desired property. A more general chain rule will be provided by Theorem 3.69.

Setting $g = G(y)$, $v = G(y+s) - G(y)$, and $w = G(y+s)$, we have

$$\begin{aligned}
&\sup_{M \in \partial^\circ \Psi(y+s)} \|\Psi(y+s) - \Psi(y) - Ms\|_{L^r} \\
&\quad \le \sup_{M_\Lambda \in \partial^\circ \Lambda(w)} \|\Lambda(w) - \Lambda(g) - M_\Lambda G'(y+s)s\|_{L^r} \\
&\quad \le \sup_{M_\Lambda \in \partial^\circ \Lambda(w)} \|\Lambda(w) - \Lambda(g) - M_\Lambda v\|_{L^r} \\
&\qquad + \sup_{M_\Lambda \in \partial^\circ \Lambda(w)} \|M_\Lambda \big(G(y+s) - G(y) - G'(y+s)s\big)\|_{L^r} \\
&\quad \overset{\text{def}}{=} \rho_\Lambda + \rho_{MG}.
\end{aligned}$$

By the local Lipschitz continuity of G and the semismoothness of Λ, we obtain

$$\rho_\Lambda = o(\|v\|_{\prod_i L^{q_i}}) = o(\|s\|_Y) \quad \text{as } s \to 0 \text{ in } Y.$$

Further, since $d \in L_\psi \bar{B}^m_{L^\infty}$ by Lemma 3.46, we have by Assumption 3.32 (a)

$$\begin{aligned}
\|\rho_{MG}\|_{L^r} &\le L_\psi \sum_i \|G_i(y+s) - G_i(y) - G_i'(y+s)s\|_{L^r} \\
&\le L_\psi \sum_i c_{r,r_i}(\Omega) \|G_i(y+s) - G_i(y) - G_i'(y+s)s\|_{L^{r_i}} \\
&= o(\|s\|_Y) \quad \text{as } s \to 0 \text{ in } Y.
\end{aligned}$$

This proves the first result.

Now let Assumptions 3.35 hold and Λ be α-order semismooth. Then ρ_Λ and ρ_{MG} are both of the order $O(\|s\|_Y^{1+\alpha})$, which implies the second assertion. □

For the proof of Theorems 3.49 and 3.50 we need, as a technical intermediate result, the Borel measurability of the function

$$\rho : \mathbb{R}^m \times \mathbb{R}^m \to \mathbb{R}, \quad \rho(x,h) = \max_{z^T \in \partial\psi(x+h)} |\psi(x+h) - \psi(x) - z^T h|. \tag{3.37}$$

We prove this by showing that ρ is upper semicontinuous. Readers familiar with this type of result might want to skip the proof of Lemma 3.60.

Recall that a function $f : \mathbb{R}^l \to \mathbb{R}$ is upper semicontinuous at x if

$$\limsup_{x' \to x} f(x') \le f(x).$$

Equivalently, f is upper semicontinuous if and only if $\{x : f(x) \ge a\}$ is closed for all $a \in \mathbb{R}$.

Lemma 3.60. *Let $f : (x,z) \in \mathbb{R}^l \times \mathbb{R}^m \mapsto \mathbb{R}$ be upper semicontinuous. Moreover, let the multifunction $\Gamma : \mathbb{R}^l \rightrightarrows \mathbb{R}^m$ be upper semicontinuous and compact-valued. Then the function*

$$g : \mathbb{R}^l \to \mathbb{R}, \quad g(x) = \max_{z\in\Gamma(x)} f(x,z),$$

is well defined and upper semicontinuous.

Proof. For $x \in \mathbb{R}^l$, let $(z_k) \subset \Gamma(x)$ be such that

$$\lim_{k\to\infty} f(x,z_k) = \sup_{z\in\Gamma(x)} f(x,z).$$

Since $\Gamma(x)$ is compact, we may assume that $z_k \to z^*(x) \in \Gamma(x)$. Now, by upper semicontinuity of f,

$$f\big(x,z^*(x)\big) \geq \limsup_{k\to\infty} f(x,z_k) = \sup_{z\in\Gamma(x)} f(x,z) \geq f\big(x,z^*(x)\big).$$

Thus, g is well defined and there exists $z^* : \mathbb{R}^l \to \mathbb{R}^m$ with $g(x) = f(x,z^*(x))$.

We now prove the upper semicontinuity of g at x. Let $(x_k) \subset \mathbb{R}^l$ tend to x in such a way that

$$\lim_{k\to\infty} g(x_k) = \limsup_{x'\to x} g(x'),$$

and set $z_k = z^*(x_k) \in \Gamma(x_k)$. By the upper semicontinuity of Γ there exists $(\hat{z}_k) \subset \Gamma(x)$ with $(\hat{z}_k - z_k) \to 0$ as $k \to \infty$.

Since $\Gamma(x)$ is compact, a subsequence can be selected such that the sequence $(\hat{z}_k)$, and thus (z_k), converges to some $\hat{z} \in \Gamma(x)$. Now, using that f is upper semicontinuous and $\hat{z} \in \Gamma(x)$,

$$\begin{aligned}\limsup_{x'\to x} g(x') &= \lim_{k\to\infty} g(x_k) = \lim_{k\to\infty} f(x_k,z_k)\\ &= \limsup_{k\to\infty} f(x_k,z_k) \leq f(x,\hat{z}) \leq g(x).\end{aligned}$$

Therefore, g is upper semicontinuous at x. □

Lemma 3.61. *Let $\psi : \mathbb{R}^m \to \mathbb{R}$ be locally Lipschitz continuous. Then the function ρ defined in* (3.37) *is well defined and upper semicontinuous.*

Proof. Since $\partial\psi$ is upper semicontinuous and compact-valued, the multifunction

$$(x,h) \in \mathbb{R}^m \times \mathbb{R}^m \mapsto \partial\psi(x+h)$$

is upper semicontinuous and compact-valued as well. Further, the mapping

$$(x,h,z) \mapsto |\psi(x+h) - \psi(x) - z^T h|$$

is continuous, and we may apply Lemma 3.60, which yields the assertion. □

Proof of Theorem 3.49. By Lemma 3.59, it suffices to prove the semismoothness of the operator

$$\Lambda : \prod_i L^{q_i}(\Omega) \ni u \mapsto \psi(u) \in L^r(\Omega). \tag{3.38}$$

In Lemma 3.61 we showed that the function

$$\rho : \mathbb{R}^m \times \mathbb{R}^m \to \mathbb{R}, \quad \rho(x,h) = \max_{z^T \in \partial\psi(x+h)} |\psi(x+h) - \psi(x) - z^T h|,$$

is upper semicontinuous and thus Borel measurable. Hence, for $u, v \in \prod_i L^{r_i}(\Omega)$, the function $\rho(u,v)$ is measurable. We define the measurable function

$$a = \frac{\rho(u,v)}{\|v\|_1 + \mathbf{1}_{\{v=0\}}}.$$

Since $\rho(u(\omega), v(\omega)) = 0$ whenever $v(\omega) = 0$, we obtain

$$\rho(u,v) = a\|v\|_1.$$

Furthermore,

$$a(\omega) = \frac{\rho\big(u(\omega), v(\omega)\big)}{\|v(\omega)\|_1 + \mathbf{1}_{\{v=0\}}(\omega)} = \frac{o\big(\|v(\omega)\|_1\big)}{\|v(\omega)\|_1 + \mathbf{1}_{\{v=0\}}(\omega)} \to 0 \quad \text{as } v(\omega) \to 0. \tag{3.39}$$

Due to the Lipschitz continuity of ψ, we have

$$\rho(x,h) \le 2L_\psi \|h\|_1, \tag{3.40}$$

which implies $a \in 2L_\psi \bar{B}_{L^\infty}$.

Now let (v_k) tend to zero in the space $\prod_i L^{q_i}(\Omega)$ and set $a_k = a|_{v=v_k}$. Then every subsequence of (v_k) contains itself a subsequence $(v_{k'})$ such that $v_{k'} \to 0$ a.e. on Ω. By (3.39), this implies $a_{k'} \to 0$ a.e. on Ω. Since $(a_{k'})$ is bounded in $L^\infty(\Omega)$, we conclude by the Lebesgue convergence theorem that

$$\lim_{k' \to \infty} \|a_{k'}\|_{L^t} = 0 \quad \forall\, t \in [1, \infty).$$

Hence, in $L^t(\Omega)$, $1 \le t < \infty$, zero is an accumulation point of every subsequence of (a_k). This proves $a_k \to 0$ in all spaces $L^t(\Omega)$, $1 \le t < \infty$.

Since the sequence (v_k), $v_k \to 0$, was arbitrary, we thus have proved that, for all $1 \le t < \infty$, there holds

$$\|a\|_{L^t} \to 0 \quad \text{as } \|v\|_{\prod_i L^{q_i}} \to 0.$$

Now we can use Hölder's inequality to obtain

$$\begin{aligned} \|\rho(u,v)\|_{L^r(\Omega)} &\le \sum_i \|a v_i\|_{L^r} \le \sum_i \|a\|_{L^{p_i}} \|v_i\|_{L^{q_i}} \\ &\le \Big(\max_{1 \le i \le m} \|a\|_{L^{p_i}}\Big) \|v\|_{\prod_i L^{q_i}} \\ &= o\big(\|v\|_{\prod_i L^{q_i}}\big) \quad \text{as } \|v\|_{\prod_i L^{q_i}} \to 0, \end{aligned} \tag{3.41}$$

where $p_i = \frac{q_i r}{q_i - r}$ if $q_i < \infty$ and $p_i = r$ if $q_i = \infty$. Note that here we exploited the fact that $r < q_i$. We recall that here and in the subsequent proofs we work with the norm $\|v\|_{\prod_i L^{q_i}} = \sum_i \|v_i\|_{L^{q_i}}$. The semismoothness of Λ is proved. $\square$

Proof of Theorem 3.50. Also here, by Lemma 3.59, it suffices to prove the β-order semismoothness of the operator Λ defined in (3.38).

We now suppose that Assumption 3.35 and (3.29) hold. First, note that for fixed $\varepsilon > 0$ the function

$$(x,h) \in \mathbb{R}^m \times \mathbb{R}^m \mapsto \rho(x,h) - \varepsilon^{-\alpha} \|h\|_1^{1+\alpha}$$

is upper semicontinuous and that the multifunction

$$x \in \mathbb{R}^m \mapsto \varepsilon \bar{B}_1^m$$

is compact-valued and upper semicontinuous. Hence, by Lemma 3.60, the function

$$x \in \mathbb{R}^m \mapsto \max_{\|h\|_1 \le \varepsilon} \left(\rho(x,h) - \varepsilon^{-\alpha} \|h\|_1^{1+\alpha} \right)$$

is upper semicontinuous and therefore Borel measurable. This proves the measurability of the set Ω_ε appearing in (3.29). For $\varepsilon > 0$ and $0 < \beta \le \alpha$ we define the set

$$\Omega_{\beta\varepsilon} = \left\{ \omega : \rho\big(u(\omega), v(\omega)\big) > \varepsilon^{-\beta} \|v(\omega)\|_1^{1+\beta} \right\},$$

and observe that

$$\Omega_{\beta\varepsilon} \subset \Omega_\varepsilon \cup \{\|v\|_1 > \varepsilon\} \stackrel{\text{def}}{=} \Omega_\varepsilon \cup \Omega'_\varepsilon.$$

In fact, let $\omega \in \Omega_{\beta\varepsilon}$ be arbitrary. The nontrivial case is $\|v(\omega)\|_1 \le \varepsilon$. We then obtain for $h = v(\omega)$

$$\begin{aligned} \rho\big(u(\omega),h\big) &> \varepsilon^{-\beta} \|h\|_1^{1+\beta} = \varepsilon^{-\alpha} \varepsilon^{\alpha-\beta} \|h\|_1^{1+\beta} \\ &\ge \varepsilon^{-\alpha} \|h\|_1^{\alpha-\beta} \|h\|_1^{1+\beta} = \varepsilon^{-\alpha} \|h\|_1^{1+\alpha}, \end{aligned}$$

and thus, since $\|h\|_1 \le \varepsilon$,

$$\max_{\|h\|_1 \le \varepsilon} \left(\rho\big(u(\omega),h\big) - \varepsilon^{-\alpha} \|h\|_1^{1+\alpha} \right) > 0,$$

showing that $\omega \in \Omega_\varepsilon$.

In the case $q_0 = \min_{1 \le i \le m} q_i < \infty$ we derive the estimate

$$\begin{aligned} \operatorname{meas}(\Omega'_\varepsilon) &= \operatorname{meas}(\{\|v\|_1 > \varepsilon\}) \le \|\varepsilon^{-1} \|v\|_1\|_{L^{q_0}(\Omega'_\varepsilon)}^{q_0} \\ &\le \varepsilon^{-q_0} \left(\max_i c_{q_0,q_i}(\Omega'_\varepsilon) \right)^{q_0} \|v\|_{\prod_i L^{q_i}}^{q_0} = \varepsilon^{-q_0} O\left(\|v\|_{\prod_i L^{q_i}}^{q_0} \right). \end{aligned}$$

If we choose $\varepsilon = \|v\|_{\prod_i L^{q_i}}^{\lambda}$, $0 < \lambda < 1$, then

$$\operatorname{meas}(\Omega_{\beta\varepsilon}) \le \operatorname{meas}(\Omega_\varepsilon) + \operatorname{meas}(\Omega'_\varepsilon) = O\left(\|v\|_{\prod_i L^{q_i}}^{\gamma\lambda} \right) + O\left(\|v\|_{\prod_i L^{q_i}}^{(1-\lambda)q_0} \right).$$

This estimate is also true in the case $q_0 = \infty$ since then $\text{meas}(\Omega'_\varepsilon) = 0$ as soon as $\|v\|_{\prod_i L^{q_i}} < 1$. This can be seen by noting that then, for a.a. $\omega \in \Omega$, there holds

$$\|v(\omega)\|_1 \le \|\|v\|_1\|_{L^\infty} \le \|v\|_{\prod_i L^{q_i}} \le \|v\|^{\lambda}_{\prod_i L^{q_i}} = \varepsilon.$$

Introducing $\nu = \frac{q_0 - r}{q_0 r}$ if $q_0 < \infty$ and $\nu = \frac{1}{r}$, otherwise, for all $0 < \beta \le \alpha$, we obtain, using (3.40) and Lemma A.5

$$\begin{aligned}
\|\rho(u,v)\|_{L^r(\Omega_{\beta\varepsilon})} &\le \|2L_\psi\|v\|_1\|_{L^r(\Omega_{\beta\varepsilon})} \le 2L_\psi c_{r,q_0}(\Omega_{\beta\varepsilon})\|v\|_{L^{q_0}(\Omega_{\beta\varepsilon})^m} \\
&\le 2L_\psi \,\text{meas}(\Omega_{\beta\varepsilon})^\nu \|v\|_{L^{q_0}(\Omega_{\beta\varepsilon})^m} \\
&= O\left(\|v\|^{1+\gamma\lambda\nu}_{\prod_i L^{q_i}}\right) + O\left(\|v\|^{1+(1-\lambda)\nu q_0}_{\prod_i L^{q_i}}\right).
\end{aligned} \tag{3.42}$$

Again, we have used here the fact that $r < q_0 \le q_i$, which allowed us to take advantage of the smallness of the set $\Omega_{\beta\varepsilon}$.

Finally, on $\Omega^c_{\beta\varepsilon}$, $(1+\beta)r \le q_0$, $0 < \beta \le \alpha$, there holds with our choice $\varepsilon = \|v\|^{\lambda}_{\prod_i L^{q_i}}$

$$\begin{aligned}
\|\rho(u,v)\|_{L^r(\Omega^c_{\beta\varepsilon})} &\le \|\varepsilon^{-\beta}\|v\|_1^{1+\beta}\|_{L^r(\Omega^c_{\beta\varepsilon})} \le c_{r,\frac{q_0}{1+\beta}}(\Omega^c_{\beta\varepsilon})\|v\|^{-\beta\lambda}_{\prod_i L^{q_i}}\|v\|^{1+\beta}_{L^{q_0}(\Omega^c_{\beta\varepsilon})^m} \\
&= O\left(\|v\|^{1+\beta(1-\lambda)}_{\prod_i L^{q_i}}\right).
\end{aligned}$$

Therefore,

$$\|\rho(u,v)\|_{L^r} = O\left(\|v\|^{1+\gamma\lambda\nu}_{\prod_i L^{q_i}}\right) + O\left(\|v\|^{1+(1-\lambda)\nu q_0}_{\prod_i L^{q_i}}\right) + O\left(\|v\|^{1+\beta(1-\lambda)}_{\prod_i L^{q_i}}\right).$$

We now choose $0 < \lambda < 1$ and $\beta > 0$ with $\beta \le \alpha$, $(1+\beta)r \le q_0$ in such a way that the order of the right-hand side is maximized. In the case $(1+\alpha)r \ge q_0$ the minimum of all three exponents is maximized for the choice $\beta = \frac{q_0 - r}{r} = \nu q_0$ and $\lambda = \frac{q_0}{\gamma + q_0}$. Then all three exponents are equal to $1 + \frac{\gamma\nu q_0}{\gamma + q_0}$ and thus

$$\|\rho(u,v)\|_{L^r} = O\left(\|v\|^{1+\frac{\gamma\nu q_0}{\gamma+q_0}}_{\prod_i L^{q_i}}\right). \tag{3.43}$$

If, on the other hand, $(1+\alpha)r < q_0$, then the third exponent is smaller than the second one for all $0 < \lambda < 1$ and $0 < \beta \le \alpha$. Further, it is not difficult to see that under these constraints the first and third exponent become maximal for $\beta = \alpha$ and $\lambda = \frac{\alpha}{\alpha + \gamma\nu}$ and attain the value $1 + \frac{\alpha\gamma\nu}{\alpha+\gamma\nu}$. Hence,

$$\|\rho(u,v)\|_{L^r} = O\left(\|v\|^{1+\frac{\alpha\gamma\nu}{\alpha+\gamma\nu}}_{\prod_i L^{q_i}}\right). \tag{3.44}$$

Combining (3.43) and (3.44) proves the β-order semismoothness of Λ with β as in (3.30). □

3.3.6 Semismooth Newton Methods for Superposition Operators

The developed semismoothness results can be used to derive superlinearly convergent Newton-type methods for the solution of the nonsmooth operator equation

$$\Psi(y) = 0 \tag{3.45}$$

with Ψ as defined in (3.20). In fact, any of the variants of the semismooth Newton method that we developed and analyzed in sections 3.2.3 and 3.2.4 can be applied, provided that the respective assumptions are satisfied. We just have to choose $Z = L^r(\Omega)$, $f \equiv \Psi$, and $\partial^* f \equiv \partial^\circ \Psi$. With these settings, Algorithms 3.10, 3.14, 3.16, 3.19, and 3.22 are applicable to (3.45) and their convergence properties are stated in Theorems 3.13, 3.15, 3.18, 3.20, and 3.24, respectively. The semismoothness requirements on Ψ are ensured by Theorems 3.49 and 3.50 under Assumptions 3.32 and 3.35, respectively. The regularity condition and the requirement on a smoothing step, i.e., Assumption 3.12, need to be specialized to the current situation; see Assumption 3.64 below.

For illustration, we restate the most general of these methods, Algorithm 3.22, when applied to reformulations (3.21) of the NCP (3.5). We also recall the local convergence properties of the resulting method. The results hold equally well for bilaterally constrained problems; the only difference is that the reformulation then requires an MCP-function instead of an NCP-function.

For the reformulation of the NCP we work with an NCP-function ϕ which, together with the operator F, satisfies Assumption 3.37. Further, we assume that we are given an admissible set

$$\mathcal{K} = \{u \in L^p(\Omega) : a_{\mathcal{K}} \le u \le b_{\mathcal{K}} \ \text{ on } \Omega\},$$

which contains the solution $\bar{u} \in L^p(\Omega)$, and in which all iterates generated by the algorithm should stay. The requirements on the bounds $a_{\mathcal{K}}$ and $b_{\mathcal{K}}$ are: There exist measurable sets $\Omega_a^{\mathcal{K}}, \Omega_b^{\mathcal{K}} \subset \Omega$ such that

$$\begin{aligned} a_{\mathcal{K}} &= -\infty \quad \text{on } \Omega \setminus \Omega_a^{\mathcal{K}}, & b_{\mathcal{K}} &= +\infty \quad \text{on } \Omega \setminus \Omega_b^{\mathcal{K}}, \\ a_{\mathcal{K}}|_{\Omega_a^{\mathcal{K}}} &\in L^p(\Omega_a^{\mathcal{K}}), & b_{\mathcal{K}}|_{\Omega_b^{\mathcal{K}}} &\in L^p(\Omega_b^{\mathcal{K}}). \end{aligned} \tag{3.46}$$

Natural choices for $\mathcal{K}$ are $\mathcal{K} = L^p(\Omega)$ or $\mathcal{K} = \mathcal{B} = \{u \in L^p(\Omega) : u \ge 0\}$.

We define the projection $P_{\mathcal{K}} : L^p(\Omega) \to \mathcal{K}$,

$$P_{\mathcal{K}}(u) = P_{[a_{\mathcal{K}}(\omega), b_{\mathcal{K}}(\omega)]}(u) = \max\{a_{\mathcal{K}}(\omega), \min\{u(\omega), b_{\mathcal{K}}(\omega)\}\},$$

which is easily seen to assign to each $u \in L^p(\Omega)$ a function $P_{\mathcal{K}}(u) \in \mathcal{K}$ that is nearest to u in L^p (for $p < \infty$, $P_{\mathcal{K}}(u)$ is the unique metric projection). Since $|P_{\mathcal{K}}(u) - P_{\mathcal{K}}(v)| \le |u - v|$ pointwise on Ω, we see that

$$\|P_{\mathcal{K}}(u) - P_{\mathcal{K}}(v)\|_{L^p} \le \|u - v\|_{L^p} \quad \forall\, u, v \in L^p(\Omega).$$

In particular, since $\bar{u} \in \mathcal{K}$, we see that

$$\|P_{\mathcal{K}}(u) - \bar{u}\|_{L^p} \le \|u - \bar{u}\|_{L^p} \quad \forall\, u \in L^p(\Omega).$$

Therefore, $\mathcal{K}$ and $P_{\mathcal{K}}$ satisfy Assumptions 3.21.

In section 3.2.3 we developed Newton-like methods that are formulated in a two-norm framework by incorporating an additional space Y_0 with $Y \hookrightarrow Y_0$. However, so far a rigorous justification for the necessity of two-norm techniques is still missing. We are now in a position to give this justification.

In the current setting, we have $Y = L^p(\Omega)$, and, as we will see, it is appropriate to choose $Y_0 = L^r(\Omega)$ (the standard situation is $r = 2$). Algorithm 3.22 then becomes the following.

Algorithm 3.62 (projected inexact Newton method for NCP).

0. Choose an initial point $u_0 \in \mathcal{K}$ and set $k = 0$.

1. Choose an invertible operator $B_k \in \mathcal{L}(L^r(\Omega), L^r(\Omega))$, compute $s_k \in L^r(\Omega)$ from
$$B_k s_k = -\Phi(u_k),$$
and set $u_{k+1}^0 = u_k + s_k$.

2. Perform a smoothing step: $L^r(\Omega) \ni u_{k+1}^0 \mapsto u_{k+1}^1 = S_k(u_{k+1}^0) \in L^p(\Omega)$.

3. Project onto $\mathcal{K}$: $u_{k+1} = P_{\mathcal{K}}(u_{k+1}^1)$.

4. If $u_{k+1} = u_k$, then STOP with result $u^* = u_{k+1}$.

5. Increment k by one and go to step 1.

To discuss the role of the two-norm technique and the smoothing step, it is convenient to consider the special case of the semismooth Newton method with smoothing step as described in Algorithm 3.10, which is obtained by choosing $\mathcal{K} = L^p(\Omega)$ and $B_k = M_k \in \partial^\circ \Phi(u_k)$.

For well-definedness of the method, it is reasonable to require that the Newton equation $M_k s_k = -\Phi(u_k)$ in step 1 always possesses a unique solution. Further, in the convergence analysis an estimate is needed that bounds the norm of s_k in terms of $\|\Phi(u_k)\|_{L^r}$. It turns out that the L^p-norm is too strong for this purpose. In fact, recall that every operator $M \in \partial^\circ \Phi(u)$ assumes the form
$$M = d_1 \cdot I + d_2 \cdot F'(u),$$
with $d \in L^\infty(\Omega)^2$, $d(\omega) \in \partial\phi(u(\omega), F(u)(\omega))$. Now define
$$\Omega_1 = \{\omega \in \Omega : d_2(\omega) = 0\}.$$
Then, for all $\omega \in \Omega_1$, there holds
$$(Mv)(\omega) = d_1(\omega)v(\omega).$$
This shows that Mv is in general not more regular (in the L^q-sense) than v and vice versa. Therefore, it is not appropriate to assume that $M \in \partial^\circ \Phi(u)$ is continuously invertible in $\mathcal{L}(L^p, L^r)$, as the norm on L^p is stronger than on L^r. However, it is reasonable to assume that M is an L^r-homeomorphism. This leads to regularity conditions of the form stated in Assumption 3.12 (a) or in Assumption 3.25 with $Y_0 = L^r(\Omega)$.

As a consequence, in the convergence analysis we only have available the uniform boundedness of $\|M_k^{-1}\|_{Z,Y_0}$, and this makes a smoothing step necessary, as can be seen from the following chain of implications that we used in the proof of Theorem 3.13 (and its generalizations). We describe it for the setting of Algorithm 3.62.

$$
\begin{aligned}
& M_k s_k = -\Phi(u_k), \quad \Phi(\bar{u}) = 0, \quad v_k = u_k - \bar{u}, \quad v_k^0 = u_k^0 - \bar{u}, \quad v_k^1 = u_k^1 - \bar{u} \\
\Longrightarrow\ & M_k v_{k+1}^0 = -\big(\Phi(\bar{u} + v_k) - \Phi(\bar{u}) - M_k v_k\big) \\
\Longrightarrow\ & \|M_k v_{k+1}^0\|_{L^r} = o(\|v_k\|_{L^p}) && \text{(semismoothness)} \\
\Longrightarrow\ & \|v_{k+1}^0\|_{L^r} \le \|M_k^{-1}\|_{L^r,L^r} \|M_k v_{k+1}^0\|_{L^r} = o(\|v_k\|_{L^p}) && \text{(regularity)} \\
\Longrightarrow\ & \|v_{k+1}^1\|_{L^p} = \|S_k(u_{k+1}^0) - \bar{u}\|_{L^p} \\
& \qquad = O(\|v_{k+1}^0\|_{L^r}) = o(\|v_k\|_{L^p}) && \text{(smoothing step)} \\
\Longrightarrow\ & \|v_{k+1}\|_{L^p} = \|P_{\mathcal{K}}(u_{k+1}^1) - \bar{u}\|_{L^p} \le \|v_{k+1}^1\|_{L^p} && \text{(nonexpansiveness} \\
& \qquad = o(\|v_k\|_{L^p}) && \text{of projection)}
\end{aligned}
$$

Therefore, we see that the two-norm framework of our abstract analysis in section 3.2.3 is fully justified.

Adapted to the current setting, Assumptions 3.17 and 3.12 required to apply Theorem 3.24 now read as follows.

Assumption 3.63 (Dennis–Moré condition for B_k).

(a) *There exist operators $M_k \in \partial^\circ \Phi(u_k + s_k)$ such that*

$$\|(B_k - M_k)s_k\|_{L^r} = o(\|s_k\|_{L^r}) \quad \textit{as } \|s_k\|_{L^p} \to 0, \tag{3.47}$$

where $s_k \in L^r(\Omega)$ is the step computed in step 1.

(b) *Condition* (a) *holds with* (3.47) *replaced by*

$$\|(B_k - M_k)s_k\|_{L^r} = O(\|s_k\|_{L^r}^{1+\alpha}) \quad \textit{as } \|s_k\|_{L^p} \to 0.$$

Assumption 3.64.

(a) *(Regularity condition) One of the following conditions holds:*

(i) *The operators M_k map $L^r(\Omega)$ continuously into itself with bounded inverses, and there exists a constant $C_{M^{-1}} > 0$ such that*

$$\|M_k^{-1}\|_{L^r,L^r} \le C_{M^{-1}}.$$

(ii) *There exist constants $\eta > 0$ and $C_{M^{-1}} > 0$ such that, for all $u \in (\bar{u} + \eta B_{L^p}) \cap \mathcal{K}$, every $M \in \partial^\circ \Phi(u)$ is an invertible element of $\mathcal{L}(L^r, L^r)$ with $\|M^{-1}\|_{L^r,L^r} \le C_{M^{-1}}$.*

(b) *(Smoothing condition) The smoothing steps in step 1 satisfy*

$$\|S_k(u_{k+1}^0) - \bar{u}\|_{L^p} \le C_S \|u_{k+1}^0 - \bar{u}\|_{L^r} \quad \forall\, k,$$

where $\bar{u} \in \mathcal{K}$ solves (3.1).

Remark 3.65. *In section* 4.3 *we develop sufficient conditions for regularity that are widely applicable and easy to apply.*

Remark 3.66. *In section* 4.1 *we discuss how smoothing steps can be computed. Further, in section* 4.2 *we propose a choice for ϕ which allows us to get rid of the smoothing step.*

Since Φ is semismooth by Theorem 3.49 and locally Lipschitz continuous by Proposition 3.36, we can apply Theorem 3.24 to the current situation and obtain the following local convergence result.

Theorem 3.67. *Denote by $\bar{u} \in \mathcal{K}$ a solution of* (3.1). *Further, let Assumptions* 3.37, 3.63 (a), *and* 3.64 *hold. Then,*

(a) *there exists $\delta > 0$ such that, for all $u_0 \in (\bar{u} + \delta B_{L^p}) \cap \mathcal{K}$, Algorithm* 3.16 *either terminates with $u_k = \bar{u}$ or generates a sequence $(u_k) \subset \mathcal{K}$ that converges q-superlinearly to $\bar{u}$ in $L^p(\Omega)$.*

(b) *if in* (a) *the mapping Φ is α-order semismooth at $\bar{u}$, $0 < \alpha \leq 1$, and if Assumption* 3.63 (b) *is satisfied, then the q-order of convergence is at least $1 + \alpha$.*

3.3.7 Semismooth Composite Operators and Chain Rules

This section considers the semismoothness of composite operators. There is a certain overlap with the result of the abstract Proposition 3.8, but we think it is helpful to study the properties of the generalized differential $\partial^\circ \Phi$ in some more detail.

We consider the scenario where $G = H_1 \circ H_2$ is a composition of the operators

$$H_1 : X \to \prod_i L^{r_i}(\Omega), \quad H_2 : Y \to X,$$

with X a Banach space, and where $\psi = \psi_1 \circ \psi_2$ is a composition of the functions

$$\psi_1 : \mathbb{R}^l \to \mathbb{R}, \quad \psi_2 : \mathbb{R}^m \to \mathbb{R}^l.$$

We impose assumptions on ψ_1, ψ_2, H_1, and H_2 to ensure that G and ψ satisfy Assumption 3.32. Here is one way to do this.

Assumption 3.68. *There are $1 \leq r \leq r_i < q_i \leq \infty$, $1 \leq i \leq m$, such that*

(a) *The operators $H_1 : X \to \prod_i L^{r_i}(\Omega)$ and $H_2 : Y \to X$ are continuously Fréchet differentiable.*

(b) *The operator H_1 maps X locally Lipschitz continuously into $L^{q_i}(\Omega)$.*

(c) *The functions ψ_1 and ψ_2 are Lipschitz continuous.*

(d) *ψ_1 and ψ_2 are semismooth.*

It is straightforward to strengthen these assumptions such that they imply Assumptions 3.35. For brevity, we will not discuss the extension of the next theorem to semismoothness of order β, which is easily obtained by slight modifications of the assumptions and the proofs.

Theorem 3.69. *Let Assumptions* 3.68 *hold and let* $G = H_1 \circ H_2$ *and* $\psi = \psi_1 \circ \psi_2$. *Then*

(a) G *and* ψ *satisfy Assumptions* 3.32.

(b) Ψ *as defined in* (3.20) *is semismooth.*

(c) *The operator* $\Psi_1 : X \ni z \mapsto \psi(H_1(z)) \in L^r(\Omega)$ *is semismooth and the following* chain rule *holds:*

$$\partial^\circ \Psi(y) = \partial^\circ \Psi_1\big(H_2(y)\big)H_2'(y) = \big\{M_1 H_2'(y) : M_1 \in \partial^\circ \Psi_1\big(H_2(y)\big)\big\}.$$

(d) *If* $l = 1$ *and* ψ_1 *is strictly differentiable* [40, p. 30] *then the operator* $\Psi_2 : Y \ni y \mapsto \psi_2(G(y)) \in L^r(\Omega)$ *is semismooth and the following* chain rule *holds:*

$$\partial^\circ \Psi(y) = \psi_1'\big(\Psi_2(y)\big)\partial^\circ \Psi_2(y) = \big\{\psi_1'\big(\Psi_2(y)\big) \cdot M_2 : M_2 \in \partial^\circ \Psi_2(y)\big\}.$$

Proof. (a) Assumption 3.68 (a) implies Assumption 3.32 (a); Assumption 3.32 (b) follows from Assumption 3.68 (a) and (b); Assumption 3.68 (c) implies Assumption 3.32 (c); and Assumption 3.32 (d) holds by Assumption 3.68 (d), since the composition of semismooth functions is semismooth.

(b) By (a), we can apply Theorem 3.49.

(c) Assumptions 3.68 imply Assumptions 3.32 with H_1 and X instead of G and Y. Hence, Ψ_1 is semismooth by Theorem 3.49.

For the proof of the "$\subset$" part of the chain rule, let $M \in \partial^\circ \Psi(y)$ be arbitrary. By definition, there exists a measurable selection d of $\partial \psi(G(y))$ such that

$$M = \sum\nolimits_i d_i \cdot G_i'(y).$$

Now, since $G_i'(y) = H_{1i}'(H_2(y))H_2'(y)$,

$$M = \sum\nolimits_i d_i \cdot H_{1i}'\big(H_2(y)\big)H_2'(y) = M_1 H_2'(y),$$

where

$$M_1 = \sum\nolimits_i d_i \cdot H_{1i}'\big(H_2(y)\big). \tag{3.48}$$

Obviously, we have $M_1 \in \partial^\circ \Psi_1(H_2(y))$.

To prove the reverse inclusion, note that any $M_1 \in \partial^\circ \Psi_1(H_2(y))$ assumes the form (3.48) with appropriate measurable selection $d \in \partial \psi(G(y))$. Then

$$M_1 H_2'(y) = \sum\nolimits_i d_i \cdot \big(H_{1i}'\big(H_2(y)\big)H_2'(y)\big) = \sum\nolimits_i d_i \cdot G_i'(y),$$

which shows $M_1 H_2'(y) \in \partial^\circ \Psi(y)$.

(d) Certainly, G and ψ_2 satisfy Assumptions 3.32 (with ψ replaced by ψ_2). Hence, Theorem 3.49 yields the semismoothness of Ψ_2. We proceed by noting that, a.e. on Ω, there holds

$$\psi_1'\big(\Psi_2(y)(\omega)\big)\partial \psi_2\big(G(y)(\omega)\big) = \partial \psi\big(G(y)(\omega)\big), \tag{3.49}$$

where we have applied the chain rule for generalized gradients [40, Thm. 2.3.9] and the identity $\partial \psi_1 = \{\psi_1'\}$; see [40, Prop. 2.2.4].

We first prove the "$\supset$" direction of the chain rule. Let $M_2 \in \partial^\circ \Psi_2$ be arbitrary. It assumes the form

$$M_2 = \sum_i \hat{d}_i \cdot G_i'(y),$$

where $\hat{d} \in L^\infty(\Omega)^m$ is a measurable selection of $\partial\psi_2(G(y))$. Now for any operator M contained in the right-hand side of the assertion we have with $d \stackrel{\text{def}}{=} \psi_1'(\Psi_2(y))\hat{d}$

$$M = \psi_1'\big(\Psi_2(y)\big) \cdot M_2 = \sum_i d_i \cdot G_i'(y).$$

Obviously, $d \in L^\infty(\Omega)^m$ and, by (3.49), d is a measurable selection of $\partial\psi(G(y))$. Hence, $M \in \partial^\circ\Psi(y)$.

Conversely, to prove "$\subset$", let $M \in \partial^\circ\Psi(y)$ be arbitrary and denote by $d \in L^\infty(\Omega)^m$ the corresponding measurable selection of $\partial\psi(G(y))$. Now let $\tilde{d} \in L^\infty(\Omega)^m$ be a measurable selection of $\partial\psi_2(G(y))$ and define $\hat{d} \in L^\infty(\Omega)^m$ by

$$\begin{aligned} \hat{d}(\omega) &= \tilde{d}(\omega) && \text{on } \Omega_0 = \big\{\omega : \psi_1'\big(\Psi_2(y)(\omega)\big) = 0\big\},\\ \hat{d}(\omega) &= \frac{d(\omega)}{\psi_1'\big(\Psi_2(y)(\omega)\big)} && \text{on } \Omega \setminus \Omega_0. \end{aligned}$$

Then $\hat{d}$ is measurable and $d = \psi_1'(\Psi_2(y))\hat{d}$. Further, $\hat{d}(\omega) = \tilde{d}(\omega) \in \partial\psi_2(G(y))$ on Ω_0 and, using (3.49),

$$\hat{d}(\omega) = \frac{d(\omega)}{\psi_1'\big(\Psi_2(y)(\omega)\big)} \in \frac{\psi_1'\big(\Psi_2(y)(\omega)\big)\partial\psi_2\big(G(y)\big)}{\psi_1'\big(\Psi_2(y)(\omega)\big)} = \partial\psi_2\big(G(y)\big) \quad \text{on } \Omega \setminus \Omega_0.$$

Thus, $\hat{d}$ is a measurable selection of $\partial\psi_2(G(y))$, and consequently also $\hat{d} \in L^\infty(\Omega)^m$ due to the Lipschitz continuity of ψ_2. Therefore,

$$M_2 = \sum_i \hat{d}_i \cdot G_i'(y) \in \partial^\circ\Psi_2(y)$$

and thus $M \in \psi_1'\big(\Psi_2(y)\big) \cdot \partial^\circ\Psi_2(y)$ as asserted. □

3.3.8 Further Properties of the Generalized Differential

We now show that our generalized differential is convex-valued, weak compact-valued, and weakly graph closed. These properties can provide a basis for future research on the connections between $\partial^\circ\Psi$ and other generalized differentials, in particular the Thibault generalized differential [186] and the Ioffe–Ralph generalized differential [118, 171]. As weak topology on $\mathcal{L}(Y, L^r)$ we use the weak operator topology, which is defined by the seminorms

$$M \mapsto |\langle w, Mv\rangle_\Omega| = \left|\int_\Omega w(\omega)(Mv)(\omega)\,d\omega\right|, \quad v \in Y, \ \ w \in L^{r'}(\Omega),$$

where $L^{r'}(\Omega)$ is the dual space of $L^r(\Omega)$.

The following result will be of importance.

Lemma 3.70. *Under Assumption* 3.32, *the set* $K(y)$ *defined in* (3.26) *is convex and weak* sequentially compact in* $L^\infty(\Omega)^m$ *for all* $y \in Y$.

Proof. From Lemma 3.46 we know that $K(y) \subset L_\psi \bar{B}^m_{L^\infty}$ is nonempty and bounded. Further, the convexity of $\partial\psi(x)$ implies the convexity of $K(y)$. Now let $s_k \in K(y)$ tend to s in $L^2(\Omega)^m$. Then for a subsequence there holds $s_{k'}(\omega) \to s(\omega)$ for a.a. $\omega \in \Omega$. Since $\partial\psi(u(\omega))$ is compact, this implies that, for a.a. $\omega \in \Omega$, there holds $s(\omega) \in \partial\psi(u(\omega))$ and thus $s \in K(y)$. Hence, $K(y)$ is a bounded, closed, and convex subset of $L^2(\Omega)^m$ and thus weakly sequentially compact in $L^2(\Omega)^m$. Therefore, $K(y)$ is also weakly* sequentially closed in $L^\infty(\Omega)^m$, for, if $(s_k) \subset K(y)$ converges weakly* to s in $L^\infty(\Omega)^m$, then $\langle w, s_k - s\rangle_\Omega \to 0$ for all $w \in L^1(\Omega)^m \supset L^2(\Omega)^m$, showing that $s_k \to s$ weakly in $L^2(\Omega)^m$. Thus, $K(y)$ is weak* sequentially closed and bounded in $L^\infty(\Omega)^m$. Since $L^1(\Omega)^m$ is separable, this yields that $K(y)$ is weakly* sequentially compact. □

Convexity and Weak Compactness

As further useful properties of $\partial^\circ\Psi$ we prove the convexity and weak compactness of its images.

Theorem 3.71. *Under Assumptions* 3.32, *the generalized differential* $\partial^\circ\Psi(y)$ *is nonempty, convex, and weakly sequentially compact for all* $y \in Y$. *If* Y *is separable, then* $\partial^\circ\Psi(y)$ *is also weakly compact for all* $y \in Y$.

Proof. The nonemptiness was already stated in Proposition 3.47. The convexity follows immediately from the convexity of the set $K(y)$ derived in Lemma 3.46. We now prove weak sequential compactness. Let $(M_k) \subset \partial^\circ\Psi(y)$ be any sequence. Then

$$M_k = \sum\nolimits_i d_{ki} \cdot G_i'(y)$$

with $d_k \in K(y)$; see (3.26). Lemma 3.70 yields that $K(y)$ is weakly* sequentially compact in $L^\infty(\Omega)^m$. Hence, we can select a subsequence such that (d_k) converges weakly* to $d^* \in K(y)$ in $L^\infty(\Omega)^m$. Define $M^* = \sum_i d_i^* \cdot G_i(y)$ and observe that $M^* \in \partial^\circ\Psi(y)$, since $d^* \in K(y)$. It remains to prove that $M_k \to M^*$ weakly. Let $w \in L^{r'}(\Omega) = L^r(\Omega)'$ and $v \in Y$ be arbitrary. We set $z_i = w \cdot G_i'(y)v$ and note that $z_i \in L^1(\Omega)$. Hence,

$$\begin{aligned} |\langle w, (M_k - M^*)v\rangle_\Omega| &\le \sum\nolimits_i |\langle w, (d_k - d^*)_i \cdot G_i'(y)v\rangle_\Omega| \\ &= \sum\nolimits_i |\langle z_i, (d_k - d^*)_i\rangle_\Omega| \longrightarrow 0 \quad \text{as } k \to \infty. \end{aligned} \tag{3.50}$$

Therefore, the weak sequential compactness is shown.

By Lemma 3.42, $\partial^\circ\Psi(y)$ is contained in a closed ball in $\mathcal{L}(Y, L^r)$, on which the weak topology is metrizable if Y is separable (note that $1 \le r < \infty$ implies that $L^r(\Omega)$ is separable). Hence, in this case the weak compactness follows from the weak sequential compactness. □

Weak Graph Closedness of the Generalized Differential

Finally, we prove that the multifunction $\partial^\circ\Psi$ is weakly graph closed.

Theorem 3.72. *Let Assumptions 3.32 be satisfied and let $(y_k)\subset Y$ and $(M_k)\subset \mathcal{L}(Y,L^r(\Omega))$ be sequences such that $M_k\in\partial^\circ\Psi(y_k)$ for all k, $y_k\to y^*$ in Y, and $M_k\to M^*$ weakly in $\mathcal{L}(Y,L^r(\Omega))$. Then, there holds $M^*\in\partial^\circ\Psi(y^*)$. If, in addition, Y is separable, then the above assertion also holds if we replace the sequences (y_k) and (M_k) by nets.*

Proof. Let $y_k\to y^*$ in Y and $\partial^\circ\Psi(y_k)\ni M_k\to M^*$ weakly. We have the representations $M_k=\sum_i d_{ki}\cdot G_i'(y_k)$ with measurable selections d_k of $\partial\psi(u_k)$, where $u_k=G(y_k)$. We also introduce $u^*=G(y^*)$. The multifunction $\omega\in\Omega\mapsto\partial\psi(u^*(\omega))$ is closed-valued (even compact-valued) and measurable. Furthermore, the function $(\omega,h)\mapsto\|d_k(\omega)-h\|_2$ is a normal integrand on $\Omega\times\mathbb{R}^m$ [177, Cor. 2P]. Hence, by [177, Thm. 2K], the multifunctions $S_k:\Omega\to\mathbb{R}^m$,

$$S_k(\omega)=\operatorname*{arg\,min}_{h\in\partial\psi(u^*(\omega))}\|d_k(\omega)-h\|_2$$

are closed-valued (even compact-valued) and measurable. We choose measurable selections s_k of S_k. The sequence (s_k) is contained by Lemma 3.70 in the sequentially weak* compact set $K(y^*)\subset L^\infty(\Omega)^m$. Further, by Lemma 3.46, we have $d_k\in L_\psi\bar{B}^m_{L^\infty}$.

Hence, by transition to subsequences we achieve $s_k\to\bar{s}\in K(y^*)$ weak* in $L^\infty(\Omega)^m$ and $d_k\to\bar{d}\in L_\psi\bar{B}^m_{L^\infty}$ weak* in $L^\infty(\Omega)^m$. Therefore, $(d_k-s_k)\to(\bar{d}-\bar{s})$ weak* in $L^\infty(\Omega)^m$ and thus also weakly in $L^2(\Omega)^m$. Since $u_k\to u^*$ in $\prod_i L^{q_i}(\Omega)$, we achieve by transition to a further subsequence that $u_k\to u^*$ a.e. on Ω. Hence, since $d_k(\omega)\in\partial\psi(u_k(\omega))$ for a.a. $\omega\in\Omega$ and $\partial\psi$ is upper semicontinuous, we obtain from the construction of s_k that $(d_k-s_k)\to 0$ a.e. on Ω. The sequence (d_k-s_k) is bounded in $L^\infty(\Omega)^m$ and thus the Lebesgue convergence theorem yields $(d_k-s_k)\to 0$ in $L^2(\Omega)^m$. From $(d_k-s_k)\to 0$ and $(d_k-s_k)\to(\bar{d}-\bar{s})$ weakly in $L^2(\Omega)^m$ we see $\bar{d}=\bar{s}$. We thus have

$$d_k\to\bar{d}=\bar{s}\in K(y^*)\quad\text{weak}^*\text{ in }L^\infty(\Omega)^m.$$

This shows that $\bar{M}\overset{\text{def}}{=}\sum_i\bar{d}_i\cdot G_i'(y^*)\in\partial^\circ\Psi(y^*)$. It remains to prove that $M_k\to\bar{M}$ weakly. To show this, let $w\in L^{r'}(\Omega)=L^r(\Omega)'$ and $v\in Y$ be arbitrary. Then with $z_{ki}=w\cdot G_i'(y_k)v$ and $z_i=w\cdot G_i'(y^*)v$ there holds $z_{ki},z_i\in L^1(\Omega)$ and

$$\|z_{ki}-z_i\|_{L^1}\le\|w\|_{L^{r'}}\|G_i'(y_k)v-G_i'(y^*)v\|_{L^r}\to 0\quad\text{as }k\to\infty.$$

Hence, we obtain similarly as in (3.50)

$$\begin{aligned}|\langle w,(M_k-\bar{M})v\rangle_\Omega|&\le\sum_i\left|\langle w,d_{ki}\cdot G_i'(y_k)v-\bar{d}_i\cdot G_i'(y^*)v\rangle_\Omega\right|\\&=\sum_i\left|\langle d_{ki},z_{ki}\rangle_\Omega-\langle\bar{d}_i,z_i\rangle_\Omega\right|\\&\le\sum_i\left(|\langle\bar{d}_i-d_{ki},z_i\rangle_\Omega|+\|d_{ki}\|_{L^\infty}\|z_i-z_{ki}\|_{L^1}\right)\to 0\quad\text{as }k\to\infty.\end{aligned}$$

This implies $M^*=\bar{M}\in\partial^\circ\Psi(y^*)$ and completes the proof of the first assertion.

Now let $(y_\kappa) \subset Y$ and $(M_\kappa) \subset \mathcal{L}(Y, L^r(\Omega))$ be nets such that $M_\kappa \in \partial^\circ \Psi(y_\kappa)$ for all κ, $y_\kappa \to y^*$ in Y, and $M_\kappa \to M$ weakly in $\mathcal{L}(Y, L^r(\Omega))$. Since (y_κ) finally stays in any neighborhood of y^* and since G' is continuous, we see from (3.25) that without loss of generality (w.l.o.g.) we may assume that (M_κ) is contained in a bounded ball $\mathcal{B} \subset \mathcal{L}(Y, L^r)$. Since, due to the assumed separability of Y, $\mathcal{B}$ is metrizable with respect to the weak topology, we see that we can work with sequences instead of nets. $\square$

Chapter 4

Smoothing Steps and Regularity Conditions

The analysis of semismooth Newton methods used three ingredients: semismoothness, a smoothing step, and a regularity condition. In this chapter we show how smoothing steps can be obtained in practice and also describe a particular method that does not require a smoothing step at all. Furthermore, we establish sufficient conditions that imply the regularity condition stated in Assumption 4.5.

4.1 Smoothing Steps

We consider the VIP (1.14) with the assumptions stated there; i.e.,

$$\begin{aligned} &u \in \mathcal{B} \stackrel{\text{def}}{=} \{v \in L^2(\Omega) : a \leq v|_{\Omega_a},\ \ v|_{\Omega_b} \leq b\}, \\ &(F(u), v-u)_{L^2} \geq 0 \quad \forall\, v \in \mathcal{B}. \end{aligned} \tag{4.1}$$

Here, $\Omega \subset \mathbb{R}^n$ is assumed to be bounded in measure (alternatively, Ω could also be a surface with bounded measure). The lower and upper bounds satisfy $a \in L^p(\Omega_a)$ and $b \in L^p(\Omega_b)$ with $p > 2$. Furthermore, $\Omega_a, \Omega_b \subset \Omega$ are measurable and there holds $a \leq b$ on $\Omega_a \cap \Omega_b$. We extend a to Ω by $a|_{\Omega\setminus\Omega_a} = -\infty$, $b|_{\Omega\setminus\Omega_b} = +\infty$. Further, $F : L^2(\Omega) \to L^2(\Omega)$ is continuous and satisfies the following assumption.

Assumption 4.1. *The operator F has the form*

$$F(u) = \lambda u + G(u),$$

where λ is positive, $G : L^2(\Omega) \to L^2(\Omega)$, and there exists $2 < p \leq \infty$ such that the operator

$$L^2(\Omega) \ni u \mapsto G(u) \in L^p(\Omega)$$

is locally Lipschitz continuous.

It was already observed that many problems of practical interest can be stated as a VIP (1.14) with the operator F satisfying the above assumption. Note that $G(u)$ lives in a smoother space than its preimage u, since $L^p(\Omega) \hookrightarrow L^2(\Omega)$ (using that Ω is bounded) with nonequivalent norms. This form of G arises, e.g., in the first-order necessary optimality conditions of a large class of optimal control problems with bounds on the control and L^2-regularization. For obtaining smoothing steps, we use an idea that goes back to Kelley and Sachs [135]. The approach was already briefly sketched at the beginning of section 3.2.3.

Since $\phi^E_{[\alpha,\beta]}(x) = x_1 - P_{[\alpha,\beta]}(x_1 - x_2)$ is an MCP-function, we know that $\bar{u} \in L^2(\Omega)$ solves the VIP (1.14) if and only if $S(\bar{u}) = \bar{u}$, where

$$S(u) \stackrel{\text{def}}{=} P_{\mathcal{B}}\left(u - \lambda^{-1} F(u)\right), \quad P_{\mathcal{B}}(u) = \max\{a, \min\{u, b\}\}. \tag{4.2}$$

Further, for all $u \in L^2(\Omega)$ we have

$$u - \lambda^{-1} F(u) = -\lambda^{-1} G(u) \in L^p(\Omega),$$

and therefore $S(u) = P_{\mathcal{B}}(-\lambda^{-1} G(u))$. We now use that for all $v, w \in L^p(\Omega)$ there holds pointwise $|P_{\mathcal{B}}(v) - P_{\mathcal{B}}(w)| \le |v - w|$, and thus $\|P_{\mathcal{B}}(v) - P_{\mathcal{B}}(w)\|_{L^p} \le \|v - w\|_{L^p}$. Further, G is Lipschitz continuous (with modulus L_G) from an $L^2(\Omega)$-neighborhood of $\bar{u}$ to $L^p(\Omega)$. Hence, for all $u \in L^2(\Omega)$ in this neighborhood, we obtain

$$\begin{aligned}\|S(u) - \bar{u}\|_{L^p} &= \|S(u) - S(\bar{u})\|_{L^p} = \|P_{\mathcal{B}}(-\lambda^{-1} G(u)) - P_{\mathcal{B}}(-\lambda^{-1} G(\bar{u}))\|_{L^p} \\ &\le \lambda^{-1} \|G(u) - G(\bar{u})\|_{L^p} \le L_G \lambda^{-1} \|u - \bar{u}\|_{L^2}.\end{aligned}$$

This shows the following.

Theorem 4.2. *Let Assumption* 4.1 *hold and define S by* (4.2). *Then in every L^2-neighborhood of $\bar{u}$ on which G is Lipschitz continuous (with modulus L_G) as an operator to $L^p(\Omega)$, the mapping*

$$L^2(\Omega) \ni u_k^0 \mapsto u_k \stackrel{\text{def}}{=} S(u_k^0) \in L^p(\Omega)$$

is a smoothing step in the sense of Assumption 3.12 (b) *with $r = 2$ and constant $C_S = L_G/\lambda$.*

Remark 4.3. *In Assumption* 4.1 *as well as the subsequent investigation of $S(u)$ defined in* (4.2), *we can replace $L^2(\Omega)$ by $L^r(\Omega)$, $1 \le r < p$. Then Theorem* 4.2 *holds with $L^r(\Omega)$ replaced by $L^2(\Omega)$. In the context of the variational inequality* (4.1), *the space $L^2(\Omega)$ is, however, by far the most natural space.*

The applicability of this approach to concrete problems is discussed in application Chapters 9 and 10. Here we only consider the introductory example control problem (1.40) of section 1.6.1. There, see Remark 3.39, we have $F(u) = \lambda u - w(u)$, where $w(u) \in H_0^1(\Omega)$ is the adjoint state, which depends continuously and affine linearly on $u \in L^2(\Omega)$. Since $H_0^1(\Omega) \hookrightarrow L^p(\Omega)$ for appropriate $p > 2$, the described scenario is given with $G(u) = -w(u)$.

4.2 A Semismooth Newton Method without Smoothing Steps

We now describe how a variant of the MCP-function ϕ^E can be used to derive a semismooth reformulation of VIP to which a semismooth Newton method without smoothing step can be applied. Due to this nice property, this approach has developed into the standard approach.

In fact, the very same idea used in the construction of smoothing steps can be adopted. Here, we assume that F has the same structure as in section 4.1. The simple idea is to reformulate (1.14) equivalently as

$$u - S(u) = 0, \tag{4.3}$$

and to show the semismoothness of the operator $L^2(\Omega) \ni u \mapsto u - S(u) \in L^2(\Omega)$. This formulation appeared first in [102], where it was observed in the context of bound-constrained linear-quadratic optimal control problems that semismooth Newton methods applied to (4.3) are identical to the class of primal dual methods developed in [20, 21]. Numerical tests in these and many other papers have proved the excellent efficiency of this class of methods, and thus underline the potential and importance of semismooth Newton methods. These positive results are confirmed by all our numerical tests; see Chapter 9.

Theorem 4.4. *Let $F : L^2(\Omega) \to L^2(\Omega)$ be continuously differentiable and let Assumption* 4.1 *hold. Define the operator*

$$\Phi : u \in L^2(\Omega) \mapsto u - S(u) \in L^2(\Omega), \tag{4.4}$$

with S as defined in (4.2). *Then Φ is locally Lipschitz continuous and $\partial^\circ\Phi$-semismooth, with $\partial^\circ\Phi(u)$ consisting of all $M \in \mathcal{L}(L^2, L^2)$ of the form*

$$M = I + \lambda^{-1} d \cdot G'(u),$$

with $d \in L^\infty(\Omega)$,

$$d(\omega) \in \partial P_{[a(\omega),b(\omega)]}(-\lambda^{-1} G(u)(\omega)), \quad \omega \in \Omega. \tag{4.5}$$

If F' is α-order Hölder continuous, $\alpha \in (0,1]$, then Φ is β-order semismooth with β as given in Theorem 3.50.

Proof. In this proof, all assertions on local Lipschitz continuity, semismoothness, etc., are meant from $L^2(\Omega)$ to $L^2(\Omega)$. We introduce the disjoint measurable partitioning $\Omega = \Omega_f \cup \Omega_l \cup \Omega_u \cup \Omega_{lu}$,

$$\Omega^f = \Omega \setminus (\Omega_a \cup \Omega_b), \quad \Omega^l = \Omega_a \setminus \Omega_b, \quad \Omega^u = \Omega_b \setminus \Omega_a, \quad \Omega^{lu} = \Omega_a \cap \Omega_b.$$

Now, set $\bar{a} = a$ on Ω_a and $\bar{a} = 0$, otherwise; $\bar{b} = b$ on Ω_b and $\bar{b} = 1$, otherwise. Since $\Psi^f(u) = -\lambda^{-1} G(u)$ maps $L^2(\Omega)$ continuously differentiable to $L^2(\Omega)$, Ψ^f is locally Lipschitz continuous and $\{-\lambda^{-1} \cdot G'\}$-semismooth. On Ω^f we have $S(u) = \Psi^f(u)$. Hence, by Proposition 3.8, $\mathbf{1}_{\Omega^f} \cdot S$ is locally Lipschitz continuous and $\{\mathbf{1}_{\Omega^f} \cdot (-\lambda^{-1} \cdot G')\}$-semismooth. Obviously, this generalized differential consists of all operators of the form

$$\mathbf{1}_{\Omega^f} \cdot (-\lambda^{-1} d \cdot G')$$

with d as in (4.5).

Next, we set $\psi^l(t) = \max\{0, t\}$ and define $\Psi^l : L^2(\Omega) \to L^2(\Omega)$,

$$\Psi^l(u) = \psi^l\big(-\lambda^{-1} G(u) - \bar{a}\big).$$

By Proposition 3.36 and Theorem 3.49, this operator is locally Lipschitz continuous and $\partial^\circ \Psi^l$-semismooth. Furthermore, there holds $S(u) = \bar{a} + \Psi^l(u)$ on Ω^l, and thus $\mathbf{1}_{\Omega^l} \cdot S$ is locally Lipschitz continuous and $(\mathbf{1}_{\Omega^l} \cdot \partial^\circ \Psi^l)$-semismooth by Propositions 3.5 and 3.8. Looking at the structure of $\partial^\circ \Psi^l$ we see that $(\mathbf{1}_{\Omega^l} \cdot \partial^\circ \Psi^l)$ is the set of all operators

$$\mathbf{1}_{\Omega^l} \cdot [-\lambda^{-1} d \cdot G'(u)],$$

where $d \in L^\infty(\Omega)$ satisfies (4.5). In fact, for $\omega \in \Omega^l$ there holds with $\alpha = \bar{a}(\omega) = a(\omega)$

$$P_{[a(\omega),b(\omega)]}(t) = \max\{\alpha, t\} = \alpha + \max\{0, t-\alpha\} = \alpha + \psi^l(t-\alpha),$$

and thus

$$\partial P_{[\alpha,\infty)}(t) = \partial \psi^l(t-\alpha).$$

In a completely analogous way, we see that $\mathbf{1}_{\Omega^u} \cdot S$ is locally Lipschitz continuous and $(\mathbf{1}_{\Omega^u} \cdot \partial^\circ \Psi^u)$-semismooth, where the latter differential is the set of all operators

$$\mathbf{1}_{\Omega^u} \cdot [-\lambda^{-1} d \cdot G'(u)]$$

with $d \in L^\infty(\Omega)$ as in (4.5).

Finally, we consider $\omega \in \Omega^{lu}$. For $\alpha = \bar{a}(\omega) = a(\omega)$, $\beta = \bar{b}(\omega) = b(\omega)$ we have

$$\begin{aligned} P_{[a(\omega),b(\omega)]}(t) &= \max\{\alpha, \min\{t,\beta\}\} = \alpha + \max\{0, \min\{t-\alpha, \beta-\alpha\}\} \\ &= \alpha + (\beta-\alpha)\psi^{lu}\left(\frac{t-\alpha}{\beta-\alpha}\right) \end{aligned}$$

with $\psi^{lu}(t) = \max\{0, \min\{t,1\}\} = P_{[0,1]}(t)$. We conclude for $\omega \in \Omega^{lu}$

$$\partial P_{[a(\omega),b(\omega)]}(t) = (\beta-\alpha)\partial_t\left[\psi^{lu}\left(\frac{t-\alpha}{\beta-\alpha}\right)\right] = \partial\psi^{lu}\left(\frac{t-\alpha}{\beta-\alpha}\right). \tag{4.6}$$

Now define

$$\Psi^{lu}(u) = \psi^{lu}\left(-\frac{\lambda^{-1}G(u)+\bar{a}}{\bar{b}-\bar{a}}\right).$$

By Proposition 3.36 and Theorem 3.49, this operator is locally Lipschitz continuous and $\partial^\circ \Psi^{lu}$-semismooth. Furthermore, there holds

$$\mathbf{1}_{\Omega^{lu}} \cdot S = \mathbf{1}_{\Omega^{lu}} \cdot [\bar{a} + (\bar{b}-\bar{a}) \cdot \Psi^{lu}].$$

We use once again Propositions 3.5 and 3.8 to conclude that $\mathbf{1}_{\Omega^{lu}} \cdot S$ is locally Lipschitz continuous and $(\mathbf{1}_{\Omega^{lu}} \cdot (\bar{b}-\bar{a}) \cdot \partial^\circ \Psi^{lu})$-semismooth. From (4.6) we see that this differential is the set of all operators

$$\mathbf{1}_{\Omega^{lu}} \cdot [-\lambda^{-1} d \cdot G'(u)],$$

where $d \in L^\infty(\Omega)$ satisfies (4.5). Now, since

$$u - S(u) = u - \mathbf{1}_{\Omega^f} \cdot S(u) - \mathbf{1}_{\Omega^l} \cdot S(u) - \mathbf{1}_{\Omega^u} \cdot S(u) - \mathbf{1}_{\Omega^{lu}} \cdot S(u),$$

we can apply Proposition 3.5 to complete the proof of the first assertion.

If F' is α-Hölder continuous, then it is straightforward to modify the proof to establish semismoothness of order $\beta > 0$. □

Therefore, we can apply the Newton methods of section 3.2.3 to solve the reformulation (4.3) of the VIP. A smoothing step is not required, since Φ is semismooth as a mapping $L^2(\Omega) \to L^2(\Omega)$, and, as we will demonstrate for NCPs in section 4.3, it is appropriate to use Assumption 3.64 (a), i.e., the uniformly bounded invertibility of the generalized differentials in $\mathcal{L}(L^2, L^2)$ as a regularity condition.

We conclude this section by showing that if F is continuously Fréchet differentiable and Assumption 4.1 holds, then choice (4.4) is, up to scaling, the only semismooth operator $\Phi : L^2(\Omega) \to L^2(\Omega)$ that can be obtained by applying NCP- or MCP-functions pointwise to the pair $(u, F(u))$. In fact, the construction of a semismooth reformulation of the VIP in the form

$$\Phi(u) = 0$$

such that $\Phi : L^2(\Omega) \to L^2(\Omega)$ is semismooth was based on the idea of using the structure $F(u) = \lambda u + G(u)$ and to find NCP- and MCP-functions in which u appears linearly, while $G(u)$ may appear nonlinearly. In fact, any nonlinear direct appearance of u in the superposition operator would destroy the semismoothness of $\Phi : L^2(\Omega) \to L^2(\Omega)$. This structural requirement means

$$\phi_{[a(\omega),b(\omega)]}(u(\omega), \lambda u(\omega) + G(u)(\omega)) = cu(\omega) - \theta(G(u)(\omega))$$

with a constant $c \neq 0$ and a suitable function $\theta : \mathbb{R} \to \mathbb{R}$. Thus, with α, β, x_1, and x_2 replacing $a(\omega)$, $b(\omega)$, $u(\omega)$, and $F(u)(\omega) = \lambda u(\omega) + G(u)(\omega)$, respectively, the term expressing $G(u)(\omega)$ is $x_2 - \lambda x_1$ and we obtain the structural requirement

$$\phi_{[\alpha,\beta]}(x_1, x_2) = cx_1 - \theta(x_2 - \lambda x_1).$$

In the following, we will only consider the case $-\infty < \alpha < \beta < +\infty$; the unilateral cases are even simpler.

From $\phi_{[\alpha,\beta]}(\alpha, x_2) = 0$ for all $x_2 \geq 0$ we see that

$$c\alpha - \theta(x_2 - \lambda\alpha) = 0 \quad \forall\, x_2 \geq 0$$

and thus

$$\theta(t) = c\alpha \quad \forall\, t \geq -\lambda\alpha.$$

Similarly, $\phi_{[\alpha,\beta]}(\beta, x_2) = 0$ for all $x_2 \leq 0$ implies

$$c\beta - \theta(x_2 - \lambda\beta) = 0 \quad \forall\, x_2 \leq 0,$$

hence

$$\theta(t) = c\beta \quad \forall\, t \leq -\lambda\beta.$$

Further, for all $x_1 \in (\alpha, \beta)$ there holds $\phi_{[\alpha,\beta]}(x_1, 0) = 0$ and thus

$$cx_1 - \theta(-\lambda x_1) = 0 \quad \forall\, x_1 \in (\alpha, \beta).$$

From this we conclude

$$\theta(t) = -\frac{c}{\lambda} t \quad \forall\, t \in (-\lambda\beta, -\lambda\alpha).$$

We thus see that necessarily there holds

$$\theta(t) = c P_{[\alpha,\beta]}\left(\frac{-t}{\lambda}\right).$$

Therefore, modulo scaling by $c \neq 0$, the function

$$\phi_{[\alpha,\beta]}(x) = x_1 - P_{[\alpha,\beta]}(x_2 - \lambda x_1)$$

is the only MCP-function that has the desired structure.

4.3 Sufficient Conditions for Regularity

In this section we establish a sufficient condition for solutions of the NCP (1.17), posed in the usual setting of (1.14), that implies the following regularity condition.

Assumption 4.5. *There exist constants $\eta > 0$ and $C_{M^{-1}} > 0$ such that, for all $u \in \bar{u} + \eta B_{L^p}$, every $M \in \partial^\circ\Phi(u)$ is an invertible element of $\mathcal{L}(L^2, L^2)$ with $\|M^{-1}\|_{L^2,L^2} \leq C_{M^{-1}}$.*

Here, $\Phi = \phi(u, F(u))$ is the superposition operator arising in the semismooth reformulation via the NCP-function ϕ. We consider problems where F has the form $F(u) = \lambda u + G(u)$, and G has a smoothing property. In this setting we show that, in broad terms, regularity is implied by L^2-coercivity of $F'(\bar{u})$ on the tangent space of the strongly active constraints.

An alternative sufficient condition for regularity, which does not require special structure of F but assumes that $F'(\bar{u})$ is L^2-coercive on the whole space, can be found in [192].

We work under the following assumptions.

Assumption 4.6. *There exist $p \in [2,\infty]$ and $p' \in (2,\infty]$ such that*

(a) $F(u) = \lambda u + G(u), \quad \lambda \in L^\infty(\Omega), \quad \lambda \geq \lambda_0 > 0.$

(b) *$G : L^2(\Omega) \to L^2(\Omega)$ is Fréchet differentiable with derivative $G'(u)$.*

(c) *$u \in L^p(\Omega) \to G'(u) \in \mathcal{L}(L^2(\Omega), L^2(\Omega))$ is continuous near $\bar{u}$.*

(d) *For u near $\bar{u}$ in $L^p(\Omega)$, the L^2-endomorphisms $G'(u)$ and $G'(u)^*$ are contained in $\mathcal{L}(L^2(\Omega), L^{p'}(\Omega))$ with their norms uniformly bounded by a constant $C_{G'}$.*

(e) *There exists a constant $\nu > 0$ such that for $F'(\bar{u}) = \lambda I + G'(u)$ there holds*

$$(v, F'(\bar{u})v)_{L^2(\Omega)} \geq \nu\|v\|^2_{L^2(\Omega)}$$

for all $v \in L^2(\Omega)$ with $v = 0$ on $\{\omega \in \Omega : F(\bar{u})(\omega) \neq 0\}$.

(f) *ϕ is Lipschitz continuous and semismooth.*

(g) *There exists a constant $\theta > 0$ such that for all $x \in \mathbb{R}^2$ and all $g \in \partial\phi(x)$ there holds*

$$g_1 g_2 \geq 0, \quad |g_1 + g_2| \geq \theta.$$

(h) *For $x \in (0,\infty) \times \{0\}$ there holds $\partial\phi(x) \subset \{0\} \times \mathbb{R}$, and for $x \in \{0\} \times (0,\infty)$ there holds $\partial\phi(x) \subset \mathbb{R} \times \{0\}$.*

Remark 4.7. *In the case of a minimization problem, i.e., $F = j'$, condition* (e) *can be interpreted as a strong second-order sufficient condition: The Hessian operator $j''(\bar{u})$ has to be coercive on the tangent space of the strongly active constraints. Similar conditions can be found in, e.g., Dunn and Tian* [62] *and Ulbrich and Ulbrich* [195]. *Strong second-order sufficient conditions are also essential for proving fast convergence of finite-dimensional algorithms; see, e.g.,* [25, 99, 149].

Observe that Assumption 4.6 with $p > 2$ implies Assumption 3.37 with $r = 2$ and $p' = \min\{p, p'\}$ on an L^p-neighborhood of $\bar{u}$. Hence, $\Phi : L^p(\Omega) \to L^2(\Omega)$ is semismooth at $\bar{u}$ by Theorem 3.53. In fact, (a)–(c) imply Assumption 3.37 (a). Further, for $u, u+v \in L^p(\Omega)$ near $\bar{u}$ there holds with $s = \min\{p, p'\}$, using (d),

$$\begin{aligned}\|F(u+v) - F(u)\|_{L^s} &\le \int_0^1 \|F'(u+tv)v\|_{L^s} dt \\ &\le c\|\lambda\|_{L^\infty}\|v\|_{L^p} + c \sup_{t\in[0,1]} \|G'(u+tv)\|_{L^p, L^{p'}}\|v\|_{L^p} \\ &\le c(\|\lambda\|_{L^\infty} + C_{G'})\|v\|_{L^p},\end{aligned}$$

where $c > 0$ is a suitable constant. This implies Assumption 3.37 (b) for $p' = s$. Finally (f) ensures Assumption 3.37 (c) and (d).

Next, we illustrate Assumptions 4.6 by verifying them for the optimal control problem (1.40). There, $F(u) = j'(u) = \lambda u - B^* w(u)$, where

$$w(u) = -A^{-1}B(B^*y - y_d) = -A^{-1}B^*(BA^{-1}Bu - y_d) \in H_0^1(\Omega) \tag{4.7}$$

is the adjoint state. Here, $A \in \mathcal{L}(H_0^1(\Omega), H^{-1}(\Omega))$ denotes the elliptic operator corresponding to $-\Delta$. Although this is often omitted in favor of compact notation, we have included the natural injection operators

$$\begin{aligned}B &= I_{L^2\to H^{-1}} : \ L^2(\Omega) \ni u \mapsto u \in H^{-1}(\Omega), \\ B^* &= I_{H_0^1\to L^2} : \ H_0^1(\Omega) \ni y \mapsto y \in L^2(\Omega)\end{aligned}$$

for the purpose of precise notation. The operator $L^2(\Omega) \ni u \mapsto w(u) \in H_0^1(\Omega)$ is continuous and affine linear. Thus, choosing $p' > 2$ such that $H_0^1(\Omega) \hookrightarrow L^{p'}(\Omega)$, F has the form as in assumption (a) with

$$G : L^2(\Omega) \ni u \mapsto -B^* w(u) \in L^2(\Omega)$$

being continuous and affine linear. In addition,

$$L^2(\Omega) \ni u \mapsto G(u) \in L^{p'}(\Omega)$$

is continuous and affine linear, too. Therefore, G is smooth and $G'(u) \in \mathcal{L}(L^2, L^{p'})$ is constant. From (4.7) we see that $G'(u) = B^*A^{-1}BB^*A^{-1}B$. Using $A^* = A$, hence $(A^{-1})^* = A^{-1}$, we conclude $G'(u)^* = G'(u)$. Further, with $v \in L^2(\Omega)$ and $z = A^{-1}Bv \in H_0^1(\Omega)$, we have

$$\begin{aligned}(F'(u)v, v)_{L^2} &= (G'(u)v, v)_{L^2} + (\lambda v, v)_{L^2} = (B^*A^{-1}Bv, B^*A^{-1}Bv)_{L^2} + (\lambda v, v)_{L^2} \\ &= \|z\|_{L^2}^2 + \lambda\|v\|_{L^2}^2 \ge \lambda\|v\|_{L^2}^2.\end{aligned}$$

Taking all together, we see that (a)–(e) are satisfied for every $p \in [2, \infty]$.

We now state and prove our sufficient condition for regularity.

Theorem 4.8. *If Assumption* 4.6 *holds at a solution* $\bar{u} \in L^p(\Omega)$ *of the NCP* (1.17), *then there exists* $\rho > 0$ *such that Assumption* 4.5 *is satisfied.*

Proof. For convenience, we set $(\cdot,\cdot) = (\cdot,\cdot)_{L^2(\Omega)}$ and $\|\cdot\| = \|\cdot\|_{L^2(\Omega)}$.

Every element $M \in \partial^\circ\Phi(u)$ can be written in the form

$$M = d_1 \cdot I + d_2 \cdot F'(u), \quad d_i \in L^\infty(\Omega), \quad (d_1, d_2) \in \partial\phi(u). \tag{4.8}$$

Due to the Lipschitz continuity of ϕ, the functions d_1, d_2 are bounded in $L^\infty(\Omega)$ uniformly in u. We define

$$c = \frac{d_2}{d_1 + \lambda d_2}, \tag{4.9}$$

which, since by assumption $d_1 d_2 \geq 0$, $\theta \leq |d_1 + d_2|$, and $\lambda \geq \lambda_0 > 0$, is well defined and uniformly bounded in $L^\infty(\Omega)$ for all $u \in L^p(\Omega)$. Furthermore, there holds $c \geq 0$. Using $F'(u) = \lambda I + G'(u)$, we see that

$$M = (d_1 + \lambda d_2) \cdot (I + c \cdot G'(u)).$$

Since $(d_1 + \lambda d_2)$ and $(d_1 + \lambda d_2)^{-1}$ are uniformly bounded in $L^\infty(\Omega)$ for all $u \in L^p(\Omega)$, the operators $M \in \partial^\circ\Phi(u)$ are continuously invertible in $\mathcal{L}(L^2(\Omega), L^2(\Omega))$ on an L^p-neighborhood of $\bar{u}$ with uniformly bounded inverses if and only if the same holds true for the operators $T = I + c \cdot G'(u)$.

Next, consider any $\bar{M} \in \partial^\circ\Phi(\bar{u})$ with corresponding functions $\bar{d}_1, \bar{d}_2, \bar{c} \in L^\infty(\Omega)$ according to (4.8) and (4.9). Define the sets

$$\Omega_1 = \{(\bar{u}, F(\bar{u})) \neq 0\}, \quad \Omega_2 = \{\bar{u} = 0, \ F(\bar{u}) = 0\},$$

and consider the function $e \in L^\infty(\Omega)$,

$$e = \bar{c} \ \text{ on } \Omega_1, \quad e = c \ \text{ on } \Omega_2. \tag{4.10}$$

From $c, \bar{c} \geq 0$ it follows that $e \geq 0$. We first prove that for arbitrary $t \in [1, \infty)$,

$$\|c - e\|_{L^t} \to 0 \quad \text{as } u \to \bar{u} \text{ in } L^p(\Omega). \tag{4.11}$$

Assume that this is not true. Then there exist $t \geq 1$, $\varepsilon > 0$, and a sequence $(u_k) \subset L^p(\Omega)$ with $u_k \to \bar{u}$ in $L^p(\Omega)$ and corresponding differentials $M_k \in \partial^\circ\Phi(u_k)$ such that

$$\|c_k - e_k\|_{L^t} \geq \varepsilon \quad \forall\, k. \tag{4.12}$$

Here, we denote by d_{1k}, d_{2k}, c_k, and e_k the associated functions defined in (4.8), (4.9), and (4.10). From $u_k \to \bar{u}$ it follows $F(u_k) \to F(\bar{u})$ in $L^{\min\{p,p'\}}(\Omega)$. Hence, there exists a subsequence such that $(u_{k'}, F(u_{k'})) \to (\bar{u}, F(\bar{u}))$ a.e. on Ω.

Since $\bar{u}F(\bar{u}) = 0$, we have the disjoint partitioning $\Omega_1 = \Omega_{11} \cup \Omega_{12}$ with

$$\Omega_{11} = \{F(\bar{u}) \neq 0\} = \{\bar{u} = 0, \ F(\bar{u}) \neq 0\},$$
$$\Omega_{12} = \{\bar{u} \neq 0\} = \{\bar{u} \neq 0, \ F(\bar{u}) = 0\}.$$

On the set Ω_{11} we have (a.e.) $u_{k'} \to 0$, $F(u_{k'}) \to F(\bar{u}) \neq 0$ and thus, by the upper semicontinuity of $\partial\phi$ and the assumptions on ϕ, $\liminf_{k'\to\infty} |d_{1k'}| \geq \theta$, $d_{2k'} \to 0$, which implies $c_{k'} \to 0 = \bar{c}$ on Ω_{11}. Since Ω has finite measure and the sequence $(c_{k'})$ is bounded in $L^\infty(\Omega)$, the Lebesgue convergence theorem implies

$$\|c_{k'} - \bar{c}\|_{L^t(\Omega_{11})} \to 0. \tag{4.13}$$

On the set Ω_{12} there holds $u_{k'} \to \bar{u} \neq 0$, $F(u_{k'}) \to F(\bar{u}) = 0$ and thus, again using the properties of $\partial\phi$, $d_{1k'} \to 0 = \bar{d}_1$, $\liminf_{k'\to\infty} |d_{2k'}| \geq \theta$, which implies $c_{k'} \to 1/\lambda = \bar{c}$. Invoking Lebesgue's convergence theorem once again we see that

$$\|c_{k'} - \bar{c}\|_{L^t(\Omega_{12})} \to 0. \tag{4.14}$$

Then it is an immediate consequence of (4.13) and (4.14) that

$$\|c_{k'} - e_{k'}\|_{L^t(\Omega)} = \|c_{k'} - \bar{c}\|_{L^t(\Omega_1)} \leq \|c_{k'} - \bar{c}\|_{L^t(\Omega_{11})} + \|c_{k'} - \bar{c}\|_{L^t(\Omega_{12})} \to 0,$$

which contradicts (4.12). Thus, (4.11) is proved.

We now consider the operators

$$T = I + c \cdot G'(u) \quad \text{and} \quad S = I + e \cdot G'(\bar{u}).$$

For all $v \in L^2(\Omega)$ there holds (with $2p'/(p'-2)$ to be interpreted as 2 if $p' = \infty$)

$$\begin{aligned}
\|Tv - Sv\| &\leq \|(c-e)\cdot G'(\bar{u})v\| + \|c\cdot(G'(u)v - G'(\bar{u})v)\| \\
&\leq \|c-e\|_{L^{\frac{2p'}{p'-2}}} \|G'(\bar{u})v\|_{L^{p'}} + \|c\|_{L^\infty}\|G'(u)v - G'(\bar{u})v\| \\
&\leq \|c-e\|_{L^{\frac{2p'}{p'-2}}} \|G'(\bar{u})\|_{L^2,L^{p'}}\|v\| + \|c\|_{L^\infty}\|G'(u) - G'(\bar{u})\|_{L^2,L^2}\|v\|.
\end{aligned}$$

This proves

$$\|T - S\|_{L^2,L^2} \to 0 \quad \text{as } u \to \bar{u} \text{ in } L^p(\Omega). \tag{4.15}$$

Next, we prove

$$\|S^*v\| \geq \gamma\|v\| \quad \forall\, v \in L^2(\Omega), \tag{4.16}$$

where $\gamma = 1$ if $G'(\bar{u}) = 0$ and

$$\gamma = \min\left\{\nu\kappa, \frac{1}{2}\right\}, \quad \kappa = \frac{1}{2\|G'(\bar{u})^*\|_{L^2,L^2}} \quad \text{if } G'(\bar{u}) \neq 0.$$

The assertion is trivial if $G'(\bar{u}) = 0$. To prove the assertion for $G'(\bar{u}) \neq 0$, we set $w = ev$ and distinguish two cases.

Case 1: $\|w\| \leq \kappa\|v\|$.

Then

$$\begin{aligned}
\|S^*v\| &= \|v + G'(\bar{u})^*(ev)\| \\
&\geq \|v\| - \|G'(\bar{u})^*w\| \\
&\geq (1 - \kappa\|G'(\bar{u})^*\|_{L^2,L^2})\|v\| \geq \frac{1}{2}\|v\| \geq \gamma\|v\|.
\end{aligned}$$

Case 2: $\|w\| > \kappa\|v\|$.

Since $w = ev$ and $e = \bar{c} = 0$ on Ω_{11}, we have $w = 0$ on Ω_{11} and thus, by (e),

$$(w, (\lambda I + G'(\bar{u})^*)w) \geq \nu\|w\|^2.$$

In the calculations to follow we will use that

$$1 - \lambda e = 1 \quad \text{on } \Omega_{11}, \qquad 1 - \lambda e = 1 - \lambda\bar{c} = 0 \quad \text{on } \Omega_{12},$$
$$1 - \lambda e = 1 - \lambda c = \frac{d_1 + \lambda d_2 - \lambda d_2}{d_1 + \lambda d_2} = \frac{d_1}{d_1 + \lambda d_2} \geq 0 \quad \text{on } \Omega_2.$$

In particular, $1 - \lambda e \geq 0$ on Ω, and thus

$$\begin{aligned}\|w\|\,\|S^*v\| &\geq (w, S^*v) = (w, v) + (w, G'(\bar{u})^*w) \\ &\geq (w, v) + \nu\|w\|^2 - (w, \lambda w) = (w, (1 - \lambda e)v) + \nu\|w\|^2 \\ &= (v, e(1 - \lambda e)v) + \nu\|w\|^2 \geq \nu\|w\|^2 \geq \nu\kappa\|w\|\,\|v\| \geq \gamma\|w\|\,\|v\|.\end{aligned}$$

Hence, (4.16) is proved. In particular, S^* is injective. Moreover, S^* has closed range. In fact, let $S^*v_k \to z$. Then

$$\|v_k - v_l\| \leq \gamma^{-1}\|S^*v_k - S^*v_l\| \to 0 \quad \text{as } k, l \to \infty.$$

Therefore, (v_k) is a Cauchy sequence and thus $v_k \to v$ for some $v \in L^2(\Omega)$. This implies $S^*v_k \to S^*v$, hence $z = S^*v$. By the closed range theorem [129, Ch. XII], the injectivity of S^* now implies the surjectivity of S.

We proceed by showing the injectivity of S. Consider any $v \in L^2(\Omega)$ with $Sv = 0$. Let us introduce the function $z \in L^{p'}(\Omega)$,

$$z = 0 \ \text{ on } \Omega_{11}, \quad z = -G'(\bar{u})v \ \text{ on } \Omega_{12} \cup \Omega_2. \tag{4.17}$$

Observing that

$$v = Sv - e \cdot G'(\bar{u})v = -e \cdot G'(\bar{u})v \quad \text{on } \Omega,$$

and $e = 0$ on Ω_{11}, we see that

$$v = ez \quad \text{on } \Omega,$$

and that v vanishes on Ω_{11}. Therefore, using (e),

$$\begin{aligned}0 = (z, Sv) &= (z, v) + (z, e \cdot G'(\bar{u})v) = (z, v) + (ez, G'(\bar{u})v) \\ &= (z, v) + (v, G'(\bar{u})v) \geq (z, v) + \nu\|v\|^2 - (v, \lambda v) = \nu\|v\|^2 + (z - \lambda ez, ez) \\ &= \nu\|v\|^2 + (z, (1 - \lambda e)ez) \geq \nu\|v\|^2,\end{aligned}$$

since $(1 - \lambda e)e \geq 0$. This implies $v = 0$, which proves the injectivity of S.

We thus have shown that $S \in \mathcal{L}(L^2(\Omega), L^2(\Omega))$ is bijective and hence, by the open mapping theorem, continuously invertible.

Furthermore, for all $v \in L^2(\Omega)$ we have

$$\|v\| = \|S^*(S^*)^{-1}v\| \geq \gamma\|(S^*)^{-1}v\|,$$

and thus

$$\|S^{-1}\|_{L^2,L^2} = \|(S^*)^{-1}\|_{L^2,L^2} \le \frac{1}{\gamma}.$$

By (4.15), there exists $\rho > 0$ such that for all $u \in L^p(\Omega)$, $\|u - \bar{u}\|_{L^p} \le \rho$, there holds $\|T - S\|_{L^2,L^2} \le \gamma/2$. Therefore, by Banach's theorem [129, Ch. V.4.6], $T \in \mathcal{L}(L^2(\Omega), L^2(\Omega))$ is invertible with

$$\|T^{-1}\|_{L^2,L^2} \le \frac{\|S^{-1}\|_{L^2,L^2}}{1 - \|S^{-1}\|_{L^2,L^2}\|T - S\|_{L^2,L^2}} \le \frac{2}{\gamma}. \qquad \square$$

The sufficient condition of Theorem 4.8 and the sufficient condition for regularity established in [192] are very helpful for proving regularity conditions in concrete applications.

Chapter 5

Variational Inequalities and Mixed Problems

So far, we have demonstrated the applicability of semismooth Newton methods mainly for the NCP (1.17). We now discuss several applications to more general classes of problems. First, we show how the semismooth reformulation approach that we investigated in detail for the NCP can be extended to the larger problem class of bound-constrained VIPs (1.14). In addition, we describe how semismooth reformulations can be obtained for even more general problems than the bound-constrained VIP. The second extension considers mixed problems consisting of VIPs and additional operator equations. In particular, the first-order necessary (Karush–Kuhn–Tucker, KKT) conditions of very general optimization problems can be written in this form.

5.1 Application to Variational Inequalities

5.1.1 Problems with Bound Constraints

We now describe how our treatment of the NCP can be carried over to the bound-constrained VIP (1.14). One possibility was already described in section 4.2, where we presented a semismooth reformulation that does not require a smoothing step. Here, we describe a similar approach for which general NCP- and MCP-functions can be used.

For the derivation of a semismooth reformulation, assume an NCP-function ϕ and MCP-functions $\phi_{[\alpha,\beta]}$ for all compact intervals. We now define the operator

$$\Phi(u)(\omega) = \begin{cases} F(u)(\omega) & \text{on } \Omega^f = \Omega \setminus (\Omega_a \cup \Omega_b), \\ \phi\big(u(\omega) - a(\omega), F(u)(\omega)\big) & \text{on } \Omega^l = \Omega_a \setminus \Omega_b, \\ -\phi\big(b(\omega) - u(\omega), -F(u)(\omega)\big) & \text{on } \Omega^u = \Omega_b \setminus \Omega_a, \\ \phi_{[a(\omega),b(\omega)]}\big(u(\omega), F(u)(\omega)\big) & \text{on } \Omega^{lu} = \Omega_a \cap \Omega_b. \end{cases} \tag{5.1}$$

It was shown in section 1.4.2 that $u \in L^2(\Omega)$ solves (1.14) if and only if

$$\Phi(u) = 0. \tag{5.2}$$

Also, it was argued that often the structure of F allows us to conclude that the solution $\bar{u}$ lives in a stronger space $L^p(\Omega)$, $p > 2$, than $L^2(\Omega)$. In the following, we will consider the superposition operator Φ as a mapping from $L^p(\Omega)$ to $L^r(\Omega)$ with $1 \le r < p \le \infty$. Our aim is to prove the semismoothness of Φ and to characterize its generalized differential. We require the following assumption.

Assumption 5.1. *There exist $r \in [1,p) \cap [1,p')$ such that*

(a) *The mapping $u \in L^p(\Omega) \mapsto F(u) \in L^r(\Omega)$ is continuously differentiable.*

(b) *The operator $F: L^p(\Omega) \to L^{p'}(\Omega)$ is locally Lipschitz continuous.*

(c) *The function $\phi: \mathbb{R}^2 \to \mathbb{R}$ is Lipschitz continuous and semismooth.*

(d) *The function $x \mapsto \psi_{[x_1,x_2]}(x_3,x_4)$ is Lipschitz continuous and semismooth.*

For semismoothness of higher order we need slightly stronger requirements.

Assumption 5.2. *There exists $r \in [1,p) \cap [1,p')$ and $\alpha \in (0,1]$ such that*

(a) *The mapping $u \in L^p(\Omega) \mapsto F(u) \in L^r(\Omega)$ is differentiable with locally α-Hölder continuous derivative.*

(b) *The operator $F: L^p(\Omega) \to L^{p'}(\Omega)$ is locally Lipschitz continuous.*

(c) *The function $\phi: \mathbb{R}^2 \to \mathbb{R}$ is Lipschitz continuous and α-order semismooth.*

(d) *The function $x \mapsto \psi_{[x_1,x_2]}(x_3,x_4)$ is Lipschitz continuous and α-order semismooth.*

Remark 5.3. *At this point it would be more convenient if we had established semismoothness results for superposition operators of the form $\psi(\omega, G(u)(\omega))$. This is certainly possible, but not really needed in this work. Instead, the trick we will use here is to build superposition operators with the inner operator given by $u \mapsto (\bar{a}, \bar{b}, u, F(u))$, where $\bar{a}$ and $\bar{b}$ are cutoff versions of a and b to make them finite.*

A different approach would be to transform the problem such that $[a,b] \to [0,1]$ on $\Omega_a \cap \Omega_b$ and $[a,b] \to [0,\infty)$ on $(\Omega_a \cup \Omega_b) \setminus (\Omega_a \cap \Omega_b)$. There is, however, a certain danger that this transformation affects the scaling of the problem in a negative way. The latter approach was implicitly used in the proof of Theorem 4.4.

Theorem 5.4. *Under Assumption* 5.1 *the operator $\Phi: L^p(\Omega) \to L^r(\Omega)$ is locally Lipschitz continuous and $\partial^\circ\Phi$-semismooth, where $\partial^\circ\Phi(u)$ consists of all operators $M \in \mathcal{L}(L^p, L^r)$ of the form*

$$M = d_1 I + d_2 \cdot F'(u),$$

with $d_1, d_2 \in L^\infty(\Omega)$,

$$(d_1,d_2)(\omega) \in \begin{cases} \{(0,1)\} & \text{on } \Omega^f, \\ \partial\phi\big(u(\omega) - a(\omega), F(u)(\omega)\big) & \text{on } \Omega^l, \\ \partial\phi\big(b(\omega) - u(\omega), -F(u)(\omega)\big) & \text{on } \Omega^u, \\ \partial\phi_{[a(\omega),b(\omega)]}\big(u(\omega), F(u)(\omega)\big) & \text{on } \Omega^{lu}. \end{cases} \tag{5.3}$$

Under Assumption 5.2 *the operator Φ is even β-order semismooth, where $\beta > 0$ is as in Theorem* 3.50.

Proof. Let us define $\bar{a}, \bar{b} \in L^p(\Omega)$ by $\bar{a} = a$ on Ω_a, $\bar{a} = 0$, otherwise; $\bar{b} = b$ on Ω_b, $\bar{b} = 0$, otherwise. Further, we introduce four functions $\mathbb{R}^4 \to \mathbb{R}$,

$$\psi^f(x) = x_4, \quad \psi^l(x) = \phi(x_3 - x_1, x_4),$$
$$\psi^u(x) = -\phi(x_2 - x_3, -x_4), \quad \psi^{lu}(x) = \phi_{[x_1,x_2]}(x_3, x_4),$$

which are Lipschitz continuous and semismooth. Define the operator

$$T : u \in L^p(\Omega) \mapsto (\bar{a}, \bar{b}, u, F(u)) \in L^r(\Omega)^4,$$

which is continuously differentiable with derivative $T'(u) = (0|0|I|F'(u))$, and locally Lipschitz continuous as a mapping $L^p(\Omega) \to L^p(\Omega)^3 \times L^{p'}(\Omega)$.

Next, for $\gamma \in \{f, l, u, lu\}$, we introduce the superposition operators

$$\Psi^\gamma : L^p(\Omega) \to L^r(\Omega), \quad \Psi^\gamma(u)(\omega) = \psi^\gamma\big(T(u)(\omega)\big).$$

By Proposition 3.36 and Theorem 3.49, these operators are $\partial^\circ\Psi^\gamma$-semismooth; here, the operator $M^\gamma \in \mathcal{L}(L^r, L^r)$ is an element of $\partial^\circ\Psi^\gamma(u)$ if and only if

$$M^\gamma = (d_a^\gamma, d_b^\gamma, d_1^\gamma, d_2^\gamma) \cdot T'(u) = d_1^\gamma I + d_2^\gamma \cdot F'(u),$$

where $d_a^\gamma, d_b^\gamma, d_1^\gamma, d_2^\gamma \in L^\infty(\Omega)$ satisfy $(d_a^\gamma, d_b^\gamma, d_1^\gamma, d_2^\gamma) \in \partial\psi^\gamma(T(u))$ on Ω. We now use [40, Prop. 2.3.16], a direct consequence of Proposition 2.3, to conclude

$$\partial_{(x_3,x_4)}\psi^\gamma(x) \subset \{g \in \mathbb{R}^2 : \exists h \in \mathbb{R}^2 : \ (h, g) \in \partial\psi^\gamma(x)\}.$$

Now let $d_1, d_2 \in L^\infty(\Omega)$ be arbitrary such that (5.3) holds. Then holds $(d_1, d_2) \in \partial_{(x_3,x_4)}\psi^\gamma(T(u))$ on Ω^γ. Therefore, using Filippov's theorem [15, Thm. 8.2.10], we conclude that there exist $d_a^\gamma, d_b^\gamma \in L^\infty(\Omega)$ with

$$(d_a^\gamma, d_b^\gamma, d_1, d_2) \in \partial\psi^\gamma(T(u)) \quad \text{on } \Omega^\gamma, \ \gamma \in \{f, l, u, lu\}.$$

This shows

$$\mathbf{1}_{\Omega^\gamma} \cdot [d_1 I + d_2 \cdot F'(u)] \in \mathbf{1}_{\Omega^\gamma} \cdot \partial^\circ\Psi^\gamma(u). \tag{5.4}$$

Finally, we define $H \in \mathcal{L}([L^r]^4, L^r)$,

$$Hv = \mathbf{1}_{\Omega^f} v_1 + \mathbf{1}_{\Omega^l} v_2 + \mathbf{1}_{\Omega^u} v_3 + \mathbf{1}_{\Omega^{lu}} v_4,$$

and observe that

$$\Phi(u) = H\big(\Psi^f(u), \Psi^l(u), \Psi^u(u), \Psi^{lu}(u)\big).$$

Thus, Φ is locally Lipschitz continuous. From application of the direct product rule and the chain rule, Propositions 3.6 and 3.8 (note that $H' \equiv H$ is bounded), we conclude that Φ is $H' \circ (\partial^\circ\Psi^f \times \partial^\circ\Psi^l \times \partial^\circ\Psi^u \times \partial^\circ\Psi^{lu})$-semismooth and that, by (5.4), this generalized differential contains all $M \in \mathcal{L}(L^r, L^r)$ of the form $M = d_1 I + d_2 \cdot F'(u)$, where $d_1, d_2 \in L^\infty(\Omega)$ satisfy (5.3).

If Assumption 5.2 holds, then it is straightforward to modify the proof to establish semismoothness of order $\beta > 0$. $\square$

It should be immediately clear from our detailed discussion of NCPs in previous sections how the semismooth reformulation (5.2) can be used to apply our class of semismooth Newton methods. The resulting algorithm looks exactly like Algorithm 3.62, with the only difference that Φ is defined by (5.1). Also the regularity condition of Assumption 3.64 is appropriate and the assertions of Theorem 3.67 can be established as well.

We now discuss ways of choosing ϕ and $\phi_{[\alpha,\beta]}$. Consider any NCP-function ϕ that is positive on $(0,\infty)^2$ and negative on $\mathbb{R}^2 \setminus [0,\infty)^2$. Then the following construction, which was proposed by Billups [24] for $\phi = \phi^{FB}$, can be used to obtain an MCP-function $\phi_{[\alpha,\beta]}$, $-\infty < \alpha < \beta < +\infty$:

$$\phi_{[\alpha,\beta]}(x) = \phi\big(x_1 - \alpha, -\phi(\beta - x_1, -x_2)\big). \tag{5.5}$$

Proposition 5.5. *Let ϕ be an NCP-function that is positive on $(0,\infty)^2$ and negative on $\mathbb{R}^2 \setminus [0,\infty)^2$. Then, for any interval $[\alpha,\beta]$, $-\infty < \alpha < \beta < \infty$, the function $\phi_{[\alpha,\beta]}(x)$ defined in* (5.5) *is an MCP-function.*

Proof. We have to show that $\phi_{[\alpha,\beta]}(x) = 0$ holds if and only if

$$\alpha \le x_1 \le \beta, \quad (x_1 - \alpha)x_2 \le 0, \quad (x_1 - \beta)x_2 \le 0. \tag{5.6}$$

To this end, observe that $\phi_{[\alpha,\beta]}(x) = 0$ is equivalent to

$$x_1 - \alpha \ge 0, \quad \phi(\beta - x_1, -x_2) \le 0, \quad (x_1 - \alpha)\phi(\beta - x_1, -x_2) = 0, \tag{5.7}$$

where we have used the fact that ϕ is an NCP-function.

For $x_1 < \alpha$, (5.6) and (5.7) are both violated. For $x_1 = \alpha$, we use the assumptions on ϕ to obtain

$$(5.6) \Longleftrightarrow x_2 \ge 0 \Longleftrightarrow \phi(\beta - \alpha, -x_2) \le 0 \Longleftrightarrow (5.7).$$

Finally, for $x_1 > \alpha$,

$$\begin{aligned}(5.6) &\Longleftrightarrow x_1 \le \beta,\ x_2 \le 0,\ (x_1 - \beta)x_2 \le 0 \\ &\Longleftrightarrow \phi(\beta - x_1, -x_2) = 0 \Longleftrightarrow (5.7). \qquad \square\end{aligned}$$

We demonstrate this construction for

$$\phi(x) = \phi^E(x) = x_1 - P_{[0,\infty)}(x_1 - x_2) = \min\{x_1, x_2\}.$$

Then

$$\begin{aligned}\phi_{[\alpha,\beta]}(x) &= \min\{x_1 - \alpha, -\min\{\beta - x_1, -x_2\}\} \\ &= \min\{x_1 - \alpha, \max\{x_1 - \beta, x_2\}\} = x_1 - P_{[\alpha,\beta]}(x_1 - x_2) = \phi^E_{[\alpha,\beta]}(x).\end{aligned}$$

Therefore, starting with the projection-based NCP-function ϕ^E, we obtain the projection-based MCP-function $\phi^E_{[\alpha,\beta]}$. Concerning the concrete calculation of $\partial\phi^E$ and $\partial\phi^E_{[\alpha,\beta]}$, we have the following.

Proposition 5.6. *The function ϕ^E is piecewise affine linear on $\mathbb{R}^2$ and affine linear on the sets $\{x : x_1 < x_2\}$, $\{x : x_1 > x_2\}$. There holds*

$$\begin{aligned} \partial\phi^E(x) = \partial_B\phi^E(x) &= \{(\phi^E)'(x)\} = \{(1,0)\} \quad \textit{for } x_1 < x_2, \\ \partial\phi^E(x) = \partial_B\phi^E(x) &= \{(\phi^E)'(x)\} = \{(0,1)\} \quad \textit{for } x_1 > x_2, \\ \partial_B\phi^E(x) = \{(1,0),(0,1)\}, \quad \partial\phi^E(x) &= \{(t,1-t) : 0 \le t \le 1\} \quad \textit{for } x_1 = x_2. \end{aligned}$$

The function $\phi^E_{[\alpha,\beta]}$ is piecewise affine linear on $\mathbb{R}^2$ and affine linear on the connected components of $\{x : x_1 - x_2 \neq \alpha,\ x_1 - x_2 \neq \beta\}$. There holds

$$\begin{aligned} &\partial\phi^E_{[\alpha,\beta]}(x) = \partial_B\phi^E_{[\alpha,\beta]}(x) = \{(\phi^E)'_{[\alpha,\beta]}(x)\} = \{(1,0)\} \quad \textit{for } x_1 - x_2 \notin [\alpha,\beta], \\ &\partial\phi^E_{[\alpha,\beta]}(x) = \partial_B\phi^E_{[\alpha,\beta]}(x) = \{(\phi^E)'_{[\alpha,\beta]}(x)\} = \{(0,1)\} \quad \textit{for } x_1 - x_2 \in (\alpha,\beta), \\ &\left.\begin{aligned} \partial_B\phi^E_{[\alpha,\beta]}(x) &= \{(1,0),(0,1)\}, \\ \partial\phi^E_{[\alpha,\beta]}(x) &= \{(t,1-t) : 0 \le t \le 1\} \end{aligned}\right\} \quad \textit{for } x_1 - x_2 \in \{\alpha,\beta\}. \end{aligned}$$

Proof. This is an immediate consequence of Proposition 2.25. □

The generalized differential of ϕ^{FB} was already derived in section 2.5.2. In a similar way, it is possible to obtain formulas for the generalized differential of $\phi^{FB}_{[\alpha,\beta]}$; see [70].

5.1.2 Pointwise Convex Constraints

More general than bound constraints, we can consider pointwise convex constraints; i.e., the feasible set $\mathcal{C}$ is given by

$$\mathcal{C} = \{u \in L^2(\Omega)^m : u(\omega) \in C \ \text{ on } \Omega\}, \tag{5.8}$$

where $C \subset \mathbb{R}^m$ is a nonempty closed convex set and, as throughout this work, Ω is bounded and measurable with $\text{meas}(\Omega) > 0$. Equally well, we could consider sets $\mathcal{C}$ consisting of all $u \in L^p(\Omega)^m$ with $u(\omega) \in C(\omega)$ on Ω, with the multifunction C having suitable properties. For convenience, however, we restrict our discussion to the case (5.8).

We wish to solve the following problem.

Variational Inequality with Pointwise Convex Constraints

$$u \in \mathcal{C}, \quad (F(u), v - u)_{L^2} \ge 0 \quad \forall\, v \in \mathcal{C}, \tag{5.9}$$

with the same assumptions as in (1.14), but F being an operator between $\mathbb{R}^m$-valued Lebesgue spaces, i.e., $F : L^2(\Omega)^m \to L^2(\Omega)^m$. Suppose that a continuous function $\pi : \mathbb{R}^m \times \mathbb{R}^m \to \mathbb{R}^m$ is available with the property

$$\pi(x_1, x_2) = 0 \Longleftrightarrow x_1 = P_C(x_1 - x_2), \tag{5.10}$$

where P_C is the Euclidean projection onto C. We will prove that (5.9) is equivalent to the operator equation

$$\Pi(u) = 0, \quad \text{where} \quad \Pi(u)(\omega) = \pi\big(u(\omega), F(u)(\omega)\big). \tag{5.11}$$

Remark 5.7. *The function*

$$\pi^E(x_1,x_2) = x_1 - P_C(x_1 - x_2) \tag{5.12}$$

satisfies (5.10). *It generalizes the projection-based NCP-function* ϕ^E.

Proposition 5.8. *Let the function* $\pi : \mathbb{R}^m \times \mathbb{R}^m \to \mathbb{R}^m$ *satisfy* (5.10) *and define* Π *by* (5.11). *Then* u *solves* (5.9) *if and only if* (5.11) *is satisfied.*

Proof. The projection $x_P = P_C(x)$ is characterized by

$$x_P \in C, \quad (x_P - x)^T(z - x_P) \ge 0 \quad \forall\, z \in C. \tag{5.13}$$

Now, if $\Pi(u) = 0$, then $u(\omega) = P_C(u(\omega) - F(u)(\omega))$ a.e. on Ω. In particular, $u(\omega) \in C$ and, by (5.13), for all $v \in \mathcal{C}$,

$$\big(u(\omega) - [u(\omega) - F(u)(\omega)]\big)^T(v(\omega) - u(\omega)) \ge 0,$$

where we have used $v(\omega) \in C$. Integrating this over Ω shows that u solves (5.9).

Conversely, assume that $\Pi(u) \ne 0$. If $u \notin \mathcal{C}$, then u does not solve (5.9). Otherwise, $u \in \mathcal{C}$ and the set

$$\Omega' = \big\{\omega : u(\omega) \ne P_C\big(u(\omega) - F(u)(\omega)\big)\big\}$$

has positive measure. Set $z = u - F(u)$ and $v = u + \sigma w$, where, for $\omega \in \Omega$,

$$w(\omega) = P_C(z(\omega)) - u(\omega), \quad \sigma(\omega) = \frac{1}{\max\{1, \|w(\omega)\|_2\}}.$$

Then holds $v \in \mathcal{C}$, $w \ne 0$, and

$$\begin{aligned}
F(u)(\omega)^T(v(\omega) - u(\omega)) &= \sigma(\omega)F(u)(\omega)^T w(\omega)\\
&= \sigma(\omega)\big(w(\omega) + F(u)(\omega)\big)^T w(\omega) - \sigma(\omega)\|w(\omega)\|_2^2\\
&= \sigma(\omega)\big(P_C(z(\omega)) - z(\omega)\big)^T\big(P_C(z(\omega)) - u(\omega)\big) - \sigma(\omega)\|w(\omega)\|_2^2\\
&\le -\sigma(\omega)\|w(\omega)\|_2^2 \le -\min\big\{\|w(\omega)\|_2, \|w(\omega)\|_2^2\big\}.
\end{aligned}$$

Integration over Ω yields

$$(F(u), v - u)_{L^2} < 0.$$

Therefore, since $v \in \mathcal{C}$, u is not a solution of (5.9). ∎

The reformulation (5.11) is an operator equation involving the superposition operator Π. The application of semismooth Newton methods is attractive if a function π can be found that is (a) Lipschitz continuous and (b) semismooth, and for which (c) π and $\partial_C \pi$ can be computed efficiently. Requirement (a) holds, e.g., for $\phi = \phi^E$, since the Euclidean projection is nonexpansive. Requirement (b) depends on the set C; if, e.g., C is a polyhedron, then P_C is piecewise affine linear, see [181, Prop. 2.4.4], and thus 1-order semismooth. Also (c) depends on the set C. We will give an example below. Requirements (a) and (b) are essential for proving the semismoothness of Π.

As a preparation for the treatment of mixed problems, we will prove the semismoothness of a slightly more general class of operators than those defined in (5.11). We consider operators $\Pi(z,u)$ that arise from the reformulation of problems (5.9) where F depends on an additional parameter $z \in Z$, where Z is a Banach space:

$$F : Z \times L^2(\Omega)^m \to L^2(\Omega)^m.$$

For $z \in Z$ we then consider the problem

$$u \in \mathcal{C}, \quad (F(z,u), v-u)_{L^2} \geq 0 \quad \forall\, v \in \mathcal{C}, \tag{5.14}$$

which can be interpreted as a class of problems (5.9) that is parameterized by z. Here, $\mathcal{C}$ is defined by (5.8).

Remark 5.9. *The problem* (5.9) *is contained in the class* (5.14) *by choosing* $Z = \{0\}$ *and* $F(0,u) = F(u)$.

By Proposition 5.8 we can use a function π satisfying (5.10) to reformulate (5.14) equivalently as

$$\Pi(z,u) = 0, \quad \text{where} \quad \Pi(z,u)(\omega) = \pi\big(u(\omega), F(z,u)(\omega)\big), \quad \omega \in \Omega. \tag{5.15}$$

Although we have formulated the problem (5.14) in an L^2-setting, as before we now investigate the semismoothness of F in a general L^q-setting. Suppose that the following holds.

Assumption 5.10. *There are* $1 \leq r < \min\{p, p'\}$ *such that*

(a) $F : Z \times L^p(\Omega)^m \to L^r(\Omega)^m$ *is continuously Fréchet differentiable.*

(b) $(z,u) \in Z \times L^p(\Omega)^m \mapsto F(z,u) \in L^{p'}(\Omega)^m$ *is locally Lipschitz continuous.*

(c) *The function* π *is Lipschitz continuous.*

(d) π *is semismooth.*

Then we obtain the following.

Theorem 5.11. *Under Assumption* 5.10 *the operator*

$$\Pi : Z \times L^p(\Omega)^m \to L^r(\Omega)^m$$

defined in (5.15) *is locally Lipschitz continuous and* $\partial_C^\circ \Pi$*-semismooth, where the generalized differential* $\partial_C^\circ \Pi(u)$ *consists of all operators* $M \in \mathcal{L}(Z \times [L^p]^m, [L^r]^m)$ *of the form*

$$M(v,w) = D_1 w + D_2(F'(z,u)(v,w)) \quad \forall\, (v,w) \in Z \times L^p(\Omega)^m, \tag{5.16}$$

where $D_i \in L^\infty(\Omega)^{m\times m}$ *and* $D = (D_1 | D_2)$ *satisfies*

$$D(\omega) \in \partial_C \pi\big(u(\omega), F(z,u)(\omega)\big), \quad \omega \in \Omega. \tag{5.17}$$

Proof. Consider the ith component $\Pi_i(z,u) = \pi_i\big(u, F(z,u)\big)$ of Π. Obviously, Assumption 5.10 implies Assumption 3.32 with $Y = Z \times L^p(\Omega)^m$, $G(z,u) = (u, F(z,u))$, $r_i = r$, $i = 1,\dots,2m$, $q_i = p$, $i = 1,\dots,m$, $q_i = p'$, $i = m+1,\dots,2m$, and $\psi = \pi_i$. Therefore, by Proposition 3.36 and Theorem 3.49, the operator $\Pi_i : Z \times L^p(\Omega)^m \to L^r(\Omega)$ is locally Lipschitz continuous and $\partial^\circ\Pi_i$-semismooth. Hence, we can apply Proposition 3.6 to conclude that

$$\Pi : Z \times L^p(\Omega)^m \to L^r(\Omega)^m$$

is $\partial^\circ_C\Pi$-semismooth, where $\partial^\circ_C\Pi = \partial^\circ\Pi_1 \times \cdots \times \partial^\circ\Pi_m$. From the definition of the C-subdifferential it is clear that $\partial^\circ_C\Pi(z,u)$ can be characterized by (5.16) and (5.17). □

We can also prove semismoothness of higher order.

Assumption 5.12. *As Assumption* 5.10, *but with (a) and (d) replaced by*
There exists $\alpha \in (0,1]$ *such that*

(a) $F : Z \times L^p(\Omega)^m \to L^r(\Omega)^m$ *is continuously Fréchet differentiable with locally* α*-Hölder continuous derivative.*

(d) π *is* α*-order semismooth.*

Under these strengthened assumptions we can use Theorem 3.50 to prove the following.

Theorem 5.13. *Under Assumption* 5.12 *the assertions of Theorem* 5.11 *hold true and, in addition, the operator* Π *is* β*-order* $\partial^\circ_C\Pi$*-semismooth, where* β *can be determined as in Theorem* 3.50.

The established semismoothness results allow us to solve problem (5.9) by applying the semismooth Newton methods of section 3.2.3 to the reformulation (5.11). The resulting methods are of the same form as Algorithm 3.62 for NCPs, only Φ has to be replaced by Π and all L^p-spaces are now m-dimensional. Smoothing steps can be obtained as described in section 4.1. An appropriate regularity condition is obtained by requiring that all M_k are elements of $\mathcal{L}([L^r]^m, [L^r]^m)$ with uniformly bounded inverses.

In section 4.2 we described a situation where, through an appropriate choice of the MCP-function, the smoothing step can be avoided. This approach can be generalized to the current situation.

Assumption 5.14. *The operator* F *has the form* $F(z,u) = \lambda u + G(z,u)$ *with* $\lambda > 0$, *and there exist* $1 \le r < p' \le \infty$ *such that*

(a) $G : Z \times L^r(\Omega)^m \to L^r(\Omega)^m$ *is continuously Fréchet differentiable.*

(b) $(z,u) \in Z \times L^r(\Omega)^m \mapsto G(z,u) \in L^{p'}(\Omega)^m$ *is locally Lipschitz continuous.*

(c) *The function* π *is defined by* $\pi(x_1,x_2) = x_1 - P_C(x_1 - \lambda^{-1}x_2)$, *where* P_C *is the projection on* C.

(d) *The projection* P_C *is semismooth.*

Under these assumptions we can prove the following theorem.

Theorem 5.15. *Let Assumption* 5.14 *hold. Then, we have*

$$\Pi(z,u)(\omega) = u(\omega) - P_C\big(-\lambda^{-1}G(z,u)(\omega)\big),$$

and $\Pi : Z \times L^r(\Omega)^m \to L^r(\Omega)^m$ *is* $\partial_C^\circ\Pi$*-semismooth. Here,* $\partial_C^\circ\Pi(z,u)$ *is the set of all* $M \in \mathcal{L}(Z \times L^r(\Omega)^m, L^r(\Omega)^m)$ *of the form*

$$M = \big(\lambda^{-1}DG_z(z,u)|I + \lambda^{-1}DG_u(z,u)\big), \tag{5.18}$$

with $D \in L^\infty(\Omega)^{m\times m}$, $D(\omega) \in \partial_C P_C\big(-\lambda^{-1}G(z,u)(\omega)\big)$ *on* Ω.

Proof. We set $T(z,u) = -\lambda^{-1}G(z,u)$, $\psi(x) = P_C(x)$. Then

$$T : Z \times L^r(\Omega)^m \to L^r(\Omega)^m$$

is continuously differentiable and maps locally Lipschitz continuous into $L^{p'}(\Omega)^m$. Further, ψ is Lipschitz continuous and semismooth. Therefore, we can apply Theorem 3.49 componentwise (with $Y = Z \times L^r(\Omega)^m, r_i = r, q_i = p'$) and obtain that $\Psi_i : (z,u) \in Z \times L^r(\Omega)^m \mapsto \psi_i(T(z,u)) \in L^r(\Omega)$ is $\partial^\circ\Psi_i$-semismooth. Therefore, by Proposition 3.6, we see that

$$\Psi : Z \times L^r(\Omega)^m \to L^r(\Omega)^m$$

is $\partial_C^\circ\Psi$-semismooth. Now, using the $(0|I)$-semismoothness of $(z,u) \mapsto u$ and the sum rule for semismooth operators, Proposition 3.5, we see that

$$\Pi : Z \times L^r(\Omega)^m \to L^r(\Omega)^m$$

is $\partial_C^\circ\Pi$-semismooth with $\partial_C^\circ\Pi = (0|I) - \partial_C^\circ\Psi$. It is straightforward to see that the elements of $\partial_C^\circ\Pi$ are characterized by (5.18). $\square$

The situation typically arising in practice is $r = 2$. Under the (reasonable) regularity requirement $M_k \in \mathcal{L}([L^r]^m, ([L^r]^m)$ with uniformly bounded inverses, superlinear convergence of the semismooth Newton method can be established as for the case of bound-constraints; see section 4.2.

Finally, we give an example of how a function π and its differential can be obtained in a concrete situation.

Example 5.16. Models for the flow of Bingham fluids [78, 79] involve VIPs of the form (5.14), where

$$C = \{x : \|x\|_2 \le 1\}.$$

We now derive explicit formulas for $\pi^E(x_1,x_2) = x_1 - P_C(x_1 - x_2)$ and its differentials $\partial_B\pi^E$, $\partial\pi^E$, and $\partial_C\pi^E$. First, observe that

$$P_C(x) = \frac{1}{\max\{1, \|x\|_2\}}x$$

is Lipschitz continuous and PC^∞ on $\mathbb{R}^m$. Further, P_C is C^∞ on $\{x : \|x\|_2 \ne 1\}$ with

$$P_C'(x) = I \quad \text{for } \|x\|_2 < 1, \qquad P_C'(x) = \frac{1}{\|x\|_2}I - \frac{xx^T}{\|x\|_2^3} \quad \text{for } \|x\|_2 > 1.$$

This shows that π^E is Lipschitz continuous and PC^∞ on $\mathbb{R}^m$. Hence, π^E is 1-order semismooth and

$$\begin{aligned}\partial_B \pi^E(x_1,x_2) &= \{(I-S|S) : S \in M_B\},\\ \partial \pi^E(x_1,x_2) &= \{(I-S|S) : S \in M\},\\ \partial_C \pi^E(x_1,x_2) &= \{(I-S|S) : S \in M_C\},\end{aligned}$$

where, with $w = x_1 - x_2$,

$$\begin{aligned}&M_B = M = M_C = \{I\} && \text{for } \|w\|_2 < 1,\\ &M_B = M = M_C = \left\{\frac{1}{\|w\|_2} I - \frac{ww^T}{\|w\|_2^3}\right\} && \text{for } \|w\|_2 > 1,\\ &\left.\begin{array}{l} M_B = \{I, I - ww^T\}, \quad M = \{I - tww^T : 0 \le t \le 1\},\\ M_C = \{I - \mathrm{diag}(t_1,\dots,t_m)ww^T : 0 \le t_1,\dots,t_m \le 1\}\end{array}\right\} && \text{for } \|w\|_2 = 1.\end{aligned}$$

■

5.2 Mixed Problems

So far we have considered variational inequalities in an L^p-setting. Often, the problem to solve is not given in this particular form, because the original problem formulation contains additional unknowns (e.g., the state) and additional operator equality constraints (e.g., the state equation). In the case of optimal control problems with unique control-to-state mapping $u \mapsto y(u)$ (induced by the state equation) we demonstrated how, by using the dependence $y = y(u)$, a reduced problem can be obtained that only depends on the control. This reduction method is called the *black-box approach*. Having the advantage of reducing the problem dimension, the black-box approach nevertheless suffers from several disadvantages. The evaluation of the objective function requires the solution of the (possibly nonlinear) state equation. Further, the black-box approach is only viable if the state equation admits a unique solution $y(u)$ for every control u.

Therefore, it can be advantageous to employ the *all-at-once approach*, i.e., to solve for u and y simultaneously. In the following we describe how the developed ideas can be extended to the all-at-once approach.

5.2.1 Karush–Kuhn–Tucker Systems

Consider the optimization problem (with control structure)

$$\text{minimize } J(y,u) \quad \text{subject to} \quad E(y,u) = 0 \quad \text{and} \quad u \in \mathcal{C}. \tag{5.19}$$

Here, let $\mathcal{C} \subset U$ be a nonempty closed convex set and assume that the operator $E : Y \times U \to W$ and the objective function $J : Y \times U \to \mathbb{R}$ are twice continuously differentiable. Further, let the control space U and the state space Y as well as W be Banach spaces.

Now consider a local solution $(\bar{y},\bar{u}) \in Y \times U$ of (5.19) at which Robinson's regularity condition [174] holds. More precisely, this means that

$$0 \in \mathrm{int}\left\{\left(E'(\bar{y},\bar{u})(h,s), \bar{u} + s - u\right) : h \in Y,\ s \in U,\ u \in \mathcal{C}\right\},$$

which can be shown to be equivalent to

$$0 \in \operatorname{int}\left\{E'(\bar{y},\bar{u})(h,u-\bar{u}) : h \in Y,\ u \in \mathcal{C}\right\}. \tag{5.20}$$

In particular, (5.20) is satisfied if $E_y(\bar{y},\bar{u})$ is onto, which holds true for many optimal control problems.

If the regularity condition (5.20) holds at a local solution $(\bar{y},\bar{u})$, then there exists a Lagrange multiplier $\bar{w} \in W^*$ such that the triple $(\bar{y},\bar{u},\bar{w})$ satisfies the KKT conditions; see, e.g., [208]:

$$\bar{u} \in \mathcal{C}, \quad \langle J_u(\bar{y},\bar{u}) + E_u(\bar{y},\bar{u})^*\bar{w}, u-\bar{u}\rangle_{U^*,U} \geq 0 \quad \forall\, u \in \mathcal{C}, \tag{5.21}$$

$$J_y(\bar{y},\bar{u}) + E_y(\bar{y},\bar{u})^*\bar{w} = 0, \tag{5.22}$$

$$E(\bar{y},\bar{u}) = 0. \tag{5.23}$$

This system consists of a variational inequality (parameterized by $\bar{z} = (\bar{y},\bar{w})$) of the form (5.14) with $F(y,u,w) = J_u(y,u) + E_u(y,u)^*w$ (except that the space U and the convex set $\mathcal{C}$ are not yet specified) and two operator equations. For convenient notation, we introduce the Lagrange function

$$L : Y \times U \times W \to \mathbb{R}, \quad L(y,u,w) = J(y,u) + \langle w, E(y,u)\rangle_{W^*,W}.$$

Then the operators appearing in (5.21)–(5.23) are $L_u(\bar{y},\bar{u},\bar{w})$, $L_y(\bar{y},\bar{u},\bar{w})$, and $L_w(\bar{y},\bar{u},\bar{w})$, respectively. Therefore, we can write (5.21)–(5.23) in the form

$$\bar{u} \in \mathcal{C}, \quad \langle L_u(\bar{y},\bar{u},\bar{w}), u-\bar{u}\rangle_{U^*,U} \geq 0 \quad \forall\, u \in \mathcal{C}, \tag{5.24}$$

$$L_y(\bar{y},\bar{u},\bar{w}) = 0, \tag{5.25}$$

$$E(\bar{y},\bar{u}) = 0. \tag{5.26}$$

Our aim is to reformulate the variational inequality as an equivalent nonsmooth operator equation. To this end, we consider $U = L^2(\Omega)^m$, Ω bounded with $\text{meas}(\Omega) > 0$, and assume that $\mathcal{C}$ has appropriate structure. In the following we analyze the case where $\mathcal{C}$ is described by pointwise convex constraints of the form (5.8) and assume that a continuous function $\pi : \mathbb{R}^m \times \mathbb{R}^m \to \mathbb{R}^m$ with the property (5.10) is available. Note that this problem class includes the NCP and the bound-constrained VIP in normal form as special cases. According to Proposition 5.8, we can reformulate (5.24) as $\Pi(\bar{y},\bar{u},\bar{w}) = 0$, where

$$\Pi(y,u,w)(\omega) = \pi\big(u(\omega), L_u(y,u,w)(\omega)\big), \quad \omega \in \Omega,$$

and thus $(\bar{y},\bar{u},\bar{w})$ is a KKT-triple if and only if it is a solution to the system

$$\Sigma(y,u,w) \stackrel{\text{def}}{=} \begin{pmatrix} L_y(y,u,w) \\ \Pi(y,u,w) \\ E(y,u) \end{pmatrix} = 0. \tag{5.27}$$

We continue by considering two approaches, parallel to the situations in Assumptions 5.10 and 5.14, respectively. Again, we formulate our assumptions in a general L^p-setting.

The first approach requires the following hypotheses.

Assumption 5.17. *There exist* $1 \le r < \min\{p, p'\} \le \infty$ *such that*

(a) $E : Y \times L^p(\Omega)^m \to W$ *and* $J : Y \times L^p(\Omega)^m \to \mathbb{R}$ *are twice continuously differentiable.*

(b) *The operator* $(y,u,w) \in Y \times L^p(\Omega)^m \times W^* \mapsto L_u(y,u,w) \in L^r(\Omega)^m$ *is well defined and continuously differentiable.*

(c) *The operator* $(y,u,w) \in Y \times L^p(\Omega)^m \times W^* \mapsto L_u(y,u,w) \in L^{p'}(\Omega)^m$ *is well defined and locally Lipschitz continuous.*

(d) π *is Lipschitz continuous and semismooth.*

Remark 5.18. *Variants of Assumption* 5.17 *are possible.*

We obtain the following.

Theorem 5.19. *Let Assumption* 5.17 *hold. Then the operator* $\Sigma : Y \times L^p(\Omega)^m \times W^* \to Y^* \times L^r(\Omega)^m \times W$ *defined in* (5.27) *is locally Lipschitz continuous and* $\partial_C \Sigma$*-semismooth with* $\partial_C \Sigma = \{L'_y\} \times \partial_C^\circ \Pi \times \{E'\}$. *More precisely,* $\partial_C \Sigma(y,u,w)$ *is the set of all* $M \in \mathcal{L}(Y \times L^p(\Omega)^m \times W^*, Y^* \times L^r(\Omega)^m \times W)$ *of the form*

$$M = \begin{pmatrix} L_{yy}(y,u,w) & L_{yu}(y,u,w) & E_y(y,u)^* \\ D_2 L_{uy}(y,u,w) & D_1 I + D_2 L_{uu}(y,u,w) & D_2 E_u(y,u)^* \\ E_y(y,u) & E_u(y,u) & 0 \end{pmatrix}, \tag{5.28}$$

where $D_i \in L^\infty(\Omega)^{m \times m}$, $(D_1|D_2)(\omega) \in \partial_C \pi(u(\omega), L_u(y,u,w)(\omega))$.

Proof. We set $Z = Y \times W^*$ and $F(y,w,u) = L_u(y,u,w)$. Assumption 5.17 then implies Assumption 5.10, and thus Π is locally Lipschitz continuous and $\partial_C^\circ \Pi$-semismooth by Theorem 5.11. From the differentiability requirements in Assumption 5.17 we obtain the local Lipschitz continuity and, by Proposition 3.4, the $\{L'_y\}$- and $\{E'\}$-semismoothness of the second and third component of Σ, respectively. Proposition 3.6 now yields the local Lipschitz continuity and the $\partial_C \Sigma$-semismoothness of Σ for $\partial_C \Sigma = \{L'_y\} \times \partial_C^\circ \Pi \times \{E'\}$. The elements of $\partial_C \Sigma(y,u,w)$ are easily seen to be given by (5.28). ☐

In Example 5.23, we apply Theorem 5.19 to an optimal control problem.

A second approach for establishing the semismoothness of Π relies on the following hypotheses.

Assumption 5.20. *There exist* $1 \le r < p' \le \infty$ *such that*

(a) $E : Y \times L^r(\Omega)^m \to W$ *and* $J : Y \times L^r(\Omega)^m \to \mathbb{R}$ *are twice continuously differentiable.*

(b) L_u *has the form* $L_u(y,u,w) = \lambda u + G(y,u,w)$ *with* $\lambda > 0$ *and*

  (i) $G : Y \times L^r(\Omega)^m \times W^* \to L^r(\Omega)^m$ *is continuously Fréchet differentiable.*

  (ii) *The operator* $(y,u,w) \in Y \times L^r(\Omega)^m \times W^* \mapsto G(y,u,w) \in L^{p'}(\Omega)^m$ *is locally Lipschitz continuous.*

(c) *The function* π *is defined by* $\pi(x_1,x_2) = x_1 - P_C(x_1 - \lambda^{-1}x_2)$ *and the projection* P_C *on* C *is semismooth.*

Theorem 5.21. *Let Assumption* 5.20 *hold. Then we have*

$$\Pi(y,u,w)(\omega) = u(\omega) - P_C\big(-\lambda^{-1}G(y,u,w)(\omega)\big),$$

and $\Sigma : Y \times L^r(\Omega)^m \times W^* \to Y^* \times L^r(\Omega)^m \times W$ *is locally Lipschitz continuous and* $\partial_C\Sigma$*-semismooth. Here,* $\partial_C\Sigma(y,u,w)$ *is the set of all* $M \in \mathcal{L}(Y \times L^r(\Omega)^m \times W^*, Y^* \times L^r(\Omega)^m \times W)$ *of the form*

$$M = \begin{pmatrix} L_{yy}(y,u,w) & L_{yu}(y,u,w) & E_y(y,u)^* \\ \lambda^{-1}DG_y(y,u,w) & I+\lambda^{-1}DG_u(y,u,w) & \lambda^{-1}DG_w(y,u,w) \\ E_y(y,u) & E_u(y,u) & 0 \end{pmatrix} \tag{5.29}$$

with

$$D \in L^\infty(\Omega)^{m\times m}, \quad D(\omega) \in \partial_C P_C\big(-\lambda^{-1}G(y,u,w)(\omega)\big) \quad \text{on } \Omega. \tag{5.30}$$

Proof. Assumption 5.20 implies Assumption 5.14 for $Z = Y \times W^*$ and $F(y,w,u) = L_u(y,u,w)$. Theorem 5.15 is applicable and yields the local Lipschitz continuity and $\partial_C^\circ\Pi$-semismoothness of $\Pi : Y \times L^r(\Omega)^m \times W^* \to L^r(\Omega)^m$, where $\partial_C^\circ\Pi(y,u,w)$ is the set of all $M_\Pi \in \mathcal{L}(Y \times L^r(\Omega)^m \times W^*, L^r(\Omega)^m)$ of the form

$$M_\Pi = \big(\lambda^{-1}DG_y(y,u,w) | I+\lambda^{-1}DG_u(y,u,w) | \lambda^{-1}DG_w(y,u,w)\big),$$

where D is as in the theorem. From Assumption 5.20 and Proposition 3.4 follow the local Lipschitz continuity as well as the $\{L'_y\}$- and $\{E'\}$-semismoothness of the second and third component of Σ, respectively. Therefore, the operator

$$\Sigma : Y \times L^r(\Omega)^m \times W^* \to Y^* \times L^r(\Omega)^m \times W$$

is locally Lipschitz continuous and, by Proposition 3.6, $\partial_C\Sigma$-semismooth with $\partial_C\Sigma = \{L'_y\} \times \partial_C^\circ\Pi \times \{E'\}$. It is straightforward to verify that the elements of $\partial_C\Sigma(y,u,w)$ are exactly the operators M in (5.29). □

Remark 5.22. *If* P_C *is* α*-order semismooth, it is easy to modify Assumption* 5.20 *and Theorem* 5.21 *such that higher-order semismoothness of* Π *can be established.*

The following example illustrates how Theorems 5.19 and 5.21 can be applied in practice.

Example 5.23. Let $\Omega \subset \mathbb{R}^n$ be a bounded Lipschitz domain and consider the optimal control problem

$$\begin{aligned} &\underset{y\in H_0^1(\Omega),u\in L^2(\Omega)}{\text{minimize}} && \frac{1}{2}\int_\Omega (y(x)-y_d(x))^2dx + \frac{\lambda}{2}\int_\Omega u(x)^2dx \\ &\text{subject to} && -\Delta y = f + gu \quad \text{on } \Omega, \quad \beta_1 \le u \le \beta_2 \quad \text{on } \Omega. \end{aligned} \tag{5.31}$$

Note that the Dirichlet boundary conditions $y = 0$ on $\partial\Omega$ are expressed by $y \in H_0^1(\Omega)$. This is a problem of the form (5.19) with $U = L^2(\Omega)$, $Y = H_0^1(\Omega)$, $W = H^{-1}(\Omega)$, $W^* = H_0^1(\Omega)$,

$C = [\beta_1, \beta_2]$, $\mathcal{C}$ defined in (5.8), and

$$J(y,u) = \frac{1}{2}\int_\Omega (y(x) - y_d(x))^2 dx + \frac{\lambda}{2}\int_\Omega u(x)^2 dx,$$
$$E(y,u) = Ay - f - gu,$$

where $A \in \mathcal{L}(H_0^1(\Omega), H^{-1}(\Omega))$ is the operator corresponding to $-\Delta$, i.e.,

$$\langle Ay, v\rangle_{H^{-1},H_0^1} = \int_\Omega \nabla y(x)^T \nabla v(x) dx.$$

We assume $-\infty < \beta_1 < \beta_2 < +\infty$, $y_d \in L^2(\Omega)$, $\lambda > 0$, $f \in H^{-1}(\Omega)$, and $g \in L^\infty(\Omega)$. Observe that

(a) J is strictly convex.

(b) $\{(y,u): Ay = f + gu,\ u \in [\beta_1,\beta_2]\} \subset H_0^1(\Omega) \times L^2(\Omega)$ is closed, convex, and bounded.

In (b) we have used that $A \in \mathcal{L}(H_0^1, H^{-1})$ is a homeomorphism. Hence, by a standard result [65, Prop. II.1.2], there exists a unique solution $(\bar{y},\bar{u}) \in H_0^1(\Omega) \times L^2(\Omega)$ to the problem.

Since $\mathcal{C} \subset \max\{|\beta_1|,|\beta_2|\}\bar{B}_{L^\infty}$, we have $\bar{u} \in L^p(\Omega)$ for all $p \in [1,\infty]$.

The continuous invertibility of $E_y(y,u) = A \in \mathcal{L}(H_0^1, H^{-1})$ guarantees that Robinson's regularity condition (5.20) is satisfied, so that the solution $(\bar{y},\bar{u})$ is characterized by (5.24)–(5.26), where $\bar{w} \in W^* = H_0^1(\Omega)$ is the Lagrange multiplier (adjoint state). Clearly, the operator A is self-adjoint, i.e., $A^* = A$, and thus the Lagrange function satisfies

$$\begin{aligned} L(y,u,w) &= J(y,u) + \langle Ay, w\rangle_{H^{-1},H_0^1} - (f + gu, w)_{L^2} \\ &= J(y,u) + \langle Aw, y\rangle_{H^{-1},H_0^1} - (f,w)_{L^2} - (gw,u)_{L^2}. \end{aligned}$$

Therefore,

$$\begin{aligned} L_y(y,u,w) &= y - y_d + Aw, \\ L_u(y,u,w) &= \lambda u - gw, \end{aligned}$$

and (5.24)–(5.26) are satisfied by the triple $(\bar{y},\bar{u},\bar{w})$ if and only if it solves the system

$$\bar{u} \in L^2(\Omega),\ \bar{u} \in C, \quad (\lambda\bar{u} - g\bar{w}, u - \bar{u})_{L^2} \geq 0 \quad \forall\, u \in L^2(\Omega),\ u \in C, \tag{5.32}$$
$$A\bar{w} = -(\bar{y} - y_d), \tag{5.33}$$
$$A\bar{y} = f + g\bar{u}. \tag{5.34}$$

Now, let q be arbitrary with $q \in (2,\infty]$ if $n = 1$, $q \in (2,\infty)$ if $n = 2$, and $q \in (2, 2n/(n-2)]$ if $n \geq 3$. Then the continuous embedding $H_0^1(\Omega) \hookrightarrow L^q(\Omega)$ implies that the operator

$$(y,u,w) \in Y \times L^p(\Omega) \times W^* \mapsto L_u(y,u,w) = \lambda u - gw \in L^q(\Omega)$$

is continuous linear and thus C^∞ for all $p \geq q$.

It is now straightforward to see that Assumption 5.17 (a)–(c) holds for any $p \in (2,\infty]$, $p' \in (2, \min\{p,q\}]$ with $q > 2$ as specified, and any $r \in [2, p')$. For π we can choose any Lipschitz continuous and semismooth MCP-function for the interval $[\beta_1,\beta_2]$ to meet Assumption 5.17 (d). This makes Theorem 5.19 applicable.

Now we turn to the situation of Assumption 5.20. Obviously, for $r = 2$ and $p' = q$, Assumptions 5.20 (a) and (b) hold with $G(y,u,w) = -gw$. Further, $P_C(x) = \max\{\beta_1, \min\{x, \beta_2\}\}$ is 1-order semismooth, so that Assumption 5.20 (c) also holds. Hence, Theorem 5.21 is applicable. ∎

Having established the semismoothness of the operator Σ, we can apply the (projected) semismooth Newton method (Algorithm 3.16 or 3.22) for the solution of (5.27). For the superlinear convergence results, Theorems 3.18 and 3.24, respectively, the regularity condition of Assumption 3.17 or one of its variants, Assumption 3.25 or 3.28, respectively, has to be satisfied. Essentially, these assumptions require the bounded invertibility of some or all elements of $\partial_C \Sigma$, viewed as operators between appropriate spaces, near the solution. In the next section we establish a relation between $\partial_C \Sigma$ and the generalized differential of the reformulated reduced problem. This relation can then be used to show that regularity conditions for the reduced problem imply regularity of the full problem (5.27). Further, we discuss how smoothing steps can be constructed for the scenario of Assumption 5.17. As we will see, in the setting of Assumption 5.20 no smoothing step is required.

5.2.2 Connections to the Reduced Problem

We consider the problem (5.19) and, in parallel, the reduced problem

$$\text{minimize} \quad j(u) \quad \text{subject to} \quad u \in \mathcal{C}, \tag{5.35}$$

where $j(u) = J(y(u),u)$ and $y(u) \in Y$ is such that

$$E(y(u),u) = 0. \tag{5.36}$$

We assume that $y(u)$ exists uniquely for all u in a neighborhood V of $\mathcal{C}$ (this can be relaxed; see Remark 5.24) and that $E_y(y(u),u)$ is continuously invertible. Then, by the implicit function theorem, the mapping $u \in U \mapsto y(u) \in Y$ is twice continuously differentiable.

The adjoint representation of the derivative $j'(u) \in U^*$ is given by

$$j'(u) = J_u(y(u),u) + E_u(y(u),u)^* w(u),$$

where $w = w(u) \in W^*$ solves the adjoint equation

$$E_y(y(u),u)^* w = -J_y(y(u),u); \tag{5.37}$$

see section A.1 in the appendix. In terms of the Lagrange function

$$L(y,u,w) = J(y,u) + \langle w, E(y,u)\rangle_{W^*,W},$$

this can be written as

$$j'(u) = L_u(y(u),u,w(u)), \tag{5.38}$$

where $w(u)$ satisfies

$$L_y(y(u),u,w(u)) = 0. \tag{5.39}$$

Any solution $\bar{u} \in U$ of (5.35) satisfies the first-order necessary optimality conditions for (5.35):

$$\bar{u} \in \mathcal{C}, \quad \langle j'(\bar{u}), u - \bar{u}\rangle_{U^*,U} \geq 0 \quad \forall\, v \in \mathcal{C}. \tag{5.40}$$

Now, setting $\bar{y} = y(\bar{u})$ and combining (5.40) with (5.38), (5.39), and (5.36), we can write (5.40) equivalently as

$$\begin{aligned} \bar{u} \in \mathcal{C}, \quad \langle L_u(\bar{y},\bar{u},\bar{w}), u - \bar{u}\rangle_{U^*,U} &\geq 0 \quad \forall\, u \in \mathcal{C}, \\ L_y(\bar{y},\bar{u},\bar{w}) &= 0, \\ E(\bar{y},\bar{u}) &= 0. \end{aligned}$$

These are exactly the KKT conditions (5.24)–(5.26) of problem (5.19).

Therefore, if $\bar{u} \in U$ is a critical point of (5.35), i.e., if $\bar{u} \in U$ satisfies (5.40), then $(\bar{y},\bar{u},\bar{w}) = (y(\bar{u}),\bar{u},w(\bar{u}))$ is a KKT-triple of (5.19); i.e., $(\bar{y},\bar{u},\bar{w})$ satisfies (5.24)–(5.26). Conversely, if $(\bar{y},\bar{u},\bar{w})$ is a KKT-triple of (5.19), then there holds $\bar{y} = y(\bar{u})$, $\bar{w} = w(\bar{u})$, and $\bar{u}$ is a critical point of (5.35).

Remark 5.24. *We have assumed that $y(u)$ exists uniquely with $E_y(y(u),u)$ being continuously invertible for all u in a neighborhood of $\mathcal{C}$. This requirement can be relaxed. In fact, let $(\bar{y},\bar{u},\bar{w})$ be a KKT-triple of* (5.19) *and assume that $E_y(\bar{y},\bar{u})$ is continuously invertible. Then, by the implicit function theorem there exist neighborhoods V_U of $\bar{u}$ and V_Y of $\bar{y}$ and a unique mapping $u \in V_U \mapsto y(u) \in V_Y$ with $y(\bar{u}) = \bar{y}$ and $E_y(y(u),u) = 0$ for all $u \in V_U$. Furthermore, $y(u)$ is twice continuously differentiable. Introducing $j(u) = J(y(u),u)$, $u \in V_U$, we see as above that* (5.24)–(5.26) *and* (5.40) *are equivalent.*

Due to this equivalence of the optimality systems for (5.19) and (5.35) we expect to find close relations between Newton methods for the solution of (5.24)–(5.26) and those for the solution of (5.40). This is the objective of the next section.

5.2.3 Relations between Full and Reduced Newton System

We now return to problem (5.19) with $U = L^2(\Omega)^m$ and

$$\mathcal{C} = \{u \in L^2(\Omega)^m : u(\omega) \in C,\ \omega \in \Omega\},$$

where $C \subset \mathbb{R}^m$ is closed and convex. As in Remark 5.24, let us suppose that $(\bar{y},\bar{u},\bar{w})$ is a KKT-triple with continuously invertible operator $E_y(\bar{y},\bar{u})$ and denote by $y(u)$ the locally unique control-to-state mapping with $y(\bar{u}) = \bar{y}$.

We consider the reformulation (5.27) of (5.24)–(5.26) under Assumption 5.17. If we work with exact elements M of the generalized differential $\partial_C \Sigma(y,u,w)$, the semismooth Newton method for the solution of (5.27) requires us to solve systems of the form $Ms = -\Sigma(y,u,w)$. According to Theorem 5.19, these systems assume the form

$$\left(\begin{array}{ccc|c} L_{yy} & L_{yu} & E_y^* & \rho_1 \\ D_2 L_{uy} & D_1 I + D_2 L_{uu} & D_2 E_u^* & \rho_2 \\ E_y & E_u & 0 & \rho_3 \end{array}\right), \tag{5.41}$$

where we have omitted the arguments (y,u,w) and (y,u). By the Banach theorem, $E_y(y,u)$ is continuously invertible in a neighborhood of $(\bar{y},\bar{u})$ with uniformly bounded inverse.

Using this, we can perform the following block elimination:

$$\left(\begin{array}{ccc|c} L_{yy} & L_{yu} & E_y^* & \rho_1 \\ D_2 L_{uy} & D_1 I + D_2 L_{uu} & D_2 E_u^* & \rho_2 \\ E_y & E_u & 0 & \rho_3 \end{array}\right)$$

$$\Updownarrow \quad (\text{Row } 1 - L_{yy} E_y^{-1} \times \text{Row } 3)$$

$$\left(\begin{array}{ccc|c} 0 & L_{yu} - L_{yy} E_y^{-1} E_u & E_y^* & \rho_1 - L_{yy} E_y^{-1} \rho_3 \\ D_2 L_{uy} & D_1 I + D_2 L_{uu} & D_2 E_u^* & \rho_2 \\ E_y & E_u & 0 & \rho_3 \end{array}\right)$$

$$\Updownarrow \quad (\text{Row } 2 - D_2 L_{uy} E_y^{-1} \times \text{Row } 3)$$

$$\left(\begin{array}{ccc|c} 0 & L_{yu} - L_{yy} E_y^{-1} E_u & E_y^* & \rho_1 - L_{yy} E_y^{-1} \rho_3 \\ 0 & D_1 I + D_2 (L_{uu} - L_{uy} E_y^{-1} E_u) & D_2 E_u^* & \rho_2 - D_2 L_{uy} E_y^{-1} \rho_3 \\ E_y & E_u & 0 & \rho_3 \end{array}\right)$$

$$\Updownarrow \quad (\text{Row } 2 - D_2 E_u^* (E_y^*)^{-1} \times \text{Row } 1)$$

$$\left(\begin{array}{ccc|c} 0 & L_{yu} - L_{yy} E_y^{-1} E_u & E_y^* & \rho_1 - L_{yy} E_y^{-1} \rho_3 \\ 0 & D_1 I + D_2 H & 0 & \rho_2' \\ E_y & E_u & 0 & \rho_3 \end{array}\right),$$

where

$$\begin{aligned} H(y,u,w) = {} & L_{uu} - L_{uy} E_y^{-1} E_u - E_u^* (E_y^*)^{-1} L_{yu} \\ & + E_u^* (E_y^*)^{-1} L_{yy} E_y^{-1} E_u, \end{aligned} \tag{5.42}$$

$$\rho_2' = \rho_2 - D_2 E_u^* (E_y^*)^{-1} \rho_1 + D_2 (E_u^* (E_y^*)^{-1} L_{yy} - L_{uy}) E_y^{-1} \rho_3.$$

The operator H can be written in the form

$$H = T^* \begin{pmatrix} L_{yy} & L_{yu} \\ L_{uy} & L_{uu} \end{pmatrix} T, \quad T(y,u) = \begin{pmatrix} -E_y^{-1} E_u \\ I \end{pmatrix}.$$

Therefore, the continuous invertibility of M is closely related to the continuous invertibility of the operator $D_1 I + D_2 H$.

We now consider the reduced objective function $j(u) = J(y(u),u)$ in a neighborhood of $\bar{u}$. It is shown in section A.1 that the Hessian $j''(u)$ can be represented in the form

$$j''(u) = T(y,u)^* \begin{pmatrix} L_{yy}(y,u,w) & L_{yu}(y,u,w) \\ L_{uy}(y,u,w) & L_{uu}(y,u,w) \end{pmatrix} T(y,u),$$

$$T(y,u) = \begin{pmatrix} -E_y(y,u)^{-1} E_u(y,u) \\ I \end{pmatrix},$$

where $y = y(u)$, and $w = w(u)$ is the adjoint state, given by the adjoint equation (5.37), which can also be written in the form (5.39). Therefore, we see that $j''(u) = H(y(u), u, w(u))$ and, hence, $j''(\bar{u}) = H(\bar{y}, \bar{u}, \bar{w})$, since $\bar{y} = y(\bar{u})$ and $\bar{w} = w(\bar{u})$. For $(y, u, w) = (y(u), u, w(u))$ we have $L_u(y(u), u, w(u)) = j'(u)$ by (5.38). Hence, with $D = (D_1|D_2)$,

$$D(\omega) \in \partial_C \pi\big(u(\omega), L_u(y(u), u, w(u))(\omega)\big) \iff D(\omega) \in \partial_C \pi\big(u(\omega), j'(u)(\omega)\big).$$

Thus, by Theorems 5.11 and 5.19, for any $(y, u, w) = (y(u), u, w(u))$ and all operators M of the form (5.28) the Schur complement satisfies

$$M^R = D_1 I + D_2 H(y(u), u, w(u)) \in \partial_C^\circ \Pi^R(u),$$

where

$$\Pi^R(u)(\omega) = \pi\big(u(\omega), j'(u)(\omega)\big).$$

For the application of the class of (projected) semismooth Newton methods to problem (5.27) we need the invertibility of $M_k \in \partial_C \Sigma(y_k, u_k, w_k)$ as an operator between appropriate spaces. We already observed that for the reduced problem it is appropriate to require the uniformly bounded invertibility of $M_k^R \in \partial_C^\circ \Pi^R(u_k)$ in $\mathcal{L}([L^r]^m, [L^r]^m)$. In agreement with this we now require the following.

Assumption 5.25. *At least one of the following conditions holds:*

(a) *The operators $M_k \in \partial_C \Sigma(y_k, u_k, w_k)$ are continuously invertible elements of $\mathcal{L}(Y \times [L^r]^m \times W^*, Y^* \times [L^r]^m \times W)$ with the norms of their inverses bounded by a constant $C_{M^{-1}}$.*

(b) *There exist constants $\eta > 0$ and $C_{M^{-1}} > 0$ such that, for all $(y, u, w) \in (\bar{y}, \bar{u}, \bar{w}) + \eta B_{Y \times [L^p]^m \times W^*}$, every $M \in \partial_C \Sigma(y_k, u_k, w_k)$ is an invertible element of $\mathcal{L}(Y \times [L^r]^m \times W^*, Y^* \times [L^r]^m \times W)$ with the norm of its inverse bounded by $C_{M^{-1}}$.*

This assumption corresponds to Assumption 3.12 (a) with $Y_0 = Y \times [L^r]^m \times W^*$.

Under Assumptions 5.17, 5.25, and 3.12 (b) (ensuring the availability of a smoothing step), we can apply Algorithm 3.10 or its projected version, Algorithm 3.22 (with $B_k = M_k$ and, e.g., $\mathcal{K} = \mathcal{C}$) for $f = \Sigma$, $\partial^* f = \partial_C \Sigma$, $Y = Y \times [L^p]^m \times W^*$, $Z = Y^* \times [L^r]^m \times W$, and $Y_0 = Y \times [L^r]^m \times W^*$. Theorems 3.13 and 3.24 then guarantee superlinear convergence since, by Theorem 5.19, Σ is $\partial_C \Sigma$-semismooth. In section 5.2.4 we will propose a way of constructing smoothing steps.

In the same way, we can consider reformulations arising under Assumption 5.20. In this case we have

$$L_u(y, u, w) = \lambda u + G(y, u, w), \quad \pi(x) = x_1 - P_C(x_1 - \lambda^{-1} x_2).$$

Further, for all $M \in \partial_C \Sigma(y, u, w)$, there exists $D \in L^\infty(\Omega)^{m \times m}$ with $D \in \partial_C P_C(-\lambda^{-1} G(y, u, w))$

such that

$$M = \begin{pmatrix} L_{yy} & L_{yu} & E_y^* \\ \lambda^{-1}DG_y & I+\lambda^{-1}DG_u & \lambda^{-1}DG_w \\ E_y & E_u & 0 \end{pmatrix}$$
$$= \begin{pmatrix} L_{yy} & L_{yu} & E_y^* \\ \lambda^{-1}DL_{uy} & I+\lambda^{-1}D(L_{uu}-\lambda I) & \lambda^{-1}DE_u^* \\ E_y & E_u & 0 \end{pmatrix}$$
$$= \begin{pmatrix} L_{yy} & L_{yu} & E_y^* \\ D_2L_{uy} & D_1I+D_2L_{uu} & D_2E_u^* \\ E_y & E_u & 0 \end{pmatrix},$$

with $D_1 = I - D$ and $D_2 = \lambda^{-1}D$. Note that $(D_1, D_2) \in \partial_C\pi(u, L_u(y,u,w))$ and, hence, for these choices of D_1 and D_2, the operator M assumes the form (5.28). Thus, we can apply the same transformations to the Newton system as before and obtain again that, for $(y,u,w) = (y(u),u,w(u))$, the generalized differentials of the reduced semismooth reformulation appear as Schur complements of the full system. We choose the following as the regularity condition.

Assumption 5.26. *At least one of the following conditions holds:*

(a) *The operators $M_k \in \partial_C\Sigma(y_k,u_k,w_k)$ are continuously invertible elements of $\mathcal{L}(Y \times [L^r]^m \times W^*, Y^* \times [L^r]^m \times W)$ with the norms of their inverses uniformly bounded by a constant $C_{M^{-1}}$.*

(b) *There exist constants $\eta > 0$ and $C_{M^{-1}} > 0$ such that, for all $(y,u,w) \in (\bar{y},\bar{u},\bar{w}) + \eta B_{Y\times[L^r]^m\times W^*}$, every $M \in \partial_C\Sigma(y_k,u_k,w_k)$ is an invertible element of $\mathcal{L}(Y \times [L^r]^m \times W^*, Y^* \times [L^r]^m \times W)$ with the norm of its inverse bounded by $C_{M^{-1}}$.*

This assumption corresponds to Assumption 3.12 (a) with $Y_0 = Y = Y \times [L^r]^m \times W^*$. Now, under Assumptions 5.20 and 5.26, we can apply Algorithm 3.10 or its projected version, Algorithm 3.22, for $f = \Sigma$, $\partial^* f = \partial_C\Sigma$, $Y = Y_0 = Y \times [L^r]^m \times W^*$, and $Z = Y^* \times [L^r]^m \times W$. Since $Y_0 = Y$, we do not need a smoothing step. Theorems 3.13 and 3.24 establish superlinear convergence since, by Theorem 5.21, Σ is $\partial_C\Sigma$-semismooth.

5.2.4 Smoothing Steps

In addition to Assumption 5.17, we require the following.

Assumption 5.27. *The derivative L_u has the form $L_u(y,u,w) = \lambda u + G(y,u,w)$, with*

$$(y,u,w) \in Y \times L^r(\Omega)^m \times W^* \mapsto G(y,u,w) \in L^p(\Omega)^m$$

being locally Lipschitz continuous.

Example 5.28. We verify this assumption for the control problem of Example 5.23. There, we had $Y = W^* = H_0^1(\Omega)$, $U = L^p(\Omega)$ with $p \geq 2$ arbitrary, and

$$L_u(y,u,w) = \lambda u - gw = \lambda u + G(y,u,w) \quad \text{with} \quad G(y,u,w) = -gw.$$

Since $g \in L^\infty$ and $w \in H_0^1(\Omega) \hookrightarrow L^q(\Omega)$ for all $q \in [1,\infty]$ if $n = 1$, all $q \in [1,\infty)$ if $n = 2$, and all $q \in [1, 2n/(n-2)]$ if $n \geq 3$, we see that G maps L^r, with $r \geq 2$ arbitrary, linear and continuous to $L^q(\Omega)$. Thus, Assumption 5.27 holds for all $p \in (2, q]$. ∎

We can show the following theorem.

Theorem 5.29. *Let Assumptions* 5.17 *and* 5.27 *hold. Then the operator*

$$S : Y \times L^r(\Omega)^m \times W^* \mapsto Y \times L^p(\Omega)^m \times W^*,$$

$$S(y,u,w) = \begin{pmatrix} y \\ P_C(u - \lambda^{-1} L_u(y,u,w)) \\ w \end{pmatrix},$$

defines a smoothing step.

Proof. We first note that

$$x_1 = P_C(x_1 - \lambda^{-1} x_2) \quad \Longleftrightarrow \quad x_1 = P_C(x_1 - x_2) \quad \Longleftrightarrow \quad \pi(x) = 0,$$

so that

$$u = P_C\big(u - \lambda^{-1} L_u(y,u,w)\big) \quad \Longleftrightarrow \quad \Pi(y,u,w) = 0.$$

Hence, for any solution $(\bar{y}, \bar{u}, \bar{w})$ of (5.27), we have

$$S(\bar{y}, \bar{u}, \bar{w}) = (\bar{y}, \bar{u}, \bar{w}).$$

Furthermore, as in section 4.1, pointwise on Ω holds

$$\begin{aligned}
&\|P_C\big(u - \lambda^{-1} L_u(y,u,w)\big) - \bar{u}\|_2 \\
&\quad = \|P_C\big(u - \lambda^{-1} L_u(y,u,w)\big) - P_C\big(\bar{u} - \lambda^{-1} L_u(\bar{y}, \bar{u}, \bar{w})\big)\|_2 \\
&\quad = \|P_C\big(-\lambda^{-1} G(y,u,w)\big) - P_C\big(-\lambda^{-1} G(\bar{y}, \bar{u}, \bar{w})\big)\|_2 \\
&\quad \leq \lambda^{-1} \|G(y,u,w) - G(\bar{y}, \bar{u}, \bar{w})\|_2,
\end{aligned}$$

and thus, with C_G denoting the local Lipschitz constant of G near $(\bar{y}, \bar{u}, \bar{w})$,

$$\begin{aligned}
&\|P_C\big(u - \lambda^{-1} L_u(y,u,w)\big) - \bar{u}\|_{[L^p]^m} \\
&\qquad \leq C_G c \lambda^{-1} \|(y,u,w) - (\bar{y}, \bar{u}, \bar{w})\|_{Y \times [L^r]^m \times W^*},
\end{aligned}$$

where c depends on m only. The proof is complete, since

$$\begin{aligned}
&\|S(y,u,w) - (\bar{y}, \bar{u}, \bar{w})\|_{Y \times [L^p]^m \times W^*} \\
&\quad \leq c\big(\|(y,w) - (\bar{y}, \bar{w})\|_{Y \times W^*} + \|P_C\big(u - \lambda^{-1} L_u(y,u,w)\big) - \bar{u}\|_{[L^p]^m}\big). \qquad \square
\end{aligned}$$

5.2.5 Regularity Conditions

We already observed that the all-at-once Newton system is closely related to the black-box Newton system. In this section we show how the regularity of the all-at-once Newton system can be reduced to regularity conditions on its Schur complement. Since, for $(y,u,w) = (y(u),u,w(u))$, this Schur complement coincides with the operator of the black-box Newton system, sufficient conditions for regularity can then be developed along the lines of section 4.3. In the following, we restrict our investigations to the situation of Assumptions 5.20 and 5.26.

Our hypothesis on the Schur complement is as follows.

Assumption 5.30. *There exist constants* $\eta > 0$ *and* $C^R_{M^{-1}} > 0$ *such that, for all* $(y,u,w) \in (\bar{y},\bar{u},\bar{w}) + \eta B_{Y\times[L^r]^m\times W^*}$ *holds*

(a) $E_y(y,u,w) \in \mathcal{L}(Y \times [L^r]^m \times W^*, Y^* \times [L^r]^m \times W)$ *is continuously invertible with uniformly bounded inverse.*

(b) *For all D satisfying* (5.30)*, the Schur complement*

$$D_1 + D_2 H,$$

with $D_1 = I - D$, $D_2 = \lambda^{-1} D$, *and H as defined in* (5.42)*, is an invertible element of* $\mathcal{L}([L^r]^m, [L^r]^m)$ *with* $\|M^{-1}\|_{[L^r]^m,[L^r]^m} \le C^R_{M^{-1}}$.

Theorem 5.31. *Let Assumptions* 5.20 *and* 5.30 *hold. Then the regularity condition of Assumption* 5.26 (b) *holds.*

Proof. Let $(y,u,w) \in (\bar{y},\bar{u},\bar{w}) + \eta B_{Y\times[L^r]^m\times W^*}$ and $M \in \partial_C \Sigma(y,u,w)$ be arbitrary. Then there exists D satisfying (5.30) such that M assumes the form (5.29). Now consider any $\rho = (\rho_1,\rho_2,\rho_3)^T \in Y^* \times [L^r]^m \times W$. Then, according to section 5.2.3, solving the system

$$M(s_y, s_u, s_w)^T = \rho$$

is equivalent to

$$\begin{aligned} (D_1 I + D_2 H)s_u &= \rho_2 - D_2 E_u^*(E_y^*)^{-1}\rho_1 \\ &\quad + D_2(E_u^*(E_y^*)^{-1}L_{yy} - L_{uy})E_y^{-1}\rho_3, && (5.43) \\ E_y s_y &= \rho_3 - E_u s_u, && (5.44) \\ E_y^* s_w &= \rho_1 - L_{yy}E_y^{-1}\rho_3 - (L_{yu} - L_{yy}E_y^{-1}E_u)s_u. && (5.45) \end{aligned}$$

The assumptions ensure twice continuous differentiability of L and uniformly bounded invertibility of E_y and $D_1 + D_2 H$. Furthermore, D and thus D_1, D_2 are uniformly bounded in $L^\infty(\Omega)^{m\times m}$ due to the Lipschitz continuity of P_C. This and (5.43)–(5.45) show that, possibly after shrinking η, there exists $C_{M^{-1}} > 0$ such that

$$\|s\|_{Y\times[L^r]^m\times W^*} \le C_{M^{-1}} \|s\|_{Y^*\times[L^r]^m\times W},$$

holds uniformly on $(\bar{y},\bar{u},\bar{w}) + \eta B_{Y\times[L^r]^m\times W^*}$. $\square$

Chapter 6

Mesh Independence

6.1 Introduction

An important motivation for investigating optimization methods in infinite dimensions is developing algorithms that are mesh independent. Here, mesh independence means the following: Suppose that for the infinite-dimensional problem (P), a local convergence theory for an abstract solution algorithm A is available. This algorithm A could, e.g., be the Newton method if (P) is an operator equation. For the numerical implementation, the problem (P) needs to be discretized, which results in a discrete problem (P_h) ($h > 0$ denoting the mesh size or, more generally, the accuracy of the discretization). The process of discretization also results in a discrete version A_h of the algorithm A under consideration. If the discretization of (P) is done appropriately, then for $h \to 0^+$, the original problem (P) is increasingly better approximated by (P_h), and the solutions $\bar{u}_h$ of (P_h) closest to the solution $\bar{u}$ of (P) converge to $\bar{u}$ as $h \to 0$. Mesh independence means that some kind of convergence also holds for the behavior of the algorithm A_h towards the behavior of algorithm A. This behavioral convergence comes with different flavors. The traditional mesh-independence results for the Newton method state that, under appropriate assumptions, there exists a neighborhood $V \subset U$ of $\bar{u}$ such that, for all $\varepsilon > 0$, there exists $h_\varepsilon > 0$ for which the following holds: If the Newton iteration for (P) is started at $u^0 \in V$ and the Newton iteration for (P_h), $0 < h \le h_\varepsilon$, is started at the discrete point $u_h^0 = \Delta_h u^0$, where $\Delta_h : U \to U_h$ is a bounded linear discretization operator, then the Newton sequences (u^k) and (u_h^k) converge to $\bar{u}$ and $\bar{u}_h$, respectively, and the indices of the first iterates lying in the ε-balls around $\bar{u}$ and $\bar{u}_h$, respectively, differ by at most one. This means that if u^k is the first iterate with $\|u^k - \bar{u}\|_U < \varepsilon$ and u_h^l is the first iterate with $\|u_h^l - \bar{u}_h\|_{U_h} < \varepsilon$, then $l \in \{k-1, k, k+1\}$. Furthermore, the distance $\|\Delta_h u^k - u_h^k\|_{U_h}$ can be estimated in terms of the discretization error. Results of this form are available for the classical Newton method and also for Josephy–Newton methods for generalized equations, which include sequential quadratic programming as a special case [111, Section 2.6]. We refer the reader to [7, 3, 61, 58, 95, 147, 200] for further details.

In this chapter, we are interested in deriving a mesh-independence result for semismooth Newton methods in connection with nonsmooth reformulations of variational inequalities. It turns out that the proof techniques that work for the aforementioned Newton methods cannot be transferred to our setting. A severe difficulty arises from the fact that, as we will see, the order of semismoothness is *not* stable with respect to perturbations of the point where they are evaluated. The first result on mesh independence of semismooth Newton methods was derived by Hintermüller and Ulbrich [106]. Subsequently, this result was shown to be applicable to mixed control-state constrained optimal control problems [100] and to regularized state-constrained optimal control problems [105].

In this chapter we extend the results of [106] by proving a mesh-independent order of q-superlinear convergence, whereas in [106] the mesh independence of any prescribed linear q-rate of convergence is shown.

To avoid unnecessary notational overhead, we do not consider bilaterally constrained variational inequalities here (which would be possible as well), but rather focus on complementarity problems CP(F) in the Lebesgue space $U = L^2(\Omega)$:

CP(F): Find $u \in U$ such that

$$u \geq 0, \quad F(u) \geq 0, \quad uF(u) = 0 \quad \text{a.e. on } \Omega.$$

Here, $\Omega \subset \mathbb{R}^n$ is a bounded open domain and

$$F : U \to U$$

is a (Fréchet) differentiable operator.

We reformulate CP(F) by using an NCP-function. As we know, the application of semismooth Newton methods requires a smoothing property. Hence, we will assume the following particular structure of F, which, as was shown before, arises frequently in PDE-constrained optimization.

Assumption 6.1. *The operator $F : U \to U$, with $U = L^2(\Omega)$, has the form*

$$F(u) = \lambda u + G(u),$$

where $\lambda \in L^\infty(\Omega)$, $\lambda \geq \lambda_0$ with a constant $\lambda_0 > 0$, and with G denoting a locally Lipschitz continuous operator

$$G : U \to L^p(\Omega)$$

with $p \in (2, \infty]$. Furthermore, the mapping $L^r(\Omega) \ni u \mapsto G(u) \in U$ is continuously Fréchet differentiable (with $r \geq 2$ specified when referring to this part of the assumption).

Remark 6.2. *It is also possible to work with a localized version of Assumption* 6.1, *i.e., to require the assumptions only to hold on an open neighborhood of a solution $\bar{u}$ of CP(F). Since it is obvious how to do such a localization, we do not include it here.*

For compact notation, we will, where required, use the following convention.

Definition 6.3. *For $p_1, p_2, q_1, q_2 \in [1, \infty]$ and an operator $Q : L^{p_1}(\Omega) \to L^{q_1}(\Omega)$, we define, if this is meaningful, the operator*

$$Q_{p_2,q_2} : L^{p_2}(\Omega) \ni u \mapsto Q(u) \in L^{q_2}(\Omega).$$

This operator is always well defined in the case $p_2 \geq p_1$, $q_2 \leq q_1$, since $L^{p_2}(\Omega) \hookrightarrow L^{p_1}(\Omega)$ and $L^{q_1}(\Omega) \hookrightarrow L^{q_2}(\Omega)$.

Remark 6.4. *The above convention is only used where required, e.g., to express relaxed differentiability or Hölder continuity requirements. However, we will not write, e.g., $\|G_{p,2}(u)\|_{L^2}$, but rather $\|G(u)\|_{L^2}$.*

We will derive nonsmooth reformulations based on the min-NCP-function. The following lemma collects several useful reformulations of CP(F).

Lemma 6.5. *Under Assumption* 6.1 *(with $r \geq 2$ arbitrary), the following assertions on $\bar{u} \in U = L^2(\Omega)$ are equivalent:*

0. *$\bar{u}$ solves CP(F).*
1. *For arbitrary, fixed $\sigma \in L^\infty(\Omega)$ with $\sigma > 0$, $\bar{u}$ solves*
$$\min\{u, \sigma F(u)\} = 0. \tag{6.1}$$
2. *$\bar{u}$ solves*
$$u + \min\left\{0, \frac{1}{\lambda} G(u)\right\} = 0. \tag{6.2}$$
3. *$\bar{u}$ satisfies $\bar{u} = -\min\{0, \bar{z}\}$ with $\bar{z} \in L^p(\Omega)$ solving*
$$z - \frac{1}{\lambda} G(-\min\{0, z\}) = 0. \tag{6.3}$$

Proof. 0. $\Longleftrightarrow$ 1.: This follows from the fact that $(x_1, x_2) \mapsto \min\{x_1, \sigma(\omega)x_2\}$ is an NCP-function for a.a. $\omega \in \Omega$.

0. $\Longleftrightarrow$ 2.: (6.2) is equivalent to (6.1) for the special choice $\sigma = 1/\lambda$. In fact,
$$\min\left\{u, \frac{1}{\lambda} F(u)\right\} = \min\left\{u, u + \frac{1}{\lambda} G(u)\right\} = u + \min\left\{0, \frac{1}{\lambda} G(u)\right\}.$$
Therefore, we can use the equivalence of 0. and 1. for this choice of σ.

2. $\Longrightarrow$ 3.: Let $\bar{u}$ solve (6.2), i.e.,
$$\bar{u} + \min\left\{0, \frac{1}{\lambda} G(\bar{u})\right\} = 0.$$
Setting $\bar{z} = (1/\lambda) G(\bar{u})$, it follows that
$$\bar{u} = -\min\{0, \bar{z}\}.$$
Hence, $\bar{z} = (1/\lambda) G(-\min\{0, \bar{z}\})$ and thus $\bar{z}$ solves (6.3) and $\bar{u} = -\min\{0, \bar{z}\}$ holds.

3. $\Longrightarrow$ 2.: Let $\bar{u} = -\min\{0, \bar{z}\}$, where $\bar{z}$ solves (6.3). Then
$$0 = \bar{u} + \min\{0, \bar{z}\} = \bar{u} + \min\left\{0, \frac{1}{\lambda} G(-\min\{0, \bar{z}\})\right\} = \bar{u} + \min\left\{0, \frac{1}{\lambda} G(\bar{u})\right\}.$$
Hence, $\bar{u} = -\min\{0, \bar{z}\}$ solves (6.2). $\square$

We recall the following consequence of Assumption 6.1 and of formulation 2 in Lemma 6.5.

Lemma 6.6. *Let Assumption* 6.1 *hold and let* $\bar{u}$ *be a solution of* $CP(F)$. *Then there holds* $\bar{u} \in L^p(\Omega)$.

Proof. By Lemma 6.5, part 2, there holds

$$\bar{u} = -\min\left\{0, \frac{1}{\lambda}G(\bar{u})\right\}.$$

Now $1/\lambda \in L^\infty(\Omega)$, $G(\bar{u}) \in L^p(\Omega)$, hence $\bar{u} \in L^p(\Omega)$. □

It will be convenient to introduce the following operators:

$$N : L^p(\Omega) \to U, \quad N(z)(\omega) = \min\{0, z(\omega)\}, \quad \omega \in \Omega. \tag{6.4}$$

$$\hat{Z} : L^p(\Omega) \to L^p(\Omega), \quad \hat{Z}(u) = \sigma F(u) - u, \tag{6.5}$$

$$Z : U \to L^p(\Omega), \quad Z(u) = \frac{1}{\lambda}G(u). \tag{6.6}$$

The well-definedness of these operators follows from Assumption 6.1 and the continuous embedding $L^p(\Omega) \hookrightarrow U$.

Since $\psi : \mathbb{R} \to \mathbb{R}$, $\psi(t) = \min\{0, t\}$, is Lipschitz continuous and strongly semismooth, Theorem 3.49 implies that the operator N defined in (6.4) is semismooth with respect to

$$\partial^\circ N(z) = \{M_N \in \mathcal{L}(L^p(\Omega), U) : M_N v = g_N v,\ g_N \in L^\infty(\Omega) \text{ satisfies (6.8)}\}, \tag{6.7}$$

where

$$g_N(\omega) \begin{cases} = 1 & \text{if } z(\omega) < 0, \\ \in [0,1] & \text{if } z(\omega) = 0, \\ = 0 & \text{if } z(\omega) > 0. \end{cases} \tag{6.8}$$

As was discussed in previous chapters, the operator

$$u \mapsto \min\{u, \sigma F(u)\} = u + \min\{0, \sigma F(u) - u\} = u + \min\{0, (\sigma\lambda - 1)u + \sigma G(u)\}$$

is, in general, only semismooth from $L^q(\Omega)$ to $U = L^2(\Omega)$ if $q > 2$ and if F is sufficiently well behaved (e.g., if $F_{q,2}$ is continuously F-differentiable). Hence, if we want to apply a semismooth Newton method to this reformulation, we have to work in $L^q(\Omega)$, $q > 2$, and to introduce a smoothing step. Suitable for a smoothing step is

$$U \ni u \mapsto -\min\left\{0, \frac{1}{\lambda}G(u)\right\} = -N(Z(u)) \in L^p(\Omega),$$

which by Assumption 6.1 is locally Lipschitz continuous and which has as fixed points exactly the solutions of CP(F).

Therefore, the rate of convergence of a semismooth Newton method for the formulation (6.1) is governed by the order of semismoothness of the mapping

$$\Phi_1 : u \in L^p(\Omega) \mapsto \min\{u, \sigma F(u)\} = u + N(\hat{Z}(u)) \in U. \tag{6.9}$$

We will now investigate the semismoothness of Φ_1 with respect to the differential

$$\begin{aligned}
\partial^\circ\Phi_1(u) &= I + \partial^\circ N(\hat{Z}(u))\hat{Z}'_{p,2}(u) \\
&= \{M_{\Phi_1} \in \mathcal{L}(L^p(\Omega),U) : M_{\Phi_1} = I + M_N(\sigma \cdot F'_{p,2}(u) - I),\ M_N \in \partial^\circ N(\hat{Z}(u))\} \\
&= \{M_{\Phi_1} \in \mathcal{L}(L^p(\Omega),U) : M_{\Phi_1} = (1-g_N)\cdot I + (g_N\sigma)\cdot F'_{p,2}(u), \\
&\qquad\qquad g_N \text{ satisfies (6.8) for } z = \hat{Z}(u)\}.
\end{aligned}$$

The next lemma proves the semismoothness of Φ_1 and provides structural information for the semismoothness remainder term of Φ_1.

Lemma 6.7. *Let Assumption* 6.1 *hold with $r = p$ and define N, $\hat{Z}$, and Φ_1 as in* (6.4), (6.5), *and* (6.9), *respectively. Furthermore, let $\hat{u} \in L^p(\Omega)$ be fixed. Then there exists an open neighborhood $V_p(\hat{u}) \subset L^p(\Omega)$ of $\hat{u}$ on which the operator $G_{p,p}$ is Lipschitz continuous with modulus $L_p > 0$ and $\hat{Z}_{p,2}$ is continuously differentiable. Furthermore, for all $u, u+d \in V_p(\hat{u})$, and all $M_{\Phi_1}(u+d) \in \partial^\circ\Phi_1(u+d)$, $M_N(\hat{Z}(u+d)) \in \partial^\circ N(\hat{Z}(u+d))$ such that $M_{\Phi_1}(u+d) = I + M_N(\hat{Z}(u+d))\hat{Z}'_{p,2}(u+d)$, there holds with $z := \hat{Z}(u)$ and $s := \hat{Z}(u+d) - \hat{Z}(u)$*

$$\begin{aligned}
&\|\Phi_1(u+d) - \Phi_1(u) - M_{\Phi_1}(u+d)d\|_{L^2} \\
&\qquad \le \|N(z+s) - N(z) - M_N(z+s)s\|_{L^2} \\
&\qquad\quad + \|\hat{Z}(u+d) - \hat{Z}(u) - \hat{Z}'_{p,2}(u+d)d\|_{L^2},
\end{aligned} \tag{6.10}$$

$$\|s\|_{L^p} = \|\hat{Z}(u+d) - \hat{Z}(u)\|_{L^p} \le L_{\hat{Z}}\|d\|_{L^p}, \tag{6.11}$$

where $L_{\hat{Z}} = \|\sigma\lambda - 1\|_{L^\infty} + \|\sigma\|_{L^\infty}L_p$. In particular, Φ_1 is semismooth on $V_p(\hat{u})$. Furthermore, Φ_1 is semismooth of order $\alpha > 0$ at $u \in V_p(\hat{u})$ if N is semismooth of order α at $z = \hat{Z}(u)$ and if $G_{p,2}$ is semismooth of order α at u.

Proof. Since $G : U \to L^p(\Omega)$ is locally Lipschitz continuous and $L^p(\Omega) \hookrightarrow U$, there exist an L^p-neighborhood $V_p(\hat{u})$ of $\hat{u}$ and $L_p > 0$ such that $G_{p,p}$ is Lipschitz continuous with modulus L_p on $V_p(\hat{u})$. Furthermore, since $G_{p,2}$ is continuously F-differentiable, $\hat{Z}_{p,2}$ is continuously F-differentiable with $\hat{Z}'_{p,2}(u) = \sigma \cdot G'_{p,2}(u) + (\sigma\lambda - 1)\cdot I$.

Now consider $u, u+d \in V_p(\hat{u})$ and set $z = \hat{Z}(u)$, $s = \hat{Z}(u+d) - z$. Then

$$\begin{aligned}
\|s\|_{L^p} &= \|\hat{Z}(u+d) - \hat{Z}(u)\|_{L^p} \le \|(\sigma\lambda - 1)d\|_{L^p} + \|\sigma(G(u+d) - G(u))\|_{L^p} \\
&\le (\|\sigma\lambda - 1\|_{L^\infty} + \|\sigma\|_{L^\infty}L_p)\|d\|_{L^p}.
\end{aligned}$$

This shows that $\hat{Z}$ is Lipschitz continuous on $V_p(\hat{u})$ with modulus

$$L_{\hat{Z}} = \|\sigma\lambda - 1\|_{L^\infty} + \|\sigma\|_{L^\infty}L_p.$$

Next, let $M_{\Phi_1}(u+d) \in \partial^\circ\Phi_1(u+d)$ be arbitrary. Then there exists $M_N(z+s) \in \partial^\circ N(z+s)$ such that

$$M_{\Phi_1}(u+d) = I + M_N(\hat{Z}(u+d))\hat{Z}'_{p,2}(u+d) = I + M_N(z+s)\hat{Z}'_{p,2}(u+d).$$

We obtain

$$\begin{aligned}
R_1(u,d) &:= \Phi_1(u+d) - \Phi_1(u) - M_{\Phi_1}(u+d)d \\
&= u+d+N(\hat{Z}(u+d)) - u - N(\hat{Z}(u)) - d - M_N(\hat{Z}(u+d))\hat{Z}'_{p,2}(u+d)d \\
&= N(z+s) - N(z) - M_N(z+s)\hat{Z}'_{p,2}(u+d)d \\
&= N(z+s) - N(z) - M_N(z+s)s \\
&\quad + M_N(z+s)(\hat{Z}(u+d) - \hat{Z}(u) - \hat{Z}'_{p,2}(u+d)d).
\end{aligned}$$

Hence, using $M_N(z+s) : v \mapsto g_N v$ with $\|g_N\|_{L^\infty} \le 1$, we arrive at

$$\begin{aligned}
\|R_1(u,d)\|_{L^2} &\le \|N(z+s) - N(z) - M_N(z+s)s\|_{L^2} \\
&\quad + \|g_N\|_{L^\infty} \|\hat{Z}(u+d) - \hat{Z}(u) - \hat{Z}'_{p,2}(u+d)d\|_{L^2} \\
&\le \|N(z+s) - N(z) - M_N(z+s)s\|_{L^2} \\
&\quad + \|\hat{Z}(u+d) - \hat{Z}(u) - \hat{Z}'_{p,2}(u+d)d\|_{L^2}.
\end{aligned}$$

Since Theorem 3.49 yields that N is semismooth and Proposition 3.4 yields that $\hat{Z}_{p,2}$ is semismooth, we conclude that Φ_1 is semismooth on $V_p(\hat{u})$. The assertion on the semismoothness of order α follows immediately from (6.10), (6.11), and the assumptions. □

Next, we consider the reformulation (6.2). We now show that, under Assumption 6.1 with $r = 2$, the operator

$$\Phi_2 : U \to U, \quad \Phi_2(u) = u + \min\left\{0, \frac{1}{\lambda}G(u)\right\} = u + N(Z(u)) \tag{6.12}$$

is semismooth with respect to

$$\begin{aligned}
\partial^\circ\Phi_2(u) &= I + \partial^\circ N(Z(u))Z'_{2,2}(u) \\
&= \{M_{\Phi_2} \in \mathcal{L}(U,U) : M_{\Phi_2} = I + M_N Z'_{2,2}(u),\ M_N \in \partial^\circ N(Z(u))\} \\
&= \{M_{\Phi_2} \in \mathcal{L}(U,U) : M_{\Phi_2} = I + g_N \cdot Z'_{2,2}(u),\ g_N \text{ satisfies (6.8) for } z = Z(u)\}.
\end{aligned}$$

The next lemma is the analogue of Lemma 6.7 for Φ_2.

Lemma 6.8. *Let Assumption* 6.1 *with $r = 2$ hold. Define N, Z, and Φ_2 according to* (6.4), (6.6), *and* (6.12), *respectively. Consider $\hat{u} \in U$ and an open neighborhood $V(\hat{u}) \subset U$ of $\hat{u}$ on which G is Lipschitz continuous with modulus L. Then Z is Lipschitz continuous on $V(\hat{u})$ with constant $\|1/\lambda\|_{L^\infty} L \le L/\lambda_0$ and $Z_{2,2}$ is continuously differentiable. Furthermore, for all $u, u+d \in V(\hat{u})$, $M_{\Phi_2}(u+d) \in \partial^\circ\Phi_2(u+d)$, and $M_N(Z(u+d)) \in \partial^\circ N(Z(u+d))$ such that $M_{\Phi_2}(u+d) = I + M_N(Z(u+d))Z'_{2,2}(u+d)$, there holds with $z := Z(u)$ and $s := Z(u+d) - Z(u)$*

$$\begin{aligned}
\|\Phi_2(u+d) - \Phi_2(u) - M_{\Phi_2}(u+d)d\|_{L^2} & \\
\le \|N(z+s) - N(z) - M_N(z+s)s\|_{L^2} & \\
+ \|Z(u+d) - Z(u) - Z'_{2,2}(u+d)d\|_{L^2}, &
\end{aligned} \tag{6.13}$$

$$\|s\|_{L^p} = \|Z(u+d) - Z(u)\|_{L^2} \le \left\|\frac{1}{\lambda}\right\|_{L^\infty} L\|d\|_{L^2} \le \frac{L}{\lambda_0}\|d\|_{L^2}. \tag{6.14}$$

In particular, Φ_2 is semismooth on $V(\hat{u})$. Furthermore, if $G_{2,2}$ is α-order semismooth at $u \in V(\hat{u})$ and if N is semismooth of order $\alpha > 0$ at $z = Z(u)$, then Φ_2 is α-order semismooth at u.

Proof. By assumption, $Z_{2,2}$ is continuously F-differentiable with $Z'_{2,2}(u) = (1/\lambda) \cdot G'_{2,2}(u)$. Consider $u, u+d \in V(\hat{u})$ and set $z = Z(u)$, $s = Z(u+d) - z$. Then

$$\begin{aligned}\|s\|_{L^p} = \|Z(u+d) - Z(u)\|_{L^p} &\leq \left\| \frac{1}{\lambda}(G(u+d) - G(u)) \right\|_{L^p} \\ &\leq \left\| \frac{1}{\lambda} \right\|_{L^\infty} L\|d\|_{L^2} \leq \frac{L}{\lambda_0}\|d\|_{L^2}.\end{aligned}$$

This shows that Z is Lipschitz continuous on $V(\hat{u})$ and proves (6.14).

Now, let $M_{\Phi_2}(u+d) \in \partial^\circ \Phi_2(u+d)$ be arbitrary. Then there exists $M_N(z+s) \in \partial^\circ N(z+s)$ such that

$$M_{\Phi_2}(u+d) = I + M_N(Z(u+d))Z'_{2,2}(u+d) = I + M_N(z+s)Z'_{2,2}(u+d).$$

As in the proof of Lemma 6.7 (essentially, $\hat{Z}$ needs to be replaced by Z), we obtain for $R_2(u,d) \stackrel{\text{def}}{=} \Phi_2(u+d) - \Phi_2(u) - M_{\Phi_2}(u+d)d$ the estimate

$$\begin{aligned}\|R_2(u,d)\|_{L^2} \leq\ &\|N(z+s) - N(z) - M_N(z+s)s\|_{L^2} \\ &+ \|Z(u+d) - Z(u) - Z'_{2,2}(u+d)d\|_{L^2}.\end{aligned}$$

From Theorem 3.49 we obtain that N is semismooth and by Proposition 3.4, $Z_{2,2}$ is semismooth. Hence, Φ_2 is semismooth at u.

The assertion on the α-order semismoothness of Φ_2 follows immediately from the corresponding assumptions, (6.13) and (6.14). □

Finally, we can consider the reformulation (6.3) under Assumption 6.1 with $r = p$. This results in analyzing the semismoothness properties of the operator

$$\Phi_3 : L^p(\Omega) \to L^p(\Omega), \quad \Phi_3(z) = z - \frac{1}{\lambda}G(-\min\{0, z\}) = z - Z(-N(z)). \tag{6.15}$$

For the reason of brevity, we do not investigate this operator further, although this would readily be possible.

Looking at Lemmata 6.7 and 6.8, we see that if G is sufficiently well behaved, then the order of semismoothness of Φ_i is governed by the order of semismoothness of the operator N.

6.2 Uniform Semismoothness

In addition to the complementarity problem CP(F), we now consider discretized complementarity problems CP(F_h) with

$$F_h : U_h \to U_h, \quad F_h = \lambda_h \cdot I + G_h, \quad G_h : U_h \to U_h \cap L^p(\Omega), \quad U_h \subset L^2(\Omega).$$

Here, $\lambda_h \in L^\infty(\Omega)$, $\lambda_h \geq \lambda_0 > 0$.

The problem CP(F_h) consists in finding $u_h \in U_h$ such that

$$u_h \geq 0, \quad F_h(u_h) \geq 0, \quad u_h F_h(u_h) = 0 \quad \text{a.e. on } \Omega.$$

The subscript $h > 0$ is a measure of the accuracy of the discretization, for instance, the grid size of a finite element mesh. The smaller h, the more accurate is the discretization. The set of discretization parameters h of interest is denoted by $\mathcal{H}$. For convenience, we let $h = 0$ correspond to the original problem, i.e., $U_0 = U = L^2(\Omega)$, $F_0 = F$, $\lambda_0 = \lambda$, and $G_0 = G$.

We will assume throughout that U_h, λ_h, and σ_h (used in the reformulation below) are such that

$$\min\{0, v_h\} \in U_h, \quad \frac{1}{\lambda_h} v_h \in U_h, \quad \sigma_h v_h \in U_h \qquad \forall\, v_h \in U_h. \tag{6.16}$$

Examples for U_h that satisfy the above requirements are the piecewise constant functions or the space $L^2(\Omega)$ itself; see Remark 6.9.

It is obvious that CP(F_h) can be reformulated equivalently by means of one of the formulations in Lemma 6.5. As before, we only consider the reformulations 1 and 2. The following assertions are equivalent:

0. $\bar{u}_h \in U_h$ solves CP(F_h).
1. For fixed $\sigma_h \in L^\infty(\Omega)$, $\sigma_h > 0$, $\bar{u}_h \in U_h$ solves

$$\Phi_{1h}(u_h) \overset{\text{def}}{=} \min\{u_h, \sigma_h F_h(u_h)\} = 0.$$

2. $\bar{u}_h \in U_h$ solves

$$\Phi_{2h}(u_h) \overset{\text{def}}{=} u_h + N\left(\frac{1}{\lambda_h} \cdot G_h(u_h)\right) = 0.$$

Due to the requirement (6.16) on U_h, λ_h, and σ_h, there holds

$$\Phi_{1h} : U_h \cap L^p(\Omega) \to U_h, \quad \Phi_{2h} : U_h \to U_h.$$

As in (6.5) and (6.6), we define

$$\hat{Z}_h : U_h \cap L^p(\Omega) \to U_h \cap L^p(\Omega), \quad \hat{Z}_h(u_h) = \sigma_h F_h(u_h) - u_h, \tag{6.17}$$

$$Z_h : U_h \to U_h \cap L^p(\Omega), \quad Z_h(u_h) = \frac{1}{\lambda_h} G_h(u_h). \tag{6.18}$$

Then we can write

$$\Phi_{1h}(u_h) = u_h + N(\hat{Z}_h(u_h)), \quad \Phi_{2h}(u_h) = u_h + N(Z_h(u_h)).$$

Remark 6.9. *The requirement* (6.16) *is satisfied in the following important cases:*

(a) *U_h is the space of piecewise constant functions on a partitioning (e.g., triangulation) of Ω and $\lambda_h \in U_h$, $\lambda_h > 0$. Furthermore, for reformulation 1, $\sigma_h \in U_h$, $\sigma_h > 0$. Then F_h, $\hat{Z}_h$, and Z_h map to U_h.*

(b) *$U_h = U = L^2(\Omega)$, i.e., U is not discretized, and $G_h : L^2(\Omega) \to \hat{U}_h$ with $\hat{U}_h \subset L^p(\Omega)$. This approach of discretizing only G corresponds to the idea of not discretizing the* *control as proposed and investigated by Hinze [108, 111]. For instance, if λ_h is piecewise constant on a triangulation and G_h maps into the space $\hat{U}_h \subset L^p(\Omega)$ of continuous piecewise linear functions on this triangulation, then the solution $\bar{u}_h$ inherits a finite representation from the structure of $\hat{U}_h$. In fact, there holds*

$$\bar{u}_h = -N(\bar{z}_h) = \max(0, -\bar{z}_h),$$

where

$$\bar{z}_h = \frac{1}{\lambda_h} \cdot G_h(\bar{u}_h)$$

is a piecewise linear, possibly discontinuous function.

We now return to CP(F), CP(F_h), and one of the two equivalent nonsmooth reformulations:

1. $\Phi_1(\bar{u}) = 0$ and $\Phi_{1h}(\bar{u}_h) = 0$,
2. $\Phi_2(\bar{u}) = 0$ and $\Phi_{2h}(\bar{u}_h) = 0$.

For proving mesh independence it will be necessary to find $h_1 > 0$, $\delta > 0$, and a uniform upper bound for

$$\|R_N(\bar{z}_h, s_h)\|_{L^2} = \|N(\bar{z}_h + s_h) - N(\bar{z}_h) - M_N(\bar{z}_h + s_h)s_h\|_{L^2}$$

for all $s_h \in U_h$, $\|s_h\|_{L^2} < \delta$, and all $h \in \mathcal{H} \cup \{0\}$, $h \leq h_1$. Here, as indicated above, $h \in \mathcal{H}$ are all discretization parameters of interest, and $h = 0$ corresponds to the original problem $\Phi(u) = 0$. Furthermore, $\bar{u}_h$, $h \in \mathcal{H}$, are the discrete solutions of $\Phi_h(u_h) = 0$ corresponding to the solution $\bar{u}_0 = \bar{u}$ of $\Phi(u) = 0$. Finally, in reformulation 2 we have $\bar{z}_h = Z_h(\bar{u}_h)$ and $\bar{z} = Z(\bar{u})$, whereas in reformulation 1 we have $\bar{z}_h = \hat{Z}_h(\bar{u}_h)$ and $\bar{z} = \hat{Z}(\bar{u})$.

In order to investigate if such a uniform semismoothness result is possible, we analyze the semismoothness properties of N in detail. An important concept in this context is the notion of strict complementarity.

Definition 6.10. *Let $\bar{u}$ be a solution of CP(F). We say that $\bar{u}$ violates strict complementarity at $\omega \in \Omega$ if $\bar{u}(\omega) = 0$ and at the same time $F(\bar{u})(\omega) = 0$. More precisely,*

- $\bar{u} \geq 0$ *satisfies strict complementarity at* $\omega \iff \bar{u}(\omega) > 0$ *or* $F(\bar{u})(\omega) > 0$.
- $\bar{u} \geq 0$ *violates strict complementarity at* $\omega \iff \bar{u}(\omega) = 0$ *and* $F(\bar{u})(\omega) = 0$.

The same terminology is used for discrete solutions $\bar{u}_h \in U_h$ of CP(F_h).

We now express strict complementarity of $\bar{u}_h$ in terms of $\bar{z}_h = Z(\bar{u}_h)$.

Lemma 6.11. *Let $\bar{u}_h \in L^2(\Omega)$ be a solution of CP(F_h) and let $\bar{z}_h = Z(\bar{u}_h)$. Then, for $\omega \in \Omega$, the following assertions are equivalent:*

(a) *$\bar{u}_h$ satisfies strict complementarity at ω.*

(b) $\hat{Z}_h(\bar{u}_h)(\omega) \neq 0$.

(c) $Z_h(\bar{u}_h)(\omega) \neq 0$.

Proof. Let $\bar{u}_h$ be a solution of CP(F_h).

Assume first that $\bar{u}_h$ satisfies strict complementarity at ω.

Case 1: $\bar{u}_h(\omega) = 0$. Then $F_h(\bar{u}_h)(\omega) > 0$ and thus

$$\hat{Z}_h(\bar{u}_h)(\omega) = \sigma_h(\omega)F_h(\bar{u}_h)(\omega) - \bar{u}_h(\omega) = \sigma_h(\omega)F_h(\bar{u}_h)(\omega) > 0.$$

For the special choice $\sigma_h = 1/\lambda_h$ we also obtain

$$Z_h(\bar{u}_h)(\omega) = \hat{Z}_h(\bar{u}_h)(\omega) > 0.$$

Case 2: $\bar{u}_h(\omega) > 0$. Then $F_h(\bar{u}_h)(\omega) = 0$ and thus

$$\hat{Z}_h(\bar{u}_h)(\omega) = \sigma_h(\omega)F_h(\bar{u}_h)(\omega) - \bar{u}_h(\omega) = -\bar{u}_h(\omega) < 0.$$

For the special choice $\sigma_h = 1/\lambda_h$ we again obtain

$$Z_h(\bar{u}_h)(\omega) = \hat{Z}_h(\bar{u}_h)(\omega) < 0.$$

Now assume that $\bar{u}_h$ violates strict complementarity at ω. Then $\bar{u}_h(\omega) = F_h(\bar{u}_h)(\omega) = 0$ and thus

$$\hat{Z}_h(\bar{u}_h)(\omega) = \sigma_h(\omega)F_h(\bar{u}_h)(\omega) - \bar{u}_h(\omega) = 0.$$

For $\sigma_h = 1/\lambda_h$ we obtain

$$Z_h(\bar{u}_h)(\omega) = \hat{Z}_h(\bar{u}_h)(\omega) = 0. \qquad \square$$

Due to this observation, we use $|\bar{z}_h| = |Z_h(\bar{u}_h)|$ as a *pointwise measure for strict complementarity* for the reformulation 2. Similarly, we uses $|\bar{z}_h| = |\hat{Z}_h(\bar{u}_h)|$ as a *pointwise measure for strict complementarity* for the reformulation 1.

For $z, \bar{z} \in L^p(\Omega)$, we now consider the remainder term

$$R_N(\bar{z}, z - \bar{z}) = N(z) - N(\bar{z}) - M_N(z)(z - \bar{z}), \tag{6.19}$$

where $M_N(z) \in \partial^\circ N(z)$.

For brevity, we will write R_N instead of $R_N(\bar{z}, z - \bar{z})$.

Lemma 6.12. *Let $z, \bar{z} \in L^p(\Omega)$ and denote by $R_N = R_N(\bar{z}, z - \bar{z})$ the remainder term defined in* (6.19)*. Then the following holds:*

(a) $|R_N| \leq |\bar{z}|$ *on* Ω.

(b) *For all $\omega \in \Omega$:* $\quad R_N(\omega) \neq 0 \implies |R_N(\omega)| \leq |\bar{z}(\omega)| \leq |z(\omega) - \bar{z}(\omega)|$.

Proof. Let $\omega \in \Omega$ be arbitrary. Then, for arbitrary $M_N(z) \in \partial^\circ N(z)$ we have

$$N(z)(\omega) = \min\{0, z(\omega)\}, \quad N(\bar{z})(\omega) = \min\{0, \bar{z}(\omega)\},$$

$$[M_N(z - \bar{z})](\omega) = \eta(z(\omega) - \bar{z}(\omega)) \quad \text{with} \quad \eta \begin{cases} = 0 & \text{if } z(\omega) > 0, \\ = 1 & \text{if } z(\omega) < 0, \\ \in [0, 1] & \text{if } z(\omega) = 0. \end{cases}$$

For the residual

$$|R_N(\bar{z}, z-\bar{z})(\omega)| = |[N(z) - N(\bar{z}) - M_N(z)(z-\bar{z})](\omega)|$$

we thus obtain the following values:

$\lvert R_N(\bar{z}, z-\bar{z})(\omega)\rvert$		$\bar{z}(\omega)$		
		< 0	$= 0$	> 0
$z(\omega)$	< 0	0	0	$\bar{z}(\omega)$
	$= 0$	$(1-\eta)\lvert\bar{z}(\omega)\rvert$	0	$\eta\bar{z}(\omega)$
	> 0	$-\bar{z}(\omega)$	0	0

Since $\eta \in [0, 1]$, we always have

$$|R_N(\bar{z}, z-\bar{z})(\omega)| \le |\bar{z}(\omega)|.$$

Furthermore, $R_N(\bar{z}, z-\bar{z})(\omega) \neq 0$ implies either $z(\omega) \le 0 \wedge \bar{z}(\omega) > 0$ or $z(\omega) \ge 0 \wedge \bar{z}(\omega) < 0$. In both cases, there holds $|R_N(\bar{z}, z-\bar{z})(\omega)| \le |\bar{z}(\omega)| \le |z(\omega) - \bar{z}(\omega)|$. □

In section 3.3.3, we observed that the order of semismoothness of nonsmooth superposition operators depends on the growth rate of a certain parametric family of subset of Ω. In the context of complementarity problems, it was shown that the set corresponding to the parameter value $t > 0$ is closely related to the set of all points where the measure of strict complementarity lies between zero and t. In our current setting we showed in Lemma 6.11 that $|\bar{z}|$, where $\bar{z} = \hat{Z}(\bar{u})$ in reformulation 1 and $\bar{z} = Z(\bar{u})$ in reformulation 2, can serve as a measure of complementarity. As we will now see, the order of semismoothness of N at $\bar{z}$ depends on the size of the parameter $\gamma > 0$ in the following growth condition:

$$\begin{aligned}&\text{There exist } \gamma > 0,\ C > 0, \text{ and } t_0 > 0 \text{ such that}\\ &\operatorname{meas}(\{\omega \in \Omega : 0 < |\bar{z}(\omega)| < t\}) \le C t^{\gamma} \quad \forall\, t \in (0, t_0].\end{aligned} \tag{6.20}$$

The following theorem is an extended version of a sharp result on the order of semismoothness of the operator N. The generalization consists of the fact that we do not impose a growth condition on the point where the order of semismoothness is considered. Rather, the growth condition is posed on a different, sufficiently nearby point. More precisely, we estimate the semismoothness residual $R_N(\hat{z}, z-\hat{z})$ at $\hat{z}$ while requiring the growth condition (6.20) for $\bar{z}$. If we choose $\hat{z} = \bar{z}$, we recover an order of semismoothness result at $\bar{z}$ with a growth condition posed at $\bar{z}$. If we choose $\hat{z} = \bar{z}_h$, we obtain an estimate that we can use for the mesh-independent semismoothness result.

Theorem 6.13. *Let $\bar{z} \in L^p(\Omega)$ satisfy the condition* (6.20) *with constants $C > 0$, $t_0 > 0$, and $\gamma > 0$. Let*

$$\delta = \min\left\{1, \left(\frac{t_0}{2}\right)^{\frac{p+\gamma}{p}}\right\} \quad \text{if } p < \infty, \qquad \delta = \min\left\{1, \frac{t_0}{2}\right\} \quad \text{if } p = \infty, \tag{6.21}$$

and consider $\hat{z} \in L^p(\Omega)$ satisfying $\|\hat{z} - \bar{z}\|_{L^p} < \delta$. Assume that

$$\operatorname{meas}(\{\omega \in \Omega : \bar{z}(\omega) = 0 \neq \hat{z}(\omega)\}) = 0. \tag{6.22}$$

Then, for any $r \in [1, p)$ and all $z \in L^p(\Omega)$ satisfying

$$\|z - \hat{z}\|_{L^p} < \delta,$$

the following holds with $s = z - \hat{z}$:

(a) *If $p = \infty$, then*

$$\|R_N(\hat{z}, s)\|_{L^r} \leq 3^{\gamma/r} C^{1/r} \max\{\|\hat{z} - \bar{z}\|_{L^p}, \|s\|_{L^p}\}^{\gamma/r} \|s\|_{L^\infty}.$$

(b) *If $p < \infty$, then*

$$\|R_N(\hat{z}, s)\|_{L^r} \leq \max\left\{1, (2^\gamma C)^{\frac{p-r}{rp}}\right\} (\max\{\|\hat{z} - \bar{z}\|_{L^p}, \|s\|_{L^p}\})^{\frac{\gamma(p-r)}{r(p+\gamma)}} \|s\|_{L^p}.$$

Proof. Let $\bar{z} \in L^p(\Omega)$ and $\delta > 0$ satisfy the conditions stated in the theorem. Note that then $\delta \leq t_0/2$, since $(p+\gamma)/p > 1$. Let $\hat{z} \in L^p(\Omega)$ satisfy $\|\hat{z} - \bar{z}\|_{L^p} < \delta$ and (6.22). Consider $z \in L^p(\Omega)$ with $\|z - \hat{z}\|_{L^p} < \delta$, let $s := z - \hat{z}$, and set $\hat{R}_N := R_N(\hat{z}, s)$.

In the case $z = \hat{z}$ there is nothing to prove since then $\hat{R}_N = 0$. Hence, we may assume $z \neq \hat{z}$ and thus $s \neq 0$.

We fix $t \leq t_0/2$ such that $\|\hat{z} - \bar{z}\|_{L^p} < t$ and $\|s\|_{L^p} < t$ (t will be adjusted later) and define

$$\begin{aligned}
\Omega_0 &= \{\omega \in \Omega : \hat{R}_N(\omega) = 0\}, \\
\Omega_1 &= \{\omega \in \Omega : \hat{R}_N(\omega) \neq 0,\ \bar{z}(\omega) = 0\}, \\
\Omega_2(t) &= \{\omega \in \Omega : \hat{R}_N(\omega) \neq 0,\ 0 < |\bar{z}(\omega)| < 2t\}, \\
\Omega_3(t) &= \{\omega \in \Omega : \hat{R}_N(\omega) \neq 0,\ 0 < |\hat{z}(\omega)| < t,\ |\bar{z}(\omega)| \geq 2t\}, \\
\Omega_4(t) &= \{\omega \in \Omega : \hat{R}_N(\omega) \neq 0,\ |\hat{z}(\omega)| \geq t,\ |\bar{z}(\omega)| \geq 2t\}.
\end{aligned}$$

Since $\hat{R}_N(\omega) = 0$ implies $\hat{z}(\omega) \neq 0$, this gives the disjoint partitioning $\Omega = \Omega_0 \cup \Omega_1 \cup \Omega_2(t) \cup \Omega_3(t) \cup \Omega_4(t)$, and thus there holds

$$\|\hat{R}_N\|_{L^r}^r = \int_{\Omega_1} |\hat{R}_N(\omega)|^r \, d\omega + \sum_{j=2}^{4} \int_{\Omega_j(t)} |\hat{R}_N(\omega)|^r \, d\omega.$$

Note that on $\Omega_0^c := \Omega \setminus \Omega_0$, we have $|s(\omega)| \geq |\hat{z}(\omega)| \geq |\hat{R}_N(\omega)| > 0$ since otherwise $\hat{R}_N(\omega) = 0$ by Lemma 6.12.

Hence, with q such that $1/q + r/p = 1$, i.e., $q = p/(p-r)$ for $p < \infty$ and $q = 1$ for $p = \infty$, there holds for any $\hat{\Omega} \subset \Omega_0^c$

$$\begin{aligned}
\int_{\hat{\Omega}} |\hat{R}_N(\omega)|^r \, d\omega &\leq \int_{\hat{\Omega}} |\hat{z}(\omega)|^r \, d\omega \leq \|1_{\hat{\Omega}} |[\hat{z}]_{\hat{\Omega}}|^r \|_{L^1} \\
&\leq \|1_{\hat{\Omega}}\|_{L^q} \|\hat{z}|_{\hat{\Omega}}\|_{L^p}^r \leq \operatorname{meas}(\hat{\Omega})^{1/q} \|\hat{z}|_{\hat{\Omega}}\|_{L^p}^r \\
&\leq \operatorname{meas}(\hat{\Omega})^{1/q} \|s|_{\hat{\Omega}}\|_{L^p}^r.
\end{aligned}$$

On the set Ω_1 we have $\bar{z}(\omega) = 0$ and $\hat{z}(\omega) \neq 0$, and hence $\operatorname{meas}(\Omega_1) = 0$ by assumption (6.22).

Furthermore, since $2t \le t_0$, there holds

$$\int_{\Omega_2(t)} |\hat{R}_N(\omega)|^r \, d\omega \le \text{meas}(\Omega_2(t))^{1/q} \|s|_{\Omega_2(t)}\|_{L^p}^r \le C^{1/q}(2t)^{\gamma/q} \|s|_{\Omega_2(t)}\|_{L^p}^r.$$

For $p = \infty$ we have $q = 1$ and thus

$$\int_{\Omega_2(t)} |\hat{R}_N(\omega)|^r \, d\omega \le C^{1/q}(2t)^{\gamma/q} \|s|_{\Omega_2(t)}\|_{L^p}^r = 2^\gamma C t^\gamma \|s|_{\Omega_2(t)}\|_{L^\infty}^r.$$

For $p < \infty$ we obtain

$$\int_{\Omega_2(t)} |\hat{R}_N(\omega)|^r \, d\omega \le C^{1/q}(2t)^{\gamma/q} \|s|_{\Omega_2(t)}\|_{L^p}^r = (2^\gamma C)^{\frac{p-r}{p}} t^{\frac{\gamma(p-r)}{p}} \|s|_{\Omega_2(t)}\|_{L^p}^r.$$

For $\omega \in \Omega_3(t)$, there holds $|\hat{z}(\omega) - \bar{z}(\omega)| > t$. Hence,

$$\Omega_3(t) \subset \{\omega \in \Omega : |\hat{z}(\omega) - \bar{z}(\omega)| > t\}.$$

For $p = \infty$ we thus have $\text{meas}(\Omega_3(t)) = 0$, since for a.a. ω there holds $|\hat{z}(\omega) - \bar{z}(\omega)| \le \|\hat{z} - \bar{z}\|_{L^\infty} < t$.

We now consider the case $p < \infty$. Then we can estimate

$$\text{meas}(\Omega_3(t)) = \int_{\Omega_3(t)} d\omega \le \int_{\Omega_3(t)} \left(\frac{|\hat{z}(\omega) - \bar{z}(\omega)|}{t}\right)^p d\omega = t^{-p} \|[\hat{z} - \bar{z}]_{\Omega_3(t)}\|_{L^p}^p.$$

Hence, we obtain for $p < \infty$, using $p/q = p - r$,

$$\int_{\Omega_3(t)} |\hat{R}_N(\omega)|^r \, d\omega \le \text{meas}(\Omega_3(t))^{1/q} \|s|_{\Omega_3(t)}\|_{L^p}^r \le t^{r-p} \|[\hat{z} - \bar{z}]_{\Omega_3(t)}\|_{L^p}^{p-r} \|s|_{\Omega_3(t)}\|_{L^p}^r.$$

For $\omega \in \Omega_4(t)$, there holds $|s(\omega)| = |z(\omega) - \hat{z}(\omega)| \ge |\hat{z}(\omega)| \ge t$, since otherwise $\hat{R}_N(\omega) = 0$ by Lemma 6.12. Therefore, $\Omega_4(t) \subset \{\omega \in \Omega : |s(\omega)| \ge t\}$.

For $p = \infty$ we thus have $\text{meas}(\Omega_4(t)) = 0$, since $|s(\omega)| \le \|s\|_{L^\infty} < t$ for a.a. $\omega \in \Omega$. We now consider the case $p < \infty$. Then we can estimate

$$\text{meas}(\Omega_4(t)) \le \int_{\Omega_4(t)} \left(\frac{|s(\omega)|}{t}\right)^p d\omega = t^{-p} \|s|_{\Omega_4(t)}\|_{L^p}^p.$$

Hence, we obtain for $p < \infty$, using $p/q = p - r$,

$$\begin{aligned}\int_{\Omega_4(t)} |\hat{R}_N(\omega)|^r \, d\omega &\le \text{meas}(\Omega_4(t))^{1/q} \|s|_{\Omega_4(t)}\|_{L^p}^r \le t^{-p/q} \|s|_{\Omega_4(t)}\|_{L^p}^{p/q} \|s|_{\Omega_4(t)}\|_{L^p}^r \\ &\le t^{r-p} \|s|_{\Omega_4(t)}\|_{L^p}^{p-r} \|s|_{\Omega_4(t)}\|_{L^p}^r = t^{r-p} \|s|_{\Omega_4(t)}\|_{L^p}^p.\end{aligned}$$

We now choose t as a suitable power of $\max\{\|\hat{z} - \bar{z}\|_{L^p}, \|s\|_{L^p}\}$ in order to balance the order of the residuals in

$$\|\hat{R}_N\|_{L^r}^r = \int_{\Omega_2(t)} |\hat{R}_N(\omega)|^r \, d\omega + \int_{\Omega_3(t)} |\hat{R}_N(\omega)|^r \, d\omega + \int_{\Omega_4(t)} |\hat{R}_N(\omega)|^r \, d\omega.$$

Note here that we proved $\text{meas}(\Omega_1) = 0$. In the case $p = \infty$, we have $\|\hat{z} - \bar{z}\|_{L^p} < \delta = \min\{1, t_0/2\}$ and $\|s\|_{L^p} < \delta$. Thus, there exists $\kappa \in (1, 3/2]$ with $t := \kappa \max\{\|\hat{z} - \bar{z}\|_{L^p}, \|s\|_{L^p}\} < \delta \le t_0/2$. By the choice of κ, we have $\|\hat{z} - \bar{z}\|_{L^p} < t$ and $\|s\|_{L^p} < t$ as required. It was shown before that then $\text{meas}(\Omega_3(t)) = 0$ and $\text{meas}(\Omega_4(t)) = 0$. Thus,

$$\begin{aligned}\|\hat{R}_N\|_{L^r}^r &= \int_{\Omega_2(t)} |\hat{R}_N(\omega)|^r \, d\omega \le 2^\gamma C t^\gamma \|s|_{\Omega_2(t)}\|_{L^\infty}^r \\ &\le (2\kappa)^\gamma C \max\{\|\hat{z} - \bar{z}\|_{L^\infty}, \|s\|_{L^\infty}\}^\gamma \|s\|_{L^\infty}^r \\ &\le 3^\gamma C \max\{\|\hat{z} - \bar{z}\|_{L^\infty}, \|s\|_{L^\infty}\}^\gamma \|s\|_{L^\infty}^r.\end{aligned}$$

Now consider the case $p < \infty$. Then $\delta = \min\{1, (t_0/2)^{\frac{p+\gamma}{p}}\}$, $\|\hat{z} - \bar{z}\|_{L^p} < \delta$, and $\|s\|_{L^p} < \delta$. Setting $t = \max\{\|\hat{z} - \bar{z}\|_{L^p}, \|s\|_{L^p}\}^{\frac{p}{p+\gamma}}$, there holds $t < 1$, and thus $\|\hat{z} - \bar{z}\|_{L^p} \le t^{\frac{p+\gamma}{p}} < t$, $\|s\|_{L^p} \le t^{\frac{p+\gamma}{p}} < t$. Furthermore, $t \le \delta^{\frac{p}{p+\gamma}} \le t_0/2$. Thus, the choice of t satisfies all requirements. We obtain

$$\begin{aligned}\int_{\Omega_2(t)} |\hat{R}_N(\omega)|^r \, d\omega &\le (2^\gamma C)^{\frac{p-r}{p}} t^{\frac{\gamma(p-r)}{p}} \|s|_{\Omega_2(t)}\|_{L^p}^r \\ &= (2^\gamma C)^{\frac{p-r}{p}} \max\{\|\hat{z} - \bar{z}\|_{L^p}, \|s\|_{L^p}\}^{\frac{\gamma(p-r)}{p+\gamma}} \|s|_{\Omega_2(t)}\|_{L^p}^r, \\ \int_{\Omega_3(t)} |\hat{R}_N(\omega)|^r \, d\omega &\le t^{r-p} \|\hat{z} - \bar{z}\|_{L^p}^{p-r} \|s|_{\Omega_3(t)}\|_{L^p}^r \\ &= \max\{\|\hat{z} - \bar{z}\|_{L^p}, \|s\|_{L^p}\}^{\frac{p(r-p)}{p+\gamma}} \|\hat{z} - \bar{z}\|_{L^p}^{p-r} \|s|_{\Omega_3(t)}\|_{L^p}^r \\ &\le \max\{\|\hat{z} - \bar{z}\|_{L^p}, \|s\|_{L^p}\}^{\frac{\gamma(p-r)}{p+\gamma}} \|s|_{\Omega_3(t)}\|_{L^p}^r, \\ \int_{\Omega_4(t)} |\hat{R}_N(\omega)|^r \, d\omega &\le t^{r-p} \|s|_{\Omega_4(t)}\|_{L^p}^p \\ &= \max\{\|\hat{z} - \bar{z}\|_{L^p}, \|s\|_{L^p}\}^{\frac{p(r-p)}{p+\gamma}} \|s|_{\Omega_4(t)}\|_{L^p}^p \\ &\le \max\{\|\hat{z} - \bar{z}\|_{L^p}, \|s\|_{L^p}\}^{\frac{\gamma(p-r)}{p+\gamma}} \|s|_{\Omega_4(t)}\|_{L^p}^r.\end{aligned}$$

Hence, in all cases, we have proved the assertions. □

We obtain the following corollary.

Corollary 6.14. *Let $\bar{z} \in L^p(\Omega)$ satisfy the condition* (6.20) *with constants $C > 0$, $t_0 > 0$, and $\gamma > 0$. Let*

$$\delta = \min\left\{1, \left(\frac{t_0}{2}\right)^{\frac{p+\gamma}{p}}\right\} \quad \text{if } p < \infty, \qquad \delta = \min\left\{1, \frac{t_0}{2}\right\} \quad \text{if } p = \infty.$$

Then, for any $r \in [1, p)$ and all $z \in L^p(\Omega)$ satisfying

$$\|z - \bar{z}\|_{L^p} < \delta,$$

the following holds with $s = z - \bar{z}$:

(a) *If* $p=\infty$, *then*

$$\|R_N(\bar{z},s)\|_{L^r} \le 3^{\gamma/r}C^{1/r}\|s\|_{L^\infty}^{1+\gamma/r}.$$

(b) *If* $p<\infty$, *then*

$$\|R_N(\bar{z},s)\|_{L^r} \le \max\{1,(2^\gamma C)^{\frac{p-r}{rp}}\}\|s\|_{L^p}^{1+\frac{\gamma(p-r)}{r(p+\gamma)}}.$$

Proof. The result follows immediately from Theorem 6.13 by choosing $\hat{z}=\bar{z}$. The condition (6.22) is then trivially satisfied. □

By a slight modification of the proof of Theorem 6.13, we can also deal with the case where no growth condition of the form (6.20) is available. In fact, we can use that the measure of the set in (6.20) tends to zero for $t\to 0^+$.

Theorem 6.15. *Let* $\eta\in(0,1)$ *be given and consider* $\bar{z}\in L^p(\Omega)$. *For any fixed* $r\in[1,p)$, *set* $\rho=\eta^{\frac{rp}{p-r}}$ *if* $p<\infty$ *and* $\rho=\eta^r$ *if* $p=\infty$. *Then there exists* $t_0>0$ *such that*

$$\operatorname{meas}(\{\omega\in\Omega : 0<|\bar{z}(\omega)|<t_0\})\le\rho. \tag{6.23}$$

Now, let

$$\delta=\min\left\{1,\rho^{1/p}\left(\frac{t_0}{2}\right)\right\} \quad \textit{if } p<\infty, \qquad \delta=\min\left\{1,\frac{t_0}{2}\right\} \quad \textit{if } p=\infty \tag{6.24}$$

and consider $\hat{z}\in L^p(\Omega)$ *satisfying* $\|\hat{z}-\bar{z}\|_{L^p}<\delta$ *and* (6.22). *Then, for all* $z\in L^p(\Omega)$ *with*

$$\|z-\hat{z}\|_{L^p}<\delta,$$

the following holds with $s=z-\hat{z}$:

$$\|R_N(\hat{z},s)\|_{L^r}\le\eta\|s\|_{L^p}.$$

Proof. From $\{0<|\bar{z}|<t\}\downarrow\emptyset$ as $t\downarrow 0$ we conclude that we can find $t_0>0$ for which (6.23) holds. Choosing $t=t_0/2$, there holds $\delta\le t$ and thus $\|\hat{z}-\bar{z}\|_{L^p}<t$ as well as $\|s\|_{L^p}<t$. We proceed as in the proof of Theorem 6.13, except for the estimate on $\Omega_2(t)$.

For Ω_1 we obtain $\operatorname{meas}(\Omega_1)=0$. In the proof of Theorem 6.13, the growth condition (6.20) was only used in connection with the set $\Omega_2(t)$. Replacing (6.20) by (6.23), we obtain

$$\int_{\Omega_2(t)}|\hat{R}_N(\omega)|^r\,d\omega\le\operatorname{meas}(\Omega_2(t))^{1/q}\|s|_{\Omega_2(t)}\|_{L^p}^r\le\rho^{1/q}\|s|_{\Omega_2(t)}\|_{L^p}^r.$$

For $p=\infty$ we have $q=1$ and thus

$$\int_{\Omega_2(t)}|\hat{R}_N(\omega)|^r\,d\omega\le\rho^{1/q}\|s|_{\Omega_2(t)}\|_{L^p}^r=\rho\|s|_{\Omega_2(t)}\|_{L^\infty}^r=\eta^r\|s|_{\Omega_2(t)}\|_{L^\infty}^r.$$

For $p < \infty$ we obtain

$$\int_{\Omega_2(t)} |\hat{R}_N(\omega)|^r \, d\omega \le \rho^{1/q} \|s|_{\Omega_2(t)}\|_{L^p}^r = \rho^{\frac{p-r}{p}} \|s|_{\Omega_2(t)}\|_{L^p}^r = \eta^r \|s|_{\Omega_2(t)}\|_{L^p}^r .$$

For $p = \infty$, a copy of the proof of Theorem 6.13 yields that $\text{meas}(\Omega_3(t)) = 0$ and $\text{meas}(\Omega_4(t)) = 0$. Furthermore, in the case $p < \infty$ we derive, exactly as in the proof of Theorem 6.13,

$$\int_{\Omega_3(t)} |\hat{R}_N(\omega)|^r \, d\omega \le t^{r-p} \|[\hat{z} - \bar{z}]_{\Omega_3(t)}\|_{L^p}^{p-r} \|s|_{\Omega_3(t)}\|_{L^p}^r ,$$

and using $t = t_0/2$, $\|\hat{z} - \bar{z}\|_{L^p} < \delta$ we can further estimate

$$\begin{aligned}
&\le \left(\frac{t_0}{2}\right)^{r-p} \delta^{p-r} \|s|_{\Omega_3(t)}\|_{L^p}^r \\
&\le \left(\frac{t_0}{2}\right)^{r-p} \rho^{\frac{p-r}{p}} \left(\frac{t_0}{2}\right)^{p-r} \|s|_{\Omega_3(t)}\|_{L^p}^r = \eta^r \|s|_{\Omega_3(t)}\|_{L^p}^r .
\end{aligned}$$

Similarly,

$$\begin{aligned}
\int_{\Omega_4(t)} |\hat{R}_N(\omega)|^r \, d\omega &\le t^{r-p} \|s|_{\Omega_4(t)}\|_{L^p}^{p-r} \|s|_{\Omega_4(t)}\|_{L^p}^r \le \left(\frac{t_0}{2}\right)^{r-p} \delta^{p-r} \|s|_{\Omega_4(t)}\|_{L^p}^r \\
&\le \eta^r \|s|_{\Omega_4(t)}\|_{L^p}^r .
\end{aligned}$$

Hence,

$$\begin{aligned}
\|\hat{R}_N(\omega)\|_{L^r}^r &= \int_{\Omega_2(t)} |\hat{R}_N(\omega)|^r \, d\omega + \int_{\Omega_3(t)} |\hat{R}_N(\omega)|^r \, d\omega + \int_{\Omega_4(t)} |\hat{R}_N(\omega)|^r \, d\omega \\
&\le \eta^r \|s|_{\Omega_2(t)}\|_{L^p}^r + \eta^r \|s|_{\Omega_3(t)}\|_{L^p}^r + \eta^r \|s|_{\Omega_4(t)}\|_{L^p}^r \le \eta^r \|s\|_{L^p}^r . \qquad \square
\end{aligned}$$

The result of Theorem 6.15 is essentially the one obtained in the proof of [106, Thm. 1], which is the main building block for the mesh-independence result established in [106]. We refer to sections 6.3.3 and 6.4.3. Theorem 6.13 extends this result significantly.

In the next two examples, we discuss the sharpness of the estimates in Theorem 6.13 and Corollary 6.14.

Example 6.16. Let $\Omega = (-1,1)$, $\bar{z}(\omega) = |\omega|^{1/\gamma}$ with $\gamma > 0$. Then, for $0 < t \le 1$, there holds

$$\{\omega : 0 < |\bar{z}(\omega)| < t\} = (-t^\gamma, 0) \cup (0, t^\gamma),$$

hence assumption (6.20) is satisfied with $t_0 = 1$, $C = 2$. Now, with arbitrary $\varepsilon \in (0,1)$ and $\delta \in (0,1)$, consider $s = -(1+\varepsilon)1_{[-\delta,\delta]}\bar{z}$. Then $\bar{z}(\omega) + s(\omega) = -\varepsilon\bar{z}(\omega)$ for all $\omega \in [-\delta,\delta]$. Thus, for all $\omega \in [-\delta,\delta] \setminus \{0\}$, there holds $[\bar{z}(\bar{z}+s)](\omega) < 0$ and thus $|R_N(\bar{z},s)(\omega)| = |\bar{z}(\omega)|$. For $|\omega| > \delta$ or $\omega = 0$ there holds $s(\omega) = 0$ and thus $R_N(\bar{z},s)(\omega) = 0$. Therefore,

$$\|R_N(\bar{z},s)\|_{L^r}^r = \|[\bar{z}]_{[-\delta,\delta]}\|_{L^r}^r = \int_{-\delta}^{\delta} |\omega|^{r/\gamma} \, d\omega = 2\frac{1}{1+r/\gamma}\delta^{1+r/\gamma} = \left(c_1 \delta^{\frac{1}{r}+\frac{1}{\gamma}}\right)^r$$

with $c_1 = \left(\frac{2}{1+r/\gamma}\right)^{1/r}$. We consider first the case $r < p < \infty$ and obtain

$$\|s\|_{L^p}^p = (1+\varepsilon)^p \int_{-\delta}^{\delta} |\omega|^{p/\gamma}\, d\omega = 2(1+\varepsilon)^p \frac{1}{1+p/\gamma} \delta^{1+p/\gamma} = \left(c_2 \delta^{\frac{1}{p}+\frac{1}{\gamma}}\right)^p$$

with $c_2 = (1+\varepsilon)\left(\frac{2}{1+p/\gamma}\right)^{1/p}$. In particular, $\|s\|_{L^p} \to 0$ as $\delta \to 0$. Now

$$\left(\frac{\|s\|_{L^p}}{c_2}\right)^{1+\frac{\gamma(p-r)}{r(p+\gamma)}} = \delta^{\left(\frac{1}{p}+\frac{1}{\gamma}\right)\left(1+\frac{\gamma(p-r)}{r(p+\gamma)}\right)} = \delta^{\frac{p+\gamma}{\gamma p}\frac{p(r+\gamma)}{r(p+\gamma)}} = \delta^{\frac{r+\gamma}{\gamma r}} = \delta^{\frac{1}{r}+\frac{1}{\gamma}}.$$

Therefore,

$$\|R_N(\bar{z},s)\|_{L^r} = c_1 \delta^{\frac{1}{r}+\frac{1}{\gamma}} = c_1 c_2^{-1-\frac{\gamma(p-r)}{r(p+\gamma)}} \|s\|_{L^p}^{1+\frac{\gamma(p-r)}{r(p+\gamma)}}.$$

We thus see that in the case $p < \infty$ the result of Corollary 6.14 is sharp.

Next, we consider the case $r < p = \infty$. Then

$$\|s\|_{L^\infty} = (1+\varepsilon)\delta^{1/\gamma}.$$

Again, $\|s\|_{L^\infty} \to 0$ as $\delta \to 0$. Now

$$\left(\frac{\|s\|_{L^\infty}}{1+\varepsilon}\right)^{1+\frac{\gamma}{r}} = \delta^{\frac{1}{\gamma}(1+\frac{\gamma}{r})} = \delta^{\frac{1}{r}+\frac{1}{\gamma}}.$$

Therefore,

$$\|R_N(\bar{z},s)\|_{L^r} = c_1 \delta^{\frac{1}{r}+\frac{1}{\gamma}} = c_1 (1+\varepsilon)^{-1-\frac{\gamma}{r}} \|s\|_{L^\infty}^{1+\frac{\gamma}{r}}.$$

Thus, the result of Corollary 6.14 is sharp also for $p = \infty$. ∎

The next example shows that the result of Theorem 6.13 is sharp. In particular, it shows that the factor involving the maximum of $\|\hat{z}-\bar{z}\|_{L^p}$ and $\|s\|_{L^p}$ cannot be avoided.

Example 6.17. As in Example 6.16, let $\Omega = (-1,1)$ and $\bar{z}(\omega) = |\omega|^{1/\gamma}$ with $\gamma > 0$. Furthermore, for $0 < \hat{\gamma} < \gamma$ and $0 < \tau < 1/2$, let $\hat{z}(\omega) \in L^p(\Omega)$,

$$\begin{aligned} &\hat{z}(\omega) = \tfrac{1}{2}|\omega|^{1/\hat{\gamma}} && \text{for } |\omega| \le \tau, \\ &\hat{z}(\omega) = |\omega|^{1/\gamma} && \text{for } |\omega| \ge 2\tau, \\ &\hat{z}(\omega) \in \left[\tfrac{1}{2}|\omega|^{1/\hat{\gamma}}, |\omega|^{1/\gamma}\right] && \text{for } \tau < |\omega| < 2\tau. \end{aligned}$$

Note here that $2\tau < 1$ and $1/\gamma < 1/\hat{\gamma}$ imply $|\omega|^{1/\gamma} \ge |\omega|^{1/\hat{\gamma}}$ for all $\omega \in (-2\tau, 2\tau)$. From Example 6.16 we know that (6.20) holds at $\bar{z}$ with $t_0 = 1$, $C = 2$.

We first consider the case $p < \infty$. Then

$$\|\hat{z}-\bar{z}\|_{L^p}^p \le \int_{-2\tau}^{2\tau} |\omega|^{p/\gamma}\, d\omega = 2\frac{1}{1+p/\gamma}(2\tau)^{1+p/\gamma} = \frac{2^{2+p/\gamma}}{1+p/\gamma}\tau^{1+p/\gamma},$$

$$\|\hat{z}-\bar{z}\|_{L^p}^p \ge \int_{-\tau}^{\tau} \left(|\omega|^{1/\gamma} - \frac{|\omega|^{1/\hat{\gamma}}}{2}\right)^p d\omega \ge \int_{-\tau}^{\tau} \left(\frac{|\omega|^{1/\gamma}}{2}\right)^p d\omega = \frac{2^{1-p}}{1+p/\gamma}\tau^{1+p/\gamma}.$$

Similarly, for $p=\infty$ there holds

$$\|\hat{z}-\bar{z}\|_{L^\infty} \le (2\tau)^{1/\gamma},$$
$$\|\hat{z}-\bar{z}\|_{L^\infty} \ge \bar{z}(\tau)-\hat{z}(\tau) = \tau^{1/\gamma} - \frac{1}{2}\tau^{1/\hat{\gamma}} \ge \tau^{1/\gamma} - \frac{1}{2}\tau^{1/\gamma} = \frac{1}{2}\tau^{1/\gamma}.$$

We thus see that for all $p \in (2,\infty]$ there exist constants $0 < c_l \le c_r$ depending only on p and γ with

$$c_l \tau^{\frac{1}{p}+\frac{1}{\gamma}} \le \|\hat{z}-\bar{z}\|_{L^p} \le c_r \tau^{\frac{1}{p}+\frac{1}{\gamma}}.$$

In particular, $\|\hat{z}-\bar{z}\|_{L^p} \to 0$ as $\tau \to 0$.

Now, choose $0<\varepsilon<1$, $0<\delta\le\tau$, and set

$$s = -(1+\varepsilon)1_{[-\delta,\delta]}\hat{z}.$$

Then there holds

$$s(\omega) = -\frac{1+\varepsilon}{2}|\omega|^{1/\hat{\gamma}} \quad \text{for } |\omega| \le \delta, \qquad s(\omega)=0 \quad \text{for } |\omega|>\delta.$$

A calculation as in Example 6.16, but with γ replaced by $\hat{\gamma}$ and taking into account the factor $1/2$ in the definition of $\hat{z}$, shows

$$\|R_N(\hat{z},s)\|_{L^r} = c_1\delta^{\frac{1}{r}+\frac{1}{\hat{\gamma}}}$$

with $c_1 = \frac{1}{2}\left(\frac{2}{1+r/\hat{\gamma}}\right)^{1/r}$. Furthermore, for $p<\infty$,

$$\|s\|_{L^p} = c_2\delta^{\frac{1}{p}+\frac{1}{\hat{\gamma}}}$$

with $c_2 = \frac{1+\varepsilon}{2}\left(\frac{2}{1+p/\hat{\gamma}}\right)^{1/p}$, and thus, as in Example 6.16,

$$\|R_N(\hat{z},s)\|_{L^r} = c_1\delta^{\frac{1}{r}+\frac{1}{\hat{\gamma}}} = c_1 c_2^{-1-\frac{\hat{\gamma}(p-r)}{r(p+\hat{\gamma})}} \|s\|_{L^p}^{1+\frac{\hat{\gamma}(p-r)}{r(p+\hat{\gamma})}}.$$

In the case $p=\infty$ we obtain

$$\|s\|_{L^\infty} = \frac{1+\varepsilon}{2}\delta^{1/\hat{\gamma}} = c_2\delta^{1/\hat{\gamma}}$$

with $c_2 = (1+\varepsilon)/2$, and thus, as in Example 6.16,

$$\|R_N(\hat{z},s)\|_{L^r} = c_1 c_2^{-1-\frac{\hat{\gamma}}{r}} \|s\|_{L^\infty}^{1+\frac{\hat{\gamma}}{r}}.$$

Now, since $\delta\le\tau$, we have in the case $p<\infty$

$$\|s\|_{L^p} = c_2\delta^{\frac{1}{p}+\frac{1}{\hat{\gamma}}} \le c_2\delta^{\frac{1}{\hat{\gamma}}-\frac{1}{\gamma}}\tau^{\frac{1}{p}+\frac{1}{\gamma}} \le \frac{c_2}{c_l}\delta^{\frac{1}{\hat{\gamma}}-\frac{1}{\gamma}}\|\hat{z}-\bar{z}\|_{L^p}.$$

In the case $p=\infty$, there holds

$$\|s\|_{L^\infty} = \frac{1+\varepsilon}{2}\delta^{1/\hat{\gamma}} \le c_2\delta^{\frac{1}{\hat{\gamma}}-\frac{1}{\gamma}}\tau^{\frac{1}{\gamma}} \le \frac{c_2}{c_l}\delta^{\frac{1}{\hat{\gamma}}-\frac{1}{\gamma}}\|\hat{z}-\bar{z}\|_{L^\infty}.$$

Hence, for sufficiently small δ, there holds $\|s\|_{L^p} \le \|\hat{z}-\bar{z}\|_{L^p}$. In particular, if $\tau > 0$ is sufficiently small, then $\delta \le \tau$ implies $\|s\|_{L^p} \le \|\hat{z}-\bar{z}\|_{L^p}$.

Hence, for $p < \infty$ and $\tau > 0$ sufficiently small, there holds with appropriate $c_m = c_m(\tau,\gamma,\hat{\gamma},\delta,\varepsilon) \in [c_l,c_r]$

$$\max\{\|\hat{z}-\bar{z}\|_{L^p},\|s\|_{L^p}\}^{\frac{\gamma(p-r)}{r(p+\gamma)}} = c_m^{\frac{\gamma(p-r)}{r(p+\gamma)}} \tau^{\left(\frac{1}{p}+\frac{1}{\gamma}\right)\frac{\gamma(p-r)}{r(p+\gamma)}} = c_m^{\frac{\gamma(p-r)}{r(p+\gamma)}} \tau^{\frac{p-r}{rp}} = c_m^{\frac{\gamma(p-r)}{r(p+\gamma)}} \tau^{\frac{1}{r}-\frac{1}{p}}.$$

Therefore,

$$\max\{\|\hat{z}-\bar{z}\|_{L^p},\|s\|_{L^p}\}^{\frac{\gamma(p-r)}{r(p+\gamma)}}\|s\|_{L^p} = c_2 c_m^{\frac{\gamma(p-r)}{r(p+\gamma)}} \tau^{\frac{1}{r}-\frac{1}{p}} \delta^{\frac{1}{p}+\frac{1}{\gamma}}.$$

For fixed $\kappa \ge 1$ and $\delta = \tau/\kappa$, this gives

$$\begin{aligned}
\max\{\|\hat{z}-\bar{z}\|_{L^p},\|s\|_{L^p}\}^{\frac{\gamma(p-r)}{r(p+\gamma)}}\|s\|_{L^p} &= c_2 c_m^{\frac{\gamma(p-r)}{r(p+\gamma)}} \kappa^{\frac{1}{r}-\frac{1}{p}} \delta^{\frac{1}{r}-\frac{1}{p}+\frac{1}{p}+\frac{1}{\gamma}} \\
&= c_2 c_m^{\frac{\gamma(p-r)}{r(p+\gamma)}} \kappa^{\frac{1}{r}-\frac{1}{p}} \delta^{\frac{1}{r}+\frac{1}{\gamma}} = \frac{c_2}{c_1} c_m^{\frac{\gamma(p-r)}{r(p+\gamma)}} \kappa^{\frac{1}{r}-\frac{1}{p}} \|R_N(\hat{z},s)\|_{L^r}.
\end{aligned}$$

Hence, the estimate in Theorem 6.13 is sharp in the case $p < \infty$.

In the case $p = \infty$, there holds with appropriate $c_m = c_m(\tau,\gamma,\hat{\gamma},\delta,\varepsilon) \in [c_l,c_r]$

$$\max\{\|\hat{z}-\bar{z}\|_{L^\infty},\|s\|_{L^\infty}\}^{\frac{\gamma}{r}} = c_m^{\frac{\gamma}{r}} \tau^{\frac{1}{\gamma}\frac{\gamma}{r}} = c_m^{\frac{\gamma}{r}} \tau^{\frac{1}{r}}.$$

Therefore,

$$\max\{\|\hat{z}-\bar{z}\|_{L^\infty},\|s\|_{L^\infty}\}^{\frac{\gamma}{r}}\|s\|_{L^\infty} = c_2 c_m^{\frac{\gamma}{r}} \tau^{\frac{1}{r}} \delta^{\frac{1}{\gamma}}.$$

For $\delta = \tau/\kappa$ with fixed $\kappa \ge 1$, this gives

$$\begin{aligned}
\max\{\|\hat{z}-\bar{z}\|_{L^\infty},\|s\|_{L^\infty}\}^{\frac{\gamma}{r}}\|s\|_{L^\infty} &= c_2 c_m^{\frac{\gamma}{r}} \kappa^{\frac{1}{r}} \delta^{\frac{1}{r}+\frac{1}{\gamma}} = c_2 c_m^{\frac{\gamma}{r}} \kappa^{\frac{1}{r}} \delta^{\frac{1}{r}+\frac{1}{\gamma}} \\
&= \frac{c_2}{c_1} c_m^{\frac{\gamma}{r}} \kappa^{\frac{1}{r}} \|R_N(\hat{z},s)\|_{L^r}.
\end{aligned}$$

Therefore, also in the case $p = \infty$, the estimate of Theorem 6.13 is sharp. ∎

6.3 Mesh-Independent Semismoothness

From the results obtained so far we can derive two types of mesh-independent order of semismoothness results for Φ_{1h} and Φ_{2h}.

The first one is based on Corollary 6.14 and poses growth conditions (6.20) on all $\bar{z}_h$, $h \in \mathcal{H} \cup \{0\}$, $h \le h_0$, where $\bar{z}_h = \hat{Z}_h(\bar{u}_h)$ for reformulation 1 and $\bar{z}_h = Z_h(\bar{u}_h)$ for reformulation 2. The second one is based on Theorem 6.13 with $\hat{z} = \bar{z}_h$ and poses the growth condition (6.20) only on $\bar{z} = \bar{z}_0$.

To state the first mesh-independent semismoothness result, we will need the following assumptions.

The first one is Assumption 6.1, but now formulated for CP(F_h) instead of CP(F).

Assumption 6.18. *The operator $F_h : U_h \to U_h$, with $U_h \subset L^2(\Omega)$, has the form*

$$F_h(u_h) = \lambda_h u_h + G_h(u_h),$$

where $\lambda_h \in L^\infty(\Omega)$, $\lambda_h \geq \lambda_0$ *with a constant* $\lambda_0 > 0$, *and with* G_h *denoting a locally Lipschitz continuous operator*

$$G_h : U_h \to U_h \cap L^p(\Omega)$$

for some $p \in (2, \infty]$. *Furthermore, the mapping* $U_h \cap L^r(\Omega) \ni u_h \mapsto G_h(u_h) \in U_h$ *is continuously Fréchet differentiable (with* $r \geq 2$ *specified when referring to this part of the assumption).*

For convenience, we will use the convention that $h = 0$ corresponds to the original problem; i.e.,

$$U_0 = U = L^2(\Omega), \quad F_0 = F, \quad G_0 = G, \quad Z_0 = Z, \quad \hat{Z}_0 = \hat{Z}, \quad \text{etc.}$$

With this convention, Assumption 6.18 with $h = 0$ is the same as Assumption 6.1.

We now relate the assumptions for the individual problems CP(F_h) as follows.

Assumption 6.19. *There exist* $h_0 > 0$, $\delta_0 \in (0, 1]$, $p > 2$, $\lambda_0 > 0$, $L_G > 0$, $C_{G'} > 0$, *and* $\kappa > 0$ *such that with* $r \geq 2$ *as specified, the following holds:*

(a) $\bar{u} = \bar{u}_0$ *solves CP(*F*) and* $\bar{u}_h$ *solves CP(*F_h*),* $h \in \mathcal{H}$, $h \leq h_0$.

(b) $\|\bar{u}_h - \bar{u}\|_{L^2} \to 0 \quad$ *as* $\mathcal{H} \ni h \to 0$.

(c) $\|G_h(\bar{u}_h) - G(\bar{u})\|_{L^p} \to 0 \quad$ *as* $\mathcal{H} \ni h \to 0$.

(d) *For* λ_0 *and* r *independent of* h, *Assumption* 6.18 *holds for all* $h \in \mathcal{H} \cup \{0\}$, $h \leq h_0$.

(e) *For all* $h \in \mathcal{H} \cup \{0\}$, $h \leq h_0$, *the operator* G_h *satisfies*

$$\|G_h(u_h) - G_h(\bar{u}_h)\|_{L^p} \leq L_G \|u_h - \bar{u}_h\|_{L^2} \quad \forall\, u_h \in B_{h,\delta_0}(\bar{u}_h),$$

where $B_{h,\delta_0}(\bar{u}_h) = \{u_h \in U_h : \|u_h - \bar{u}_h\|_{L^2} < \delta_0\}$.
Furthermore, for all $h \in \mathcal{H} \cup \{0\}$, $h \leq h_0$, *and all* $u_h \in B_{h,\delta_0}(\bar{u}_h)$, *there holds*

$$\|[G_h]'_{r,2}(u_h) - [G_h]'_{r,2}(\bar{u}_h)\|_{L^r, L^2} \leq C_{G'} \|u_h - \bar{u}_h\|^\kappa_{L^r}.$$

6.3.1 Mesh-Independent Semismoothness under Uniform Growth Conditions

We consider first the case where growth conditions of the form (6.20) are posed on all $\bar{z}_h$, $h \in \mathcal{H} \cup \{0\}$, $h \leq h_1 \leq h_0$. We start by considering Φ_{1h}.

Theorem 6.20. *Let Assumption* 6.19 *hold with* $r = p$. *Furthermore, assume that there exist constants* $0 < h_1 \leq h_0$, $\lambda_1 > 0$, $\sigma_1 > 0\, \gamma > 0$, $C > 0$, *and* $t_0 > 0$ *such that with* $\bar{z}_0 = \bar{z} = \hat{Z}(\bar{u})$ *and* $\bar{z}_h = \hat{Z}_h(\bar{u}_h)$ *there holds*

$$\begin{aligned} &\|\lambda_h\|_{L^\infty} \leq \lambda_1 \quad \text{and} \quad \|\sigma_h\|_{L^\infty} \leq \sigma_1 && \forall\, h \in \mathcal{H} \cup \{0\},\ h \leq h_1, \\ &\operatorname{meas}(\{\omega : 0 < |\bar{z}_h(\omega)| < t\}) \leq C t^\gamma && \forall\, t \in (0, t_0],\ \forall\, h \in \mathcal{H} \cup \{0\},\ h \leq h_1. \end{aligned}$$

Then there exist $\delta_0' \ge \delta_0/\text{meas}(\Omega)^{\frac{p-2}{2p}}$ *and* $L_p \le \text{meas}(\Omega)^{\frac{p-2}{2p}} L_G$ *(with* $(p-2)/(2p) = 1/2$ *if* $p = \infty$*) such that, for all* $h \in \mathcal{H} \cup \{0\}$, $h \le h_1$,

$$B_{h,p,\delta_0'}(\bar{u}_h) := \{u_h \in U_h : \|u_h - \bar{u}_h\|_{L^p} < \delta_0'\} \subset B_{h,\delta_0}(\bar{u}_h)$$

and the operators $[G_h]_{p,p}$ *are Lipschitz continuous on* $B_{h,p,\delta_0'}(\bar{u}_h)$ *with modulus* L_p. *Let*

$$C_{\hat{Z}} = \frac{\sigma_1 C_{G'}}{1+\kappa}, \qquad L_{\hat{Z}} = \sup_{h \in \mathcal{H} \cup \{0\},\, h \le h_1} (\|\sigma_h \lambda_h - 1\|_{L^\infty} + \|\sigma_h\|_{L^\infty} L_p),$$

$$\theta = \begin{cases} \frac{\gamma(p-2)}{2(p+\gamma)} & \text{if } p < \infty, \\ \frac{\gamma}{2} & \text{if } p = \infty, \end{cases} \qquad C_N = \begin{cases} \max\{1, (2^\gamma C)^{\frac{p-2}{2p}}\} L_{\hat{Z}}^{1+\frac{\gamma(p-2)}{2(p+\gamma)}} & \text{if } p < \infty, \\ \sqrt{3^\gamma C} L_{\hat{Z}}^{1+\frac{\gamma}{2}} & \text{if } p = \infty. \end{cases}$$

Then, with $\delta = \min\{\delta'/L_{\hat{Z}}, \delta_0'\}$, *where* $\delta' = \min\{1, (t_0/2)^{\frac{p+\gamma}{p}}\}$ *if* $p < \infty$ *and* $\delta' = \min\{1, t_0/2\}$ *if* $p = \infty$, *there holds*

$$\begin{aligned} &\forall\, u_h \in U_h,\ \|u_h - \bar{u}_h\|_{L^p} < \delta,\ h \in \mathcal{H} \cup \{0\},\ h \le h_1 : \\ &\|\Phi_{1h}(u_h) - \Phi_{1h}(\bar{u}_h) - M_{\Phi_{1h}}(u_h)(u_h - \bar{u}_h)\|_{L^2} && (6.25) \\ &\qquad \le C_N \|u_h - \bar{u}_h\|_{L^p}^{1+\theta} + C_{\hat{Z}} \|u_h - \bar{u}_h\|_{L^p}^{1+\kappa}. && (6.26) \end{aligned}$$

Proof. For $v \in L^p(\Omega)$ there holds by Hölder's inequality

$$\|v\|_{L^2} \le \|1\|_{L^{2p/(p-2)}} \|v\|_{L^p} = \text{meas}(\Omega)^{\frac{p-2}{2p}} \|v\|_{L^p}.$$

Thus the assertions on δ_0', $B_{h,p,\delta_0'}(\bar{u}_h)$, and L_p hold true.

Let δ', $L_{\hat{Z}}$, and δ be defined as in the theorem. Consider $h \in \mathcal{H} \cup \{0\}$, $h \le h_1$, and $u_h \in U_h \cap L^p(\Omega)$, $\|u_h - \bar{u}_h\|_{L^p} < \delta$. From Lemma 6.7 we then obtain

$$\begin{aligned} &\|\Phi_{1h}(u_h) - \Phi_{1h}(\bar{u}_h) - M_{\Phi_{1h}}(u_h)(u_h - \bar{u}_h)\|_{L^2} \\ &\quad \le \|N(z_h) - N(\bar{z}_h) - M_N(z_h)(z_h - \bar{z}_h)\|_{L^2} && (6.27) \\ &\qquad + \|\hat{Z}_h(u_h) - \hat{Z}_h(\bar{u}_h) - [\hat{Z}_h]'_{p,2}(u_h)(u_h - \bar{u}_h)\|_{L^2} \\ &\quad \stackrel{\text{def}}{=} \|R_N(\bar{z}_h, z_h - \bar{z}_h)\|_{L^2} + \|R_{\hat{Z}_h}(\bar{u}_h, u_h - \bar{u}_h)\|_{L^2}, \\ &\|z_h - \bar{z}_h\|_{L^p} \le L_{\hat{Z}} \|u_h - \bar{u}_h\|_{L^p} < L_{\hat{Z}} \delta \le \delta'. && (6.28) \end{aligned}$$

Now, by Corollary 6.14, there holds

$$\begin{aligned} \|R_N(\bar{z}_h, z_h - \bar{z}_h)\|_{L^2} &\le \begin{cases} \max\{1, (2^\gamma C)^{\frac{p-2}{2p}}\} \|z_h - \bar{z}_h\|_{L^p}^{1+\frac{\gamma(p-2)}{2(p+\gamma)}} & \text{if } p < \infty, \\ \sqrt{3^\gamma C} \|z_h - \bar{z}_h\|_{L^\infty}^{1+\frac{\gamma}{2}} & \text{if } p = \infty \end{cases} \\ &\le \begin{cases} \max\{1, (2^\gamma C)^{\frac{p-2}{2p}}\} L_{\hat{Z}}^{1+\frac{\gamma(p-2)}{2(p+\gamma)}} \|u_h - \bar{u}_h\|_{L^p}^{1+\frac{\gamma(p-2)}{2(p+\gamma)}} & \text{if } p < \infty, \\ \sqrt{3^\gamma C} L_{\hat{Z}}^{1+\frac{\gamma}{2}} \|u_h - \bar{u}_h\|_{L^\infty}^{1+\frac{\gamma}{2}} & \text{if } p = \infty. \end{cases} \end{aligned}$$

Furthermore,

$$
\begin{aligned}
&\|R_{\hat{Z}_h}(\bar{u}_h, u_h - \bar{u}_h)\|_{L^2} \\
&\le \int_0^1 \|([\hat{Z}_h]'_{p,2}(tu_h + (1-t)\bar{u}_h) - [\hat{Z}_h]'_{p,2}(u_h))(u_h - \bar{u}_h)\|_{L^2}\, dt \\
&\le \int_0^1 \|\sigma_h\|_{L^\infty} \|[G_h]'_{p,2}(tu_h + (1-t)\bar{u}_h) - [G_h]'_{p,2}(u_h)\|_{L^p, L^2}\, dt\, \|u_h - \bar{u}_h\|_{L^p} \\
&\le \sigma_1 C_{G'} \int_0^1 (1-t)^\kappa\, dt \|u_h - \bar{u}_h\|_{L^p}^{1+\kappa} = \frac{\sigma_1 C_{G'}}{1+\kappa} \|u_h - \bar{u}_h\|_{L^p}^{1+\kappa}. \qquad \square
\end{aligned} \tag{6.29}
$$

Next, we derive a similar result for Φ_{2h}.

Theorem 6.21. *Let Assumption* 6.19 *hold with* $r = 2$. *Assume further that there exist constants* $0 < h_1 \le h_0$, $\gamma > 0$, $t_0 > 0$, *and* $C > 0$ *such that for* $\bar{z}_0 = \bar{z} = Z(\bar{u})$ *and* $\bar{z}_h = Z_h(\bar{u}_h)$ *there holds*

$$
\operatorname{meas}(\{\omega : 0 < |\bar{z}_h(\omega)| < t\}) \le C t^\gamma \quad \forall\, t \in (0, t_0],\ \forall\, h \in \mathcal{H} \cup \{0\},\ h \le h_1.
$$

Let

$$
C_Z = \frac{C_{G'}}{\lambda_0(1+\kappa)}, \qquad L_Z = \frac{L_G}{\lambda_0},
$$

$$
\theta = \begin{cases} \frac{\gamma(p-2)}{2(p+\gamma)} & \text{if } p < \infty, \\ \frac{\gamma}{2} & \text{if } p = \infty, \end{cases} \qquad C_N = \begin{cases} \max\{1, (2^\gamma C)^{\frac{p-2}{2p}}\} L_Z^{1+\frac{\gamma(p-2)}{2(p+\gamma)}} & \text{if } p < \infty, \\ \sqrt{3^\gamma C}\, L_Z^{1+\frac{\gamma}{2}} & \text{if } p = \infty. \end{cases}
$$

Then, with $\delta = \min\{\delta'/L_Z, \delta_0\}$, *where* $\delta' = \min\{1, (t_0/2)^{\frac{p+\gamma}{p}}\}$ *if* $p < \infty$ *and* $\delta' = \min\{1, t_0/2\}$ *if* $p = \infty$, *there holds*

$$
\begin{aligned}
&\forall\, u_h \in B_{h,\delta}(\bar{u}_h),\ h \in \mathcal{H} \cup \{0\},\ h \le h_1 : \\
&\|\Phi_{2h}(u_h) - \Phi_{2h}(\bar{u}_h) - M_{\Phi_{2h}}(u_h)(u_h - \bar{u}_h)\|_{L^2} \\
&\qquad \le C_N \|u_h - \bar{u}_h\|_{L^2}^{1+\theta} + C_Z \|u_h - \bar{u}_h\|_{L^2}^{1+\kappa}.
\end{aligned} \tag{6.30}
$$

Proof. Let δ', L_Z, and δ be defined as in the theorem. Consider $h \in \mathcal{H} \cup \{0\}$, $h \le h_1$, and $u_h \in U_h$, $\|u_h - \bar{u}_h\|_{L^2} < \delta$. From Lemma 6.8 we then obtain

$$
\begin{aligned}
&\|\Phi_{2h}(u_h) - \Phi_{2h}(\bar{u}_h) - M_{\Phi_{2h}}(u_h)(u_h - \bar{u}_h)\|_{L^2} \\
&\quad \le \|N(z_h) - N(\bar{z}_h) - M_N(z_h)(z_h - \bar{z}_h)\|_{L^2} \\
&\qquad + \|Z_h(u_h) - Z_h(\bar{u}_h) - [Z_h]'_{2,2}(u_h)(u_h - \bar{u}_h)\|_{L^2} \\
&\quad \stackrel{\text{def}}{=} \|R_N(\bar{z}_h, z_h - \bar{z}_h)\|_{L^2} + \|R_{Z_h}(\bar{u}_h, u_h - \bar{u}_h)\|_{L^2},
\end{aligned} \tag{6.31}
$$

$$
\|z_h - \bar{z}_h\|_{L^p} \le L_Z \|u_h - \bar{u}_h\|_{L^2} < L_Z \delta \le \delta'. \tag{6.32}
$$

Now, by Corollary 6.14, there holds

$$\|R_N(\bar{z}_h, z_h - \bar{z}_h)\|_{L^2} \le \begin{cases} \max\{1, (2^\gamma C)^{\frac{p-2}{2p}}\}\|z_h - \bar{z}_h\|_{L^p}^{1+\frac{\gamma(p-2)}{2(p+\gamma)}} & \text{if } p < \infty, \\ \sqrt{3^\gamma C}\|z_h - \bar{z}_h\|_{L^\infty}^{1+\frac{\gamma}{2}} & \text{if } p = \infty \end{cases}$$

$$\le \begin{cases} \max\{1, (2^\gamma C)^{\frac{p-2}{2p}}\}L_Z^{1+\frac{\gamma(p-2)}{2(p+\gamma)}}\|u_h - \bar{u}_h\|_{L^2}^{1+\frac{\gamma(p-2)}{2(p+\gamma)}} & \text{if } p < \infty, \\ \sqrt{3^\gamma C}L_Z^{1+\frac{\gamma}{2}}\|u_h - \bar{u}_h\|_{L^2}^{1+\frac{\gamma}{2}} & \text{if } p = \infty. \end{cases}$$

Furthermore,

$$\begin{aligned} &\|R_{Z_h}(\bar{u}_h, u_h - \bar{u}_h)\|_{L^2} \\ &\le \int_0^1 \|([Z_h]'_{2,2}(tu_h + (1-t)\bar{u}_h) - [Z_h]'_{2,2}(u_h))(u_h - \bar{u}_h)\|_{L^2}\,dt \\ &\le \int_0^1 \left\|\frac{1}{\lambda_h}\right\|_{L^\infty} \|[G_h]'_{2,2}(tu_h + (1-t)\bar{u}_h) - [G_h]'_{2,2}(u_h)\|_{L^2,L^2}\,dt\,\|u_h - \bar{u}_h\|_{L^2} \\ &\le \frac{C_{G'}}{\lambda_0}\int_0^1 (1-t)^\kappa\,dt\,\|u_h - \bar{u}_h\|_{L^2}^{1+\kappa} = \frac{C_{G'}}{\lambda_0(1+\kappa)}\|u_h - \bar{u}_h\|_{L^2}^{1+\kappa}. \quad \square \end{aligned} \tag{6.33}$$

6.3.2 Mesh-Independent Semismoothness without Uniform Growth Conditions

Now we consider the case where the growth condition (6.20) is only posed on $\bar{z}$, but not on the discrete functions $\bar{z}_h$, $h \in \mathcal{H}$, $h \le h_1 \le h_0$.

For the operators Φ_{1h} we obtain the following.

Theorem 6.22. *Let Assumption* 6.19 *hold with* $r = p$. *Furthermore, assume that there exist constants* $0 < h_1' \le h_0$, $\lambda_1 > 0$, $\sigma_1 > 0$ $\gamma > 0$, $C > 0$, *and* $t_0 > 0$ *such that with* $\bar{z}_0 = \bar{z} = \hat{Z}(\bar{u})$ *there holds*

$$\begin{aligned} &\|\lambda_h\|_{L^\infty} \le \lambda_1 \quad \text{and} \quad \|\sigma_h\|_{L^\infty} \le \sigma_1 \quad \forall\, h \in \mathcal{H} \cup \{0\},\ h \le h_1', \\ &\text{meas}(\{\omega : 0 < |\bar{z}(\omega)| < t\}) \le Ct^\gamma \qquad \forall\, t \in (0, t_0]. \end{aligned}$$

Furthermore, let strict complementarity hold at $\bar{u}$*:*

$$\text{meas}(\{\omega : \bar{z}(\omega) = 0\}) = 0.$$

Then there exists $0 < h_1 \le h_1'$ *such that for* $\bar{z}_h = \hat{Z}_h(\bar{u}_h)$ *there holds*

$$\|\bar{z}_h - \bar{z}\|_{L^p} < \delta' \quad \forall\, h \in \mathcal{H},\ h \le h_1, \tag{6.34}$$

where $\delta' = \min\{1, (t_0/2)^{\frac{p+\gamma}{p}}\}$ *if* $p < \infty$ *and* $\delta' = \min\{1, t_0/2\}$ *if* $p = \infty$. *Furthermore, there exist* $\delta_0' \ge \delta_0/\text{meas}(\Omega)^{\frac{p-2}{2p}}$ *and* $L_p \le \text{meas}(\Omega)^{\frac{p-2}{2p}} L_G$ *(with* $(p-2)/(2p) = 1/2$ *if* $p = \infty$*) such that, for all* $h \in \mathcal{H} \cup \{0\}$, $h \le h_1$,

$$B_{h,p,\delta_0'}(\bar{u}_h) := \{u_h \in U_h : \|u_h - \bar{u}_h\|_{L^p} < \delta_0'\} \subset B_{h,\delta_0}(\bar{u}_h)$$

and the operators $[G_h]_{p,p}$ are Lipschitz continuous on $B_{h,p,\delta_0'}(\bar{u}_h)$ with modulus L_p. Let

$$C_{\hat{Z}} = \frac{\sigma_1 C_{G'}}{1+\kappa}, \qquad L_{\hat{Z}} = \sup_{h\in\mathcal{H}\cup\{0\},\, h\le h_1} (\|\sigma_h\lambda_h - 1\|_{L^\infty} + \|\sigma_h\|_{L^\infty} L_p),$$

$$\theta = \begin{cases} \frac{\gamma(p-2)}{2(p+\gamma)} & \text{if } p<\infty, \\ \frac{\gamma}{2} & \text{if } p=\infty, \end{cases} \qquad C_N = \begin{cases} \max\{1,(2^\gamma C)^{\frac{p-2}{2p}}\} L_{\hat{Z}} & \text{if } p<\infty, \\ \sqrt{3^\gamma C} L_{\hat{Z}} & \text{if } p=\infty. \end{cases}$$

Then, with $\delta = \min\{\delta'/L_{\hat{Z}}, \delta_0'\}$, there holds

$$\begin{aligned} &\forall\, u_h \in U_h,\ \|u_h - \bar{u}_h\|_{L^p} < \delta,\ h\in\mathcal{H}\cup\{0\},\ h\le h_1: \\ &\|\Phi_{1h}(u_h) - \Phi_{1h}(\bar{u}_h) - M_{\Phi_{1h}}(u_h)(u_h-\bar{u}_h)\|_{L^2} \\ &\qquad \le C_N \max\{\|\bar{z}_h - \bar{z}\|_{L^p}, L_{\hat{Z}}\|u_h - \bar{u}_h\|_{L^p}\}^\theta \|u_h-\bar{u}_h\|_{L^p} \\ &\qquad\quad + C_{\hat{Z}}\|u_h - \bar{u}_h\|_{L^p}^{1+\kappa}. \end{aligned} \tag{6.35}$$

Proof. With δ' as defined in the theorem, Assumption 6.19 (c) yields $h_1 > 0$ such that (6.34) is satisfied.

Exactly as in the proof of Theorem 6.20 we can now proceed to derive (6.27) and (6.28). Now, by Theorem 6.13, there holds in the case $p<\infty$

$$\begin{aligned} &\|R_N(\bar{z}_h, z_h - \bar{z}_h)\|_{L^2} \\ &\quad \le \max\left\{1,(2^\gamma C)^{\frac{p-2}{2p}}\right\} \max\{\|\bar{z}_h - \bar{z}\|_{L^p}, \|z_h - \bar{z}_h\|_{L^p}\}^{\frac{\gamma(p-2)}{2(p+\gamma)}} \|z_h - \bar{z}_h\|_{L^p} \\ &\quad \le \max\left\{1,(2^\gamma C)^{\frac{p-2}{2p}}\right\} \max\{\|\bar{z}_h - \bar{z}\|_{L^p}, L_{\hat{Z}}\|u_h - \bar{u}_h\|_{L^p}\}^{\frac{\gamma(p-2)}{2(p+\gamma)}} L_{\hat{Z}}\|u_h - \bar{u}_h\|_{L^p}. \end{aligned}$$

In the case $p=\infty$ we obtain

$$\begin{aligned} &\|R_N(\bar{z}_h, z_h - \bar{z}_h)\|_{L^2} \le \\ &\quad \le \sqrt{3^\gamma C}\max\{\|\bar{z}_h - \bar{z}\|_{L^\infty}, \|z_h - \bar{z}_h\|_{L^\infty}\}^{\frac{\gamma}{2}} \|z_h - \bar{z}_h\|_{L^\infty} \\ &\quad \le \sqrt{3^\gamma C}\max\{\|\bar{z}_h - \bar{z}\|_{L^\infty}, L_{\hat{Z}}\|u_h - \bar{u}_h\|_{L^\infty}\}^{\frac{\gamma}{2}} L_{\hat{Z}}\|u_h - \bar{u}_h\|_{L^\infty}. \end{aligned}$$

Furthermore, we can derive (6.29) as in the proof of Theorem 6.20. Combining these estimates completes the proof. □

In the next theorem we consider the operators Φ_{2h}.

Theorem 6.23. *Let Assumption 6.19 hold with $r=2$. Assume further that there exist constants $\gamma>0$, $t_0>0$, and $C>0$ such that with $\bar{z}_0 = \bar{z} = Z(\bar{u})$ there hold strict complementarity*

$$\text{meas}(\{\omega : \bar{z}(\omega) = 0\}) = 0$$

as well as the growth condition

$$\text{meas}(\{\omega : 0 < |\bar{z}(\omega)| < t\}) \le Ct^\gamma \quad \forall\, t\in(0,t_0].$$

Then there exists $0<h_1\le h_0$ such that for $\bar{z}_h = Z_h(\bar{u}_h)$ there holds

$$\|\bar{z}_h - \bar{z}\|_{L^p} < \delta' \quad \forall\, h\in\mathcal{H},\ h\le h_1, \tag{6.36}$$

where $\delta' = \min\{1,(t_0/2)^{\frac{p+\gamma}{p}}\}$ *if* $p < \infty$ *and* $\delta' = \min\{1,t_0/2\}$ *if* $p = \infty$. *Let*

$$C_Z = \frac{C_{G'}}{\lambda_0(1+\kappa)}, \qquad L_Z = \frac{L_G}{\lambda_0},$$

$$\theta = \begin{cases} \frac{\gamma(p-2)}{2(p+\gamma)} & \text{if } p<\infty, \\ \frac{\gamma}{2} & \text{if } p=\infty, \end{cases} \qquad C_N = \begin{cases} \max\{1,(2^\gamma C)^{\frac{p-2}{2p}}\}L_Z & \text{if } p<\infty, \\ \sqrt{3^\gamma C}L_Z & \text{if } p=\infty. \end{cases}$$

Then, with $\delta = \min\{\delta'/L_Z, \delta_0\}$, *there holds*

$$\begin{aligned} &\forall\, u_h \in U_h,\ \|u_h - \bar{u}_h\|_{L^2} < \delta,\ h \in \mathcal{H} \cup \{0\},\ h \le h_1 : \\ &\|\Phi_{2h}(u_h) - \Phi_{2h}(\bar{u}_h) - M_{\Phi_{2h}}(u_h)(u_h - \bar{u}_h)\|_{L^2} \\ &\quad \le C_N \max\{\|\bar{z}_h - \bar{z}\|_{L^p}, L_Z\|u_h - \bar{u}_h\|_{L^2}\}^\theta \|u_h - \bar{u}_h\|_{L^2} \\ &\quad\quad + C_Z\|u_h - \bar{u}_h\|_{L^2}^{1+\kappa}. \end{aligned} \tag{6.37}$$

Proof. With δ' as defined in the theorem, Assumption 6.19 (c) ensures the existence of $0 < h_1 \le h_0$ such that (6.36) is satisfied. We now proceed as in the proof of Theorem 6.21 to derive (6.31) and (6.32). Now, by Theorem 6.13, there holds in the case $p < \infty$

$$\begin{aligned} &\|R_N(\bar{z}_h, z_h - \bar{z}_h)\|_{L^2} \\ &\le \max\left\{1,(2^\gamma C)^{\frac{p-2}{2p}}\right\} \max\{\|\bar{z}_h - \bar{z}\|_{L^p}, \|z_h - \bar{z}_h\|_{L^p}\}^{\frac{\gamma(p-2)}{2(p+\gamma)}} \|z_h - \bar{z}_h\|_{L^p} \\ &\le \max\left\{1,(2^\gamma C)^{\frac{p-2}{2p}}\right\} \max\{\|\bar{z}_h - \bar{z}\|_{L^p}, L_Z\|u_h - \bar{u}_h\|_{L^2}\}^{\frac{\gamma(p-2)}{2(p+\gamma)}} L_Z\|u_h - \bar{u}_h\|_{L^2}. \end{aligned}$$

In the case $p = \infty$ we obtain

$$\begin{aligned} &\|R_N(\bar{z}_h, z_h - \bar{z}_h)\|_{L^2} \\ &\quad \le \sqrt{3^\gamma C} \max\{\|\bar{z}_h - \bar{z}\|_{L^\infty}, \|z_h - \bar{z}_h\|_{L^\infty}\}^{\frac{\gamma}{2}} \|z_h - \bar{z}_h\|_{L^\infty} \\ &\quad \le \sqrt{3^\gamma C} \max\{\|\bar{z}_h - \bar{z}\|_{L^\infty}, L_Z\|u_h - \bar{u}_h\|_{L^2}\}^{\frac{\gamma}{2}} L_Z\|u_h - \bar{u}_h\|_{L^2}. \end{aligned}$$

Furthermore, we can derive (6.33) as in the proof of Theorem 6.21. Combining these estimates completes the proof. □

6.3.3 Mesh-Independent Semismoothness without Growth Conditions

Based on Theorem 6.15, which just requires strict complementarity at $\bar{u}$, i.e.,

$$\text{meas}(\{\omega : \bar{z}(\omega) = 0\}) = 0,$$

but no growth condition on $\bar{z}$, it is possible, by very similar arguments as in the previous two subsections, to derive linear mesh-independent estimates of the residual. Under suitable assumptions, given $\eta > 0$, there exist $\delta > 0$ and $h_1 > 0$ such that for all $u_h \in U_h \cap L^p(\Omega)$, $\|u_h - \bar{u}_h\|_{L^p} < \delta$, and all $h \in \mathcal{H} \cup \{0\}$, $h \le h_1$, the following holds:

$$\|\Phi_{1h}(u_h) - \Phi_{1h}(\bar{u}_h) - M_{\Phi_{1h}}(u_h)(u_h - \bar{u}_h)\|_{L^2} \le \eta\|u_h - \bar{u}_h\|_{L^p}.$$

Similarly, given $\eta > 0$, there exist $\delta > 0$ and $h_1 > 0$ such that for all $u_h \in U_h$, $\|u_h - \bar{u}_h\|_{L^2} < \delta$, and all $h \in \mathcal{H} \cup \{0\}$, $h \le h_1$, the following holds:

$$\|\Phi_{2h}(u_h) - \Phi_{2h}(\bar{u}_h) - M_{\Phi_{2h}}(u_h)(u_h - \bar{u}_h)\|_{L^2} \le \eta \|u_h - \bar{u}_h\|_{L^2}.$$

Estimates of this form were first derived in [106] and used to prove the first mesh-independence result for semismooth Newton methods. We refer to [106] for details. Alternatively, the reader is encouraged to make the required (quite obvious) changes to the proofs of Theorems 6.22 and 6.23 to obtain the above estimates.

6.4 Mesh Independence of the Semismooth Newton Method

We can apply the mesh-independent semismoothness theory obtained in the previous section to prove mesh-independence results for semismooth Newton methods applied to CP(F_h).

For the reformulation based on Φ_{1h}, we need a smoothing step satisfying the following assumptions.

Assumption 6.24. *There exist $h_s > 0$, $\delta_s \in (0,1]$, $L_S > 0$, and operators $S_h : U_h \to U_h \cap L^p(\Omega)$ such that $S_h(\bar{u}_h) = \bar{u}_h$ and*

$$\forall\, u_h \in U_h,\ \|u_h - \bar{u}_h\|_{L^2} < \delta_s,\ h \in \mathcal{H} \cup \{0\},\ h \le h_s :$$
$$\|S_h(u_h) - \bar{u}_h\|_{L^p} \le L_S \|u_h - \bar{u}_h\|_{L^2}.$$

We already mentioned a standard way to construct such smoothing steps. This is made precise in the following lemma.

Lemma 6.25. *Let Assumption 6.19 for $r = p$ hold. Then the smoothing steps $S_h(u_h) = -N(Z_h(u_h))$ satisfy Assumption 6.24 with $h_s = h_0$, $\delta_s = \delta_0$, and $L_S = L_G/\lambda_0$.*

Proof. From Lemma 6.5 we know that $\bar{u}_h$ solves CP(F_h) if and only if $\Phi_{2h}(\bar{u}_h) = 0$. Since $\Phi_{2h}(u_h) = u_h + N(Z_h(u_h)) = u_h - S_h(u_h)$, we see that $S_h(\bar{u}_h) = 0$ is satisfied. Furthermore, using

$$|N(z_1)(\omega) - N(z_2)(\omega)| = |\min\{z_1(\omega), 0\} - \min\{z_2(\omega), 0\}| \le |z_1(\omega) - z_2(\omega)|,$$

we obtain

$$\begin{aligned}
\|S_h(u_h) - \bar{u}_h\|_{L^p} &= \|S_h(u_h) - S_h(\bar{u}_h)\|_{L^p} \\
&= \left\| N\left(\left(\frac{1}{\lambda_h}\right)\cdot G_h(u_h)\right) - N\left(\left(\frac{1}{\lambda_h}\right)\cdot G_h(\bar{u}_h)\right)\right\|_{L^p} \\
&\le \left\|\left(\frac{1}{\lambda_h}\right)\cdot (G_h(u_h) - G_h(\bar{u}_h))\right\|_{L^p} \\
&\le \left\|\frac{1}{\lambda_h}\right\|_{L^\infty} \|G_h(u_h) - G_h(\bar{u}_h)\|_{L^p} \\
&\le \frac{L_G}{\lambda_0}\|u_h - \bar{u}_h\|_{L^2}. \qquad \square
\end{aligned}$$

The semismooth Newton iteration for Φ_{1h} including a smoothing step looks as follows.

Algorithm 6.26 (semismooth Newton method with smoothing step for Φ_{1h}).

0. Choose $u_h^0 \in U_h \cap L^p(\Omega)$.

For $k = 0, 1, 2, \dots$:

1. Choose $M_{hk} \in \partial^\circ \Phi_{1h}(u_h^k)$ and obtain $s_h^k \in U_h$ by solving
$$M_{hk} s_h^k = -\Phi_{1h}\left(u_h^k\right).$$
2. Set $u_h^{k+1} = S_h(u_h^k + s_h^k)$.

The semismooth Newton method for Φ_{2h} does not require a smoothing step and can be cast in the following way.

Algorithm 6.27 (semismooth Newton method for Φ_{2h}).

0. Choose $u_h^0 \in U_h$.

For $k = 0, 1, 2, \dots$:

1. Choose $M_{hk} \in \partial^\circ \Phi_{2h}(u_h^k)$ and obtain $s_h^k \in U_h$ by solving
$$M_{hk} s_h^k = -\Phi_{2h}\left(u_h^k\right).$$
2. Set $u_h^{k+1} = u_h^k + s_h^k$.

In a practical application of both algorithms, a suitable stopping criterion would be included. In the present form the sequence of iterates would become stationary if u_h^k solves the problem. Furthermore, since we will show that for Algorithm 6.26 the sequence $\|u_h^k - \bar{u}_h\|_{L^p}$ is strictly monotonically decreasing if $\|u_h^0 - \bar{u}_h\|_{L^p}$ is sufficiently small, stationarity for $k \geq l$ can only occur if $u_h^k = \bar{u}_h$ for all $k \geq l$. In the same way, we will show that for Algorithm 6.27 the sequence $\|u_h^k - \bar{u}_h\|_{L^2}$ is strictly monotonically decreasing if $\|u_h^0 - \bar{u}_h\|_{L^2}$ is sufficiently small. Hence, stationarity for $k \geq l$ can only occur if $u_h^k = \bar{u}_h$ for all $k \geq l$. This can be used for finite termination; see section 3.2.3.

There holds for both iterations

$$\begin{aligned} M_{hk}(u_h^k + s_h^k - \bar{u}_h) &= M_{hk}(u_h^k - \bar{u}_h) - \Phi_{ih}\left(u_h^k\right) \\ &= -\left[\Phi_{ih}\left(u_h^k\right) - \Phi_{ih}(\bar{u}_h) - M_{hk}(u_h^k - \bar{u}_h)\right] \\ &= -R_{\Phi_{ih}}\left(\bar{u}_h, u_h^k - \bar{u}_h\right), \end{aligned}$$

where $R_{\Phi_{ih}}$ is the semismoothness residual of Φ_{ih}.

Therefore, in order to prove fast local convergence, it is appropriate for both algorithms to require the following regularity condition.

Assumption 6.28. *There exist $h_r > 0$ and $C_M > 0$ such that the operators $M_{hk} \in \mathcal{L}(U_h, U_h)$ chosen in step 2 of the semismooth Newton algorithm satisfy*

$$\|M_{hk}^{-1}\|_{U_h,U_h} \leq C_M \quad \forall\, k \geq 0,\ h \in \mathcal{H} \cup \{0\},\ h \leq h_r.$$

Remark 6.29. *The conditions in Assumption 6.28 are, e.g., ensured if all elements of $\partial^\circ \Phi_{ih}(u_h)$ are uniformly bounded invertible for $u_h \in U_h$ in a neighborhood of $\bar{u}_h$. Here, in the case $i = 1$ an $(U_h \cap L^p)$-neighborhood is required, and in the case $i = 2$ an U_h-neighborhood is needed.*

6.4.1 Mesh-Independent Convergence under Uniform Growth Conditions

We now use the uniform order of semismoothness estimates of section 6.3.1 to derive estimates on the mesh-independent rate of convergence of semismooth Newton methods. We begin with the result for the operator Φ_{2h}, since no smoothing step is required there, which makes this case a bit easier than for the operator Φ_{1h}.

Theorem 6.30. *We consider the semismooth Newton method, Algorithm 6.27, applied to the equation $\Phi_{2h}(u_h) = 0$. Let Assumption 6.19 hold with $r = 2$ and let Assumption 6.28 hold. Assume further that there exist constants $0 < h_1 \leq \min\{h_0, h_r\}$, $\gamma > 0$, $t_0 > 0$, and $C > 0$ such that with $\bar{z}_0 = \bar{z} = Z(\bar{u})$ and $\bar{z}_h = Z_h(\bar{u}_h)$ there holds*

$$\operatorname{meas}(\{\omega : 0 < |\bar{z}_h(\omega)| < t\}) \leq C t^\gamma \quad \forall\, t \in (0, t_0],\ \forall h \in \mathcal{H} \cup \{0\},\ h \leq h_1.$$

Let

$$C_Z = \frac{C_{G'}}{\lambda_0(1+\kappa)}, \qquad L_Z = \frac{L_G}{\lambda_0},$$

$$\theta = \begin{cases} \frac{\gamma(p-2)}{2(p+\gamma)} & \text{if } p < \infty, \\ \frac{\gamma}{2} & \text{if } p = \infty, \end{cases} \qquad C_N = \begin{cases} \max\{1, (2^\gamma C)^{\frac{p-2}{2p}}\} L_Z^{1+\frac{\gamma(p-2)}{2(p+\gamma)}} & \text{if } p < \infty, \\ \sqrt{3^\gamma C} L_Z^{1+\frac{\gamma}{2}} & \text{if } p = \infty. \end{cases}$$

Then, with $\delta'' = \min\{\delta'/L_Z, \delta_0\}$, where $\delta' = \min\{1, (t_0/2)^{\frac{p+\gamma}{p}}\}$ if $p < \infty$ and $\delta' = \min\{1, t_0/2\}$ if $p = \infty$, there holds

$$\begin{aligned} &\forall\, u_h^k \in U_h,\ \|u_h^k - \bar{u}_h\|_{L^2} < \delta'',\ h \in \mathcal{H} \cup \{0\},\ h \leq h_1 : \\ &\|u_h^{k+1} - \bar{u}_h\|_{L^2} \leq C_M C_N \|u_h^k - \bar{u}_h\|_{L^2}^{1+\theta} + C_M C_Z \|u_h^k - \bar{u}_h\|_{L^2}^{1+\kappa}. \end{aligned} \tag{6.38}$$

In particular, there exists $0 < \delta \leq \delta''$ such that for all $u_h^0 \in U_h$, $\|u_h^0 - \bar{u}_h\|_{L^2} < \delta$, and all $h \in \mathcal{H} \cup \{0\}$, $h \leq h_1$, (u_h^k) converges q-superlinearly with at least order $1 + \min\{\theta, \kappa\}$ to $\bar{u}_h$.

Proof. From Theorem 6.21 we obtain the estimate (6.30) for all $u_h \in U_h$, $\|u_h - \bar{u}_h\|_{L^2} < \delta''$, and all $h \in \mathcal{H} \cup \{0\}$, $h \leq h_1$. Using

$$\begin{aligned} \|u_h^{k+1} - \bar{u}_h\|_{L^2} &= \|u_h^k + s_h^k - \bar{u}_h\|_{L^2} = \|M_{hk}^{-1} R_{\Phi_{2h}}(\bar{u}_h, u_h^k - \bar{u}_h)\|_{L^2} \\ &\leq C_M \|R_{\Phi_{2h}}(\bar{u}_h, u_h^k - \bar{u}_h)\|_{L^2} \end{aligned}$$

and (6.30) with $u_h = u_h^k$ yields (6.38). Given $\eta \in (0,1)$, we choose $0 < \delta \le \delta''$ such that

$$C_M C_N \|u_h^k - \bar{u}_h\|_{L^2}^{\theta} + C_M C_Z \|u_h^k - \bar{u}_h\|_{L^2}^{\kappa} \le \eta$$

for all $u_h \in U_h$, $\|u_h - \bar{u}_h\|_{L^2} < \delta$, and all $h \in \mathcal{H} \cup \{0\}$, $h \le h_1$. This gives

$$\|u_h^{k+1} - \bar{u}_h\|_{L^2} \le \eta \|u_h^k - \bar{u}_h\|_{L^2}$$

for all $u_h^k \in U_h$, $\|u_h^k - \bar{u}_h\|_{L^2} < \delta$, and all $h \in \mathcal{H} \cup \{0\}$, $h \le h_1$. Thus, for arbitrary $u_h^0 \in U_h$, $\|u_h^0 - \bar{u}_h\|_{L^2} < \delta$, we obtain convergence of (u_h^k) to $\bar{u}_h$ for all $h \in \mathcal{H} \cup \{0\}$, $h \le h_1$. By the convergence estimate (6.38), the order of q-superlinear convergence is at least $1 + \min\{\kappa, \theta\}$. □

In the next theorem we investigate the rate of local convergence of Algorithm 6.26.

Theorem 6.31. *We consider the semismooth Newton method with smoothing step, Algorithm 6.26, applied to $\Phi_{1h}(u_h) = 0$. Let Assumption 6.19 hold with $r = p$. Furthermore, let Assumptions 6.24 and 6.28 hold. Assume that there exist constants $0 < h_1 \le \min\{h_0, h_r, h_s\}$, $\lambda_1 > 0$, $\sigma_1 > 0$ $\gamma > 0$, $C > 0$, and $t_0 > 0$ such that with $\bar{z}_0 = \bar{z} = Z(\bar{u})$ and $\bar{z}_h = Z_h(\bar{u}_h)$ there holds*

$$\begin{aligned} &\|\lambda_h\|_{L^\infty} \le \lambda_1 \quad \text{and} \quad \|\sigma_h\|_{L^\infty} \le \sigma_1 && \forall\, h \in \mathcal{H} \cup \{0\},\ h \le h_1, \\ &\operatorname{meas}(\{\omega : 0 < |\bar{z}_h(\omega)| < t\}) \le C t^{\gamma} && \forall\, t \in (0, t_0],\ \forall\, h \in \mathcal{H} \cup \{0\},\ h \le h_1. \end{aligned}$$

Then there exist $\delta_0' \ge \delta_0 / \operatorname{meas}(\Omega)^{\frac{p-2}{2p}}$ and $L_p \le \operatorname{meas}(\Omega)^{\frac{p-2}{2p}} L_G$ (with $(p-2)/(2p) = 1/2$ if $p = \infty$) such that, for all $h \in \mathcal{H} \cup \{0\}$, $h \le h_1$,

$$B_{h,p,\delta_0'}(\bar{u}_h) := \{u_h \in U_h : \|u_h - \bar{u}_h\|_{L^p} < \delta_0'\} \subset B_{h,\delta_0}(\bar{u}_h)$$

and the operators $[G_h]_{p,p}$ are Lipschitz continuous on $B_{h,p,\delta_0'}(\bar{u}_h)$ with modulus L_p. Let

$$C_{\hat{Z}} = \frac{\sigma_1 C_{G'}}{1+\kappa}, \qquad L_{\hat{Z}} = \sup_{h \in \mathcal{H} \cup \{0\},\ h \le h_1} (\|\sigma_h \lambda_h - 1\|_{L^\infty} + \|\sigma_h\|_{L^\infty} L_p),$$

$$\theta = \begin{cases} \frac{\gamma(p-2)}{2(p+\gamma)} & \text{if } p < \infty, \\ \frac{\gamma}{2} & \text{if } p = \infty, \end{cases} \qquad C_N = \begin{cases} \max\{1, (2^{\gamma} C)^{\frac{p-2}{2p}}\} L_{\hat{Z}}^{1+\frac{\gamma(p-2)}{2(p+\gamma)}} & \text{if } p < \infty, \\ \sqrt{3^{\gamma} C}\, L_{\hat{Z}}^{1+\frac{\gamma}{2}} & \text{if } p = \infty. \end{cases}$$

Then, with $\delta'' = \min\{\delta'/L_{\hat{Z}}, \delta_0'\}$, where $\delta' = \min\{1, (t_0/2)^{\frac{p+\gamma}{p}}\}$ if $p < \infty$ and $\delta' = \min\{1, t_0/2\}$ if $p = \infty$, there holds

$$\begin{aligned} &\forall\, u_h^k \in U_h,\ \|u_h^k - \bar{u}_h\|_{L^p} < \delta'',\ h \in \mathcal{H} \cup \{0\},\ h \le h_1 : \\ &\|u_h^{k+1} - \bar{u}_h\|_{L^p} \le C_N C_M L_S \|u_h - \bar{u}_h\|_{L^p}^{1+\theta} + C_{\hat{Z}} C_M L_S \|u_h - \bar{u}_h\|_{L^p}^{1+\kappa}. \end{aligned} \tag{6.39}$$

In particular, there exists $0 < \delta \le \delta''$ such that for all $u_h^0 \in U_h$, $\|u_h^0 - \bar{u}_h\|_{L^p} < \delta$, and all $h \in \mathcal{H} \cup \{0\}$, $h \le h_1$, (u_h^k) converges q-superlinearly in $(U_h \cap L^p(\Omega), \|\cdot\|_{L^p})$ to $\bar{u}_h$ with order at least $1 + \min\{\theta, \kappa\}$.

Proof. Except for (6.39), all the assertions follow immediately from Theorem 6.20, which also provides the estimate (6.26) for all $u_h \in U_h$, $\|u_h - \bar{u}_h\|_{L^p} < \delta''$, and all $h \in \mathcal{H} \cup \{0\}$, $h \le h_1$. We estimate

$$\begin{aligned}\|u_h^{k+1} - \bar{u}_h\|_{L^p} &= \left\| S_h\left(u_h^k + s_h^k\right) - \bar{u}_h \right\|_{L^p} \le L_S \|u_h^k + s_h^k - \bar{u}_h\|_{L^2} \\ &= L_S \left\| M_{hk}^{-1} R_{\Phi_{1h}}\left(\bar{u}_h, u_h^k - \bar{u}_h\right) \right\|_{L^2} \le C_M L_S \left\| R_{\Phi_{1h}}\left(\bar{u}_h, u_h^k - \bar{u}_h\right) \right\|_{L^2}.\end{aligned}$$

Using (6.26) with $u_h = u_h^k$, we obtain (6.39). Next, for given $\eta \in (0,1)$, we choose $0 < \delta \le \delta''$ such that

$$C_N C_M L_S \|u_h - \bar{u}_h\|_{L^p}^{\theta} + C_{\hat{Z}} C_M L_S \|u_h - \bar{u}_h\|_{L^p}^{\kappa} \le \eta$$

for all $u_h \in U_h$, $\|u_h - \bar{u}_h\|_{L^p} < \delta$, and all $h \in \mathcal{H} \cup \{0\}$, $h \le h_1$. This gives

$$\|u_h^{k+1} - \bar{u}_h\|_{L^p} \le \eta \|u_h^k - \bar{u}_h\|_{L^p}$$

for all $u_h^k \in U_h$, $\|u_h^k - \bar{u}_h\|_{L^p} < \delta$, and all $h \in \mathcal{H} \cup \{0\}$, $h \le h_1$. Thus, for arbitrary $u_h^0 \in U_h$, $\|u_h^0 - \bar{u}_h\|_{L^p} < \delta$, (u_h^k) converges in $(U_h \cap L^p(\Omega), \|\cdot\|_{L^p})$ to $\bar{u}_h$ for all $h \in \mathcal{H} \cup \{0\}$, $h \le h_1$. The estimate (6.39) shows that the local rate of convergence is q-superlinear with order at least $1 + \min\{\theta, \kappa\}$. □

6.4.2 Mesh-Independent Convergence without Uniform Growth Conditions

The next result investigates the local convergence properties of Algorithm 6.27 under a growth condition on $\bar{z}$ only.

Theorem 6.32. *We consider the semismooth Newton method, Algorithm* 6.27, *applied to the equation* $\Phi_{2h}(u_h) = 0$. *Let Assumption* 6.19 *hold with* $r = 2$ *and let Assumption* 6.28 *hold. Assume further that there exist constants* $\gamma > 0$, $t_0 > 0$, *and* $C > 0$ *such that with* $\bar{z}_0 = \bar{z} = Z(\bar{u})$ *there holds*

$$\text{meas}(\{\omega : 0 < |\bar{z}(\omega)| < t\}) \le C t^{\gamma} \quad \forall\, t \in (0, t_0].$$

Furthermore, let strict complementarity hold at $\bar{u}$*:*

$$\text{meas}(\{\omega : \bar{z}(\omega) = 0\}) = 0.$$

Then there exists $0 < h_1 \le \min\{h_0, h_r\}$ *such that for* $\bar{z}_h = Z_h(\bar{u}_h)$ *there holds*

$$\|\bar{z}_h - \bar{z}\|_{L^p} < \delta' \quad \forall\, h \in \mathcal{H},\ h \le h_1, \tag{6.40}$$

where $\delta' = \min\{1, (t_0/2)^{\frac{p+\gamma}{p}}\}$ *if* $p < \infty$ *and* $\delta' = \min\{1, t_0/2\}$ *if* $p = \infty$. *Let*

$$C_Z = \frac{C_{G'}}{\lambda_0(1+\kappa)}, \qquad L_Z = \frac{L_G}{\lambda_0},$$

$$\theta = \begin{cases} \frac{\gamma(p-2)}{2(p+\gamma)} & \text{if } p < \infty, \\ \frac{\gamma}{2} & \text{if } p = \infty, \end{cases} \qquad C_N = \begin{cases} \max\{1, (2^{\gamma} C)^{\frac{p-2}{2p}}\} L_Z & \text{if } p < \infty, \\ \sqrt{3^{\gamma} C} L_Z & \text{if } p = \infty. \end{cases}$$

Then, with $\delta'' = \min\{\delta'/L_Z, \delta_0\}$, *there holds*

$$\forall\, u_h^k \in U_h,\ \ \|u_h^k - \bar{u}_h\|_{L^2} < \delta'',\ h \in \mathcal{H} \cup \{0\},\ h \le h_1:$$
$$\begin{aligned}\|u_h^{k+1} - \bar{u}_h\|_{L^2} \le C_M C_N \max\left\{\|\bar{z}_h - \bar{z}\|_{L^p}, L_Z\|u_h^k - \bar{u}_h\|_{L^2}\right\}^{\theta} \|u_h^k - \bar{u}_h\|_{L^2} \\ + C_M C_Z \|u_h^k - \bar{u}_h\|_{L^2}^{1+\kappa}.\end{aligned} \tag{6.41}$$

From this convergence estimate, it follows in particular that, given any $\eta \in (0,1)$, *there exist* $0 < \delta \le \delta''$ *and* $0 < h_2 \le h_1$ *such that for all* $u_h^0 \in U_h$, $\|u_h^0 - \bar{u}_h\|_{L^2} < \delta$, *and all* $h \in \mathcal{H} \cup \{0\}$, $h \le h_2$, (u_h^k) *converges at least q-linearly with rate* η *to* $\bar{u}_h$. *A detailed convergence estimate is provided by* (6.41).

Proof. From Theorem 6.23 we obtain (6.40) as well as the estimate (6.37) for all $u_h \in U_h$, $\|u_h - \bar{u}_h\|_{L^2} < \delta''$, and all $h \in \mathcal{H} \cup \{0\}$, $h \le h_1$. Using

$$\begin{aligned}\|u_h^{k+1} - \bar{u}_h\|_{L^2} &= \|u_h^k + s_h^k - \bar{u}_h\|_{L^2} = \left\| M_{hk}^{-1} R_{\Phi_{2h}}\left(\bar{u}_h, u_h^k - \bar{u}_h\right)\right\|_{L^2} \\ &\le C_M \left\| R_{\Phi_{2h}}\left(\bar{u}_h, u_h^k - \bar{u}_h\right)\right\|_{L^2}\end{aligned}$$

and (6.37) with $u_h = u_h^k$, we arrive at (6.41). Given $\eta \in (0,1)$, we choose $0 < h_2 \le h_1$ and $0 < \delta \le \delta''$ such that

$$C_M C_N \max\{\|\bar{z}_h - \bar{z}\|_{L^p}, L_Z\|u_h - \bar{u}_h\|_{L^2}\}^{\theta} + C_M C_Z \|u_h - \bar{u}_h\|_{L^2}^{\kappa} \le \eta$$

for all $u_h \in U_h$, $\|u_h - \bar{u}_h\|_{L^2} < \delta$, and all $h \in \mathcal{H} \cup \{0\}$, $h \le h_2$. This gives

$$\|u_h^{k+1} - \bar{u}_h\|_{L^2} \le \eta \|u_h^k - \bar{u}_h\|_{L^2}$$

for all $u_h^k \in U_h$, $\|u_h^k - \bar{u}_h\|_{L^2} < \delta$, and all $h \in \mathcal{H} \cup \{0\}$, $h \le h_2$. Thus, for arbitrary $u_h^0 \in U_h$, $\|u_h^0 - \bar{u}_h\|_{L^2} < \delta$, we have convergence of (u_h^k) to $\bar{u}_h$ for all $h \in \mathcal{H} \cup \{0\}$, $h \le h_2$. The rate of convergence is at least η and the convergence estimate (6.41) holds. □

We now consider the semismooth Newton method with smoothing step applied to Φ_{1h}.

Theorem 6.33. *We consider the semismooth Newton method with smoothing step, Algorithm* 6.26, *applied to* $\Phi_{1h}(u_h) = 0$. *Let Assumption* 6.19 *hold with* $r = p$. *Furthermore, let Assumptions* 6.28 *and* 6.24 *hold. Assume that there exist constants* $0 < h_1' \le \min\{h_0, h_r, h_s\}$, $\lambda_1 > 0$, $\sigma_1 > 0$ $\gamma > 0$, $C > 0$, *and* $t_0 > 0$ *such that with* $\bar{z}_0 = \bar{z} = \hat{Z}(\bar{u})$ *there holds*

$$\begin{aligned}&\|\lambda_h\|_{L^\infty} \le \lambda_1 \quad \text{and} \quad \|\sigma_h\|_{L^\infty} \le \sigma_1 && \forall\, h \in \mathcal{H} \cup \{0\},\ h \le h_1', \\ &\operatorname{meas}(\{\omega : 0 < |\bar{z}(\omega)| < t\}) \le C t^{\gamma} && \forall\, t \in (0, t_0].\end{aligned}$$

Furthermore, let strict complementarity hold at $\bar{u}$:

$$\operatorname{meas}(\{\omega : \bar{z}(\omega) = 0\}) = 0.$$

Then there exists $0 < h_1 \le h_1'$ *such that for* $\bar{z}_h = \hat{Z}_h(\bar{u}_h)$ *there holds*

$$\|\bar{z}_h - \bar{z}\|_{L^p} < \delta' \quad \forall\, h \in \mathcal{H},\ h \le h_1, \tag{6.42}$$

where $\delta' = \min\{1, (t_0/2)^{\frac{p+\gamma}{p}}\}$ *if* $p < \infty$ *and* $\delta' = \min\{1, t_0/2\}$ *if* $p = \infty$. *Furthermore, there exist* $\delta_0' \geq \delta_0/\mathrm{meas}(\Omega)^{\frac{p-2}{2p}}$ *and* $L_p \leq \mathrm{meas}(\Omega)^{\frac{p-2}{2p}} L_G$ *(with* $(p-2)/(2p) = 1/2$ *if* $p = \infty$*) such that, for all* $h \in \mathcal{H} \cup \{0\}$, $h \leq h_1$,

$$B_{h,p,\delta_0'}(\bar{u}_h) := \{u_h \in U_h : \|u_h - \bar{u}_h\|_{L^p} < \delta_0'\} \subset B_{h,\delta_0}(\bar{u}_h)$$

and the operators $[G_h]_{p,p}$ *are Lipschitz continuous on* $B_{h,p,\delta_0'}(\bar{u}_h)$ *with modulus* L_p. *Let*

$$C_{\hat{Z}} = \frac{\sigma_1 C_{G'}}{1+\kappa}, \qquad L_{\hat{Z}} = \sup_{h \in \mathcal{H}\cup\{0\},\, h \leq h_1} (\|\sigma_h \lambda_h - 1\|_{L^\infty} + \|\sigma_h\|_{L^\infty} L_p),$$

$$\theta = \begin{cases} \frac{\gamma(p-2)}{2(p+\gamma)} & \text{if } p < \infty, \\ \frac{\gamma}{2} & \text{if } p = \infty, \end{cases} \qquad C_N = \begin{cases} \max\{1, (2^\gamma C)^{\frac{p-2}{2p}}\} L_{\hat{Z}} & \text{if } p < \infty, \\ \sqrt{3^\gamma C} L_{\hat{Z}} & \text{if } p = \infty. \end{cases}$$

Then, with $\delta'' = \min\{\delta'/L_{\hat{Z}}, \delta_0'\}$, *there holds*

$$\begin{aligned} &\forall\, u_h^k \in U_h,\ \|u_h^k - \bar{u}_h\|_{L^p} < \delta'',\ h \in \mathcal{H} \cup \{0\},\ h \leq h_1 : \\ &\|u_h^{k+1} - \bar{u}_h\|_{L^p} \leq C_N C_M L_S \max\{\|\bar{z}_h - \bar{z}\|_{L^p}, L_{\hat{Z}}\|u_h - \bar{u}_h\|_{L^p}\}^\theta \|u_h - \bar{u}_h\|_{L^p} \\ &\qquad\qquad + C_{\hat{Z}} C_M L_S \|u_h - \bar{u}_h\|_{L^p}^{1+\kappa}. \end{aligned} \tag{6.43}$$

From this convergence estimate, it follows in particular that, given any $\eta \in (0,1)$, *there exist* $0 < \delta \leq \delta''$ *and* $0 < h_2 \leq h_1$ *such that for all* $u_h^0 \in U_h \cap L^p(\Omega)$, $\|u_h^0 - \bar{u}_h\|_{L^p} < \delta$, *and all* $h \in \mathcal{H} \cup \{0\}$, $h \leq h_2$, (u_h^k) *converges in* $(U_h \cap L^p(\Omega), \|\cdot\|_{L^p})$ *at least q-linearly with rate* η *to* $\bar{u}_h$.

Proof. Except for (6.43), all the assertions follow immediately from Theorem 6.22, which also provides the estimate (6.35) for all $u_h \in U_h$, $\|u_h - \bar{u}_h\|_{L^p} < \delta''$, and all $h \in \mathcal{H} \cup \{0\}$, $h \leq h_1$. We estimate

$$\begin{aligned} \|u_h^{k+1} - \bar{u}_h\|_{L^p} &= \left\| S_h\left(u_h^k + s_h^k\right) - \bar{u}_h \right\|_{L^p} \leq L_S \|u_h^k + s_h^k - \bar{u}_h\|_{L^2} \\ &= L_S \left\| M_{hk}^{-1} R_{\Phi_{1h}}\left(\bar{u}_h, u_h^k - \bar{u}_h\right)\right\|_{L^2} \leq C_M L_S \left\| R_{\Phi_{1h}}\left(\bar{u}_h, u_h^k - \bar{u}_h\right)\right\|_{L^2}. \end{aligned}$$

Using (6.35) with $u_h = u_h^k$, we obtain (6.43). Next, for given $\eta \in (0,1)$, we choose $0 < h_2 \leq h_1$ and $0 < \delta \leq \delta''$ such that

$$C_N C_M L_S \max\{\|\bar{z}_h - \bar{z}\|_{L^p}, L_{\hat{Z}}\|u_h - \bar{u}_h\|_{L^p}\}^\theta + C_{\hat{Z}} C_M L_S \|u_h - \bar{u}_h\|_{L^p}^\kappa \leq \eta$$

for all $u_h \in U_h$, $\|u_h - \bar{u}_h\|_{L^p} < \delta$, and all $h \in \mathcal{H} \cup \{0\}$, $h \leq h_2$. This gives

$$\|u_h^{k+1} - \bar{u}_h\|_{L^p} \leq \eta \|u_h^k - \bar{u}_h\|_{L^p}$$

for all $u_h^k \in U_h$, $\|u_h^k - \bar{u}_h\|_{L^p} < \delta$, and all $h \in \mathcal{H} \cup \{0\}$, $h \leq h_2$. Thus, for arbitrary $u_h^0 \in U_h$, $\|u_h^0 - \bar{u}_h\|_{L^p} < \delta$, (u_h^k) converges in $(U_h \cap L^p(\Omega), \|\cdot\|_{L^p})$ to $\bar{u}_h$ for all $h \in \mathcal{H} \cup \{0\}$, $h \leq h_2$. The rate of convergence is at least η and the convergence estimate (6.43) holds. ∎

6.4.3 Mesh-Independent Convergence without Growth Conditions

Using the mesh-independent semismoothness results sketched in section 6.3.3 for the case without any growth condition, just strict complementarity, a mesh-independent linear rate of convergence can be shown for any prescribed linear rate of convergence. In fact, under suitably adjusted assumptions, a modification of the proofs of Theorems 6.32 and 6.33 shows that for any $\eta \in (0,1)$ there exist $\delta > 0$ and $h_1 > 0$ such that for all $u_h^0 \in U_h \cap L^p(\Omega)$, $\|u_h^0 - \bar{u}_h\|_{L^p} < \delta$, and all $h \in \mathcal{H} \cup \{0\}$, $h \le h_1$, the semismooth Newton method with smoothing step, Algorithm 6.26, generates a sequence that satisfies

$$\|u_h^{k+1} - \bar{u}_h\|_{L^p} \le \eta \|u_h^k - \bar{u}_h\|_{L^p} \le \eta^{k+1} \|u_h^0 - \bar{u}_h\|_{L^p} \quad \forall\, k \ge 0.$$

Similarly, under suitable assumptions, given any $\eta \in (0,1)$, there exist $\delta > 0$ and $h_1 > 0$ such that for all $u_h^0 \in U_h$, $\|u_h^0 - \bar{u}_h\|_{L^2} < \delta$, and all $h \in \mathcal{H} \cup \{0\}$, $h \le h_1$, the semismooth Newton method, Algorithm 6.27, generates a sequence that satisfies

$$\|u_h^{k+1} - \bar{u}_h\|_{L^2} \le \eta \|u_h^k - \bar{u}_h\|_{L^2} \le \eta^{k+1} \|u_h^0 - \bar{u}_h\|_{L^2} \quad \forall\, k \ge 0.$$

The first mesh-independence results for semismooth Newton methods had this form and were derived in [106]. We refer to [106] for details. A rigorous proof can also be obtained by straightforward modifications of the proofs given for Theorems 6.33 and 6.32. We do not carry out the proofs here in detail since the results in the sections 6.4.1 and 6.4.2 are deeper than those sketched here.

6.4.4 An Application

In this section we briefly sketch how the mesh-independence results can be applied to control-constrained semilinear elliptic optimal control problems. We use the same setting as in [14, 106] and refer to [106] for the details. Since in [14, 106] bilaterally constrained problems are considered and we want to apply the investigations of [106], there are two options: On the one hand, it can be shown that the developed mesh-independence theory can be extended to bilaterally constrained problems. In this case, the operator N would be $N(z) = -P_{[\alpha,\beta]}(-z)$, where $P_{[\alpha,\beta]}(t) = \min\{\max\{a,t\},b\}$. Alternatively, we can assume that the continuous and discrete optimal controls are pointwise bounded above by a constant $\beta > 0$. The continuous and discrete solutions are then the same if we add the constraint $u \le 2\beta$, and we can then use the error estimates for the bilaterally constrained case. For simplicity, we follow this second approach. A third (more laborious) option would be to develop an analysis similar to [106] based on error estimates for unilateral optimal control problems.

We consider the following problem:

$$\begin{aligned} &\min_{y \in H^1(\Omega), u \in L^2(\Omega)} && J(y,u) = \frac{1}{2}\|y - y_d\|_{L^2}^2 + \frac{\lambda}{2}\|u\|_{L^2}^2 \\ &\text{subject to} && Ay + f(y) = u \text{ in } \Omega, \quad y = 0 \text{ on } \Gamma = \partial\Omega, \\ &&& u \ge \alpha \text{ in } \Omega, \end{aligned} \tag{6.44}$$

where $y_d \in L^4(\Omega)$, $\lambda > 0$, $\alpha \in \mathbb{R}$, and A denotes a second-order elliptic operator of the form

$$[Ay](x) = -\sum_{i,j=1}^{n} (a_{ij}(x) y_{x_i}(x))_{x_j}.$$

The coefficients are supposed to be Lipschitz continuous functions in $\bar{\Omega}$ satisfying the ellipticity condition

$$\sum_{i,j=2}^{n} a_{ij}(x)\xi_i\xi_j \geq \gamma_a \|\xi\|^2 \quad \forall\, (\xi, x) \in \mathbb{R}^n \times \bar{\Omega}, \quad \gamma_a > 0.$$

We assume that the domain $\Omega \subset \mathbb{R}^d$, with $d = 2,3$, is convex and bounded with sufficiently smooth boundary Γ. The function $f : \mathbb{R} \to \mathbb{R}$ is assumed to be of class C^3, f' is nonnegative, and $f(0) = 0$. In addition, we require that there exist constants c_1, c_2 such that

$$|f'''(u)| \leq c_1 + c_2|u|^{\frac{p-6}{2}} \quad \forall\, u \in \mathbb{R},$$

where $p \in [6,\infty)$ for $n = 2$ and $p = 6$ for $n = 3$. Then we have the continuous embedding

$$H_0^1(\Omega) \subset L^p(\Omega).$$

Under the above assumptions one can show that the semilinear elliptic PDE

$$Ay + f(y) = u \quad \text{in } \Omega, \qquad y = 0 \quad \text{on } \Gamma \tag{6.45}$$

admits a unique solution $y(u) \in H_0^1(\Omega)$ for every $u \in L^2(\Omega)$. Further, by classical arguments, one can show that (6.44) admits at least one solution.

We construct a triangulation of Ω as follows, where we consider the case $d = 2$. An extension to the case $d = 3$ is possible. $\Omega \subset \mathbb{R}^2$ is triangulated regularly and quasi-uniformly. Denoting by $\bar{\Omega}_h$ the union of all triangles $T \in \mathcal{T}_h$ and by Ω_h its interior, we assume $\Omega_h \subset \Omega$. Furthermore, all nodes lying on the boundary Γ_h of Ω_h are assumed to lie also on Γ. From this triangulation, we derive a boundary fitted triangulation $\hat{\mathcal{T}}_h$ by replacing the edges of Γ_h by the corresponding boundary curve connecting the two boundary nodes.

U_h is now defined as the space of all functions $u : \Omega \to \mathbb{R}$ that are constant on $\operatorname{int}(\hat{T})$ for all $\hat{T} \in \hat{\mathcal{T}}_h$. U_h is equipped with the $L^2(\Omega)$-norm and identified with its dual space. The state space $Y = H_0^1(\Omega)$ is discretized by the set of all continuous functions $y : \bar{\Omega} \to \mathbb{R}$ such that $y|_T$ is affine linear for all $T \in \mathcal{T}_h$ and $y|_{\bar{\Omega}\setminus\Omega_h} = 0$. Y_h is equipped with the $H^1(\Omega)$-norm. For convenience, let us assume $u_d \in U_h$. The discrete optimal control problem then is given by

$$\begin{aligned} &\min_{y_h \in Y_h, u_h \in U_h} && J(y_h, u_h) \\ &\text{subject to} && \langle Ay_h + f(y_h), v_h\rangle_{H^{-1},H_0^1} = (u_h, v_h)_{L^2} \quad \forall\, v_h \in U_h, \\ &&& u_h \geq \alpha. \end{aligned} \tag{6.46}$$

For any $u_h \in U_h$, the discrete state equation possesses a unique solution $y_h(u_h) \in U_h$. Furthermore, the problem (6.46) possesses at least one solution; see [14].

We can use the solution operator to consider the reduced problem, which is given by

$$\begin{aligned} &\min\nolimits_{u_h \in U_h} && j_h(u_h) \stackrel{\text{def}}{=} J(y_h(u_h), u_h) \\ &\text{subject to} && u_h \geq \alpha. \end{aligned} \tag{6.47}$$

A corresponding reduced formulation can also be derived for the original problem (6.44). It is now standard to write the first-order optimality conditions as complementarity problems involving the derivative $j_h'(u_h)$. It has been shown earlier that the adjoint representation of $j_h'(u_h)$ implies the structure required from F_h in Assumption 6.18 (which is referenced in Assumption 6.19). Following the arguments in [106], Assumption 6.19 (with $\kappa = 1$ and $r = 2$ or $r = p$) as well as Assumptions 6.24 and 6.28 can be verified. We refer to [106] for the details.

Chapter 7

Trust-Region Globalization

So far, we have concentrated on locally convergent Newton-type methods. We now propose a class of trust-region algorithms which are globally convergent and use (projected) Newton steps as candidates for trial steps. We restrict ourselves to the case where the problem is posed in Hilbert space, which, from a practical point of view, is not very restrictive.

To motivate our approach, we consider (1.14) with $U = L^2(\Omega)$ and a continuously (Fréchet) differentiable function $F : U \to U$. Using an MCP/NCP-function ϕ, we reformulate the problem in the form

$$\Phi(u) = 0. \tag{7.1}$$

Let Assumption 5.1 hold with $r = 2$ and some $p, p' \in (2, \infty]$. Then the operator $\Phi : L^p(\Omega) \to L^2(\Omega)$ is semismooth by Theorem 5.4. Alternatively, if F assumes the form $F(u) = \lambda u + G(u)$ and G has the smoothing property of section 4.2, and if $\Phi(u) = u - P_{\mathcal{B}}(u - \lambda^{-1}G(u))$ is chosen, then by Theorem 4.4, $\Phi : L^2(\Omega) \to L^2(\Omega)$ is locally Lipschitz continuous and semismooth.

For globalization, we need a minimization problem whose solutions or critical points correspond to solutions of (7.1). We propose three different approaches to obtain these minimization reformulations.

Most naturally, we can choose the squared residual

$$h(u) = \frac{1}{2}\|\Phi(u)\|_{L^2}^2$$

as the objective function. In fact, if the operator equation (7.1) possesses a solution, then every global solution of h is a solution to $\Phi(u) = 0$ and vice versa. Therefore, (7.1) is equivalent to the minimization problem

$$\underset{u \in L^2(\Omega)}{\text{minimize}} \;\; h(u). \tag{7.2}$$

We will show that, for appropriate choices of ϕ, the function $h(u) = \|\Phi(u)\|_{L^2}^2/2$ is continuously differentiable. This makes (7.2) a C^1 problem posed in the Hilbert space $L^2(\Omega)$.

As was discussed in the context of the projected semismooth Newton method (Algorithm 3.22), it is often desirable that the algorithm stay feasible with respect to a given closed convex set $\mathcal{K} \subset L^p(\Omega)$ which contains the solution $\bar{u} \in L^p(\Omega)$. Usually $\mathcal{K} = \mathcal{B}$ is chosen. We consider sets of the general form $\mathcal{K} = \{a_{\mathcal{K}} \le u \le b_{\mathcal{K}}\}$ with lower and upper bound functions satisfying the conditions (3.46). Then the constrained minimization problem

$$\underset{u \in L^2(\Omega)}{\text{minimize}} \ h(u) \quad \text{subject to} \quad u \in \mathcal{K} \tag{7.3}$$

is equivalent to (7.1) in the sense that every global solution $\bar{u} \in \mathcal{K}$ of (7.3) solves (7.1) and vice versa.

Finally, we come to a third possibility of globalization, which can be used if the VIP is obtained from the first-order necessary optimality conditions of the constrained minimization problem

$$\text{minimize} \ j(u) \quad \text{subject to} \quad u \in \mathcal{B} \tag{7.4}$$

with $\mathcal{B} = \{u \in L^2(\Omega) : a \le u \le b\}$ as in (1.14). Then we can use problem (7.4) itself for the purpose of globalization.

In all three approaches, (7.2), (7.3), and (7.4), we obtain a minimization problem of the form

$$\underset{u \in L^2(\Omega)}{\text{minimize}} \ f(u) \quad \text{subject to} \quad u \in \mathcal{K}. \tag{7.5}$$

For the development and analysis of the trust-region method, rather than working in L^2, we prefer to choose a general Hilbert space setting. This has the advantage of covering also the finite-dimensional case, and many other situations, e.g., the reformulation of mixed problems; see section 5.2. Therefore, in the following we consider the problem

$$\underset{u \in U}{\text{minimize}} \ f(u) \quad \text{subject to} \quad u \in \mathcal{K}, \tag{7.6}$$

where $f : U \to \mathbb{R}$ is a continuously differentiable function that is defined on the Hilbert space U. The feasible set $\mathcal{K} \subset U$ is assumed to be nonempty, closed, and convex. In particular, there exists a unique metric projection

$$P_{\mathcal{K}} : U \to \mathcal{K}, \quad P_{\mathcal{K}}(u) = \underset{v \in \mathcal{K}}{\operatorname{argmin}} \|v - u\|_U.$$

We identify the dual U^* of U with U; i.e., we use $\langle \cdot, \cdot \rangle_{U^*,U} = (\cdot, \cdot)_U$.

Our idea is to use projected semismooth Newton steps as trial step candidates for a trust-region globalization based on (7.6). In general, the presence of the smoothing step in the semismooth Newton method makes it difficult to prove rigorously transition to fast local convergence. There are ways to do this, but the approach would be highly technical, and thus we will prove transition to fast local convergence only for the case where the semismooth Newton method converges superlinearly without a smoothing step. This is justified for two reasons: As we will see in our numerical tests, experience shows that we usually observe fast convergence without incorporating a smoothing step in the algorithm. One reason for this is that a discretization would have to be very fine to resolve functions that yield an excessively big $\|\cdot\|_{L^p}/\|\cdot\|_{L^2}$-ratio. Second, in section 4.2 we developed a reformulation to which the semismooth Newton method is applicable without a smoothing step.

For unconstrained problems, global convergence usually means that the method "converges" to a critical point, i.e., a point $u \in U$ such that $f'(u) = 0$ in the sense that at least

$\liminf_{k\to\infty} \|f'(u_k)\|_U = 0$. In the constrained context, we have to clarify what we mean by a critical point.

Definition 7.1. *We call $u \in U$ a* critical point *of* (7.6) *if*

$$u \in \mathcal{K} \quad \textit{and} \quad (f'(u), v-u)_U \ge 0 \quad \forall\, v \in \mathcal{K}. \tag{7.7}$$

The following result is important.

Lemma 7.2.

(a) *Let u be a local solution of* (7.6)*; more precisely, $u \in \mathcal{K}$ and there exists $\delta > 0$ such that $f(v) \ge f(u)$ for all $v \in (u+\delta B_U)\cap\mathcal{K}$. Then u is a critical point of* (7.6).

(b) *The following statements are equivalent:*

 (i) *u is a critical point of* (7.6).

 (ii) $u - P_{\mathcal{K}}(u - f'(u)) = 0$.

 (iii) $u - P_{\mathcal{K}}(u - tf'(u)) = 0$ *for some $t > 0$.*

 (iv) $u - P_{\mathcal{K}}(u - tf'(u)) = 0$ *for all $t > 0$.*

Proof. (see also [85, sect. 8]).

(a) For every $v \in \mathcal{K}$, there holds $v(t) = u + t(v-u) \in (u+\delta B_U)\cap\mathcal{K}$ for sufficiently small $t > 0$, and thus

$$0 \le [f(v(t)) - f(u)]/t \to (f'(u), v-u)_U \quad \text{as } t \to 0^+.$$

(b) Let $t > 0$ be arbitrary. Condition (7.7) is equivalent to

$$u \in \mathcal{K}, \quad (u - (u - tf'(u)), v-u)_U \ge 0 \quad \forall\, v \in \mathcal{K},$$

which is the same as $u = P_{\mathcal{K}}(u - tf'(u))$. This proves the equivalence of (i)–(iv). □

Next, we introduce the concept of criticality measures.

Definition 7.3. *A continuous function $\chi : \mathcal{K} \to [0,\infty)$ with the property*

$$\chi(u) = 0 \iff u \textit{ is a critical point of problem } (7.6) \tag{7.8}$$

is called a criticality measure *for* (7.6).

Example 7.4. By Lemma 7.2, for every $t > 0$, the function

$$\chi^{P,t}(u) = \|u - P_{\mathcal{K}}(u - tf'(u))\|_U$$

is a criticality measure for (7.6). For $t = 1$, the resulting criticality measure

$$\chi^{P}(u) = \chi^{P,1}(u) = \|u - P_{\mathcal{K}}(u - f'(u))\|_U$$

is the norm of the *projected gradient*. ∎

The algorithm that we present in this chapter uses ideas developed in the author's paper [190] on trust-region methods for finite-dimensional semismooth equations. Other trust-region approaches for the solution of finite-dimensional NCPs and VIPs can be found in, e.g., [126, 133, 130, 131, 167, 203]. Trust-region algorithms for infinite-dimensional constrained optimization problems are investigated in, e.g., [97, 136, 187, 198]. The method we propose allows for nonmonotonicity of the sequence of generated function values. This has proven advantageous to avoid convergence to local but nonglobal solutions of the problem [36, 84, 133, 188, 190, 205].

Before we describe the trust-region algorithm, we show that for appropriate choice of ϕ the function $h(u) = \|\Phi(u)\|_{L^2}^2/2$ is continuously differentiable. We begin with the following result.

Lemma 7.5. *Let $\psi : V \to \mathbb{R}$ be locally Lipschitz continuous on the nonempty open set $V \subset \mathbb{R}^m$. Assume that ψ is continuously differentiable on $V \setminus \psi^{-1}(0)$. Then the function ψ^2 is continuously differentiable on V. Moreover, $(\psi^2)'(x) = 2\psi(x)g$ for all $g \in \partial\psi(x)$ and all $x \in V$.*

The simple proof can be found in [190].

Lemma 7.6. *Let $\psi : \mathbb{R}^m \to \mathbb{R}$ be Lipschitz continuous on $\mathbb{R}^m$ and continuously differentiable on $\mathbb{R}^m \setminus \psi^{-1}(0)$. Further, let $G : U \mapsto L^2(\Omega)^m$ be continuously differentiable. Then the function*

$$h : u \in U \mapsto \frac{1}{2}\|\Psi(u)\|_{L^2(\Omega)^m}^2 \quad \textit{with} \quad \Psi(u)(\omega) = \psi(G(u)(\omega)), \quad \omega \in \Omega,$$

is continuously differentiable with

$$h'(u) = M^*\Psi(u) \quad \forall\, M \in \partial^\circ\Psi(u).$$

Remark 7.7. *Note that $\partial^\circ\Psi(u) \subset \mathcal{L}(U, L^2)$ by Lemma* 3.42.

Proof. Using Lemma 7.5, $\eta = \psi^2/2$ is continuously differentiable with $\eta'(x) = \psi(x)g$ for all $g \in \partial\psi(x)$. The Lipschitz continuity of ψ implies

$$\|\eta'(x)\|_2 = |\psi(x)|\|g\|_2 \le L(|\psi(0)| + |\psi(x) - \psi(0)|) \le L|\psi(0)| + L^2\|x\|_2.$$

Hence, by Proposition A.11, the superposition operator

$$T : w \in L^2(\Omega)^m \mapsto \eta(w) \in L^1(\Omega)^m$$

is continuously differentiable with derivative

$$(T'(w)v)(\omega) = \eta'(w(\omega))v(\omega) = \psi(w(\omega))g^T v(\omega) \quad \forall\, g^T \in \partial\psi(w(\omega)).$$

From this and the chain rule we see that $H : u \in U \mapsto T(G(u)) \in L^1(\Omega)^m$ is continuously differentiable with

$$\begin{aligned}(H'(u)v)(\omega) &= \eta'(G(u)(\omega))(G'(u)v)(\omega)\\ &= \psi(G(u)(\omega))g^T(G'(u)v)(\omega) \quad \forall\, g^T \in \partial\psi(G(u)(\omega)).\end{aligned}$$

Hence,

$$H'(u) = \Psi(u) \cdot M \quad \forall\, M \in \partial^{\circ}\Psi(u).$$

Thus, we see that $h : u \in U \mapsto \int_{\Omega} H(u)(\omega)d\omega$ is continuously differentiable with

$$(h'(u), v)_U = \int_{\Omega} H'(u)(\omega)v(\omega)d\omega = \int_{\Omega} \Psi(u)(\omega)(Mv)(\omega)d\omega = (M^*\Psi(u), v)_U$$

for all $M \in \partial^{\circ}\Psi(u)$. □

Remark 7.8. *The Fischer–Burmeister function ϕ^{FB} meets all requirements of Lemma* 7.6. *Hence, if $F : L^2(\Omega) \to L^2(\Omega)$ is continuously differentiable, then $h(u) = \|\Phi(u)\|_{L^2}^2/2$ with $\Phi(u) = \phi(u, F(u))$ continuously differentiable. The same holds true for the MCP-function $\phi_{[\alpha,\beta]}^{FB}$ defined in* (5.5).

7.1 The Trust-Region Algorithm

We use the continuous differentiability of f to build an at least first-order accurate quadratic model

$$q_k(s) = (g_k, s)_U + \frac{1}{2}(s, B_k s)_U$$

of $f(u_k + s) - f(u_k)$ at the current iterate u_k, where $g_k \overset{\text{def}}{=} f'(u_k) \in U$ is the gradient of f at u_k. The self-adjoint operator $B_k \in \mathcal{L}(U, U)$ can be viewed as an approximation of the Hessian operator of f (if it exists). We stress, however, that the proposed trust-region method is globally convergent for very general choices of B_k, including $B_k = 0$.

In each iteration of the trust-region algorithm, a trial step s_k is computed as an approximate solution of the following.

Trust-Region Subproblem:

$$\text{minimize } q_k(s) \quad \text{subject to} \quad u_k + s \in \mathcal{K}, \ \|s\|_U \le \Delta_k. \tag{7.9}$$

We will assume that the trial steps meet the following two requirements:

Feasibility Condition:

$$u_k + s_k \in \mathcal{K} \quad \text{and} \quad \|s_k\|_U \le \beta_1 \Delta_k, \tag{7.10}$$

Reduction Condition:

$$\text{pred}_k(s_k) \overset{\text{def}}{=} -q_k(s_k) \ge \beta_2 \chi(u_k) \min\{\Delta_k, \chi(u_k)\} \tag{7.11}$$

with constants $\beta_1 \ge 1$ and $\beta_2 > 0$ independent of k. Here, χ is a suitably chosen criticality measure, see Definition 7.3. Usually, the update of the trust-region radius Δ_k is controlled by the ratio of *actual reduction*

$$\text{ared}_k(s) \overset{\text{def}}{=} f(u_k) - f(u_k + s)$$

and *predicted reduction* $\text{pred}_k(s) \overset{\text{def}}{=} -q_k(s)$.

It has been observed [36, 84, 133, 188, 205] that the performance of nonlinear programming algorithms can be significantly improved by using nonmonotone line-search or trust-region techniques. Here, in contrast to the traditional approach, the monotonicity $f(u_{k+1}) \le f(u_k)$ of the function values is not enforced in every iteration. To achieve this, we generalize a nonmonotone trust-region technique that was introduced by the author [190] in the context of finite-dimensional semismooth equations. For this algorithm all global convergence results for monotone, finite-dimensional trust-region methods remain valid. However, the decrease requirement is significantly relaxed. Before we describe this approach and the corresponding reduction ratio $\rho_k(s)$ in detail, we first state the basic trust-region algorithm.

Algorithm 7.9 (trust-region algorithm).

1. Initialization: Choose $\eta_1 \in (0,1)$, $\Delta_{\min} \ge 0$, and a criticality measure χ. Choose $u_0 \in \mathcal{K}$, $\Delta_0 > 0$ such that $\Delta_0 \ge \Delta_{\min}$, and a model Hessian $B_0 \in \mathcal{L}(U,U)$. Choose an integer $m \ge 0$ and fix $\lambda > 0$ with $m\lambda \le 1$ for the computation of ρ_k. Set $k := 0$ and $i := -1$.
2. Compute $\chi_k := \chi(u_k)$. If $\chi_k = 0$, then STOP.
3. Compute a trial step s_k satisfying the conditions (7.10) and (7.11).
4. Compute the reduction ratio $\rho_k := \rho_k(s_k)$ by calling Algorithm 7.11 with $m_k := \min\{i+1, m\}$.
5. Compute the new trust-region radius Δ_{k+1} by invoking Algorithm 7.10.
6. If $\rho_k \le \eta_1$, then reject the step s_k; i.e., set $u_{k+1} := u_k$, $B_{k+1} := B_k$, increment k by 1, and go to step 3.
7. Accept the step: Set $u_{k+1} := u_k + s_k$ and choose a new model Hessian $B_{k+1} \in \mathcal{L}(U,U)$. Set $j_{i+1} := k$, increment k and i by 1 and go to step 2.

The increasing sequence $(j_i)_{i\ge 0}$ enumerates all indices of accepted steps. Moreover,

$$u_k = u_{j_i} \quad \forall\, j_{i-1} < k \le j_i, \quad \forall\, i \ge 1. \tag{7.12}$$

Conversely, if $k \ne j_i$ for all i, then s_k was rejected. In the following we denote the set of all these "successful" indices j_i by $\mathcal{S}$:

$$\mathcal{S} \stackrel{\text{def}}{=} \{j_i : i \ge 0\} = \{k : \text{trial step } s_k \text{ is accepted}\}.$$

Sometimes, accepted steps will also be called successful. We will repeatedly use the fact that

$$\{u_k : k \ge 0\} = \{u_k : k \in \mathcal{S}\}.$$

The trust-region updates are implemented as usual. We deal with two different flavors of update rules simultaneously by introducing a nonnegative parameter $\Delta_{\min}$. We require that after successful steps $\Delta_{k+1} \ge \Delta_{\min}$ holds. If $\Delta_{\min} = 0$ is chosen, this is automatically satisfied. For $\Delta_{\min} > 0$, however, it is an additional feature that allows for special proof techniques.

Algorithm 7.10 (update of the trust-region radius).

$\Delta_{\min} \geq 0$ and $\eta_1 \in (0,1)$ are the constants defined in step 1 of Algorithm 7.9. Let $\eta_1 < \eta_2 < 1$, and let $0 \leq \gamma_0 < \gamma_1 < 1 < \gamma_2$ be fixed.

1. If $\rho_k \leq \eta_1$, then choose
$$\Delta_{k+1} \in (\gamma_0 \Delta_k, \gamma_1 \Delta_k].$$
2. If $\rho_k \in (\eta_1, \eta_2)$, then choose
$$\Delta_{k+1} \in [\gamma_1 \Delta_k, \max\{\Delta_{\min}, \Delta_k\}] \cap [\Delta_{\min}, \infty).$$
3. If $\rho_k \geq \eta_2$, then choose
$$\Delta_{k+1} \in (\Delta_k, \max\{\Delta_{\min}, \gamma_2 \Delta_k\}] \cap [\Delta_{\min}, \infty).$$

We still have to describe how the reduction ratios $\rho_k(s)$ are defined. Here is a detailed description.

Algorithm 7.11 (computation of relaxed reduction ratio).

$\lambda > 0$ with $m\lambda \leq 1$ is the constant defined in step 1 of Algorithm 7.9 and $m_k \in \{0, \ldots, m\}$ is the value passed from Algorithm 7.9.

1. Choose scalars
$$\lambda_{kr} \geq \lambda, \quad r = 0, \ldots, m_k - 1, \quad \sum_{r=0}^{m_k-1} \lambda_{kr} = 1.$$
2. Compute the *relaxed actual reduction* $\mathrm{rared}_k := \mathrm{rared}_k(s_k)$, where
$$\mathrm{rared}_k(s) \stackrel{\text{def}}{=} \begin{cases} \max\left\{ f(u_k), \sum_{r=0}^{m_k-1} \lambda_{kr} f(u_{j_{i-r}}) \right\} - f(u_k + s) & (m_k \geq 1), \\ f(u_k) - f(u_k + s) & (m_k = 0). \end{cases} \tag{7.13}$$
3. Compute the reduction ratio $\rho_k := \rho_k(s_k)$ according to
$$\rho_k(s) \stackrel{\text{def}}{=} \frac{\mathrm{rared}_k(s)}{\mathrm{pred}_k(s)}.$$

Remark 7.12. *At the very beginning of Algorithm* 7.9, *step* 4 *invokes Algorithm* 7.11 *with* $m_k = 0$. *In this case*
$$\mathrm{rared}_k(s) = f(u_k) - f(u_k + s) = \mathrm{ared}_k(s).$$
Furthermore, if $m = 0$ *is chosen, then there always holds* $m_k = 0$ *and thus always* $\mathrm{rared}_k(s) = \mathrm{ared}_k(s)$, *which corresponds to the traditional monotone trust-region method.*

The idea behind the above update rule is the following: Instead of requiring that $f(u_k + s_k)$ be smaller than $f(u_k)$, it is only required that $f(u_k + s_k)$ is either less than $f(u_k)$ or less than the weighted mean of the function values at the last $m_k = \min\{i+1, m\}$

successful iterates. Our approach is a slightly stronger requirement than the straightforward idea to replace ared_k with

$$\text{rared}_k^\infty(s) = \max\{f(u_k), f(u_{j_i}), \dots, f(u_{j_{i-m_k+1}})\} - f(u_k + s).$$

Unfortunately, for this latter choice it does not seem to be possible to establish all the global convergence results that are available for the monotone case. For our approach, however, this is possible without making the theory substantially more difficult. Moreover, we can approximate rared_k^∞ arbitrarily accurately by rared_k if we choose λ sufficiently small; in each iteration select $0 \le r_k < m_k$ satisfying $f(u_{j_{i-r_k}}) = \max_{0 \le r < m_k} f(u_{j_{i-r}})$, and set

$$\lambda_{kr} = \lambda \quad \text{if } r \neq r_k, \qquad \lambda_{kr_k} = 1 - (m_k - 1)\lambda. \tag{7.14}$$

7.2 Global Convergence

For the global convergence analysis we rely on the following.

Assumption 7.13.

(a) *The objective function f is continuously differentiable on an open neighborhood of the nonempty closed convex set $\mathcal{K}$.*

(b) *The function f is bounded below on $\mathcal{K}$.*

(c) *The norms of the model Hessians are uniformly bounded:*

$$\|B_k\|_{U,U} \le C_B \quad \forall\, k.$$

Throughout this section, Assumption 7.13 is required to hold. We first prove an important decrease property of the function values $f(u_k)$.

Lemma 7.14. *Let u_k, s_k, Δ_k, j_i, etc., be generated by Algorithm 7.9. Then for all computed indices $i \ge 1$ the following holds:*

$$f(u_{j_i}) < f(u_0) - \eta_1 \lambda \sum_{r=0}^{i-2} \text{pred}_{j_r}(s_{j_r}) - \eta_1 \text{pred}_{j_{i-1}}(s_{j_{i-1}}) < f(u_0). \tag{7.15}$$

Proof. We will use the short notations $\text{ared}_k = \text{ared}_k(s_k)$, $\text{rared}_k = \text{rared}_k(s_k)$, and $\text{pred}_k = \text{pred}_k(s_k)$. First, let us note that (7.11) implies $\text{pred}_k > 0$ whenever u_k is not critical. Therefore, the second inequality holds.

The proof of the first inequality is by induction. For $i = 1$ we have by (7.12) and using $\rho_{j_0}(s_{j_0}) > \eta_1$

$$f(u_{j_1}) = f(u_{j_0+1}) = f(u_{j_0}) - \text{ared}_{j_0} < f(u_{j_0}) - \eta_1 \text{pred}_{j_0} = f(u_0) - \eta_1 \text{pred}_{j_0}.$$

Now assume that (7.15) holds for $1, \dots, i$.

If $\text{rared}_{j_i} = \text{ared}_{j_i}$ then, using (7.15) and $\lambda \le 1$,

$$\begin{aligned}
f(u_{j_{i+1}}) = f(u_{j_i+1}) &= f(u_{j_i}) - \text{ared}_{j_i} = f(u_{j_i}) - \text{rared}_{j_i} \\
&< f(u_0) - \eta_1\lambda \sum_{r=0}^{i-2} \text{pred}_{j_r} - \eta_1 \text{pred}_{j_{i-1}} - \eta_1 \text{pred}_{j_i} \\
&\le f(u_0) - \eta_1\lambda \sum_{r=0}^{i-1} \text{pred}_{j_r} - \eta_1 \text{pred}_{j_i}.
\end{aligned}$$

If $\text{rared}_{j_i} \neq \text{ared}_{j_i}$ then $\text{rared}_{j_i} > \text{ared}_{j_i}$, and with $q = \min\{i, m-1\}$ we obtain

$$\begin{aligned}
f(u_{j_{i+1}}) = f(u_{j_i+1}) &= \sum_{p=0}^{q} \lambda_{j_i p} f(u_{j_{i-p}}) - \text{rared}_{j_i} \\
&< \sum_{p=0}^{q} \lambda_{j_i p}\left(f(u_0) - \eta_1\lambda \sum_{r=0}^{i-p-2} \text{pred}_{j_r} - \eta_1 \text{pred}_{j_{i-p-1}} \right) - \eta_1 \text{pred}_{j_i}.
\end{aligned}$$

Using $\lambda_{j_i 0} + \cdots + \lambda_{j_i q} = 1$, $\lambda_{j_i p} \ge \lambda$, and

$$\{0, \dots, q\} \times \{0, \dots, i-q-2\} \subset \{(p,r) : 0 \le p \le q,\ 0 \le r \le i-p-2\},$$

we can proceed to

$$\begin{aligned}
f(u_{j_{i+1}}) &< f(u_0) - \eta_1\lambda \sum_{r=0}^{i-q-2} \left(\sum_{p=0}^{q} \lambda_{j_i p} \right) \text{pred}_{j_r} \\
&\quad - \eta_1\lambda \sum_{p=0}^{q} \text{pred}_{j_{i-p-1}} - \eta_1 \text{pred}_{j_i} \\
&\le f(u_0) - \eta_1\lambda \sum_{r=0}^{i-q-2} \text{pred}_{j_r} - \eta_1\lambda \sum_{r=i-q-1}^{i-1} \text{pred}_{j_r} - \eta_1 \text{pred}_{j_i} \\
&= f(u_0) - \eta_1\lambda \sum_{r=0}^{i-1} \text{pred}_{j_r} - \eta_1 \text{pred}_{j_i}. \qquad \square
\end{aligned}$$

Lemma 7.15. *Let u_k, s_k, Δ_k, etc., be generated by Algorithm 7.9. Then for arbitrary $u \in \mathcal{K}$ with $\chi(u) \neq 0$ and $0 < \eta < 1$ there exist $\Delta > 0$ and $\delta > 0$ such that*

$$\rho_k \ge \eta$$

holds whenever $\|u_k - u\|_U \le \delta$ and $\Delta_k \le \Delta$ are satisfied.

Proof. Since $\chi(u) \neq 0$, by continuity there exist $\delta > 0$ and $\varepsilon > 0$ such that $\chi(u_k) \ge \varepsilon$ for all k with $\|u_k - u\|_U \le \delta$. Now, for $0 < \Delta \le \varepsilon$ and every k with $\|u_k - u\|_U \le \delta$ and $0 < \Delta_k \le \Delta$, we obtain from the decrease condition (7.11):

$$\text{pred}_k(s_k) = -q_k(s_k) \ge \beta_2 \chi(u_k) \min\{\Delta_k, \chi(u_k)\} \ge \beta_2 \varepsilon \Delta_k.$$

In particular, by (7.10)

$$\|s_k\|_U \le \beta_1 \Delta_k \le \frac{\beta_1}{\beta_2 \varepsilon} \operatorname{pred}_k(s_k). \tag{7.16}$$

Further, with appropriate $y_k = u_k + \tau_k s_k$, $\tau_k \in [0,1]$, by the mean value theorem

$$\begin{aligned} \operatorname{ared}_k(s_k) &= f(u_k) - f(u_k + s_k) = -(f'(y_k), s_k)_U \\ &= -q_k(s) + (g_k - f'(y_k), s_k)_U + \frac{1}{2}(s_k, B_k s_k)_U \\ &\ge \operatorname{pred}_k(s_k) - \left(\|g_k - f'(y_k)\|_U + \frac{1}{2}\|B_k s_k\|_U \right) \|s_k\|_U. \end{aligned}$$

Since f' is continuous, there exists $\delta' > 0$ such that

$$\|f'(u') - f'(u)\|_U \le (1-\eta)\frac{\beta_2 \varepsilon}{4\beta_1}$$

for all $u' \in \mathcal{K}$ with $\|u' - u\|_U < \delta'$. Further, since $\|B_k\|_{U,U} \le C_B$ by Assumption 7.13 (c), choosing Δ sufficiently small yields

$$\frac{1}{2}\|B_k s_k\|_U \le (1-\eta)\frac{\beta_2 \varepsilon}{2\beta_1}$$

for all k with $\Delta_k \le \Delta$. By reducing Δ and δ, if necessary, such that $\delta + \beta_1 \Delta < \delta'$ we achieve, using (7.10), that for all k with $\|u_k - u\|_U \le \delta$ and $0 < \Delta_k \le \Delta$

$$\|y_k - u\|_U \le \|u_k - u\|_U + \tau_k \|s_k\|_U \le \delta + \beta_1 \Delta < \delta', \quad \|u_k - u\|_U \le \delta < \delta'.$$

Hence, for all these indices k,

$$\|g_k - f'(y_k)\|_U \le \|g_k - f'(u)\|_U + \|f'(u) - f'(y_k)\|_U \le (1-\eta)\frac{\beta_2 \varepsilon}{2\beta_1},$$

and thus by (7.16)

$$\begin{aligned} \left(\|g_k - f'(y_k)\|_U + \frac{1}{2}\|B_k s_k\|_U \right) \|s_k\|_U &\le (1-\eta)\frac{\beta_2 \varepsilon}{\beta_1} \|s_k\|_U \\ &\le (1-\eta)\operatorname{pred}_k(s_k). \end{aligned}$$

This implies that for all these k there holds

$$\begin{aligned} \operatorname{rared}_k(s_k) \ge \operatorname{ared}_k(s_k) &\ge \operatorname{pred}_k(s_k) - \left(\|g_k - f'(y_k)\|_U + \frac{1}{2}\|B_k s_k\|_U \right) \|s_k\|_U \\ &\ge \eta \operatorname{pred}_k(s_k). \end{aligned}$$

The proof is complete. □

Lemma 7.16. *Algorithm 7.9 either terminates after finitely many steps with a critical point u_k of (7.6) or generates an infinite sequence (s_{j_i}) of accepted steps.*

Proof. Assume that Algorithm 7.9 neither terminates nor generates an infinite sequence (s_{j_i}) of accepted steps. Then there exists a smallest index k_0 such that all steps s_k are rejected for $k \geq k_0$. In particular, $u_k = u_{k_0}$, $k \geq k_0$, and the sequence of trust-region radii Δ_k tends to zero as $k \to \infty$, because

$$\Delta_{k_0+j} \leq \gamma_1^j \Delta_{k_0}.$$

Since the algorithm does not terminate, we know that $\chi(u_{k_0}) \neq 0$. But now Lemma 7.15 with $u = u_{k_0}$ yields that s_k is accepted as soon as Δ_k becomes sufficiently small. This contradicts our assumption. Therefore, the assertion of the lemma is true. □

Lemma 7.17. *Assume that Algorithm 7.9 generates infinitely many successful steps s_{j_i} and that there exists $\mathcal{S}' \subset \mathcal{S}$ with*

$$\sum_{k\in\mathcal{S}'} \Delta_k = \infty. \tag{7.17}$$

Then $\liminf_{\mathcal{S}'\ni k\to\infty} \chi(u_k) = 0$.

Proof. Let the assumptions of the lemma hold and assume that the assertion is wrong. Then there exists $\varepsilon > 0$ such that $\chi(u_k) \geq \varepsilon$ for all $k \in \mathcal{S}' \subset \mathcal{S}$. From (7.17) it follows that $\mathcal{S}'$ is not finite. For all $k \in \mathcal{S}'$ there holds by (7.11)

$$\operatorname{pred}_k(s_k) \geq \beta_2 \chi(u_k) \min\{\Delta_k, \chi(u_k)\} \geq \beta_2 \varepsilon \min\{\Delta_k, \varepsilon\}.$$

From this estimate, the fact that f is bounded below on $\mathcal{K}$ (see Assumption 7.13 (b)) and Lemma 7.14 we obtain for all $j \in \mathcal{S}'$, using $\lambda \leq 1$,

$$\begin{aligned} f(u_0) - f(u_j) &> \eta_1 \lambda \sum_{\substack{k\in\mathcal{S}\\ k<j}} \operatorname{pred}_k(s_k) \geq \eta_1 \lambda \sum_{\substack{k\in\mathcal{S}'\\ k<j}} \operatorname{pred}_k(s_k) \\ &\geq \eta_1 \lambda \beta_2 \varepsilon \sum_{\substack{k\in\mathcal{S}'\\ k<j}} \min\{\Delta_k, \varepsilon\} \to \infty \qquad (\text{as } j \to \infty). \end{aligned}$$

This is a contradiction. Therefore, the assumption was wrong, and thus the lemma is proved. □

We now have everything at hand that we need to establish our first global convergence result. It is applicable in the case $\gamma_0 > 0$, $\Delta_{\min} > 0$ and says that accumulation points are critical points of (7.6).

Theorem 7.18. *Let $\gamma_0 > 0$ and $\Delta_{min} > 0$. Assume that Algorithm 7.9 does not terminate after finitely many steps with a critical point u_k of (7.6). Then the algorithm generates infinitely many accepted steps (s_{j_i}). Moreover, every accumulation point of (u_k) is a critical point of (7.6).*

Proof. Suppose that Algorithm 7.9 does not terminate after a finite number of steps. Then according to Lemma 7.16 infinitely many successful steps (s_{j_i}) are generated. Assume that $\bar{u}$ is an accumulation point of (u_k) that is not a critical point of (7.6). Since $\chi(\bar{u}) \neq 0$, invoking Lemma 7.15 with $u = \bar{u}$ yields $\Delta > 0$ and $\delta > 0$ such that $k \in \mathcal{S}$ holds for all k

with $\|u_k - \bar{u}\|_U \le \delta$ and $\Delta_k \le \Delta$. Since $\bar{u}$ is an accumulation point, there exists an infinite increasing sequence $j_i' \in \mathcal{S}$, $i \ge 0$, of indices such that $\|u_{j_i'} - \bar{u}\|_U \le \delta$ and $u_{j_i'} \to \bar{u}$.

If $(j_i' - 1) \in \mathcal{S}$, then $\Delta_{j_i'} \ge \Delta_{\min}$. Otherwise, $s_{j_i'-1}$ was rejected, which, since then $u_{j_i'-1} = u_{j_i'}$, is only possible if $\Delta_{j_i'-1} > \Delta$, and therefore $\Delta_{j_i'} \ge \gamma_0 \Delta_{j_i'-1} > \gamma_0 \Delta$. We conclude that for all i there holds $\Delta_{j_i'} \ge \min\{\Delta_{\min}, \gamma_0 \Delta\}$. Now Lemma 7.17 is applicable with $\mathcal{S}' = \{j_i' : i \ge 0\}$ and yields

$$0 \ne \chi(\bar{u}) = \lim_{i \to \infty} \chi(u_{j_i'}) = \liminf_{i \to \infty} \chi(u_{j_i'}) = 0,$$

where we have used the continuity of χ. This is a contradiction. Therefore, the assumption $\chi(\bar{u}) \ne 0$ was wrong. □

Next, we prove a result that holds also for $\Delta_{\min} = 0$. Moreover, the existence of accumulation points is not required.

Theorem 7.19. *Let $\gamma_0 > 0$ or $\Delta_{min} = 0$ hold. Assume that Algorithm* 7.9 *does not terminate after finitely many steps with a critical point u_k of* (7.6). *Then the algorithm generates infinitely many accepted steps (s_{j_i}). Moreover,*

$$\liminf_{k \to \infty} \chi(u_k) = 0. \tag{7.18}$$

In particular, if u_k converges to $\bar{u}$, then $\bar{u}$ is a critical point of (7.6).

Proof. By Lemma 7.16, infinitely many successful steps (s_{j_i}) are generated. Now assume that (7.18) is wrong; i.e.,

$$\liminf_{k \to \infty} \chi(u_k) > 0. \tag{7.19}$$

Then we obtain from Lemma 7.17 that

$$\sum_{k \in \mathcal{S}} \Delta_k < \infty. \tag{7.20}$$

In particular, (u_{j_i}) is a Cauchy sequence by (7.10) and (7.12). Therefore, (u_k) converges to some limit $\bar{u}$, at which according to (7.19) and the continuity of χ there holds $\chi(\bar{u}) \ne 0$.

Case 1: $\Delta_{\min} > 0$.

Then by assumption also $\gamma_0 > 0$, and Theorem 7.18 yields $\chi(\bar{u}) = 0$, which is a contradiction.

Case 2: $\Delta_{\min} = 0$.

Lemma 7.15 with $u = \bar{u}$ and $\eta = \eta_2$ yields $\Delta > 0$ and $\delta > 0$ such that $k \in \mathcal{S}$ and $\Delta_{k+1} \ge \Delta_k$ hold for all k with $\|u_k - \bar{u}\|_U \le \delta$ and $\Delta_k \le \Delta$. Since $u_k \to \bar{u}$, there exists $k' \ge 0$ with $\|u_k - \bar{u}\|_U \le \delta$ for all $k \ge k'$.

Case 2.1: There exists $k'' \ge k'$ with $\Delta_k \le \Delta$ for all $k \ge k''$.

Then $k \in \mathcal{S}$ and (inductively) $\Delta_k \ge \Delta_{k''}$ for all $k \ge k''$. This contradicts (7.20).

Case 2.2: For infinitely many k there holds $\Delta_k > \Delta$.

By (7.20) there exists $k'' \ge k'$ with $\Delta_{j_i} \le \Delta$ for all $j_i \ge k''$. Now, for each $j_i \ge k''$, there exists an index $k_i > j_i$ such that $\Delta_k \le \Delta$, $j_i \le k < k_i$, and $\Delta_{k_i} > \Delta$. If $k_i \in \mathcal{S}$, set $j_i' = k_i$, thus obtaining $j_i' \in \mathcal{S}$ with $\Delta_{j_i'} > \Delta$. If $k_i \notin \mathcal{S}$, we have $j_i' \stackrel{\text{def}}{=} k_i - 1 \ge j_i \ge k'$, and thus $j_i' \in \mathcal{S}$, since by construction $\Delta_{j_i'} \le \Delta$. Moreover, $\Delta < \Delta_{k_i} \le \gamma_2 \Delta_{j_i'}$ (here $\Delta_{\min} = 0$ is used) implies that $\Delta_{j_i'} > \Delta/\gamma_2$. By this construction, we obtain an infinitely increasing sequence $(j_i') \subset \mathcal{S}$ with $\Delta_{j_i'} > \Delta/\gamma_2$. Again, this yields a contradiction to (7.20).

Therefore, in all cases we obtain a contradiction. Thus, the assumption was wrong and the proof of (7.18) is complete.

Finally, if $u_k \to \bar{u}$, the continuity of χ and (7.18) imply $\chi(\bar{u}) = 0$. Therefore, $\bar{u}$ is a critical point of (7.6). $\square$

The next result shows that under appropriate assumptions the "lim inf" in (7.18) can be replaced by "lim".

Theorem 7.20. *Let $\gamma_0 > 0$ or $\Delta_{min} = 0$ hold. Assume that Algorithm* 7.9 *does not terminate after finitely many steps with a critical point u_k of* (7.6). *Then the algorithm generates infinitely many accepted steps (s_{j_i}). Moreover, if there exists a set $\mathcal{O}$ that contains (u_k) and on which χ is uniformly continuous, then*

$$\lim_{k\to\infty} \chi(u_k) = 0. \tag{7.21}$$

Proof. In view of Theorem 7.19 we only have to prove (7.21). Thus, let us assume that (7.21) is not true. Then there exists $\varepsilon > 0$ such that $\chi(u_k) \ge 2\varepsilon$ for infinitely many $k \in \mathcal{S}$. Since (7.18) holds, we thus can find increasing sequences $(j_i')_{i\ge 0}$ and $(k_i')_{i\ge 0}$ with $j_i' < k_i' < j_{i+1}'$ and

$$\chi(u_{j_i'}) \ge 2\varepsilon, \qquad \chi(u_k) > \varepsilon \quad \forall\, k \in \mathcal{S} \text{ with } j_i' < k < k_i', \qquad \chi(u_{k_i'}) \le \varepsilon.$$

Setting $\mathcal{S}' = \bigcup_{i=0}^{\infty} \mathcal{S}_i'$ with $\mathcal{S}_i' = \{k \in \mathcal{S} : j_i' \le k < k_i'\}$, we have

$$\liminf_{\mathcal{S}' \ni k \to \infty} \chi(u_k) \ge \varepsilon.$$

Therefore, with Lemma 7.17

$$\sum_{k \in \mathcal{S}'} \Delta_k < \infty.$$

In particular, $\sum_{k \in \mathcal{S}_i'} \Delta_k \to 0$ as $i \to \infty$, and thus, using (7.10) and (7.12),

$$\|u_{k_i'} - u_{j_i'}\|_U \le \sum_{k \in \mathcal{S}_i'} \|s_k\|_U \le \beta_1 \sum_{k \in \mathcal{S}_i'} \Delta_k \to 0 \quad (\text{as } i \to \infty).$$

This is a contradiction to the uniform continuity of χ, since

$$\lim_{i\to\infty} (u_{k_i'} - u_{j_i'}) = 0, \quad \text{but} \quad |\chi(u_{k_i'}) - \chi(u_{j_i'})| \ge \varepsilon \quad \forall\, i \ge 0.$$

Therefore, the assumption was wrong and the assertion is proved. $\square$

7.3 Implementable Decrease Conditions

Algorithm 7.9 requires the computation of trial steps that satisfy the conditions (7.10) and (7.11). We now describe how these conditions can be implemented by means of a generalized Cauchy point which is based on the projected gradient path. As a criticality measure we can use any criticality measure χ that is majorized by the projected gradient in the following sense:

$$\theta\chi(u) \le \chi^P(u) \stackrel{\text{def}}{=} \|u - P_{\mathcal{K}}(u - f'(u))\|_U \tag{7.22}$$

with a fixed parameter $\theta > 0$. For $u_k \in \mathcal{K}$ and $t \ge 0$, we introduce the projected gradient path

$$\pi_k(t) = P_{\mathcal{K}}(u_k - tg_k) - u_k$$

and define the **generalized Cauchy point** s_k^c as follows:

$s_k^c = \pi_k(\sigma_k)$, with $\sigma_k \in \{1, 2^{-1}, 2^{-2}, \dots\}$ chosen maximal such that

$$q_k(\pi_k(\sigma_k)) \le \gamma(g_k, \pi_k(\sigma_k))_U, \tag{7.23}$$

$$\|\pi_k(\sigma_k)\|_U \le \Delta_k, \tag{7.24}$$

where $\gamma \in (0,1)$ is a fixed parameter.

Our aim is to show that the following condition ensures that (7.11) is satisfied with a constant β_2 independent of u_k.

Fraction of Cauchy Decrease Condition:

$$\text{pred}_k(s_k) \ge \beta_3 \text{pred}_k(s_k^c), \tag{7.25}$$

where $\beta_3 \in (0,1]$ is fixed.

We first establish several useful properties of the projected gradient path.

Lemma 7.21. *Let $u_k \in \mathcal{K}$. Then for all $t \in (0,1]$ and all $s \ge 1$ holds*

$$\|\pi_k(t)\|_U \le \|\pi_k(st)\|_U \le s\|\pi_k(t)\|_U, \tag{7.26}$$

$$-(g_k, \pi_k(t))_U \ge \frac{1}{t}\|\pi_k(t)\|_U^2 \ge \chi^P(u_k)\|\pi_k(t)\|_U \ge t\chi^P(u_k)^2. \tag{7.27}$$

Proof. The first inequality in (7.26) is well known; see, e.g., [187, Lem. 2]. The second inequality is proved in [34]. For (7.27), we use that

$$(P_{\mathcal{K}}(v) - v, u - P_{\mathcal{K}}(v))_U \ge 0 \quad \forall\, u \in \mathcal{K},\ v \in U, \tag{7.28}$$

since $w = P_{\mathcal{K}}(v)$ minimizes $\|w - v\|_U^2$ on $\mathcal{K}$. We set $v_k(t) = u_k - tg_k$ and derive

$$\begin{aligned} -(tg_k, \pi_k(t))_U &= (\pi_k(t) + [v_k(t) - P_{\mathcal{K}}(v_k(t))], \pi_k(t))_U \\ &= \|\pi_k(t)\|_U^2 + (v_k(t) - P_{\mathcal{K}}(v_k(t)), P_{\mathcal{K}}(v_k(t)) - u_k)_U \ge \|\pi_k(t)\|_U^2, \end{aligned}$$

where we have used (7.28) in the last step. From $\chi^P(u_k) = \pi_k(1)$ and (7.26) follow the remaining assertions. □

This allows us to prove the well-definedness of the generalized Cauchy point.

Lemma 7.22. *For all $u_k \in \mathcal{K}$, the condition* (7.23) *is satisfied whenever*

$$0 < \sigma_k \le \hat{\sigma} \stackrel{\text{def}}{=} \min\left\{1, \frac{2(1-\gamma)}{C_B}\right\}.$$

Furthermore, the condition (7.24) *holds for all $\sigma_k \in (0,1]$ with*

$$\sigma_k \|g_k\|_U \le \Delta_k.$$

Proof. For all $0 < t \le \hat{\sigma}$ there holds by Assumption 7.13 (c) and (7.27) that

$$\begin{aligned} q_k(\pi_k(t)) &= (g_k, \pi_k(t))_U + \frac{1}{2}(\pi_k(t), B_k \pi_k(t))_U \le (g_k, \pi_k(t))_U + \frac{C_B}{2}\|\pi_k(t)\|_U^2 \\ &\le \left(1 - \frac{C_B t}{2}\right)(g_k, \pi_k(t))_U \le \gamma (g_k, \pi_k(t))_U. \end{aligned}$$

Furthermore, (7.24) is met by all $\sigma_k \in (0,1]$ satisfying $\sigma_k \|g_k\|_U \le \Delta_k$, since

$$\|\pi_k(t)\|_U \le t\|g_k\|_U$$

holds for all $t \in [0,1]$; see (7.27). □

Lemma 7.23. *Let s_k satisfy the feasibility condition* (7.10) *and the fraction of Cauchy decrease condition* (7.25). *Then s_k satisfies the reduction condition* (7.11) *for every criticality measure χ verifying* (7.22) *and every*

$$0 < \beta_2 \le \frac{1}{2}\beta_3 \gamma \theta^2 \min\left\{1, \frac{2(1-\gamma)}{C_B}\right\}.$$

Proof. 1. If $\sigma_k = 1$, then by (7.23) and (7.27)

$$\text{pred}_k(s_k^c) = -q_k(\pi_k(\sigma_k)) \ge -\gamma(g_k, \pi_k(1))_U \ge \gamma \chi^P(u_k)^2.$$

2. If $\sigma_k < 1$, then for $\tau_k = 2\sigma_k$ there either holds $\|\pi_k(\tau_k)\|_U > \Delta_k$ or

$$q_k(\pi_k(\tau_k)) > \gamma(g_k, \pi_k(\tau_k))_U.$$

In the second case we must have $\tau_k > \hat{\sigma}$ by Lemma 7.22, and thus, using (7.26),

$$\|\pi_k(\tau_k)\|_U \ge \tau_k \|\pi_k(1)\|_U \ge \hat{\sigma}\chi^P(u_k).$$

Therefore, in both cases,

$$\|\pi_k(\sigma_k)\|_U = \left\|\pi_k\left(\frac{\tau_k}{2}\right)\right\|_U \ge \frac{1}{2}\|\pi_k(\tau_k)\|_U \ge \frac{1}{2}\min\{\hat{\sigma}\chi^P(u_k), \Delta_k\}.$$

Now, we obtain from (7.23) and (7.27) that

$$\begin{aligned} \text{pred}_k(s_k^c) &= -q_k(\pi_k(\sigma_k)) \ge -\gamma(g_k, \pi_k(\sigma_k))_U \ge \gamma \chi^P(u_k)\|\pi_k(\sigma_k)\|_U \\ &\ge \frac{\gamma}{2}\chi^P(u_k)\min\{\hat{\sigma}\chi^P(u_k), \Delta_k\}. \end{aligned}$$

As shown in 1, this also holds for the case $\sigma_k = 1$. The proof is completed by using (7.22) and (7.25). □

Remark 7.24. *Obviously, the generalized Cauchy point s_k^c satisfies* (7.10) *and* (7.25). *Since s_k^c is computed by an Armijo-type projected line search, we thus have an easily implementable way of computing an admissible trial step by choosing $s_k = s_k^c$.*

7.4 Transition to Fast Local Convergence

We now return to the problem of solving the semismooth operator equation

$$\Phi(u) = 0.$$

We assume that every $\bar{u} \in U$ with $\Phi(\bar{u}) = 0$ is a critical point of the minimization problem (7.6). Especially the smoothing step makes it theoretically difficult to prove that close to a regular solution projected semismooth Newton steps satisfy the reduction condition (7.11) (or (7.25)). In order to prevent our discussion from becoming too technical, we avoid the consideration of smoothing steps by assuming that $\Phi : U \to U$ is $\partial\Phi$-semismooth. In the framework of MCPs this is, e.g., satisfied for $U = L^2(\Omega)$ and $\Phi(u) = u - P_{\mathcal{B}}(u - \lambda^{-1}F(u))$ if F has the form $F(u) = \lambda u + G(u)$ and $G : L^2(\Omega) \mapsto L^{p'}(\Omega)$ is locally Lipschitz continuous; see section 4.2.

Therefore, the assumptions of this section are as follows.

Assumption 7.25. *In addition to Assumption* 7.13, *let the following hold:*

(a) *The operator $\Phi : U \to U$ is continuous with generalized differential $\partial\Phi$.*

(b) *The criticality measure χ satisfies*

$$v_k \in \mathcal{K}, \quad \lim_{k\to\infty} \|\Phi(v_k)\|_U = 0 \quad \Longrightarrow \quad \lim_{k\to\infty} \chi(v_k) = 0.$$

Remark 7.26. *Assumption* (b) *implies that every $u \in U$ with $\Phi(u) = 0$ is a critical point of* (7.6).

In order to cover the different variants (7.2)–(7.4) of minimization problems that can be used for globalization of (1.14), we propose the following hybrid method.

Algorithm 7.27 (trust-region projected Newton algorithm).

1. Initialization: Choose $\eta_1 \in (0,1)$, $\Delta_{\min} \geq 0$, $\nu \in (0,1)$, and a criticality measure χ. Choose $u_0 \in \mathcal{K}$, $\Delta_0 > \Delta_{\min}$, and a model Hessian $B_0 \in \mathcal{L}(U,U)$. Choose an integer $m \geq 1$ and fix $\lambda \in (0, 1/m]$ for the computation of ρ_k. Compute $\zeta_{-1} := \|\Phi(u_0)\|_U$ and set $l_{-1} := -1$, $r := -1$, $k := 0$, $i := -1$, and $i^n := -1$.
2. Compute $\chi_k := \chi(u_k)$. If $\chi_k = 0$, then STOP.
3. Compute a model Hessian $B_k \in \mathcal{L}(U,U)$ and a differential $M_k \in \partial\Phi(u_k)$.
4. Try to compute $s_k^{n,1} \in U$ by solving

$$M_k s_k^{n,1} = -\Phi(u_k).$$

If this fails, then go to step 11.
Otherwise, set $s_k^{n,2} := P_{\mathcal{K}}(u_k + s_k^{n,1}) - u_k$.

5. Compute $s_k^n := \min\left\{1, \dfrac{\Delta_k}{\|s_k^{n,2}\|_U}\right\} s_k^{n,2}$ and $\zeta_k := \|\Phi(u_k + s_k^n)\|_U$.

6. If $\zeta_k \le \nu\zeta_{l_r}$, then set $s_k := s_k^n$.
Otherwise, go to step 10.

7. If s_k fails to satisfy (7.11), then go to step 9.

8. Call Algorithm 7.11 with $m_k = \min\{i - i^n, m\}$ to compute $\rho_k := \rho_k(s_k)$. If $\rho_k \le \eta_1$, then go to step 9.
Otherwise, obtain a new trust-region radius Δ_{k+1} by invoking Algorithm 7.10, set $l_{r+1} := k$, increment r by 1 and go to step 15.

9. Set $u_{k+1} := u_k + s_k$, $\Delta_{k+1} := \max\{\Delta_{\min}, \Delta_k\}$, $j_{i+1} := k$, $l_{r+1} := k$, and $i^n := i + 1$.
Increment k, r, and i by 1 and go to step 2.

10. If $s_k = s_k^n$ satisfies (7.11), then set $s_k := s_k^n$ and go to step 12.

11. Compute a trial step s_k satisfying the conditions (7.10) and (7.11).

12. Compute the reduction ratio $\rho_k := \rho_k(s_k)$ by calling Algorithm 7.11 with $m_k = \min\{i - i^n, m\}$.

13. Compute the new trust-region radius Δ_{k+1} by invoking Algorithm 7.10.

14. If $\rho_k \le \eta_1$, then reject the step s_k: Set $u_{k+1} := u_k$, $B_{k+1} := B_k$, and $M_{k+1} := M_k$. If the computation of $s_k^{n,2}$ was successful, then set $s_{k+1}^{n,2} := s_k^{n,2}$, increment k by 1, and go to step 5.
Otherwise, increment k by 1 and go to step 11.

15. Accept the step: Set $u_{k+1} := u_k + s_k$ and $j_{i+1} := k$. Increment k and i by 1 and go to step 2.

In each iteration, a semismooth Newton step $s_k^{n,1}$ for the equation $\Phi(u) = 0$ is computed. This step is projected onto $\mathcal{K}$ and scaled to lie in the trust-region; the resulting step is s_k^n. In step 6 a test is performed to decide whether s_k^n can be accepted right away or not. If the outcome is positive, the step s_k^n is accepted in any case (either in step 9 or, via step 8, in step 15, see below), the index k is stored in l_{r+1}, and r is incremented. Therefore, the sequence $l_0 < l_2 < \cdots$ lists all iterations at which the test in step 6 was successful and, thus, the semismooth Newton step was accepted. The resulting residual $\zeta_{l_r} = \|\Phi(u_{l_r} + s_{l_r}^n)\|_U$ is stored in ζ_{l_r}, and $\zeta_{l_{-1}}$ holds the initial residual $\|\Phi(u_0)\|_U$. The test in step 6 ensures that

$$\zeta_{l_r} \le \nu\zeta_{l_{r-1}} \le \cdots \le \nu^{r+1}\zeta_{l_{-1}} = \nu^{r+1}\|\Phi(u_0)\|_U.$$

After a positive outcome of the test in step 6, it is first checked if the step $s_k = s_k^n$ also passes the "ordinary" (relaxed) reduction-ratio-based acceptance test. This is done to embed the new acceptance criterion as smoothly as possible in the trust-region framework. If $s_k = s_k^n$ satisfies the reduction-ratio-based test, then s_k is treated as every other step that is accepted by the trust-region mechanism. If it does not, the step is nevertheless accepted (in step 9), but now i'' is set to $i+1$, which has the consequence that in the next iteration we have $m_k = 0$, which results in a restart of the rared-nonmonotonicity mechanism. If the test $\zeta_k \le \nu\zeta_{l_r}$ in step 6 fails, then s_k^n is chosen as the "ordinary" trial step if it satisfies the condition (7.11); note that (7.10) is satisfied automatically. Otherwise, a different trial step is computed.

The global convergence result of Theorem 7.19 can now easily be generalized to Algorithm 7.27.

Theorem 7.28. *Let Assumption 7.25 hold and let $\gamma_0 > 0$ or $\Delta_{min} = 0$. Assume that Algorithm 7.27 does not terminate after finitely many steps with a critical point u_k of (7.6). Then the algorithm generates infinitely many accepted steps (s_{j_i}). Moreover,*

$$\liminf_{k\to\infty} \chi(u_k) = 0.$$

In particular, if u_k converges to $\bar{u}$, then $\bar{u}$ is a critical point of (7.6).

Proof. The well-definedness of Algorithm 7.27 follows immediately from the well-definedness of Algorithm 7.9, which was established in Lemma 7.16. Therefore, if Algorithm 7.27 does not terminate finitely, the sequences (s_{j_i}) of accepted steps is infinite. If r remains bounded during the algorithm, i.e., if only finitely many steps s_k^n pass the test in step 6, then Algorithm 7.27 eventually turns into Algorithm 7.9. In fact, if step 9 is never entered, then all accepted steps pass the reduction-ratio-based test and thus Algorithm 7.27 behaves like Algorithm 7.9 from the very beginning. Otherwise, let $k' = j_{i'}$ be the last iteration at which step 9 is entered. Then $\Delta_{k'+1} \ge \Delta_{\min}$ and $i'' = i'+1$ for all $k > k'$. In particular, $m_k = 0$ for all $j_{i'} < k \le j_{i'+1}$. Thus, Algorithm 7.27 behaves like an instance of Algorithm 7.9 started at $u_0 = u_{k'+1}$ with $\Delta_0 = \Delta_{k'+1}$. Hence, the assertion follows from Theorem 7.19. If, on the other hand, $r \to \infty$ during the algorithm, then we have inductively

$$\|\Phi(u_{l_r+1})\|_U = \zeta_{l_r} \le \nu\zeta_{l_{r-1}} \le \cdots \le \nu^{r+1}\|\Phi(u_0)\|_U \to 0 \quad \text{as } r \to \infty.$$

By Assumption 7.25 (b) this implies $\chi(u_{l_r+1}) \to 0$. Since χ is continuous, we see that $u_k \to \bar{u}$ implies that $\bar{u}$ is a critical point of (7.6). □

Remark 7.29. *Various generalizations can be incorporated. For instance, it is possible not to reset m_k to zero after acceptance of s_k^n in step 9. This can be achieved by generalizing Lemma 7.14 along the lines of [196]. Further, we could allow for nonmonotonicity of the residuals ζ_{l_r} in a similar way as for the function values $f(u_{j_i})$.*

We now come to the proof of transition to fast local convergence.

Theorem 7.30. *Let Assumption 7.25 hold and let $\Delta_{min} > 0$. Assume that Algorithm 7.27 generates an infinite sequence (u_k) of iterates that converges to a point $\bar{u} \in U$ with $\Phi(\bar{u}) = 0$. Let Φ be $\partial\Phi$-semismooth at $\bar{u}$ and Lipschitz continuous near $\bar{u}$. Further, assume that M_k is invertible with $\|M_k^{-1}\|_{U,U} \le C_{M^{-1}}$, whenever u_k is sufficiently close to $\bar{u}$. Then (u_k) converges q-superlinearly to $\bar{u}$. If Φ is even α-order semismooth at $\bar{u}$, $0 < \alpha \le 1$, then the q-rate of convergence is at least $1+\alpha$.*

Proof. Using the assumptions, the abstract local convergence result of Theorem 3.24 for projected semismooth Newton methods is applicable with $S_k(u) = u$ and yields

$$\|u_k + s_k^{n,2} - \bar{u}\|_U = o(\|u_k - \bar{u}\|_U) \quad (\text{as } u_k \to \bar{u}). \tag{7.29}$$

Therefore,

$$\|s_k^{n,2}\|_U \le \|u_k - \bar{u}\|_U + \|u_k + s_k^{n,2} - \bar{u}\|_U \le \frac{3}{2}\|u_k - \bar{u}\|_U, \tag{7.30}$$

$$\|s_k^{n,2}\|_U \ge \|u_k - \bar{u}\|_U - \|u_k + s_k^{n,2} - \bar{u}\|_U \ge \frac{1}{2}\|u_k - \bar{u}\|_U \tag{7.31}$$

for all u_k in a neighborhood of $\bar{u}$, and thus

$$\frac{1}{2}\|u_k - \bar{u}\|_U \le \|s_k^{n,2}\|_U \le \|s_k^{n,1}\|_U = \|M_k^{-1}\Phi(u_k)\|_U \le C_{M^{-1}}\|\Phi(u_k)\|_U.$$

We conclude that for u_k near $\bar{u}$ holds

$$\|\Phi(u_k + s_k^{n,2})\|_U \le L\|u_k + s_k^{n,2} - \bar{u}\|_U = o(\|u_k - \bar{u}\|_U) = o(\|\Phi(u_k)\|_U), \tag{7.32}$$

where L is the Lipschitz constant of Φ near $\bar{u}$.

Since $u_k \to \bar{u}$, we see from (7.30) and (7.32) that there exists K with

$$\|s_k^{n,2}\|_U \le \Delta_{\min}, \quad \left\|\Phi\left(u_k + s_k^{n,2}\right)\right\|_U \le \nu\|\Phi(u_k)\|_U \quad \forall\, k \ge K.$$

The mechanism of updating Δ_k implies $\Delta_k \ge \Delta_{\min}$ whenever $k-1 \in \mathcal{S}$. Hence, for all $k \ge K$ with $k-1 \in \mathcal{S}$ we have $s_k^n = s_k^{n,2}$ and thus $\zeta_k \le \nu\|\Phi(u_k)\|_U$.

Now assume that none of the steps s_k^n, $k \ge K$, passes the test in step 6. Then r and thus $\zeta_{l_r} > 0$ remain unchanged for all $k \ge K$. But since $\Phi(u_k) \to 0$ as $k \to \infty$, there exists $k \ge K$ with $k-1 \in \mathcal{S}$ and $\|\Phi(u_k)\|_U \le \zeta_{l_r}$. Thus s_k^n would satisfy the test in step 6, which is a contradiction. Hence, there exists $k' \ge K$ for which $s_{k'}^n$ satisfies the test in step 6 and thus is accepted. Then, in iteration $k = k'+1$, we have $\Delta_k \ge \Delta_{\min}$, $s_k^n = s_k^{n,2}$, and $\zeta_k \le \nu\|\Phi(u_k)\|_U = \nu\zeta_{k'}$, so that s_k^n again passes the test in step 6 and therefore is accepted. Inductively, all steps $s_k^n = s_k^{n,2}$, $k \ge k'$, are accepted. The superlinear convergence now follows from (7.29). If Φ is α-order semismooth, then (7.29) holds with $o(\|u_k - \bar{u}\|_U)$ replaced by $O(\|u_k - \bar{u}\|_U^{1+\alpha})$, and the rate of convergence is thus at least $1+\alpha$. $\square$

The reason why we require convergence $u_k \to \bar{u}$ instead of considering an accumulation point $\bar{u}$ is that, although we can show that $\zeta_k = o(\|\Phi(u_k)\|_U)$ for $k-1 \in \mathcal{S}$ and u_k close to $\bar{u}$, it could be that ζ_{l_r} is so small that nevertheless $\zeta_k > \nu\zeta_{l_r}$.

However, depending on the choice of the objective function f, it is often easy to establish that there exists a constant $C_\Phi > 0$ with

$$\|\Phi(u_k)\|_U \le C_\Phi\|\Phi(u_{l_r})\|_U \quad \forall \text{ iterations } k \text{ and corresponding } r. \tag{7.33}$$

This holds, e.g., for $f(u) = \|\Phi(u)\|_U^2/2$ if the amount of nonmonotonicity of $f(u_{l_r})$ is slightly restricted. If (7.33) holds, we can prove the following more general result.

Theorem 7.31. *Let Assumption 7.25 hold and let $\Delta_{min} > 0$. Assume that Algorithm 7.27 generates an infinite sequence (u_k) of iterates that has an accumulation point $\bar{u} \in U$ with $\Phi(\bar{u}) = 0$. Let Φ be $\partial\Phi$-semismooth at $\bar{u}$ and Lipschitz continuous near $\bar{u}$. Further, assume that M_k is invertible with $\|M_k^{-1}\|_{U,U} \leq C_{M^{-1}}$, whenever u_k is sufficiently close to $\bar{u}$. Finally, assume that (7.33) holds. Then (u_k) converges q-superlinearly to $\bar{u}$. If Φ is even α-order semismooth at $\bar{u}$, $0 < \alpha \leq 1$, then the q-rate of convergence is at least $1+\alpha$.*

Proof. As in the proof of Theorem 7.30 we can show that (7.29) holds. We then can proceed in a way similar to the above to show that there exists $\delta > 0$ such that for all k with $k-1 \in \mathcal{S}$ and $u_k \in \bar{u} + \delta B_U$ holds

$$
\begin{aligned}
&s_k^n = s_k^{n,2}, \quad u_k + s_k^n \in \bar{u} + \delta B_U, \\
&\zeta_k = \|\Phi(u_k + s_k^n)\|_U \leq \frac{\nu}{C_\Phi}\|\Phi(u_k)\|_U \leq \nu\|\Phi(u_{l_r})\|_U = \nu\zeta_{l_r},
\end{aligned}
$$

where we have used (7.33). Let k' be any of those k. Then the step $s_{k'}^n$ satisfies the test in step 6 and hence is accepted. Furthermore, $k = k'+1$ again satisfies $k-1 \in \mathcal{S}$ and $u_k \in \bar{u} + \delta B_U$, so that also s_k^n is accepted. Inductively, s_k^n is accepted for all $k \geq k'$. Superlinear convergence to $\bar{u}$ and convergence with rate $\geq 1+\alpha$ now follow as in the proof of Theorem 7.30. $\square$

Chapter 8

State-Constrained and Related Problems

In this chapter we consider problems that result in complementarity conditions that are *not* posed in an L^p-space, but rather include measures. The prototypes of such problems are optimal control problems with state constraints, on which we mainly focus here. Since the treatment of state constraints is very challenging, we restrict our attention to convex optimization problems to avoid additional complications.

8.1 Problem Setting

We consider the problem

$$\min_{y,u} \ J(y,u) \quad \text{subject to} \quad Ay - Bu = f, \quad y \le b. \tag{8.1}$$

Here, $y \in Y$ is the state and $u \in U$ is the control. For brevity, we will set $Z = Y \times U$ with norm $\|(y,u)\|_Z = (\|y\|_Y^2 + \|u\|_U^2)^{1/2}$.

The constraint $y \le b$ with $b \in Y_0$ is meant pointwise a.e. on Ω, where Y and Y_0 are appropriate function spaces on the domain Ω; see below. We require the following.

Assumption 8.1.

- $\Omega \subset \mathbb{R}^d$ *is a bounded open domain with sufficiently nice boundary.*
- $b \in Y_0$, *where* $Y_0 \hookrightarrow L^2(\Omega)$ *is a Banach space such that* $\max\{0,v\} \in Y_0$ *holds for all* $v \in Y_0$.
- $Y \hookrightarrow Y_0$ *is a reflexive Banach space such that* $Y \hookrightarrow L^2(\Omega)$ *is compact. Furthermore,* U *is a Hilbert space.*
- $A \in \mathcal{L}(Y,W)$ *and* $B \in \mathcal{L}(U,W)$, *with* W *a reflexive Banach space, and* $f \in W$.
- *The bounded linear operator*

$$C : (y,u) \in Y \times U \mapsto \begin{pmatrix} Ay - Bu \\ y \end{pmatrix} \in W \times Y$$

is surjective.

- *Either* $Y_0 = Y$ *(and thus* $b \in Y$*) or there exists* $\tilde{y} \in Y \cap \mathrm{int}_{Y_0}(\{y \in Y_0 : y \le b\})$.
- *The objective function* $J : Y \times U \to \mathbb{R}$ *is twice continuously differentiable.*
- *There exists* $\alpha > 0$ *such that for all* $z_i = (y_i, u_i) \in Z = Y \times U$ *with* $Ay_i - Bu_i = f$, $i = 1,2$, *there holds*

$$\langle J'(z_1) - J'(z_2), z_1 - z_2 \rangle_{Z^*,Z} \ge \alpha \|z_1 - z_2\|_Z^2. \tag{8.2}$$

- $\|J'(z)\|_{Z^*}$ *is bounded on bounded subsets of* $Z = Y \times U$.

For achieving maximum regularity of the multiplier corresponding to the state constraint $y \le b$, we consider this constraint in the space Y_0 rather than Y. The space Y_0 needs to be sufficiently strong such that a suitable constraint qualification holds. To be more precise, we then should write $Ty \le b$ instead of $y \le b$ with $T \in \mathcal{L}(Y, Y_0)$ denoting the injection $y \in Y \mapsto y \in Y_0$. Since, however, T acts like the identity, we will not always write the operator T explicitly.

Lemma 8.2. *The surjectivity of* C *is equivalent to* B *being surjective.*

Proof. Let B be surjective. Then, given any $v \in W$ and $y \in Y$, there exists u with $Bu = Ay - v$, and thus

$$C\begin{pmatrix} y \\ u \end{pmatrix} = \begin{pmatrix} v \\ y \end{pmatrix}.$$

Conversely, let C be surjective and consider any $v \in W$. Then there exist $y \in Y$ and $u \in U$ with $C\binom{y}{u} = \binom{-v}{0}$. In particular, $y = 0$ and $Ay - Bu = -v$, hence $Bu = v$. $\square$

Lemma 8.3. *The surjectivity of* C *and the existence of* $\tilde{y} \in Y$ *with* $\tilde{y} \le b$ *imply that the problem* (8.1) *possesses feasible points. More precisely, for every* $y \in Y$ *with* $y \le b$ *there exists* $u \in U$ *with* $Ay - Bu = f$.

Proof. Let $\hat{y} \in Y$ satisfy $\hat{y} \le b$ (such a $\hat{y}$ exists; e.g., $\hat{y} = b$ if $Y_0 = Y$ or $\hat{y} = \tilde{y}$, otherwise). Solving

$$C\begin{pmatrix} y \\ u \end{pmatrix} = \begin{pmatrix} f \\ \hat{y} \end{pmatrix}$$

results in $(y, u) \in Y \times U$ such that $Ay - Bu = f$ and $y = \hat{y} \le b$. $\square$

For applying optimality theory, we need to show that a constraint qualification is satisfied. We work with Robinson's constraint qualification:

$$0 \in \mathrm{int}\left(\begin{pmatrix} A\bar{y} - B\bar{u} - f \\ T\bar{y} - b \end{pmatrix} + \begin{pmatrix} A & -B \\ T & 0 \end{pmatrix}(Y \times U) - \begin{pmatrix} \{0\} \\ \{h \in Y_0 : h \le 0\} \end{pmatrix}\right) \subset W \times Y_0.$$

Lemma 8.4. *Let Assumption* 8.1 *hold and let* $(\bar{y}, \bar{u})$ *be feasible for* (8.1). *Then Robinson's constraint qualification is satisfied at* $(\bar{y}, \bar{u})$.

Proof. By assumption, either $Y_0 = Y$ or there exists $\tilde{y} \in \mathrm{int}_{Y_0}(\{y \in Y_0 : y \le b\})$.

1. Consider first the case $Y_0 = Y$. Since C is surjective, given an arbitrary $v \in W$ and $s \in Y_0 = Y$, we can find $y \in Y$ and $u \in U$ with $C\binom{y}{u} = \binom{v}{s-\bar{y}+b}$. Thus, using $T = I_Y$,

$$\begin{pmatrix} A\bar{y} - B\bar{u} - f \\ T\bar{y} - b \end{pmatrix} + \begin{pmatrix} A & -B \\ T & 0 \end{pmatrix}\begin{pmatrix} y \\ u \end{pmatrix} = \begin{pmatrix} 0 \\ \bar{y} - b \end{pmatrix} + C\begin{pmatrix} y \\ u \end{pmatrix}$$
$$= \begin{pmatrix} v \\ \bar{y} - b + s - \bar{y} + b \end{pmatrix} = \begin{pmatrix} v \\ s \end{pmatrix}.$$

Thus, Robinson's constraint qualification is satisfied.

2. Now consider the case $Y_0 \ne Y$. Then there exists $\tilde{y} \in Y \cap \mathrm{int}_{Y_0}(\{y \in Y_0 : y \le b\})$. Since C is surjective, given an arbitrary $v \in W$ and setting $y = \tilde{y} - \bar{y} \in Y$, we can find $u \in U$ with $C\binom{y}{u} = \binom{v}{y}$. Hence, there holds

$$\begin{pmatrix} A\bar{y} - B\bar{u} - f \\ T\bar{y} - b \end{pmatrix} + \begin{pmatrix} A & -B \\ T & 0 \end{pmatrix}\begin{pmatrix} y \\ u \end{pmatrix} = \begin{pmatrix} v \\ T\bar{y} - b + T(\tilde{y} - \bar{y}) \end{pmatrix} = \begin{pmatrix} v \\ T\tilde{y} - b \end{pmatrix}.$$

Now, $T\tilde{y} = \tilde{y}$ is an interior point of $\{y \in Y_0 : y \le b\}$. Thus, there exists $\varepsilon > 0$ such that for every $s \in Y_0$ with $\|s\|_{Y_0} < \varepsilon$ we have $T\tilde{y} - s \le b$. Hence, $h := T\tilde{y} - b - s \in Y_0$ satisfies $h \le 0$ and there holds

$$\begin{pmatrix} v \\ s \end{pmatrix} = \begin{pmatrix} A\bar{y} - B\bar{u} - f \\ T\bar{y} - b \end{pmatrix} + \begin{pmatrix} A & -B \\ T & 0 \end{pmatrix}\begin{pmatrix} y \\ u \end{pmatrix} - \begin{pmatrix} 0 \\ h \end{pmatrix}.$$

This shows that Robinson's constraint qualification is satisfied. □

Example 8.5. Consider the optimal control problem

$$\begin{aligned} &\min_{y,u} \ \frac{1}{2}\|y - y_d\|_{L^2}^2 + \frac{\lambda}{2}\|u\|_{L^2}^2 \\ &\text{subject to} \quad \begin{array}{rl} -\Delta y = f + u & \text{in } \Omega, \\ y = 0 & \text{in } \partial\Omega, \\ y \le b & \text{in } \Omega \end{array} \end{aligned}$$

with $y \in Y := H_0^1(\Omega) \cap H^2(\Omega)$ (i.e., the boundary values are included in the choice of Y), $u \in U = L^2(\Omega)$, $f \in W = L^2(\Omega)$, $y_d \in L^2(\Omega)$, $\lambda > 0$, and $b \in Y_0 := C(\bar{\Omega})$, $b|_{\partial\Omega} \ge \nu_0 > 0$. We assume that the open bounded domain $\Omega \subset \mathbb{R}^d$, $1 \le d \le 3$, is sufficiently well shaped (in terms of available regularity results) such that the equation $-\Delta y = v$ possesses a unique weak solution $y \in Y$ for all $v \in L^2(\Omega)$ and that, by the Sobolev embedding theorem, $Y \hookrightarrow C(\bar{\Omega})$. We then can choose $A = -\Delta$ and $B : U = L^2(\Omega) \ni u \mapsto u \in L^2(\Omega) = W$. Since B is surjective, the operator C is also surjective according to Lemma 8.2. Furthermore, b is continuous on the compact set $\bar{\Omega}$, and thus there exists a radius $\varepsilon > 0$ such that $b \ge \frac{3}{4}\nu_0$ on $\{x \in \bar{\Omega} : x = x_0 + s,\ x_0 \in \partial\Omega,\ \|s\|_2 \le \varepsilon\}$. It is now possible to construct a function $\tilde{y} \in Y = H_0^1(\Omega) \cap H^2(\Omega)$ such that $\tilde{y} \le b - \nu_0/4$ on Ω. This function then satisfies $\tilde{y} \in Y \cap \mathrm{int}_{Y_0}(\{y \in Y_0 : y \le b\})$.

We choose the objective function

$$J(y,u) = \frac{1}{2}\|y - y_d\|_{L^2}^2 + \frac{\lambda}{2}\|u\|_{L^2}^2.$$

This functional is continuous and quadratic, hence infinitely F-differentiable. In order to verify the monotonicity assumption (8.2), we consider $(y_i, u_i) \in Y \times U$ with $Ay_i - Bu_i = f$. Using $u_1 - u_2 = A(y_1 - y_2)$, we obtain

$$\begin{aligned}
&\langle J'(y_1,u_1) - J'(y_2,u_2), (y_1,u_1) - (y_2,u_2)\rangle_{Z^*,Z} \\
&\quad = (y_1 - y_2, y_1 - y_2)_{L^2} + \lambda(u_1 - u_2, u_1 - u_2)_{L^2} \\
&\quad = \|y_1 - y_2\|_{L^2}^2 + \frac{\lambda}{2}\|A(y_1 - y_2)\|_{L^2}^2 + \frac{\lambda}{2}\|u_1 - u_2\|_{L^2}^2.
\end{aligned}$$

Since $A \in \mathcal{L}(Y, W) = \mathcal{L}(Y, L^2(\Omega))$ is an isomorphism, there holds for all $y \in Y$

$$\|y\|_Y = \|A^{-1}Ay\|_Y \le \|A^{-1}\|_{W,Y}\|Ay\|_{L^2}$$

and thus

$$\begin{aligned}
&\langle J'(y_1,u_1) - J'(y_2,u_2), (y_1,u_1) - (y_2,u_2)\rangle_{Z^*,Z} \\
&\quad = \|y_1 - y_2\|_{L^2}^2 + \frac{\lambda}{2}\|A(y_1 - y_2)\|_{L^2}^2 + \frac{\lambda}{2}\|u_1 - u_2\|_{L^2}^2 \\
&\quad \ge \|y_1 - y_2\|_{L^2}^2 + \frac{\lambda}{2\|A^{-1}\|_{W,Y}^2}\|y_1 - y_2\|_Y^2 + \frac{\lambda}{2}\|u_1 - u_2\|_{L^2}^2,
\end{aligned}$$

which shows (8.2) with $\alpha = \min\{\lambda/2, \lambda/(2\|A^{-1}\|_{W,Y}^2)\}$. The linearity and continuity of J' implies that $\|J'(y,u)\|_{Y^*\times U^*}$ is bounded on bounded subsets of $Y \times U$. ∎

Example 8.6. We consider the elliptic obstacle problem

$$\min_{y\in Y} \frac{1}{2}\langle Ay, y\rangle_{Y^*,Y} - \langle f, y\rangle_{Y^*,Y} \quad \text{subject to} \quad y \le b \tag{8.3}$$

with $Y := H_0^1(\Omega)$, where $A \in \mathcal{L}(Y, Y^*) = \mathcal{L}(H_0^1(\Omega), H^{-1}(\Omega))$ is a symmetric second-order linear elliptic operator, $b \in Y_0 := Y = H_0^1(\Omega)$, and $f \in Y^* = H^{-1}(\Omega)$. We have chosen $b \in H_0^1(\Omega)$ instead of $b \in H^1(\Omega)$, $b \ge 0$ on $\partial\Omega$, to ease the verification of the constraint qualification. We briefly argue that this is no real restriction. In fact, let $\bar{y}$ be a solution of problem (8.3) with b replaced by $\tilde{b} \in H^1(\Omega)$, $\tilde{b} \ge 0$ on $\partial\Omega$. Let $\varphi \in H_0^1(\Omega) \cap C^\infty(\bar{\Omega})$ be arbitrary with $\varphi \ge 0$ and set $b = \bar{y} + \varphi(\tilde{b} - \bar{y})$.

We show that after replacing $\tilde{b}$ with b, $\bar{y}$ is still a solution of (8.3). First of all, $\varphi \ge 0$ and $\bar{y} \le \tilde{b}$ imply

$$b - \bar{y} = \varphi(\tilde{b} - \bar{y}) \ge 0.$$

Due to convexity (see below), it will be sufficient to show that for all $h \in Y$ with $\bar{y} + h \le b$ there exists $t \in (0,1]$ with $\bar{y} + th \le \tilde{b}$. To this end, given any such h, let $t = \min\{1, 1/\|\varphi\|_{C(\bar{\Omega})}\}$. Then

$$\bar{y} + th = (1-t)\bar{y} + t(\bar{y} + h) \le (1-t)\bar{y} + tb = (1-t)\bar{y} + t\bar{y} + t\varphi(\tilde{b} - \bar{y}) \le \bar{y} + \tilde{b} - \bar{y} = \tilde{b}.$$

Thus, the assertion that $\bar{y}$ remains a solution is shown.

There are now several ways to proceed. One possibility would be to introduce the "state equation" $Ay - u = f$ with an artificial "control" $u \in Y^*$ to mimic the form of (8.1) as much as possible. Alternatively, we can just choose the state equation and the control void (which equivalently could be viewed as choosing $U = \{0\}$, $W = \{0\}$). The objective function

$$J(y) = \frac{1}{2}\langle Ay, y\rangle_{Y^*,Y} - \langle f, y\rangle_{Y^*,Y}$$

then is quadratic and thus infinitely F-differentiable. Also, $J'(y) = Ay - f \in Y^*$ is linear and thus bounded on bounded subsets of Y. For verifying the monotonicity assumption (8.2), we consider $y_i \in Y$, $i = 1,2$. We obtain

$$\langle J'(y_1) - J'(y_2), y_1 - y_2\rangle_{Y^*,Y} = \langle A(y_1 - y_2), y_1 - y_2\rangle_{Y^*,Y} \geq \alpha\|y_1 - y_2\|_Y^2,$$

where $\alpha > 0$ is the coercivity constant of the elliptic operator A.

The surjectivity of $C = I$ is clear. Further, there holds $Y_0 = Y$. ■

We introduce the closed affine subspace

$$Z_f = \{(y,u) : Ay - Bu = f\} \subset Z = Y \times U.$$

Lemma 8.7. *From Assumptions* 8.1 *(especially* (8.2)*) it follows that* J *satisfies*

$$J(z_1) - J(z_2) \geq \langle J'(z_2), z_1 - z_2\rangle_{Z^*,Z} + \frac{\alpha}{2}\|z_1 - z_2\|_Z^2 \quad \forall\, z_1, z_2 \in Z_f. \tag{8.4}$$

In particular, J *is uniformly convex on* Z_f.

Proof. Let $z_1 = (y_1,u_1) \in Z_f$, $z_2 = (y_2,u_2) \in Z_f$, and $z_t = tz_1 + (1-t)z_2$, $0 \leq t \leq 1$. We then have $z_t \in Z_f$ and

$$\langle J'(z_t) - J'(z_2), z_1 - z_2\rangle_{Z^*,Z} \geq t\alpha\|z_1 - z_2\|_{Z^*,Z}^2.$$

In fact, for $t = 0$ this is trivial, since then $z_t = z_2$. For $0 < t \leq 1$ we use $z_t - z_2 = t(z_1 - z_2)$ to obtain

$$\begin{aligned}\langle J'(z_t) - J'(z_2), z_1 - z_2\rangle_{Z^*,Z} &= \frac{1}{t}\langle J'(z_t) - J'(z_2), z_t - z_2\rangle_{Z^*,Z}\\ &\geq \frac{1}{t}\alpha\|z_t - z_2\|_Z^2 = t\alpha\|z_1 - z_2\|_Z^2.\end{aligned}$$

We conclude

$$\begin{aligned}J(z_1) - J(z_2) &= \int_0^1 \langle J'(z_t), z_1 - z_2\rangle_{Z^*,Z}\, dt\\ &= \langle J'(z_2), z_1 - z_2\rangle_{Z^*,Z} + \int_0^1 \langle J'(z_t) - J'(z_2), z_1 - z_2\rangle_{Z^*,Z}\, dt\\ &\geq \langle J'(z_2), z_1 - z_2\rangle_{Z^*,Z} + \int_0^1 t\alpha\|z_1 - z_2\|_Z^2\, dt\\ &= \langle J'(z_2), z_1 - z_2\rangle_{Z^*,Z} + \frac{\alpha}{2}\|z_1 - z_2\|_Z^2.\end{aligned}$$

Therefore, (8.4) is shown. This proves the uniform convexity of J on Z_f. □

Now fix $z_f = (y_f, u_f) \in Z_f$ with $y_f \le b$. Such a feasible point exists by Lemma 8.3. For arbitrary $(y,u) \in Z_f$ we then have

$$J(y,u) - J(y_f,u_f) \ge \langle J'(y_f,u_f),(y,u)-(y_f,u_f)\rangle_{Z^*,Z} + \frac{\alpha}{2}\|(y,u)-(y_f,u_f)\|_Z^2.$$

Hence, there exists $R > 0$ such that

$$J(y,u) \ge \frac{\alpha}{4}\|(y,u)\|_Z^2 \quad \forall\,(y,u) \in Z_f,\ \|(y,u)\|_Z \ge R.$$

Since Z_f is a closed affine subspace of $Y \times U$ and thus the set

$$N_J = \{(y,u) : Ay - Bu = f,\ J(y,u) \le J(y_f,y_f),\ y \le b\}$$

is nonempty (note that $z_f \in N_J$), closed, convex, and bounded, there exists a weakly convergent minimizing sequence $(y_k,u_k) \rightharpoonup (\bar{y},\bar{u}) \in N_J$ for (8.1). Since J is convex and continuous on $Z_f \supset N_J$, it is also weakly sequentially lower semicontinuous so that the weak limit $(\bar{y},\bar{u})$ of the minimizing sequence is a solution of the problem. Since J is uniformly convex on Z_f, the solution $(\bar{y},\bar{u})$ is unique. We thus have proved the following.

Lemma 8.8. *Under Assumption* 8.1, *the problem* (8.1) *possesses a unique solution* $(\bar{y},\bar{u}) \in Y \times U$.

We next consider first-order optimality conditions and the uniqueness of Lagrange multipliers.

Lemma 8.9. *Under Assumption* 8.1, *the following holds:*

(a) *Problem* (8.1) *has a unique solution* $(\bar{y},\bar{u}) \in Y \times U$. *Furthermore, there exist Lagrange multipliers* $\bar{w} \in W^*$, $\bar{\mu} \in Y_0^*$ *such that the following first-order optimality conditions (KKT conditions) are satisfied.*

$$J_y(\bar{y},\bar{u}) + T^*\bar{\mu} + A^*\bar{w} = 0, \tag{8.5}$$
$$J_u(\bar{y},\bar{u}) - B^*\bar{w} = 0, \tag{8.6}$$
$$A\bar{y} - B\bar{u} = f, \tag{8.7}$$
$$\bar{y} \le b, \qquad \langle \bar{\mu}, v\rangle_{Y_0^*,Y_0} \ge 0 \quad \forall\, v \in Y_0,\ v \ge 0, \qquad \langle \bar{\mu}, T\bar{y} - b\rangle_{Y_0^*,Y_0} = 0. \tag{8.8}$$

Furthermore, $(\bar{w}, T^*\bar{\mu}) \in W^* \times Y^*$ *is unique.*

(b) *Let* $(\bar{y},\bar{u},\bar{w},\bar{\mu}) \in Y \times U \times W^* \times Y_0^*$ *satisfy the KKT conditions* (8.5)–(8.8). *Then for all* $(y,u) \in Z_f$ *there holds*

$$\begin{aligned} J(y,u) - J(\bar{y},\bar{u}) - \frac{\alpha}{2}\|(y,u)-(\bar{y},\bar{u})\|_Z^2 &\ge \langle J'(\bar{y},\bar{u}),(y,u)-(\bar{y},\bar{u})\rangle_{Z^*,Z} \\ &= \langle \bar{\mu}, T(\bar{y}-y)\rangle_{Y_0^*,Y_0} = \langle \bar{\mu}, b - Ty\rangle_{Y_0^*,Y_0}. \end{aligned}$$

Furthermore, if $(y,u) \in Z_f$ *satisfies* $y \le b$, *then*

$$J(y,u) - J(\bar{y},\bar{u}) \ge \frac{\alpha}{2}\|(y,u)-(\bar{y},\bar{u})\|_Z^2.$$

In particular, $(\bar{y},\bar{u})$ *is the unique solution of* (8.1).

(c) *If* $(\bar{y},\bar{u},\bar{w},\bar{\mu}) \in Y \times U \times W^* \times Y_0^*$ *satisfies the KKT conditions* (8.5)–(8.8), *then* $(\bar{y},\bar{u},\bar{w},T^*\bar{\mu}) \in Y \times U \times W^* \times Y^*$ *is uniquely determined.*

Remark 8.10. *Note that the first two equations in the KKT conditions can be written as*

$$J'(\bar{y},\bar{u})+C^*\begin{pmatrix}\bar{w}\\ T^*\bar{\mu}\end{pmatrix}=0.$$

Proof. (a) The existence of a unique solution $(\bar{y},\bar{u}) \in Y \times U$ of (8.1) follows from Lemma 8.8. We apply Lemma 8.4 and obtain that Robinson's constraint qualification is satisfied at $(\bar{x},\bar{u})$. Hence, by abstract optimality theory, see [156, 208], there exist $(\bar{w},\bar{\mu}) \in W^* \times Y_0^*$ such that the KKT conditions hold. Since C is surjective, we obtain that C^* is injective and thus, using Remark 8.10, the multiplier $\bar{w} \in W^*$ as well as $T^*\bar{\mu} \in Y^*$ are uniquely determined.

(b) Let $(\bar{y},\bar{u},\bar{w},\bar{\mu})$ satisfy the KKT conditions. Due to the uniform convexity of J on Z_f, we obtain for all $(y,u) \in Z_f$, using (8.4) and the optimality conditions,

$$\begin{aligned}
J(y,u)-J(\bar{y},\bar{u})-\frac{\alpha}{2}\|(y,u)-(\bar{y},\bar{u})\|_Z^2 &\geq \langle J'(\bar{y},\bar{u}),(y,u)-(\bar{y},\bar{u})\rangle_{Z^*,Z}\\
&= \langle T^*\bar{\mu}+A^*\bar{w},\bar{y}-y\rangle_{Y^*,Y}-\langle B^*\bar{w},\bar{u}-u\rangle_{U^*,U}\\
&= \langle \bar{\mu},T(\bar{y}-y)\rangle_{Y_0^*,Y_0}+\langle \bar{w},A(\bar{y}-y)-B(\bar{u}-u)\rangle_{W^*,W}\\
&= \langle \bar{\mu},T(\bar{y}-y)\rangle_{Y_0^*,Y_0}=\langle \bar{\mu},T\bar{y}-b\rangle_{Y_0^*,Y_0}+\langle \bar{\mu},b-Ty\rangle_{Y_0^*,Y_0}=\langle \bar{\mu},b-Ty\rangle_{Y_0^*,Y_0}.
\end{aligned}$$

In the last step, the complementarity condition (8.8) was used. Now, if $Ty=y\leq b$, we obtain from the nonnegativity of $\bar{\mu}$

$$\langle \bar{\mu},b-Ty\rangle_{Y^*,Y}\geq 0.$$

Hence, $(\bar{y},\bar{u})$ is the unique solution of (8.1).

(c) By (a), the solvability of the KKT system follows. Now consider a KKT tuple $(\bar{y},\bar{u},\bar{w},\bar{\mu})$. Then, by (b), $(\bar{y},\bar{u})$ is the unique solution of (8.1) and, by (a), the corresponding tuple $(\bar{w},T^*\bar{\mu})$ is unique. □

8.2 A Regularization Approach

We use the idea of approximating the problem by handling the state constraint by a penalty-barrier term of the form $\frac{1}{\gamma}\int_\Omega \phi(\gamma(y-b))dx$ with $\gamma \in (0,\infty)$ and a suitable continuously differentiable and convex function $\phi:\mathbb{R}\to\mathbb{R}$. We require the following.

Assumption 8.11. *The function* $\phi:\mathbb{R}\to\mathbb{R}$ *is convex, continuously differentiable, and satisfies*

$$\begin{aligned}
&\phi(0)=0, \quad \phi'(0)=\sigma\geq 0, \quad \phi'(t)\geq 0 \quad \forall\, t\in\mathbb{R},\\
&\lim_{t\to\infty}\phi'(t)=+\infty, \quad \lim_{t\to-\infty}\phi'(t)=0.
\end{aligned}$$

Remark 8.12. *We note that the convexity of ϕ implies that ϕ' is monotonically increasing or, equivalently, that ϕ' is monotone; i.e.,*

$$(\phi'(t_1)-\phi'(t_2))(t_1-t_2)\geq 0 \quad \forall\, t_1,t_2\in\mathbb{R}.$$

This follows by adding the inequalities $\phi(t_1)-\phi(t_2)\geq\phi'(t_2)(t_1-t_2)$ and $\phi(t_2)-\phi(t_1)\geq\phi'(t_1)(t_2-t_1)$.

We now consider the regularized problem

$$\min_{y,u}\ J_\gamma(y,u):=J(y,u)+\frac{1}{\gamma}\int_\Omega\phi(\gamma(y-b))dx \quad \text{subject to} \quad Ay-Bu=f. \tag{8.9}$$

Example 8.13. A suitable choice is, e.g., $\phi(t)=(1/2)\max^2\{0,t\}$. We then obtain the regularization term $\frac{1}{2\gamma}\int_\Omega\max^2\{0,\gamma(y-b)\}dx$, which is the Moreau–Yosida regularization [103, 104]. This choice for ϕ is the one found most frequently in the literature, but our more general setting might open the door for new developments. The function $\phi'(t)=\max\{0,t\}$ is nonsmooth, but Lipschitz continuous and piecewise smooth, hence semismooth. Since superposition operators involving ϕ' will occur in the optimality conditions, the Moreau–Yosida regularization results in an optimality system that contains nonsmooth superposition operators, and thus semismooth Newton methods are a suitable choice.

It should be mentioned that the Moreau–Yosida regularization sometimes also includes a shift $\hat{\mu}\in L^2(\Omega)$, $\hat{\mu}\geq 0$, in the penalization functional, i.e.,

$$\frac{1}{2\gamma}\int_\Omega\max^2\{0,\hat{\mu}+\gamma(y-b)\}dx;$$

see [103, 104]. For brevity, we work without shift in the following. ∎

The proposed regularization is a generalization of the Moreau–Yosida regularization, which again is related to augmented Lagrangian methods. For more details on the Moreau–Yosida regularization and its connections to augmented Lagrangian methods, we refer to [103, 104, 119, 121, 122]. The generalization considered here is partially inspired by penalty-barrier multiplier methods [17]. However, we do not include a multiplier update, although this might be possible.

We require the following.

Assumption 8.14. *The operator $Y_0\ni v\mapsto\phi(v)\in L^1(\Omega)$ is continuously F-differentiable with derivative $Y_0\ni h\mapsto\phi'(v)h\in L^1(\Omega)$ at $v\in Y_0$. Furthermore, $\phi'(v)\in L^2(\Omega)$ for all $v\in Y_0$.*

Example 8.15. For the Moreau–Yosida regularization we have $\phi(t)=\max^2\{0,t\}/2$. Since $\phi'(t)=\max\{0,t\}$ is Lipschitz continuous, Proposition A.11 shows that $v\mapsto\phi(v)$ is continuously differentiable from $L^2(\Omega)$ to $L^1(\Omega)$ and there holds $\phi'(v)\in L^2(\Omega)$. ∎

Example 8.16. Consider $Y_0=C(\bar{\Omega})$. Then, by Proposition A.13, if ϕ is k times continuously differentiable, then $C(\bar{\Omega})\ni v\mapsto\phi(v)\in C(\bar{\Omega})$ is k times continuously F-differentiable with rth derivative

$$(h_1,\dots,h_r)\mapsto\phi^{(r)}(v)h_1\cdots h_r. \quad ∎$$

From Assumption 8.14 on ϕ, we have that $\Phi : Y \ni y \mapsto \phi(y) \in L^1(\Omega)$ is continuously differentiable with $\Phi'(y)v = \phi'(y)v$. Hence, the function J_γ is continuously differentiable with

$$\langle J_\gamma'(y,u),(v_y,v_u)\rangle_{Z^*,Z} = \langle J'(y,u),(v_y,v_u)\rangle_{Z^*,Z} + \int_\Omega \phi'(\gamma(y-b))v_y\,dx.$$

Concerning uniform convexity of J_γ, we have the following result.

Lemma 8.17. *Let Assumptions* 8.1 *and* 8.11, *and* 8.14 *hold. Then the function J_γ satisfies for all $z_1, z_2 \in Z_f$*

$$\langle J_\gamma'(z_1) - J_\gamma'(z_2), z_1 - z_2\rangle_{Z^*,Z} \geq \alpha\|z_1 - z_2\|_Z^2$$

and, as a consequence,

$$J_\gamma(z_1) - J_\gamma(z_2) \geq \langle J_\gamma'(z_2), z_1 - z_2\rangle_{Z^*,Z} + \frac{\alpha}{2}\|z_1 - z_2\|_Z^2 \quad \forall\, z_1, z_2 \in Z_f. \tag{8.10}$$

Proof. We will use that, as shown in Remark 8.12, for all $t_1, t_2 \in \mathbb{R}$, there holds

$$(\phi'(t_1) - \phi'(t_2))(t_1 - t_2) \geq 0.$$

Hence, for all $(y_1,u_1),(y_2,u_2) \in Z_f$,

$$\begin{aligned}
&\langle J_\gamma'(y_1,u_1) - J_\gamma'(y_2,u_2),(y_1,u_1)-(y_2,u_2)\rangle_{Z^*,Z}\\
&= \langle J'(y_1,u_1) - J'(y_2,u_2),(y_1,u_1)-(y_2,u_2)\rangle_{Z^*,Z}\\
&\quad + \int_\Omega (\phi'(\gamma(y_1-b)) - \phi'(\gamma(y_2-b)))(y_1-y_2)\,dx\\
&\geq \alpha\|(y_1,u_1)-(y_2,u_2)\|_Z^2,
\end{aligned}$$

where we have used assumption (8.2) and

$$\begin{aligned}
&(\phi'(\gamma(y_1-b)) - \phi'(\gamma(y_2-b)))(y_1-y_2)\\
&= \frac{1}{\gamma}(\phi'(\gamma(y_1-b)) - \phi'(\gamma(y_2-b)))(\gamma(y_1-b) - \gamma(y_2-b)) \geq 0.
\end{aligned}$$

The uniform convexity estimate (8.10) follows by applying Lemma 8.7 to J_γ instead of to J. □

From this, we see as for the original problem that there exists a unique solution (y_γ, u_γ) of the regularized problem (8.9).

The problem (8.9) is equality constrained and the derivative of the constraint, $(A,-B) \in \mathcal{L}(Y \times U, W)$, is by assumption surjective, since it is the first row of the block operator C and C is surjective. The surjectivity of $(A,-B)$ is a constraint qualification. Hence, the solution $(y_\gamma, u_\gamma) \in Y \times U$ of (8.9) satisfies the following KKT conditions.

There exists a Lagrange multiplier $w_\gamma \in W^*$ (the adjoint state) such that

$$\begin{aligned} J_y(y_\gamma, u_\gamma) + \phi'(\gamma(y_\gamma - b)) + A^* w_\gamma &= 0 \\ J_u(y_\gamma, u_\gamma) - B^* w_\gamma &= 0, \\ A y_\gamma - B u_\gamma &= f. \end{aligned}$$

Here and further on, the function $\phi'(\gamma(y_\gamma - b)) \in L^2(\Omega)$ (see Assumption 8.14) is interpreted as an element of Y^*, which is appropriate since $Y \hookrightarrow L^2(\Omega) \hookrightarrow Y^*$ via

$$v \in L^2(\Omega) \mapsto \ell_v \in Y^*, \quad \ell_v(y) = \int_\Omega v y \, dx \quad \forall\, y \in Y.$$

Setting $\mu_\gamma = \phi'(\gamma(y_\gamma - b))$, we obtain the following.

Lemma 8.18. *Let Assumptions* 8.1, 8.11, *and* 8.14 *hold. Then there exist unique Lagrange multipliers $w_\gamma \in W^*$ (the adjoint state) and $\mu_\gamma \in Y^*$ such that the following optimality conditions hold.*

$$J_y(y_\gamma, u_\gamma) + \mu_\gamma + A^* w_\gamma = 0, \tag{8.11}$$

$$J_u(y_\gamma, u_\gamma) - B^* w_\gamma = 0, \tag{8.12}$$

$$A y_\gamma - B u_\gamma = f, \tag{8.13}$$

$$\mu_\gamma - \phi'(\gamma(y_\gamma - b)) = 0. \tag{8.14}$$

Furthermore, there holds $\mu_\gamma \in L^2(\Omega)$.

Remark 8.19. *Using the operator C, we can write* (8.11) *and* (8.12) *as*

$$J'(y_\gamma, u_\gamma) + C^* \begin{pmatrix} w_\gamma \\ \mu_\gamma \end{pmatrix} = 0. \tag{8.15}$$

Proof. The optimality conditions (8.11)–(8.14) follow immediately from the KKT conditions above. The uniqueness of μ_γ is ensured, since the last equation implies $\mu_\gamma = \phi'(\gamma(y_\gamma - b))$. From $\gamma(y_\gamma - b) \in Y_0$ and $\phi'(Y_0) \subset L^2(\Omega)$ we see that $\mu_\gamma \in L^2(\Omega)$. Further, using (8.15) and the fact that C is surjective by assumption, w_γ and μ_γ are uniquely determined (we know this already for μ_γ). □

8.2.1 Convergence of the Path

We now prove the convergence of the path $(y_\gamma, u_\gamma, w_\gamma, \mu_\gamma)$ to $(\bar{y}, \bar{u}, \bar{w}, T^*\bar{\mu})$. We will need the following consequence of the open mapping theorem.

Lemma 8.20. *If $M \in \mathcal{L}(Z, X)$ is a surjective operator between Banach spaces, then there exists a constant $c > 0$ such that $\|x'\|_{X^*} \le c\|M^* x'\|_{Z^*}$ for all $x' \in X^*$.*

Proof. Let $B_Z = \{z \in Z : \|z\|_Z < 1\}$ and $\bar{B}_X = \{x \in X : \|x\|_X \le 1\}$. Consider an arbitrary $x' \in X^*$ with $\|x'\|_{X^*} = 1$. Then there exists $x \in X$ with $\|x\|_X = 1$ and $\langle x', x\rangle_{X^*,X} \ge 1/2$.

The open mapping theorem yields that MB_Z is open in X and thus contains the closed δ-ball $\delta\bar{B}_X$ for suitable $\delta > 0$. Hence, there exists $\hat{z} \in B_Z$ with $M\hat{z} = \delta x$. Therefore,

$$\|M^*x'\|_{Z^*} \geq \langle M^*x', \hat{z}\rangle_{Z^*,Z} = \langle x', M\hat{z}\rangle_{X^*,X} = \delta\langle x', x\rangle_{X^*,X} \geq \delta/2.$$

This shows $\|M^*x'\|_{Z^*} \geq (\delta/2)\|x'\|_{X^*}$ for all $x' \in X^*$. We thus can choose $c = 2/\delta$. □

Theorem 8.21. *Under Assumptions* 8.1, 8.11, *and* 8.14, *the path of unique solutions* $(y_\gamma, u_\gamma, w_\gamma, \mu_\gamma)$ *of* (8.9) *stays in a bounded subset of* $Y \times U \times W^* \times Y^*$ *for all* $\gamma > 0$ *and converges for* $\gamma \to \infty$ *strongly in* $Y \times U \times W^* \times Y^*$ *to the unique tuple* $(\bar{y}, \bar{u}, \bar{w}, T^*\bar{\mu})$, *with* $(\bar{y}, \bar{u}, \bar{w}, \bar{\mu}) \in Y \times U \times W^* \times Y_0^*$ *denoting a KKT tuple of* (8.1).

Proof. We have, using the KKT conditions (8.11)–(8.14) of (8.9),

$$\begin{aligned}
&\langle J'(y_\gamma, u_\gamma), (y_\gamma, u_\gamma) - (\bar{y}, \bar{u})\rangle_{Z^*,Z} \\
&\quad = \langle \mu_\gamma + A^*w_\gamma, \bar{y} - y_\gamma\rangle_{Y^*,Y} - \langle B^*w_\gamma, \bar{u} - u_\gamma\rangle_{U^*,U} \\
&\quad = \langle \mu_\gamma, \bar{y} - y_\gamma\rangle_{Y^*,Y} + \langle w_\gamma, A(\bar{y} - y_\gamma) - B(\bar{u} - u_\gamma)\rangle_{W^*,W} \\
&\quad = \langle \mu_\gamma, \bar{y} - y_\gamma\rangle_{Y^*,Y}.
\end{aligned} \tag{8.16}$$

This implies

$$\begin{aligned}
&\langle J'(y_\gamma, u_\gamma) - J'(\bar{y}, \bar{u}), (y_\gamma, u_\gamma) - (\bar{y}, \bar{u})\rangle_{Z^*,Z} \\
&\quad = \langle J'(\bar{y}, \bar{u}), (\bar{y}, \bar{u}) - (y_\gamma, u_\gamma)\rangle_{Z^*,Z} + \langle \mu_\gamma, \bar{y} - y_\gamma\rangle_{Y^*,Y}.
\end{aligned}$$

Therefore, by (8.2),

$$\alpha\|(y_\gamma, u_\gamma) - (\bar{y}, \bar{u})\|_Z^2 \leq \langle J'(\bar{y}, \bar{u}), (\bar{y}, \bar{u}) - (y_\gamma, u_\gamma)\rangle_{Z^*,Z} + \langle \mu_\gamma, \bar{y} - y_\gamma\rangle_{Y^*,Y}. \tag{8.17}$$

Let $\Omega_\gamma^+ = \{y_\gamma > b\}$. Then, using $\mu_\gamma \geq 0$ and $\bar{y} \leq b$, there follows on Ω_γ^+ that (note that μ_γ is a function)

$$\mu_\gamma(\bar{y} - y_\gamma) \leq \mu_\gamma(b - y_\gamma) \leq 0.$$

On $\Omega_\gamma^- = \{y_\gamma \leq b\}$ we obtain, using once again $\mu_\gamma \geq 0$ and $\bar{y} \leq b$,

$$\mu_\gamma(\bar{y} - y_\gamma) \leq \mu_\gamma(b - y_\gamma) = \phi'(\gamma(y_\gamma - b))(b - y_\gamma) \leq \sigma|y_\gamma - b|.$$

Hence,

$$\begin{aligned}
\langle \mu_\gamma, \bar{y} - y_\gamma\rangle_{Y^*,Y} &\leq \int_{\Omega_\gamma^+} \mu_\gamma(b - y_\gamma)dx + \int_{\Omega_\gamma^-} \mu_\gamma(b - y_\gamma)dx \\
&\leq \int_{\Omega_\gamma^-} \mu_\gamma(b - y_\gamma)dx \leq \sigma\|y_\gamma - b\|_{L^1} \leq c_{1Y_0}\sigma\|y_\gamma - b\|_{Y_0}
\end{aligned} \tag{8.18}$$

with $c_{1Y_0} > 0$ such that $\|\cdot\|_{L^1} \leq c_{1Y_0}\|\cdot\|_{Y_0}$. Therefore,

$$\begin{aligned}
\alpha\|(y_\gamma, u_\gamma) - (\bar{y}, \bar{u})\|_Z^2 &\leq \langle J'(\bar{y}, \bar{u}), (\bar{y}, \bar{u}) - (y_\gamma, u_\gamma)\rangle_{Z^*,Z} + \langle \mu_\gamma, \bar{y} - y_\gamma\rangle_{Y^*,Y} \\
&\leq \langle J'(\bar{y}, \bar{u}), (\bar{y}, \bar{u}) - (y_\gamma, u_\gamma)\rangle_{Z^*,Z} + c_{1Y_0}\sigma\|y_\gamma - b\|_{Y_0}.
\end{aligned}$$

From this and $Y \hookrightarrow Y_0$ we conclude that $(y_\gamma, u_\gamma) \in Y \times U$ is bounded independently of γ. Hence, by the boundedness of J' on bounded sets, see Assumption 8.1, it follows that $\|J'(y_\gamma, u_\gamma)\|_{Z^*}$ is bounded independently of γ. From the surjectivity of C and (8.15) we obtain by applying Lemma 8.20 to C that $(w_\gamma) \subset W^*$ and $(\mu_\gamma) \subset Y^*$ are bounded independently of γ.

Therefore, there exists a sequence $\gamma_k \to \infty$ such that

$$(y_{\gamma_k}, u_{\gamma_k}, w_{\gamma_k}, \mu_{\gamma_k}) \rightharpoonup (\hat{y}, \hat{u}, \hat{w}, \hat{\mu}) \quad \text{in } Y \times U \times W^* \times Y^*.$$

From $(y_{\gamma_k}, u_{\gamma_k}) \in Z_f$ and the weak sequential closedness of the closed affine subspace Z_f, we obtain $(\hat{y}, \hat{u}) \in Z_f$. Furthermore, for all $y \in Y$ with $y \geq 0$ there holds, using $\mu_\gamma \geq 0$,

$$\langle \mu_\gamma, y \rangle_{Y^*,Y} = \int_\Omega \mu_\gamma y \, dx \geq 0$$

and thus, due to weak convergence, $\langle \hat{\mu}, y \rangle_{Y^*,Y} \geq 0$. From the compact embedding $Y \hookrightarrow L^2(\Omega)$ we obtain $y_{\gamma_k} \to \hat{y}$ in $L^2(\Omega)$ (strongly). We use this to prove $\hat{y} \leq b$.

Assume that $\hat{y} \leq b$ does not hold. Then for $\hat{v}^+ = \max\{\hat{y} - b, 0\}$ there holds $\|\hat{v}^+\|_{L^1} > 0$. Setting $v_\gamma^+ = \max\{y_\gamma - b, 0\}$, the sequence $v_{\gamma_k}^+$ converges in $L^1(\Omega)$ (even in $L^2(\Omega)$) to $\hat{v}^+$. On $\Omega_\gamma^- = \{y_\gamma \leq b\}$ we obtain, using the monotonicity of ϕ', $\mu_\gamma \geq 0$, and $\bar{y} \leq b$,

$$\mu_\gamma(\bar{y} - y_\gamma) \leq \mu_\gamma |y_\gamma - \bar{y}| = \phi'(\gamma(y_\gamma - b))|y_\gamma - \bar{y}| \leq \phi'(0)|y_\gamma - \bar{y}| = \sigma |y_\gamma - \bar{y}|.$$

This implies

$$\int_{\Omega_\gamma^-} \mu_\gamma(\bar{y} - y_\gamma) dx \leq \int_{\Omega_\gamma^-} \sigma |y_\gamma - \bar{y}| dx \leq \sigma \|y_\gamma - \bar{y}\|_{L^1} \leq c_{1Y} \sigma \|y_\gamma - \bar{y}\|_Y,$$

where $\|\cdot\|_{L^1} \leq c_{1Y} \|\cdot\|_Y$. Together with (8.17), we obtain on $\Omega_\gamma^+ = \{y_\gamma > b\}$

$$\begin{aligned}\int_{\Omega_\gamma^+} \mu_\gamma(y_\gamma - \bar{y}) dx &\leq \langle J'(\bar{y}, \bar{u}), (\bar{y}, \bar{u}) - (y_\gamma, u_\gamma) \rangle_{Z^*,Z} + \int_{\Omega_\gamma^-} \mu_\gamma(\bar{y} - y_\gamma) dx \\ &\leq \langle J'(\bar{y}, \bar{u}), (\bar{y}, \bar{u}) - (y_\gamma, u_\gamma) \rangle_{Z^*,Z} + \sigma \|y_\gamma - \bar{y}\|_{L^1} \\ &\leq \langle J'(\bar{y}, \bar{u}), (\bar{y}, \bar{u}) - (y_\gamma, u_\gamma) \rangle_{Z^*,Z} + c_{1Y} \sigma \|y_\gamma - \bar{y}\|_Y.\end{aligned} \tag{8.19}$$

The right-hand side is bounded independently of γ. Now, on Ω_γ^+, there holds

$$\mu_\gamma(y_\gamma - \bar{y}) \geq \mu_\gamma(y_\gamma - b) = \phi'(\gamma(y_\gamma - b))(y_\gamma - b) = \phi'(\gamma v_\gamma^+) v_\gamma^+.$$

Choose $\varepsilon > 0$ such that $M_\varepsilon = \{\hat{v}^+ \geq 2\varepsilon\}$ has nonzero measure. Such an ε exists since $\hat{v}^+ \geq 0$, $\|\hat{v}^+\|_{L^1} > 0$ and thus $1_{\{\hat{v}^+ \geq \delta\}} \hat{v}^+$ converges to $\hat{v}^+$ in $L^1(\Omega)$ as $\delta \to 0^+$. Then $v_{\gamma_k}^+ \to \hat{v}^+$ in L^1 implies that for k sufficiently large, there holds $\text{meas}(M_k) \geq \text{meas}(M_\varepsilon)/2$, where $M_k := \{v_{\gamma_k}^+ \geq \varepsilon\}$. Note here that $\text{meas}(M_k) < \text{meas}(M_\varepsilon)/2$ would imply $\|\hat{v}^+|_{M_\varepsilon} - v_{\gamma_k}^+|_{M_\varepsilon}\|_{L^1} \geq \text{meas}(M_\varepsilon \setminus M_k)\varepsilon \geq \varepsilon \, \text{meas}(M_\varepsilon)/2$. On M_k we can estimate

$$\phi'(\gamma_k v_{\gamma_k}^+) v_{\gamma_k}^+ \geq \phi'(\gamma_k \varepsilon)\varepsilon \to \infty \quad (k \to \infty),$$

where we have used $\gamma_k \to \infty$ and $\lim_{t\to\infty} \phi'(t) = \infty$. Hence, using $v_{\gamma_k}^+ = y_{\gamma_k} - b$ on $\Omega_{\gamma_k}^+$, $v_{\gamma_k}^+ = 0$ else, we obtain for large k,

$$\int_{\Omega_{\gamma_k}^+} \mu_{\gamma_k}(y_{\gamma_k} - \bar{y})\,dx = \int_\Omega \phi'\left(\gamma_k v_{\gamma_k}^+\right) v_{\gamma_k}^+\,dx \geq \int_{M_k} \phi'(\gamma_k \varepsilon)\varepsilon\,dx$$
$$\geq \phi'(\gamma_k\varepsilon)\varepsilon \frac{1}{2}\,\mathrm{meas}(M_\varepsilon) \to \infty \quad (k \to \infty).$$

This is a contradiction to the uniform boundedness of the left-hand side. Therefore, $\hat{y} \leq b$ is proved. Together with $(\hat{y},\hat{u}) \in Z_f$ this yields that $(\hat{y},\hat{u})$ is feasible for (8.1).

Next, we show

$$\limsup_{k\to\infty} \langle \mu_{\gamma_k}, \bar{y} - y_{\gamma_k}\rangle_{Y^*,Y} \leq 0. \tag{8.20}$$

We already have shown $\mu_\gamma(\bar{y} - y_\gamma) \leq 0$ on Ω_γ^+. Hence, we can focus our investigation on Ω_γ^-. Setting $v_\gamma^- := \max\{b - y_\gamma, 0\}$, we obtain on Ω_γ^-

$$\mu_\gamma(\bar{y} - y_\gamma) \leq \mu_\gamma(b - y_\gamma) = \phi'(\gamma(y_\gamma - b))(b - y_\gamma) = \phi'(-\gamma v_\gamma^-)v_\gamma^-.$$

Since $(b - y_\gamma)_\gamma$ is bounded in Y_0 and therefore also in L^1 independently of γ, there exists $c_v > 0$ such that $\|v_\gamma^-\|_{L^1} \leq c_v$ for all γ. Thus, for arbitrary $\delta > 0$

$$\begin{aligned}
\langle \mu_\gamma, \bar{y} - y_\gamma\rangle_{Y^*,Y} &\leq \int_{\Omega_\gamma^-} \phi'(-\gamma v_\gamma^-)v_\gamma^-\,dx \\
&= \int_{\{0<v_\gamma^-\leq\delta\}} \phi'(-\gamma v_\gamma^-)v_\gamma^-\,dx + \int_{\{v_\gamma^->\delta\}} \phi'(-\gamma v_\gamma^-)v_\gamma^-\,dx \\
&\leq \int_{\{0<v_\gamma^-\leq\delta\}} \sigma\delta\,dx + \int_{\{v_\gamma^->\delta\}} \phi'(-\gamma\delta)v_\gamma^-\,dx \\
&\leq \mathrm{meas}(\Omega)\sigma\delta + \phi'(-\gamma\delta)c_v \leq 2\,\mathrm{meas}(\Omega)\sigma\delta \quad \forall\, \gamma \geq \hat{\gamma}(\delta),
\end{aligned}$$

where $\hat{\gamma}(\delta)$ is chosen so large that $\phi'(-\hat{\gamma}(\delta)\delta) \leq \mathrm{meas}(\Omega)\sigma\delta/c_v$. This is possible as $\phi'(t) \to 0$ for $t \to -\infty$. Since $\delta > 0$ was arbitrary, (8.20) is proved.

As in (8.16), we obtain from the optimality conditions (8.5)–(8.8) of (8.1)

$$\begin{aligned}
&\langle J'(\bar{y},\bar{u}), (y_\gamma,u_\gamma) - (\bar{y},\bar{u})\rangle_{Z^*,Z} \\
&\quad = \langle T^*\bar{\mu} + A^*\bar{w}, \bar{y} - y_\gamma\rangle_{Y^*,Y} - \langle B^*\bar{w}, \bar{u} - u_\gamma\rangle_{U^*,U} \\
&\quad = \langle T^*\bar{\mu}, \bar{y} - y_\gamma\rangle_{Y^*,Y} + \langle \bar{w}, A(\bar{y} - y_\gamma) - B(\bar{u} - u_\gamma)\rangle_{W^*,W} \\
&\quad = \langle T^*\bar{\mu}, \bar{y} - y_\gamma\rangle_{Y^*,Y}.
\end{aligned} \tag{8.21}$$

Combining this with (8.16) gives

$$\langle J'(y_\gamma,u_\gamma) - J'(\bar{y},\bar{u}), (y_\gamma,u_\gamma) - (\bar{y},\bar{u})\rangle_{Z^*,Z} = \langle \mu_\gamma - T^*\bar{\mu}, \bar{y} - y_\gamma\rangle_{Y^*,Y}.$$

Using the monotonicity of J' on Z_f, we obtain

$$\alpha\|(y_\gamma,u_\gamma) - (\bar{y},\bar{u})\|_Z^2 \leq \langle \mu_\gamma, \bar{y} - y_\gamma\rangle_{Y^*,Y} + \langle T^*\bar{\mu}, y_\gamma - \bar{y}\rangle_{Y^*,Y} =: r_\gamma.$$

Now, since $\langle T^*\bar{\mu}, y_{\gamma_k} - \bar{y}\rangle_{Y^*,Y} \to \langle T^*\bar{\mu}, \hat{y} - \bar{y}\rangle_{Y^*,Y}$, (8.20) yields

$$\begin{aligned}\limsup_{k\to\infty} r_{\gamma_k} &\le \langle T^*\bar{\mu}, \hat{y} - \bar{y}\rangle_{Y^*,Y} = \langle \bar{\mu}, T\hat{y} - b\rangle_{Y_0^*,Y_0} + \langle \bar{\mu}, b - T\bar{y}\rangle_{Y_0^*,Y_0} \\ &= -\langle \bar{\mu}, b - T\hat{y}\rangle_{Y_0^*,Y_0} \le 0,\end{aligned}$$

where we have used (8.8) and $T\hat{y} = \hat{y} \le b$. Therefore,

$$\lim_{k\to\infty} \|(y_{\gamma_k}, u_{\gamma_k}) - (\bar{y}, \bar{u})\|_Z = 0.$$

In particular, $\hat{y} = \bar{y}$ and $\hat{u} = \bar{u}$.

Using the KKT conditions of $(8.9)_{\gamma=\gamma_k}$ and of (8.1), we obtain

$$\begin{aligned} J'(y_{\gamma_k}, u_{\gamma_k}) + C^*\begin{pmatrix} w_{\gamma_k} \\ \mu_{\gamma_k}\end{pmatrix} &= 0, \\ J'(\bar{y}, \bar{u}) + C^*\begin{pmatrix} \bar{w} \\ T^*\bar{\mu}\end{pmatrix} &= 0.\end{aligned}$$

Subtracting the two equations gives

$$\left\| C^*\left(\begin{pmatrix} w_{\gamma_k} \\ \mu_{\gamma_k}\end{pmatrix} - \begin{pmatrix} \bar{w} \\ T^*\bar{\mu}\end{pmatrix}\right)\right\|_{Z^*} = \|J'(\bar{y}, \bar{u}) - J'(y_{\gamma_k}, u_{\gamma_k})\|_{Z^*} \to 0 \quad (k \to \infty),$$

since $J'(y_{\gamma_k}, u_{\gamma_k}) \to J'(\bar{y}, \bar{u})$ in Z^*. By Lemma 8.20 there exists $c > 0$ with

$$\left\|\begin{pmatrix} w_{\gamma_k} \\ \mu_{\gamma_k}\end{pmatrix} - \begin{pmatrix} \bar{w} \\ T^*\bar{\mu}\end{pmatrix}\right\|_{W^*\times Y^*} \le c\left\| C^*\left(\begin{pmatrix} w_{\gamma_k} \\ \mu_{\gamma_k}\end{pmatrix} - \begin{pmatrix} \bar{w} \\ T^*\bar{\mu}\end{pmatrix}\right)\right\|_{Z^*} \to 0 \quad (k \to \infty).$$

We thus have proved $y_{\gamma_k} \to \bar{y}$ in Y, $u_{\gamma_k} \to \bar{u}$ in U, $w_{\gamma_k} \to \bar{w}$ in W^*, and $\mu_{\gamma_k} \to T^*\bar{\mu}$ in Y^*.

Since we considered an arbitrary weakly convergent subsequence of the bounded family $(y_\gamma, u_\gamma, w_\gamma, \mu_\gamma)$ as $\gamma \to \infty$, we conclude that

$$y_\gamma \to \bar{y}, \quad u_\gamma \to \bar{u}, \quad w_\gamma \to \bar{w}, \quad \mu_\gamma \to T^*\bar{\mu} \qquad \text{for } \gamma \to \infty. \qquad \square$$

8.2.2 Hölder Continuity of the Path

We next consider Hölder continuity of the path.

Lemma 8.22. *Let Assumptions* 8.1, 8.11, *and* 8.14 *hold, and assume in addition that ϕ' is Hölder continuous of order $\zeta \in (0,1]$ with rank $L_{\phi'}$. Then the path of solutions*

$$\gamma \in (0,\infty] \mapsto (y_\gamma, u_\gamma) \in Y \times U$$

is Hölder continuous of order ζ.

Proof. With $\gamma_1 \geq \gamma_2 > 0$ there holds, using the KKT conditions of (8.9) with $\gamma = \gamma_{1/2}$

$$\begin{aligned}\alpha\|(y_{\gamma_1},u_{\gamma_1})-(y_{\gamma_2},u_{\gamma_2})\|_Z^2 &\leq \langle J'(y_{\gamma_1},u_{\gamma_1})-J'(y_{\gamma_2},u_{\gamma_2}),(y_{\gamma_1},u_{\gamma_1})-(y_{\gamma_2},u_{\gamma_2})\rangle_{Z^*,Z}\\ &= \langle w_{\gamma_2}-w_{\gamma_1},A(y_{\gamma_1}-y_{\gamma_2})-B(u_{\gamma_1}-u_{\gamma_2})\rangle\\ &\quad+\langle \mu_{\gamma_2}-\mu_{\gamma_1},y_{\gamma_1}-y_{\gamma_2}\rangle_{Y^*,Y}\\ &= \langle \mu_{\gamma_2}-\mu_{\gamma_1},y_{\gamma_1}-y_{\gamma_2}\rangle_{Y^*,Y}.\end{aligned}$$

We calculate

$$\begin{aligned}(\mu_{\gamma_2}-\mu_{\gamma_1})(y_{\gamma_1}-y_{\gamma_2}) &= (\phi'(\gamma_2(y_{\gamma_2}-b))-\phi'(\gamma_1(y_{\gamma_1}-b)))(y_{\gamma_1}-y_{\gamma_2})\\ &= (\phi'(\gamma_2(y_{\gamma_2}-b))-\phi'(\gamma_1(y_{\gamma_2}-b)))(y_{\gamma_1}-y_{\gamma_2})\\ &\quad+(\phi'(\gamma_1(y_{\gamma_2}-b))-\phi'(\gamma_1(y_{\gamma_1}-b)))(y_{\gamma_1}-y_{\gamma_2}).\end{aligned}$$

Now, since ϕ' is increasing, we see that

$$(\phi'(\gamma_1(y_{\gamma_2}-b))-\phi'(\gamma_1(y_{\gamma_1}-b)))(y_{\gamma_1}-y_{\gamma_2}) \leq 0.$$

Hence

$$(\mu_{\gamma_2}-\mu_{\gamma_1})(y_{\gamma_1}-y_{\gamma_2}) \leq (\phi'(\gamma_2(y_{\gamma_2}-b))-\phi'(\gamma_1(y_{\gamma_2}-b)))(y_{\gamma_1}-y_{\gamma_2}).$$

Next, assume that ϕ' is ζ-Hölder continuous with modulus $L_{\phi'}$. Then

$$\begin{aligned}(\mu_{\gamma_2}-\mu_{\gamma_1})(y_{\gamma_1}-y_{\gamma_2}) &\leq (\phi'(\gamma_2(y_{\gamma_2}-b))-\phi'(\gamma_1(y_{\gamma_2}-b)))(y_{\gamma_1}-y_{\gamma_2})\\ &\leq L_{\phi'}|\gamma_2-\gamma_1|^\zeta\,|y_{\gamma_2}-b|^\zeta\,|y_{\gamma_1}-y_{\gamma_2}|.\end{aligned}$$

Therefore,

$$\begin{aligned}\alpha\|(y_{\gamma_1},u_{\gamma_1})-(y_{\gamma_2},u_{\gamma_2})\|_Z^2 &\leq \langle \mu_{\gamma_2}-\mu_{\gamma_1},y_{\gamma_1}-y_{\gamma_2}\rangle_{Y^*,Y}\\ &\leq L_{\phi'}|\gamma_1-\gamma_2|^\zeta\,\|y_{\gamma_2}-b\|_{L^{2\zeta}}^\zeta\,\|y_{\gamma_1}-y_{\gamma_2}\|_{L^2}\\ &\leq \text{const}\,\|y_{\gamma_1}-y_{\gamma_2}\|_Y|\gamma_1-\gamma_2|^\zeta,\end{aligned}$$

where we used the uniform boundedness of y_γ in Y for all $\gamma > 0$ and the embeddings $Y \hookrightarrow L^2(\Omega) \hookrightarrow L^{2\zeta}(\Omega)$. □

8.2.3 Rate of Convergence

Here, we give results on the rate of convergence as $\gamma \to \infty$. Our first aim is to estimate $\|\max\{y_\gamma - b, 0\}\|$ in different norms. The L^∞-norm is especially important. In fact, then we know how much y_γ violates the bound.

We now start our analysis. We make the following assumption.

Assumption 8.23. *There exist constants $c_{\phi 1} > 0$ and $\theta > 0$ such that*

$$\phi'(t) \geq c_{\phi 1} t^\theta \quad \forall\, t \geq 0. \tag{8.22}$$

Lemma 8.24. *Assumptions* 8.14 *and* 8.23 *imply* $Y_0 \hookrightarrow L^{1+\theta}(\Omega)$.

Proof. There holds, for all $t \geq 0$, using $\phi(0) = 0$,

$$\phi(t) = \int_0^t \phi'(\tau)d\tau \geq c_{\phi 1} \int_0^t \tau^\theta \, d\tau = \frac{c_{\phi 1}}{1+\theta} t^{1+\theta}.$$

Now, for $v \in Y_0$ there also holds $-v \in Y_0$ and thus $\phi(\pm v) \in L^1(\Omega)$ by Assumption 8.14. Hence,

$$\int_\Omega \max{}^{1+\theta}\{v,0\}\,dx \leq \frac{1+\theta}{c_{\phi 1}} \int_{\{v \geq 0\}} \phi(v)\,dx \leq \frac{1+\theta}{c_{\phi 1}} \|\phi(v)\|_{L^1},$$
$$\int_\Omega \max{}^{1+\theta}\{-v,0\}\,dx \leq \frac{1+\theta}{c_{\phi 1}} \int_{\{v \leq 0\}} \phi(-v)\,dx \leq \frac{1+\theta}{c_{\phi 1}} \|\phi(-v)\|_{L^1}.$$

This shows $v \in L^{1+\theta}(\Omega)$. ◻

Example 8.25. For the Moreau–Yosida regularization we have $\phi(t) = \frac{1}{2}\max^2\{0,t\}$ and $\phi'(t) = \max\{0,t\}$. Hence, $\phi'(t) = t$ for all $t \geq 0$, and thus Assumption 8.23 holds with $c_{\phi 1} = 1$ and $\theta = 1$. ∎

Lemma 8.26. *Let Assumptions* 8.1, 8.11, 8.14, *and* 8.23 *hold and denote by* $\bar{z} = (\bar{y},\bar{u})$ *and* $z_\gamma = (y_\gamma, u_\gamma)$, $\gamma > 0$, *the unique solutions of* (8.1) *and* (8.9), *respectively. Then* $v_\gamma^+ = \max\{y_\gamma - b, 0\}$ *satisfies the following estimate:*

$$\begin{aligned} \|v_\gamma^+\|_{L^{1+\theta}}^{1+\theta} &\leq \frac{1}{c_{\phi 1}\gamma^\theta}\left(\langle J'(\bar{z}), \bar{z} - z_\gamma\rangle_{Z^*,Z} + \sigma\|y_\gamma - \bar{y}\|_{L^1}\right) \\ &= O(\gamma^{-\theta}\|z_\gamma - \bar{z}\|_Z) = o(\gamma^{-\theta}) \quad (\gamma \to \infty). \end{aligned} \tag{8.23}$$

If, in addition, there exist $r \in [1+\theta, \infty]$, $p \in (rd/(r+d), \infty)$ *(with* $rd/(r+d) = d$ *if* $r = \infty$*) such that* $v_\gamma^+ = \max\{y_\gamma - b, 0\} \in Y_0$ *is bounded in* $W^{1,p}(\Omega)$ *independently of* γ, *then with* $\tau = \frac{pd(r-1-\theta)}{prd+(1+\theta)r(p-d)}$ *and a suitable constant* $C > 0$, *there holds*

$$\begin{aligned} \|v_\gamma^+\|_{L^r} &\leq C\|v_\gamma^+\|_{W^{1,p}}^\tau \|v_\gamma^+\|_{L^{1+\theta}}^{1-\tau} \\ &= O\left(\gamma^{-\frac{\theta(1-\tau)}{1+\theta}}\|z_\gamma - \bar{z}\|_Z^{\frac{1-\tau}{1+\theta}}\right) = o\left(\gamma^{-\frac{\theta(1-\tau)}{1+\theta}}\right) \quad (\gamma \to \infty). \end{aligned} \tag{8.24}$$

Before we prove this Lemma, we make two remarks.

Remark 8.27. *It is important to point out that the order of* $\|v_\gamma^+\|$ *in terms of* γ *will be improved later since, based on the above estimates, we will be able to show an order in terms of* γ *for the term* $\|z_\gamma - \bar{z}\|_Z$.

Remark 8.28. *There are situations where the problem setting allows us to derive better estimates than* (8.23)*; see Example* 8.39. *As we will show in Theorem* 8.40, *this improved knowledge then can be used to derive better order estimates for* $\|z_\gamma - \bar{z}\|_Z$ *than the one obtained in Theorem* 8.35 *on the basis of* (8.23).

Proof. Setting $\Omega_\gamma^+ = \{y_\gamma > b\}$, $v_\gamma^+ = \max\{y_\gamma - b, 0\}$, $\mu_\gamma = \phi'(\gamma(y_\gamma - b))$, and using $\bar{y} \le b$, we obtain from (8.19)

$$\begin{aligned}\int_\Omega \mu_\gamma v_\gamma^+ \, dx &= \int_{\Omega_\gamma^+} \mu_\gamma (y_\gamma - b) \, dx \le \int_{\Omega_\gamma^+} \mu_\gamma (y_\gamma - \bar{y}) \, dx \\ &\le \langle J'(\bar{y},\bar{u}), (\bar{y},\bar{u}) - (y_\gamma, u_\gamma)\rangle_{Z^*,Z} + \sigma \|y_\gamma - \bar{y}\|_{L^1}.\end{aligned}$$

The right-hand side converges to zero as $\gamma \to \infty$, since $(y_\gamma, u_\gamma) \to (\bar{y},\bar{u})$ in $Y \times U$. Furthermore,

$$\int_\Omega \mu_\gamma v_\gamma^+ \, dx = \int_\Omega \phi'(\gamma v_\gamma^+) v_\gamma^+ \, dx \ge c_{\phi 1} \int_\Omega \gamma^\theta (v_\gamma^+)^{1+\theta} \, dx = c_{\phi 1} \gamma^\theta \|v_\gamma^+\|_{L^{1+\theta}}^{1+\theta}.$$

Therefore,

$$\begin{aligned}\|v_\gamma^+\|_{L^{1+\theta}}^{1+\theta} &\le \frac{1}{c_{\phi 1}\gamma^\theta} \int_\Omega \mu_\gamma v_\gamma^+ \, dx \\ &\le \frac{1}{c_{\phi 1}\gamma^\theta} \left(\langle J'(\bar{y},\bar{u}), (\bar{y},\bar{u}) - (y_\gamma, u_\gamma)\rangle_{Z^*,Z} + \sigma \|y_\gamma - \bar{y}\|_{L^1} \right) \\ &= O(\gamma^{-\theta} \|(y_\gamma, u_\gamma) - (\bar{y},\bar{u})\|_{Y \times U}) = o(\gamma^{-\theta}) \quad (\gamma \to \infty).\end{aligned}$$

The last step follows from $(y_\gamma, u_\gamma) \to (\bar{y},\bar{u})$ in $Y \times U$. This proves (8.23).

Now let the additional assumptions hold. Then $\|v_\gamma^+\|_{W^{1,p}}$ is bounded independently of γ. By our assumptions, there holds $r = \infty$ and $p > d$ or $1 + \theta \le r < \infty$ and $p > rd/(r+d)$. We next use the following interpolation inequality [2, 164]. There holds for $1 \le q \le r$, $1 < p < \infty$,

$$\|u\|_{L^r} \le K \|u\|_{W^{1,p}}^\tau \|u\|_{L^q}^{1-\tau}$$

with $0 \le \tau \le 1$ satisfying

$$\frac{1}{r} = \tau \left(\frac{1}{p} - \frac{1}{d} \right) + (1-\tau)\frac{1}{q}.$$

We obtain

$$\left(\frac{1}{q} + \frac{1}{d} - \frac{1}{p} \right) \tau = \frac{1}{q} - \frac{1}{r}.$$

Using $p > rd/(r+d)$ yields

$$\frac{1}{q} + \frac{1}{d} - \frac{1}{p} > \frac{1}{q} + \frac{1}{d} - \frac{r+d}{rd} = \frac{1}{q} - \frac{1}{r} \ge 0,$$

and solving for τ yields

$$\tau = \frac{pd(r-q)}{prd + qr(p-d)} \in [0,1).$$

We use this inequality for the choice $q = 1 + \theta$. Then there holds

$$\|u\|_{L^r} \le K \|u\|_{W^{1,p}}^\tau \|u\|_{L^{1+\theta}}^{1-\tau},$$

where

$$\tau = \frac{pd(r-1-\theta)}{prd+(1+\theta)r(p-d)} \in [0,1).$$

Thus, with this choice of τ, we obtain, using (8.23),

$$\begin{aligned}\|v_\gamma^+\|_{L^r} &\le C\|v_\gamma^+\|_{W^{1,p}}^\tau \|v_\gamma^+\|_{L^{1+\theta}}^{1-\tau} \\ &= O\left(\gamma^{-\frac{\theta(1-\tau)}{1+\theta}} \|(y_\gamma,u_\gamma)-(\bar{y},\bar{u})\|_{Y\times U}^{\frac{1-\tau}{1+\theta}}\right) \\ &= o\left(\gamma^{-\frac{\theta(1-\tau)}{1+\theta}}\right) \quad (\gamma\to\infty). \qquad \square\end{aligned}$$

Next, we derive an order of convergence for $\|(y_\gamma,u_\gamma)-(\bar{y},\bar{u})\|_Z$ in terms of γ. We need two further assumptions, the following one and Assumption 8.33 below.

Assumption 8.29. *There exist $r \in [1+\theta,\infty]$, $p \in (rd/(r+d),\infty)$ (with $rd/(r+d) = d$ if $r=\infty$), and $C_{r,p}(\bar{\mu}) > 0$ such that $v_\gamma^+ = \max\{y_\gamma - b, 0\} \in Y_0$ is bounded in $W^{1,p}(\Omega)$ independently of γ and*

$$\langle \bar{\mu}, v_\gamma^+\rangle_{Y_0^*,Y_0} \le C_{r,p}(\bar{\mu})\|v_\gamma^+\|_{L^r}. \tag{8.25}$$

Remark 8.30. *The choice of r and p is as in Lemma 8.26 and allows us to estimate the L^r-norm in terms of the $W^{1,p}$- and $L^{1+\theta}$-norm via the interpolation inequality as done in the proof of Lemma 8.26. In particular, there holds $W^{1,p}(\Omega) \hookrightarrow L^r(\Omega)$.*

The smaller r can be chosen, the better the regularity of $\bar{\mu}$ is. On the other hand, larger values of p correspond to better regularity of v_γ^+.

Lemma 8.31. *Let Assumptions 8.1, 8.11, and 8.14 hold and suppose that there exists $p \in (d,\infty)$ such that $Y \hookrightarrow W^{1,p}(\Omega)$ and $b \in Y_0 \cap W^{1,p}(\Omega)$. Further, let $\bar{\mu}$ satisfy $\bar{\mu} \in C(\bar{\Omega})^*$. Then Assumption 8.29 holds for this choice of p and $r=\infty$.*

Proof. By Theorem 8.21, (y_γ) is uniformly bounded in Y. Thus, from

$$\|v_\gamma^+\|_{W^{1,p}} = \|\max\{y_\gamma - b, 0\}\|_{W^{1,p}} \le \|y_\gamma - b\|_{W^{1,p}} \le \text{const}(\|y_\gamma\|_Y + \|b\|_{W^{1,p}})$$

we obtain that $\|v_\gamma^+\|_{W^{1,p}}$ is bounded independently of γ. Furthermore, since $p>d$ implies $W^{1,p}(\Omega) \hookrightarrow C(\bar{\Omega})$, the requirement $\bar{\mu} \in C(\bar{\Omega})^*$ yields

$$\langle \bar{\mu}, v_\gamma^+\rangle_{Y_0^*,Y_0} = \langle \bar{\mu}, v_\gamma^+\rangle_{C(\bar{\Omega})^*,C(\bar{\Omega})} \le \|\bar{\mu}\|_{C(\bar{\Omega})^*}\|v_\gamma^+\|_{C(\bar{\Omega})} = \|\bar{\mu}\|_{C(\bar{\Omega})^*}\|v_\gamma^+\|_{L^\infty},$$

and thus (8.25) holds true for $r=\infty$. $\square$

Example 8.32. We consider the optimal control problem of Example 8.5 with $d \in \{1,2,3\}$. The corresponding choice of spaces is $Y = H_0^1(\Omega)\cap H^2(\Omega)$, $Y_0 = C(\bar{\Omega})$. By the embedding theorems, with $p \in (d,\infty)$ for $d \le 2$ and $p \in (3,6]$ for $d=3$, there holds $Y \hookrightarrow W^{1,p}(\Omega) \hookrightarrow Y_0$. ■

Assumption 8.33. *There exist constants $c_{\phi 2} > 0$ and $\kappa \in [0,1)$ such that*

$$\phi'(t) \le c_{\phi 2}|t|^{\kappa-1} \quad \forall\, t < 0. \tag{8.26}$$

Example 8.34. For the Moreau–Yosida regularization we have $\phi(t) = (1/2)\max^2\{0,t\}$ and $\phi'(t) = \max\{0,t\}$. Hence, $\phi'(t) = 0$ for all $t < 0$ and thus Assumption 8.33 holds for all $c_{\phi 2} > 0$ and all $\kappa \in [0,1)$. ∎

Theorem 8.35. *Let Assumptions* 8.1, 8.11, 8.14, 8.23, 8.29, *and* 8.33 *hold and denote by* $\bar{z} = (\bar{y},\bar{u})$ *and* $z_\gamma = (y_\gamma, u_\gamma)$, $\gamma > 0$, *the unique solutions of* (8.1) *and* (8.9), *respectively. Then with* $\tau = \frac{pd(r-1-\theta)}{prd+(1+\theta)r(p-d)}$ *and a suitable constant* $C > 0$ *there holds*

$$\|z_\gamma - \bar{z}\|_Z \le C\left[\alpha^{\frac{-1}{2-\kappa}}\gamma^{\frac{\kappa-1}{2-\kappa}} + \alpha^{-\frac{1}{2}}\gamma^{-\frac{1}{2}} + \alpha^{\frac{-(2\theta+\tau+1)}{4\theta+3\tau+1}}\gamma^{\frac{-2\theta(1-\tau)}{4\theta+3\tau+1}}\right].$$

Proof. We use (8.2), (8.16), and (8.21) to derive

$$\begin{aligned}
\alpha\|z_\gamma - \bar{z}\|_Z^2 &\le \langle J'(z_\gamma) - J'(\bar{z}), z_\gamma - \bar{z}\rangle_{Z^*,Z} \\
&= \langle \mu_\gamma, \bar{y} - y_\gamma\rangle_{Y^*,Y} + \langle \bar{\mu}, T(y_\gamma - \bar{y})\rangle_{Y_0^*,Y_0} \\
&= \langle \mu_\gamma, \bar{y} - y_\gamma\rangle_{Y^*,Y} + \langle \bar{\mu}, T y_\gamma - b\rangle_{Y_0^*,Y_0} \\
&\le \langle \mu_\gamma, \bar{y} - y_\gamma\rangle_{Y^*,Y} + \langle \bar{\mu}, v_\gamma^+\rangle_{Y_0^*,Y_0},
\end{aligned}$$

where $v_\gamma^+ = \max\{y_\gamma - b, 0\}$. In the third step, $\langle \bar{\mu}, T\bar{y} - b\rangle_{Y_0^*,Y_0} = 0$ was used and in the last (fourth) step we used that $Ty_\gamma - b = y_\gamma - b$ can be written as $y_\gamma - b = v_\gamma^+ - v_\gamma^-$ with $v_\gamma^+ \in Y_0$ as just defined and $v_\gamma^- = \max\{b - y_\gamma, 0\} \in Y_0$. Since $v_\gamma^- \ge 0$, there holds $\langle \bar{\mu}, v_\gamma^-\rangle_{Y_0^*,Y_0} \ge 0$.

Further, we have

$$\langle \mu_\gamma, \bar{y} - y_\gamma\rangle_{Y^*,Y} = \int_\Omega \phi'(\gamma(y_\gamma - b))(\bar{y} - y_\gamma)\,dx.$$

Consider first the set $\{y_\gamma \ge b\}$. On this set there holds $y_\gamma - \bar{y} \ge y_\gamma - b \ge 0$ and thus

$$\phi'(\gamma(y_\gamma - b))(\bar{y} - y_\gamma) \le -c_{\phi 1}\gamma^\theta (y_\gamma - b)^\theta (y_\gamma - \bar{y}) \le -c_{\phi 1}\gamma^\theta (y_\gamma - b)^{1+\theta}.$$

Next, consider the set $\{\bar{y} \le y_\gamma < b\}$. Then it follows from $\phi' \ge 0$ that

$$\phi'(\gamma(y_\gamma - b))(\bar{y} - y_\gamma) \le 0.$$

Now, we address the set $\{b - 1/\gamma \le y_\gamma < \bar{y}\}$. Then $y_\gamma < \bar{y} \le b$ yields

$$\phi'(\gamma(y_\gamma - b))(\bar{y} - y_\gamma) \le \phi'(0)(\bar{y} - y_\gamma) \le \sigma(b - y_\gamma) \le \sigma\frac{1}{\gamma}.$$

The last case we have to consider is the set $\{y_\gamma < b - 1/\gamma$ and $y_\gamma < \bar{y}\}$. It then follows that

$$\phi'(\gamma(y_\gamma - b))(\bar{y} - y_\gamma) \le c_{\phi 2}\gamma^{\kappa-1}(b - y_\gamma)^{\kappa-1}(\bar{y} - y_\gamma) \le c_{\phi 2}\gamma^{\kappa-1}(\bar{y} - y_\gamma)^\kappa.$$

Thus, setting $\Omega_\gamma^1 = \{y_\gamma \geq \bar{y}\}$ and $\Omega_\gamma^2 = \{y_\gamma < \bar{y}\}$, we have

$$\begin{aligned}
&\int_{\Omega_\gamma^1} \phi'(\gamma(y_\gamma - b))(\bar{y} - y_\gamma)\,dx \\
&\quad = \int_{\{y_\gamma \geq b\}} \phi'(\gamma(y_\gamma - b))(\bar{y} - y_\gamma)\,dx + \int_{\{\bar{y} \leq y_\gamma < b\}} \phi'(\gamma(y_\gamma - b))(\bar{y} - y_\gamma)\,dx \\
&\quad \leq -c_{\phi 1}\gamma^\theta \int_{\{y_\gamma \geq b\}} (y_\gamma - b)^{1+\theta}\,dx = -c_{\phi 1}\gamma^\theta \|v_\gamma^+\|_{L^{1+\theta}}^{1+\theta},
\end{aligned}$$

where $v_\gamma^+ = \max\{y_\gamma - b, 0\}$. Furthermore,

$$\int_{\Omega_\gamma^2} \phi'(\gamma(y_\gamma - b))(\bar{y} - y_\gamma)\,dx \leq \frac{\sigma}{\gamma}\text{meas}(\Omega) + c_{\phi 2}\gamma^{\kappa-1}\||\bar{y} - y_\gamma|^\kappa\|_{L^1}.$$

Combining these estimates with (8.25), we obtain

$$\begin{aligned}
\alpha\|z_\gamma - \bar{z}\|_Z^2 &\leq \langle \mu_\gamma, \bar{y} - y_\gamma\rangle_{Y^*,Y} + \langle \bar{\mu}, v_\gamma^+\rangle_{Y_0^*,Y} \\
&\leq -c_{\phi 1}\gamma^\theta \|v_\gamma^+\|_{L^{1+\theta}}^{1+\theta} + \frac{\sigma}{\gamma}\text{meas}(\Omega) \\
&\quad + c_{\phi 2}\gamma^{\kappa-1}\||\bar{y} - y_\gamma|^\kappa\|_{L^1} + C_{r,p}(\bar{\mu})\|v_\gamma^+\|_{L^r}.
\end{aligned}$$

If $\kappa > 0$, we can estimate

$$\||\bar{y} - y_\gamma|^\kappa\|_{L^1} \leq \|1\|_{L^{1/(1-\kappa)}}\||\bar{y} - y_\gamma|^\kappa\|_{L^{1/\kappa}} = \text{meas}(\Omega)^{1-\kappa}\|\bar{y} - y_\gamma\|_{L^1}^\kappa.$$

This also holds true for $\kappa = 0$ with the usual definition $a^0 = 1$ for all $a \geq 0$, which will be used in the following.

In the following, we write "const" to denote generic constants that depend on the context. From the previous estimate and the estimate (8.24) of Lemma 8.26 we conclude

$$\begin{aligned}
&\alpha\|z_\gamma - \bar{z}\|_Z^2 + c_{\phi 1}\gamma^\theta \|v_\gamma^+\|_{L^{1+\theta}}^{1+\theta} \\
&\quad \leq \text{const}\,\gamma^{\kappa-1}\|z_\gamma - \bar{z}\|_Z^\kappa + \text{const}\,\gamma^{-1} + C_{r,p}(\bar{\mu})\|v_\gamma^+\|_{L^r} \\
&\quad \leq \text{const}\,\gamma^{\kappa-1}\|z_\gamma - \bar{z}\|_Z^\kappa + \text{const}\,\gamma^{-1} + \text{const}\,\|v_\gamma^+\|_{L^{1+\theta}}^{1-\tau}.
\end{aligned} \tag{8.27}$$

We now use the following estimate (8.23) from Lemma 8.26:

$$\|v_\gamma^+\|_{L^{1+\theta}} \leq \text{const}\,\gamma^{\frac{-\theta}{1+\theta}}\|z_\gamma - \bar{z}\|_Z^{\frac{1}{1+\theta}}. \tag{8.28}$$

Hence,

$$\text{const}\,\|v_\gamma^+\|_{L^{1+\theta}}^{1-\tau} \leq \text{const}\,\|v_\gamma^+\|_{L^{1+\theta}}^{\frac{1-\tau}{2}}\gamma^{\frac{-\theta(1-\tau)}{2(1+\theta)}}\|z_\gamma - \bar{z}\|_Z^{\frac{1-\tau}{2(1+\theta)}}.$$

Next, we apply the following generalization of Young's inequality, which follows from the weighted arithmetic mean and geometric mean inequality: For $a_1, \ldots, a_k \geq 0$, $q_1, \ldots, q_k > 1$, $\frac{1}{q_1} + \cdots + \frac{1}{q_k} = 1$ there holds

$$\prod_{j=1}^k a_j \leq \sum_{j=1}^k \frac{a_j^{q_j}}{q_j}.$$

With $\beta_1, \beta_2, \beta_3 > 0$ such that $\beta_1\beta_2\beta_3 = \text{const}$ and $q_1, q_2, q_3 > 1$ such that $1/q_1 + 1/q_2 + 1/q_3 = 1$ to be determined below, and setting $k = 3$, $a_1 = \beta_1 \|v_\gamma^+\|_{L^{1+\theta}}^{\frac{1-\tau}{2}}$, $a_2 = \beta_2 \|z_\gamma - \bar{z}\|_Z^{\frac{1-\tau}{2(1+\theta)}}$, and $a_3 = \beta_3 \gamma^{\frac{-\theta(1-\tau)}{2(1+\theta)}}$, we obtain

$$\begin{aligned}\text{const}\,\|v_\gamma^+\|_{L^{1+\theta}}^{1-\tau} &\le \text{const}\,\|v_\gamma^+\|_{L^{1+\theta}}^{\frac{1-\tau}{2}} \gamma^{\frac{-\theta(1-\tau)}{2(1+\theta)}} \|z_\gamma - \bar{z}\|_Z^{\frac{1-\tau}{2(1+\theta)}} \\ &\le \frac{1}{q_1}\beta_1^{q_1} \|v_\gamma^+\|_{L^{1+\theta}}^{\frac{(1-\tau)q_1}{2}} + \frac{1}{q_2}\beta_2^{q_2} \|z_\gamma - \bar{z}\|_Z^{\frac{(1-\tau)q_2}{2(1+\theta)}} + \frac{1}{q_3}\beta_3^{q_3} \gamma^{\frac{-\theta(1-\tau)q_3}{2(1+\theta)}}.\end{aligned}$$

We choose the parameters such that the first and second summand generate the terms $c_{\phi 1}\gamma^\theta \|v_\gamma^+\|_{L^{1+\theta}}^{1+\theta}$ and $\frac{\alpha}{3}\|z_\gamma - \bar{z}\|_Z^2$, respectively, which requires

$$\frac{(1-\tau)q_1}{2} = 1+\theta, \quad \frac{1}{q_1}\beta_1^{q_1} = c_{\phi 1}\gamma^\theta, \quad \frac{(1-\tau)q_2}{2(1+\theta)} = 2, \quad \frac{1}{q_2}\beta_2^{q_2} = \frac{\alpha}{3}.$$

This is achieved by the choice

$$\begin{aligned} &q_1 = \frac{2(1+\theta)}{1-\tau}, \quad q_2 = \frac{4(1+\theta)}{1-\tau}, \quad q_3 = \frac{1}{1-\frac{1}{q_1}-\frac{1}{q_2}} = \frac{4(1+\theta)}{4\theta + 3\tau + 1}, \\ &\beta_1 = (c_{\phi 1} q_1 \gamma^\theta)^{\frac{1}{q_1}} = \text{const}\,\gamma^{\frac{\theta(1-\tau)}{2(1+\theta)}}, \quad \beta_2 = \left(\frac{\alpha q_2}{3}\right)^{\frac{1}{q_2}} = \text{const}\,\alpha^{\frac{1-\tau}{4(1+\theta)}}, \\ &\beta_3 = \frac{\text{const}}{\beta_1\beta_2} = \text{const}\,\alpha^{\frac{-(1-\tau)}{4(1+\theta)}} \gamma^{\frac{-\theta(1-\tau)}{2(1+\theta)}}. \end{aligned}$$

With this choice, we obtain

$$\begin{aligned} &\alpha\|z_\gamma - \bar{z}\|_Z^2 + c_{\phi 1}\gamma^\theta \|v_\gamma^+\|_{L^{1+\theta}}^{1+\theta} \\ &\quad \le \text{const}\,\gamma^{\kappa-1}\|z_\gamma - \bar{z}\|_Z^\kappa + \text{const}\,\gamma^{-1} + \text{const}\,\|v_\gamma^+\|_{L^{1+\theta}}^{1-\tau} \\ &\quad \le \text{const}\,\gamma^{\kappa-1}\|z_\gamma - \bar{z}\|_Z^\kappa + \text{const}\,\gamma^{-1} + c_{\phi 1}\gamma^\theta \|v_\gamma^+\|_{L^{1+\theta}}^{1+\theta} \\ &\qquad + \frac{\alpha}{3}\|z_\gamma - \bar{z}\|_Z^2 + \frac{1}{q_3}\beta_3^{q_3}\gamma^{\frac{-\theta(1-\tau)q_3}{2(1+\theta)}}. \end{aligned}$$

We calculate

$$\frac{1}{q_3}\beta_3^{q_3}\gamma^{\frac{-\theta(1-\tau)q_3}{2(1+\theta)}} = \text{const}\,\alpha^{\frac{-(1-\tau)q_3}{4(1+\theta)}} \gamma^{\frac{-\theta(1-\tau)q_3}{2(1+\theta)}} \gamma^{\frac{-\theta(1-\tau)q_3}{2(1+\theta)}} = \text{const}\,\alpha^{\frac{-(1-\tau)}{4\theta+3\tau+1}} \gamma^{\frac{-4\theta(1-\tau)}{4\theta+3\tau+1}}.$$

In the case $\kappa > 0$, where we again apply Young's inequality with $k = 2$, $q_1 = \frac{2}{\kappa}$, $a_1 = \left(\frac{q_1\alpha}{3}\right)^{\frac{1}{q_1}} \|z_\gamma - \bar{z}\|_Z^\kappa$, $q_2 = \frac{2}{2-\kappa}$, and $a_2 = \text{const}\left(\frac{q_1\alpha}{3}\right)^{-\frac{1}{q_1}}\gamma^{\kappa-1}$, we have

$$\text{const}\,\gamma^{\kappa-1}\|z_\gamma - \bar{z}\|_Z^\kappa \le \frac{\alpha}{3}\|z_\gamma - \bar{z}\|_Z^2 + \text{const}\,\alpha^{\frac{-\kappa}{2-\kappa}}\gamma^{\frac{2(\kappa-1)}{2-\kappa}}.$$

This is also true for $\kappa = 0$, since the derived inequality then becomes

$$\text{const}\,\gamma^{-1}\|z_\gamma - \bar{z}\|_Z^0 \le \frac{\alpha}{3}\|z_\gamma - \bar{z}\|_Z^2 + \text{const}\,\alpha^0\gamma^{-1},$$

which is trivially satisfied.

Hence, taking all together,

$$\frac{\alpha}{3}\|z_\gamma - \bar{z}\|_Z^2 \leq \operatorname{const}\alpha^{\frac{-\kappa}{2-\kappa}}\gamma^{\frac{2(\kappa-1)}{2-\kappa}} + \operatorname{const}\gamma^{-1} + \operatorname{const}\alpha^{\frac{-(1-\tau)}{4\theta+3\tau+1}}\gamma^{\frac{-4\theta(1-\tau)}{4\theta+3\tau+1}}.$$

Since, for $a_1, a_2, a_3 \geq 0$ there holds $(a_1+a_2+a_3)^{1/2} \leq a_1^{1/2} + a_2^{1/2} + a_3^{1/2}$, we obtain

$$\|z_\gamma - \bar{z}\|_Z \leq \operatorname{const}\alpha^{\frac{-1}{2-\kappa}}\gamma^{\frac{\kappa-1}{2-\kappa}} + \operatorname{const}\alpha^{-\frac{1}{2}}\gamma^{-\frac{1}{2}} + \operatorname{const}\alpha^{\frac{-(2\theta+\tau+1)}{4\theta+3\tau+1}}\gamma^{\frac{-2\theta(1-\tau)}{4\theta+3\tau+1}}. \qquad \square$$

Remark 8.36. *A similar result for the special case of the Moreau–Yosida regularization was recently presented in* [101] *in a finite element framework.*

As already mentioned in Remark 8.27, we now can improve our estimate for v_γ^+ by inserting the estimates of Theorem 8.35 for $\|z_\gamma - \bar{z}\|_Z$ into (8.23) and (8.24). Since the resulting formulas become quite lengthy, we do not reformulate Lemma 8.26 with these improved results. Rather, we address this issue as part of the following examples.

Example 8.37. For the Moreau–Yosida regularization $\phi(t) = \frac{1}{2}\max^2\{0,t\}$ we can choose $c_{\phi 1} = 1$, $\theta = 1$, $c_{\phi 2} > 0$ arbitrarily small and $\kappa = 0$. Thus, the estimate in Theorem 8.35 becomes

$$\|z_\gamma - \bar{z}\|_Z \leq C\left[\alpha^{-\frac{1}{2}}\gamma^{-\frac{1}{2}} + \alpha^{\frac{-(\tau+3)}{3\tau+5}}\gamma^{\frac{-2(1-\tau)}{3\tau+5}}\right]. \qquad \blacksquare$$

Example 8.38. We continue Example 8.37 and combine it with the elliptic optimal control setting of Example 8.5. Thus, we have $1 \leq d \leq 3$, $Y = H_0^1(\Omega) \cap H^2(\Omega)$, and $Y_0 = C(\bar{\Omega})$. Then the Sobolev embedding theorem yields

$$Y \hookrightarrow H^2(\Omega) \hookrightarrow W^{1,p}(\Omega) \quad \text{with} \quad 2 \leq p < \infty \ \text{if } d \leq 2, \ 2 \leq p \leq \frac{2d}{d-2} = 6 \ \text{if } d = 3.$$

According to Lemma 8.31, we then can verify Assumption 8.29 with p as above and $r = \infty$.

Thus, for $d = 2$, we can choose p arbitrarily large and obtain

$$\tau = \frac{pd}{pd + (1+\theta)(p-d)} = \frac{2p}{2p + 2(p-2)} = \frac{1}{2 - 2/p} \to \frac{1}{2} \quad (p \to \infty).$$

Hence, in this case, $\tau \in (1/2, 1]$ can be chosen arbitrarily close to $1/2$. We have

$$\frac{-(\tau+3)}{3\tau+5} \downarrow \frac{-7/2}{13/2} = -\frac{7}{13}, \quad \frac{-2(1-\tau)}{3\tau+5} \downarrow \frac{-1}{13/2} = -\frac{2}{13} \quad \text{as } \tau \downarrow \frac{1}{2}.$$

Thus, there holds

$$\|z_\gamma - \bar{z}\|_Z \leq C\alpha^{-\frac{7}{13}}\gamma^{-\frac{2}{13}+\varepsilon},$$

where $\varepsilon > 0$ can be chosen arbitrarily small.

For the feasibility violation $v_\gamma^+ = \max\{y_\gamma - b, 0\}$ we obtain

$$\|v_\gamma^+\|_{L^2} \le C\gamma^{-\frac{1}{2}}\|z_\gamma - \bar{z}\|_Z^{\frac{1}{2}} \le C\alpha^{-\frac{7}{26}}\gamma^{-\frac{15}{26}+\varepsilon},$$

where $\varepsilon > 0$ can be chosen arbitrarily small.

From (8.24) and

$$\frac{1-\tau}{2}\left(\frac{-(\tau+3)}{3\tau+5}\right) \downarrow -\frac{7}{52}, \quad -\frac{1-\tau}{2} + \frac{1-\tau}{2}\left(\frac{-2(1-\tau)}{3\tau+5}\right) \downarrow = -\frac{15}{52} \quad \text{as } \tau \downarrow \frac{1}{2},$$

we obtain

$$\|v_\gamma^+\|_{C(\bar{\Omega})} \le C\alpha^{-\frac{7}{52}}\gamma^{-\frac{15}{52}+\varepsilon},$$

where $\varepsilon > 0$ can be chosen arbitrarily small.

Now consider the same situation, but with $d = 3$. Then we can choose $p = 6$ and therefore

$$\tau = \frac{pd}{pd + (1+\theta)(p-d)} = \frac{18}{18 + 2(6-3)} = \frac{3}{4}.$$

Hence,

$$\frac{-(\tau+3)}{3\tau+5} = \frac{-15/4}{29/4} = -\frac{15}{29}, \quad \frac{-2(1-\tau)}{3\tau+5} = \frac{-1/2}{29/4} = -\frac{2}{29},$$

and therefore

$$\|z_\gamma - \bar{z}\|_Z \le C\alpha^{-\frac{15}{29}}\gamma^{-\frac{2}{29}}.$$

For the feasibility violation we obtain

$$\|v_\gamma^+\|_{L^2} \le C\gamma^{-\frac{1}{2}}\|z_\gamma - \bar{z}\|_Z^{\frac{1}{2}} \le C\alpha^{-\frac{15}{58}}\gamma^{-\frac{31}{58}}.$$

From (8.24) and

$$\frac{1-\tau}{2}\left(\frac{-(\tau+3)}{3\tau+5}\right) = -\frac{15}{232}, \quad -\frac{1-\tau}{2} + \frac{1-\tau}{2}\left(\frac{-2(1-\tau)}{3\tau+5}\right) = -\frac{31}{232},$$

we obtain

$$\|v_\gamma^+\|_{C(\bar{\Omega})} \le C\alpha^{-\frac{15}{232}}\gamma^{-\frac{31}{232}}. \qquad \blacksquare$$

From the previous examples it can also be seen that the larger θ is, the better the order is. To this end, consider again the case $d = 3$ and $Y \hookrightarrow H^2(\Omega) \hookrightarrow W^{1,6}(\Omega)$ and $Y_0 = C(\bar{\Omega})$. Then $p = 6$, $r = \infty$, and

$$\tau = \frac{pd}{pd + (1+\theta)(p-d)} = \frac{18}{21 + 3\theta} \downarrow 0 \quad \text{as } \theta \to \infty.$$

We then obtain

$$\frac{-2(2\theta + \tau + 1)}{4\theta + 3\tau + 1} \uparrow -\frac{1}{2}, \quad \frac{-2\theta(1-\tau)}{4\theta + 3\tau + 1} \downarrow -\frac{1}{2} \quad \text{as } \theta \to \infty.$$

Hence, for every $\varepsilon > 0$ there exists $\theta > 0$ such that

$$\|z_\gamma - \bar{z}\|_Z \le C\left[\alpha^{\frac{-1}{2-\kappa}}\gamma^{\frac{\kappa-1}{2-\kappa}} + \alpha^{-\frac{1}{2}-\varepsilon}\gamma^{-\frac{1}{2}+\varepsilon}\right].$$

Thus, if, e.g., we generalize the Moreau–Yosida regularization to

$$\phi(t) = \frac{1}{\theta+1}\max{}^{\theta+1}\{0,t\}$$

with $\theta \ge 1$ fixed, then we can choose $c_{\phi 1} = 1$, $c_{\phi 2} > 0$ arbitrary, and $\kappa = 0$. We then can omit the terms involving κ.

Example 8.39. We consider the obstacle problem of Example 8.6,

$$\min_{y\in Y} \; J(y) := \frac{1}{2}\langle Ay, y\rangle_{Y^*,Y} - \langle f, y\rangle_{Y^*,Y} \quad \text{subject to} \quad y \le b \tag{8.29}$$

with $Y := H_0^1(\Omega)$, where $A \in \mathcal{L}(Y, Y^*)$ is given by

$$\langle Ay, v\rangle_{Y^*,Y} = \int_\Omega \nabla y^T \mathbf{A} \nabla v \, dx$$

and $\mathbf{A} \in \mathbb{R}^{d\times d}$ is a positive definite symmetric matrix such that there exists $\alpha > 0$ with

$$\langle Av, v\rangle_{Y^*,Y} \ge \alpha\|v\|_Y^2 \quad \forall\, v \in Y.$$

As shown in Example 8.6, Assumption 8.1 is then satisfied.

Further, we assume that H^2 regularity holds, i.e., that there exists a constant $C_{H^2} > 0$ such that for every $z \in L^2(\Omega)$ the unique solution $v \in Y$ of $Av = z$ satisfies $v \in Y \cap H^2(\Omega)$ and

$$\|v\|_{H^2} \le C_{H^2}\|z\|_{L^2}.$$

We assume that $b \in Y_0 := Y = H_0^1(\Omega)$ has the additional regularity $b \in H^2(\Omega)$. Further, let $f \in L^2(\Omega)$.

The Moreau–Yosida regularization results in the problem

$$\min_{y\in Y} \; J_\gamma(y) := \frac{1}{2}\langle Ay, y\rangle_{Y^*,Y} - \langle f, y\rangle_{Y^*,Y} + \frac{\gamma}{2}\|\max\{y-b, 0\}\|_{L^2}^2.$$

Then $\phi(t) = \max^2\{0,t\}/2$ satisfies Assumptions 8.11, 8.14, 8.23 with $\sigma = 0$, $c_{\phi 1} = 1$, $\theta = 1$, $c_{\phi 2} > 0$ arbitrarily small, and $\kappa = 0$. The corresponding optimality condition is

$$Ay_\gamma + \mu_\gamma = f, \quad \mu_\gamma = \gamma\max\{y_\gamma - b, 0\}.$$

The uniform boundedness of $(y_\gamma, \mu_\gamma) \in Y \times Y^*$ and the convergence $(y_\gamma, \mu_\gamma) \to (\bar{y}, \bar{\mu})$ in $Y \times Y^*$ as $\gamma \to \infty$ is shown in Theorem 8.21. Also, Lemma 8.26 yields for $v_\gamma^+ = \max\{y_\gamma - b, 0\} \in Y$ the estimate (note that $\theta = 1$)

$$\|v_\gamma^+\|_{L^2} = o\left(\gamma^{-1/2}\right) \quad (\gamma \to \infty).$$

As an alternative to applying Lemma 8.26 and as an additional illustration, the boundedness of $\|y_\gamma\|_Y$ and of $\sqrt{\gamma}\|v_\gamma^+\|_{L^2}$ can also be derived directly.

For $y^0 = \min\{b, 0\}$ there holds $y^0 \in Y$ and $y^0 - b \le 0$. Hence,

$$\frac{1}{2}\langle Ay_\gamma, y_\gamma\rangle_{Y^*,Y} - \langle f, y_\gamma\rangle_{Y^*,Y} + \frac{\gamma}{2}\|v_\gamma^+\|_{L^2}^2 = J_\gamma(y_\gamma) \le J_\gamma(y^0) = J(y^0) =: C_0.$$

This implies

$$\frac{\alpha}{2}\|y_\gamma\|_Y^2 + \frac{\gamma}{2}\|v_\gamma^+\|_{L^2}^2 \le C_0 + \|f\|_{Y^*}\|y_\gamma\|_Y \le C_0 + \frac{\alpha}{4}\|y_\gamma\|_Y^2 + \frac{1}{\alpha}\|f\|_{Y^*}^2,$$

where Young's inequality was used. Hence,

$$\alpha\|y_\gamma\|_Y^2 + 2\gamma\|v_\gamma^+\|_{L^2}^2 \le 4C_0 + \frac{4}{\alpha}\|f\|_{Y^*}^2.$$

In particular, $\|y_\gamma\|_Y$ and $\sqrt{\gamma}\|v_\gamma^+\|_{L^2}$ are bounded independently of γ.

Our aim is now to find best possible values for p and r such that Assumption 8.29 is satisfied.

First, we note that

$$Ay_\gamma = f - \gamma v_\gamma^+ \in L^2(\Omega).$$

We test this equation with $v_\gamma^+ \in Y$ and obtain

$$\langle Ay_\gamma, v_\gamma^+\rangle_{Y^*,Y} + \gamma\|v_\gamma^+\|_{L^2}^2 = \langle f, v_\gamma^+\rangle_{Y^*,Y} \le \|f\|_{Y^*}\|v_\gamma^+\|_Y.$$

Now there holds

$$\begin{aligned}
\langle Av_\gamma^+, v_\gamma^+\rangle_{Y^*,Y} &= \int_{\{y_\gamma > b\}} \nabla(y_\gamma - b)^T \mathbf{A}\nabla v_\gamma^+\, dx \\
&= \int_{\{y_\gamma > b\}} \nabla y_\gamma^T \mathbf{A}\nabla v_\gamma^+\, dx - \int_{\{y_\gamma > b\}} \nabla b^T \mathbf{A}\nabla v_\gamma^+\, dx \\
&= \int_\Omega \nabla y_\gamma^T \mathbf{A} v_\gamma^+\, dx - \int_\Omega \nabla b^T \mathbf{A}\nabla v_\gamma^+\, dx \\
&= \langle Ay_\gamma, v_\gamma^+\rangle_{Y^*,Y} + \int_\Omega \operatorname{div}(\mathbf{A}\nabla b) v_\gamma^+\, dx.
\end{aligned}$$

Hence,

$$\begin{aligned}
\langle Av_\gamma^+, v_\gamma^+\rangle_{Y^*,Y} + \gamma\|v_\gamma^+\|_{L^2}^2 &= \langle Ay_\gamma, v_\gamma^+\rangle_{Y^*,Y} + \gamma\|v_\gamma^+\|_{L^2}^2 + \int_\Omega \operatorname{div}(\mathbf{A}\nabla b) v_\gamma^+\, dx \\
&= \langle f, v_\gamma^+\rangle_{Y^*,Y} + \int_\Omega \operatorname{div}(\mathbf{A}\nabla b) v_\gamma^+\, dx \\
&\le \|f\|_{L^2}\|v_\gamma^+\|_{L^2} + C\|b\|_{H^2}\|v_\gamma^+\|_{L^2}.
\end{aligned}$$

Therefore,

$$\gamma\|v_\gamma^+\|_{L^2} \le \|f\|_{L^2} + C\|b\|_{H^2}.$$

Hence, $\gamma \|v_\gamma^+\|_{L^2}$ is bounded independently of γ. Since $\mu_\gamma = \gamma v_\gamma^+$, we see that $\|\mu_\gamma\|_{L^2}$ is bounded independently of γ. From $\mu_\gamma \to \bar{\mu}$ in $Y^* = H^{-1}(\Omega)$ and the boundedness of μ_γ in $L^2(\Omega)$ we conclude $\bar{\mu} \in L^2(\Omega)$. This can also be obtained from regularity results for elliptic obstacle problems; see, e.g., [29].

Although the boundedness of (y_γ) in $H^1(\Omega)$ will be sufficient for our purposes, see below, we use the derived results to briefly obtain the H^2-regularity of the states. Concerning the regularity of y_γ, we obtain

$$\|Ay_\gamma\|_{L^2} \le \|f\|_{L^2} + \|\mu_\gamma\|_{L^2} \le 2\|f\|_{L^2} + C\|b\|_{H^2}.$$

Hence, using the H^2-regularity of the elliptic operator A, we see that $\|y_\gamma\|_{H^2}$ is bounded independently of γ. Since $y_\gamma \to \bar{y}$ in $Y = H_0^1(\Omega)$, this also shows $\bar{y} \in H^2(\Omega)$. Again, this can also be obtained directly from regularity results for elliptic obstacle problems [29].

Since $\bar{\mu} \in L^2(\Omega)$, we can choose $r = 2 = 1+\theta$ in Assumption 8.29. For this r, the choice $p = 2$ is sufficient so that we do not need the H^2-regularity of y_γ here. In fact, independently of the choice of $p > rd/(r+d) = 2d/(2+d)$, which includes $p = 2$, we have

$$\tau = \frac{pd(r-1-\theta)}{prd + (1+\theta)r(p-d)} = 0.$$

Hence,

$$\frac{-(2\theta+\tau+1)}{4\theta+3\tau+1} = -\frac{3}{5}, \quad \frac{-2\theta(1-\tau)}{4\theta+3\tau+1} = -\frac{2}{5},$$

and thus

$$\|z_\gamma - \bar{z}\|_Z \le C\left[\alpha^{-\frac{1}{2}}\gamma^{-\frac{1}{2}} + \alpha^{-\frac{3}{5}}\gamma^{-\frac{2}{5}}\right] \le C\left[\alpha^{-\frac{3}{5}} + \gamma^{-\frac{2}{5}}\right].$$

As we will see, this estimate is not sharp. In fact, Theorem 8.35 builds on the estimate (8.23). But as we showed in our context, there even holds $\|v_\gamma^+\|_{L^2} = O(\gamma^{-1})$. ∎

As we found out in the previous example, there are situations where a better estimate for $\|v_\gamma^+\|_{L^{1+\theta}}$ is available than (8.23). We now prove a version of Theorem 8.35 that takes advantage of such an improved estimate.

Theorem 8.40. *Let Assumptions* 8.1, 8.11, 8.14, 8.23, 8.29, *and* 8.33 *hold and denote by* $\bar{z} = (\bar{y}, \bar{u})$ *and* $z_\gamma = (y_\gamma, u_\gamma)$, $\gamma > 0$, *the unique solutions of* (8.1) *and* (8.9), *respectively. Further, assume that there exist* $\eta > \theta/(1+\theta)$ *and* $\rho_\eta > 0$ *such that for all* $\gamma > 0$ *there holds*

$$\|v_\gamma^+\|_{L^{1+\theta}} \le \rho_\eta \gamma^{-\eta}. \tag{8.30}$$

Then with $\tau = \frac{pd(r-1-\theta)}{prd+(1+\theta)r(p-d)}$ *and a suitable constant* $C > 0$ *there holds*

$$\|z_\gamma - \bar{z}\|_Z \le C\left[\alpha^{\frac{-1}{2-\kappa}}\gamma^{\frac{\kappa-1}{2-\kappa}} + \alpha^{-\frac{1}{2}}\gamma^{-\frac{1}{2}} + \alpha^{-\frac{1}{2}}\rho_\eta^{\frac{(1-\tau)(1+\theta)}{2(2\theta+\tau+1)}}\gamma^{\frac{-(1-\tau)(\theta+\eta(1+\theta))}{2(2\theta+\tau+1)}}\right].$$

Proof. The beginning of this proof is identical to that of Theorem 8.35 until (8.27) is derived, which we recall here:

$$\alpha\|z_\gamma - \bar{z}\|_Z^2 + c_{\phi 1}\gamma^\theta \|v_\gamma^+\|_{L^{1+\theta}}^{1+\theta} \le \text{const}\,\gamma^{\kappa-1}\|z_\gamma - \bar{z}\|_Z^\kappa + \text{const}\,\gamma^{-1} + \text{const}\,\|v_\gamma^+\|_{L^{1+\theta}}^{1-\tau}.$$

Here and in the following, "const" again denotes a generic constant depending on the context. Now instead of (8.28), we use

$$\|v_\gamma^+\|_{L^{1+\theta}} \leq \rho_\eta \gamma^{-\eta}.$$

Hence,

$$\text{const}\, \|v_\gamma^+\|_{L^{1+\theta}}^{1-\tau} \leq \text{const}\, \|v_\gamma^+\|_{L^{1+\theta}}^{\frac{1-\tau}{2}} \rho_\eta^{\frac{1-\tau}{2}} \gamma^{\frac{-\eta(1-\tau)}{2}}.$$

Next, we apply Young's inequality with suitable factors $\beta_1, \beta_2 > 0$, $\beta_1\beta_2 = \text{const}$, $a_1 = \beta_1 \|v_\gamma^+\|_{L^{1+\theta}}^{\frac{1-\tau}{2}}$, $a_2 = \beta_2 \rho_\eta^{\frac{1-\tau}{2}} \gamma^{\frac{-\eta(1-\tau)}{2}}$, $q_1, q_2 > 1$, and $1/q_1 + 1/q_2 = 1$ such that

$$\begin{aligned}
\text{const}\, \|v_\gamma^+\|_{L^{1+\theta}}^{\frac{1-\tau}{2}} \rho_\eta^{\frac{1-\tau}{2}} \gamma^{\frac{-\eta(1-\tau)}{2}} &= a_1 a_2 \leq \frac{1}{q_1} a_1^{q_1} + \frac{1}{q_2} a_2^{q_2} \\
&= c_{\phi 1} \gamma^\theta \|v_\gamma^+\|_{L^{1+\theta}}^{1+\theta} + \frac{1}{q_2} \beta_2^{q_2} \rho_\eta^{\frac{(1-\tau)q_2}{2}} \gamma^{\frac{-\eta(1-\tau)q_2}{2}}.
\end{aligned}$$

This is achieved with

$$q_1 = \frac{2(1+\theta)}{1-\tau}, \quad q_2 = \frac{2(1+\theta)}{2\theta+\tau+1}, \quad \beta_1^{q_1} = q_1 c_{\phi 1} \gamma^\theta = \frac{2(1+\theta)}{1-\tau} c_{\phi 1} \gamma^\theta.$$

We obtain

$$\begin{aligned}
\text{const}\, \|v_\gamma^+\|_{L^{1+\theta}}^{\frac{1-\tau}{2}} \rho_\eta^{\frac{1-\tau}{2}} \gamma^{\frac{-\eta(1-\tau)}{2}} &\leq c_{\phi 1} \gamma^\theta \|v_\gamma^+\|_{L^{1+\theta}}^{1+\theta} + \text{const}\, \gamma^{-\frac{\theta q_2}{q_1}} \rho_\eta^{\frac{(1-\tau)q_2}{2}} \gamma^{\frac{-\eta(1-\tau)q_2}{2}} \\
&= c_{\phi 1} \gamma^\theta \|v_\gamma^+\|_{L^{1+\theta}}^{1+\theta} + \text{const}\, \rho_\eta^{\frac{(1-\tau)(1+\theta)}{2\theta+\tau+1}} \gamma^{\frac{-(1-\tau)(\theta+\eta(1+\theta))}{2\theta+\tau+1}}.
\end{aligned}$$

Hence,

$$\alpha \|z_\gamma - \bar{z}\|_Z^2 \leq \text{const}\, \gamma^{\kappa-1} \|z_\gamma - \bar{z}\|_Z^\kappa + \text{const}\, \gamma^{-1} + \text{const}\, \rho_\eta^{\frac{(1-\tau)(1+\theta)}{2\theta+\tau+1}} \gamma^{\frac{-(1-\tau)(\theta+\eta(1+\theta))}{2\theta+\tau+1}}.$$

In the case $\kappa > 0$, we again apply Young's inequality to estimate

$$\text{const}\, \gamma^{\kappa-1} \|z_\gamma - \bar{z}\|_Z^\kappa \leq \frac{\alpha}{2} \|z_\gamma - \bar{z}\|_Z^2 + \text{const}\, \alpha^{\frac{-\kappa}{2-\kappa}} \gamma^{\frac{2(\kappa-1)}{2-\kappa}}.$$

This is also true for $\kappa = 0$. Hence,

$$\frac{\alpha}{2} \|z_\gamma - \bar{z}\|_Z^2 \leq \text{const}\, \alpha^{\frac{-\kappa}{2-\kappa}} \gamma^{\frac{2(\kappa-1)}{2-\kappa}} + \text{const}\, \gamma^{-1} + \text{const}\, \rho_\eta^{\frac{(1-\tau)(1+\theta)}{2\theta+\tau+1}} \gamma^{\frac{-(1-\tau)(\theta+\eta(1+\theta))}{2\theta+\tau+1}}.$$

From this we obtain

$$\|z_\gamma - \bar{z}\|_Z \leq \text{const}\, \alpha^{\frac{-1}{2-\kappa}} \gamma^{\frac{\kappa-1}{2-\kappa}} + \text{const}\, \alpha^{-\frac{1}{2}} \gamma^{-\frac{1}{2}} + \text{const}\, \rho_\eta^{\frac{(1-\tau)(1+\theta)}{2(2\theta+\tau+1)}} \alpha^{-\frac{1}{2}} \gamma^{\frac{-(1-\tau)(\theta+\eta(1+\theta))}{2(2\theta+\tau+1)}}. \qquad \square$$

Example 8.41. We return to the obstacle problem with Moreau–Yosida regularization; see Example 8.39. Then (8.30) is satisfied with $\eta = 1$ and a constant $\rho_\eta > 0$. Now, we have

$$\frac{-(1-\tau)(\theta+\eta(1+\theta))}{2(2\theta+\tau+1)} = \frac{-(1-0)(1+1\cdot(1+1))}{2(2\cdot 1+0+1)} = -\frac{1}{2},$$

and thus the estimate of Theorem 8.40 becomes

$$\|z_\gamma - \bar{z}\|_Z \le \text{const}\,\alpha^{-\frac{1}{2}}\gamma^{-\frac{1}{2}}. \quad \blacksquare$$

Example 8.42. We consider Example 8.38 with $\theta = 1$, $\kappa = 0$, $d = 3$, and $p = 6$. We calculated in Example 8.38 that then $\tau = 3/4$ and derived the estimate

$$\|v_\gamma^+\|_{L^{1+\theta}} = \|v_\gamma^+\|_{L^2} \le C\alpha^{-\frac{15}{58}}\gamma^{-\frac{31}{58}}.$$

Hence, assumption (8.30) holds with $\eta = \frac{31}{58}$ and $\rho_\eta = C\alpha^{-\frac{15}{58}}$. From

$$\frac{-(1-\tau)(\theta+\eta(1+\theta))}{2(2\theta+\tau+1)} = \frac{-\left(1-\frac{3}{4}\right)\left(1+\frac{31}{58}\cdot(1+1)\right)}{2\left(2\cdot 1+\frac{3}{4}+1\right)} = -\frac{2}{29},$$

$$-\frac{15}{58}\cdot\frac{(1-\tau)(1+\theta)}{2(2\theta+\tau+1)} - \frac{1}{2} = -\frac{15}{58}\cdot\frac{\left(1-\frac{3}{4}\right)(1+1)}{2\left(2\cdot 1+\frac{3}{4}+1\right)} - \frac{1}{2} = -\frac{15}{29},$$

we conclude that the estimate of Theorem 8.40 becomes

$$\|z_\gamma - \bar{z}\|_Z \le \text{const}\,\alpha^{-\frac{15}{29}}\gamma^{-\frac{2}{29}}.$$

This is the same result as we obtained in Example 8.38, which shows the compatibility between Theorems 8.35 and 8.40. ■

8.2.4 Interpretation as a Dual Regularization

We now give a second view of the investigated penalization technique. Although the following considerations can be extended to a more general setting, we focus on the Moreau–Yosida regularization, i.e., the case $\phi(t) = (1/2)\max^2\{0,t\}$.

We will show that the Moreau–Yosida regularization is equivalent to an L^2-regularization for the multiplier μ in the Lagrange function. In fact, based on the Lagrange function $L : Y \times U \times W^* \times Y_0^* \to \mathbb{R}$,

$$L(y,u,w,\mu) = J(y,u) + \langle w, Ay - Bu - f\rangle_{W^*,W} + \langle \mu, Ty - b\rangle_{Y_0^*,Y_0},$$

the KKT conditions (8.5)–(8.8) can be written as

$$\begin{aligned} L_y(\bar{y},\bar{u},\bar{w},\bar{\mu}) &= 0,\\ L_u(\bar{y},\bar{u},\bar{w},\bar{\mu}) &= 0,\\ L_w(\bar{y},\bar{u},\bar{w},\bar{\mu}) &= 0, \end{aligned}$$

$$L_\mu(\bar{y},\bar{u},\bar{w},\bar{\mu}) \le 0, \quad \langle \bar{\mu}, v\rangle_{Y_0^*,Y_0} \ge 0 \ \forall\, v \in Y_0,\ v \ge 0, \quad \langle \bar{\mu}, L_\mu(\bar{y},\bar{u},\bar{w},\bar{\mu})\rangle_{Y_0^*,Y_0} = 0.$$

These conditions are first-order conditions for $((\bar{y},\bar{\mu}),(\bar{w},\bar{\mu}))$ being a saddle point of L on $(Y\times U)\times(W^*\times(Y_0^*)_+)$, where

$$(Y_0^*)_+ = \{\mu \in Y_0^* : \langle \mu, v\rangle_{Y_0^*,Y_0} \geq 0 \;\; \forall\, v \in Y_0,\; v \geq 0\}.$$

Here, the saddle point is a minimizer with respect to $(y,u)\in Y\times U$ and a maximizer with respect to $(w,\mu)\in W^*\times(Y_0^*)_+$. As explained, the difficulty of state constraints is that the corresponding multiplier is quite irregular, since it is only an element of the dual space $\mu\in Y_0^*$ (e.g., a regular Borel measure on $\bar{\Omega}$ in the case $Y_0 = C(\bar{\Omega})$). A stable related saddle point problem where Y_0^* can be replaced by $L^2(\Omega)$ is obtained by adding an L^2-regularization term for μ to the Lagrange function. We obtain the regularized Lagrange function

$$Y\times U\times W^*\times L^2(\Omega)\ni(y,u,w,\mu)\mapsto \mathcal{L}(\gamma;y,u,w,\mu) = L(y,u,w,\mu) - \frac{1}{2\gamma}\|\mu\|_{L^2}^2 \in \mathbb{R}.$$

For the regularization we have chosen the minus sign since L is maximized with respect to μ and thus the regularization term should be concave rather than convex. If we now write down the first-order saddle point conditions for $\mathcal{L}(\gamma;\cdot)$ on $(Y\times U)\times(W^*\times L^2(\Omega)_+)$, where $L^2(\Omega)_+ = \{v\in L^2(\Omega): v\geq 0\}$, we obtain for the saddle point $(y_\gamma,u_\gamma,w_\gamma,\mu_\gamma)$ the following conditions:

$$J_y(y_\gamma,u_\gamma) + A^* w_\gamma + \mu_\gamma = 0, \tag{8.31}$$

$$J_u(y_\gamma,u_\gamma) - B^* w_\gamma = 0, \tag{8.32}$$

$$Ay_\gamma - Bu_\gamma = f, \tag{8.33}$$

$$y_\gamma - \frac{1}{\gamma}\mu_\gamma \leq b, \quad \mu_\gamma \geq 0, \quad \mu_\gamma\left(y_\gamma - \frac{1}{\gamma}\mu_\gamma - b\right) = 0. \tag{8.34}$$

The complementarity condition (8.34) can equivalently be written with the min-NCP-function as follows:

$$\min\left\{\mu_\gamma, -\gamma\left(y_\gamma - \frac{1}{\gamma}\mu_\gamma - b\right)\right\} = 0.$$

This gives

$$\mu_\gamma + \min\{0, -\gamma(y_\gamma - b)\} = 0,$$

and thus

$$\mu_\gamma - \max\{0, \gamma(y_\gamma - b)\} = 0.$$

With $\phi(t) = (1/2)\max^2\{0,t\}$ there holds $\phi'(t) = \max\{0,t\}$ and, therefore, (8.34) is equivalent to

$$\mu_\gamma - \phi'(\gamma(y_\gamma - b)) = 0. \tag{8.35}$$

This is exactly the condition (8.14). Since the conditions (8.11)–(8.14) and (8.31)–(8.34) are exactly the same, we see that dual L^2-regularization of μ results in the optimality system (8.31)–(8.34) which via the equivalence of (8.34) and (8.35) is equivalent to (8.11)–(8.14) for the choice $\phi(t) = (1/2)\max^2\{0,t\}$ corresponding to Moreau–Yosida regularization. Thus, the Moreau–Yosida regularization can equivalently be viewed as an L^2-regularization for μ in the Lagrange function. We consider this dual regularization approach in section 9.2 for an elliptic obstacle problem.

8.2.5 Related Approaches

As was seen in the examples, the described approach includes the Moreau–Yosida regularization as a special case, which in a similar context was investigated in, e.g., [101, 103, 104]. A different method is Lavrentiev regularization, where the state constraint $y \le b$ is regularized via $y + u/\gamma \le b$. This approach is restricted to the situation where y and u are functions on the same domain Ω, which, e.g., does not apply to boundary control. Investigations of the Lavrentiev regularization approach can be found in, e.g., [158, 159]. The Lavrentiev regularization can be extended to more general situations by the virtual control Lavrentiev regularization. Here, a new, virtual control $v \in L^2(\Omega)$ is introduced and the constraint $y \le b$ is regularized via $y + v/\gamma \le b$. In addition, a regularization $\frac{\psi(\gamma)}{2}\|v\|_{L^2}^2$ is added to the objective function; see, e.g., [142]. This concept is very closely related to the Moreau–Yosida regularization.

Chapter 9

Several Applications

9.1 Distributed Control of a Semilinear Elliptic Equation

Let $\Omega \subset \mathbb{R}^n$ be a nonempty and bounded open domain with sufficiently smooth boundary, and consider the semilinear optimal control problem

$$\begin{aligned}
&\underset{y\in H_0^1(\Omega),u\in L^2(\Omega)}{\text{minimize}} && \frac{1}{2}\int_\Omega (y(x)-y_d(x))^2dx + \frac{\lambda}{2}\int_\Omega (u(x)-u_d(x))^2dx \\
&\text{subject to} && -\Delta y + \varphi(y) = f + gu \quad \text{on } \Omega, \\
& && \beta_1 \le u \le \beta_2 \quad \text{on } \Omega.
\end{aligned} \tag{9.1}$$

Note that the boundary conditions $y = 0$ on $\partial\Omega$ are included in the requirement $y \in H_0^1(\Omega)$. We assume $y_d \in L^2(\Omega)$, $u_d \in L^\infty(\Omega)$ (L^q with $q > 2$ would also be possible), $f \in L^2(\Omega)$, $g \in L^\infty(\Omega)$, $-\infty \le \beta_1 < \beta_2 \le +\infty$; $\lambda > 0$ is the regularization parameter. Further, let $\varphi : \mathbb{R} \to \mathbb{R}$ be nondecreasing and twice continuously differentiable with

$$|\varphi''(\tau)| \le c_1 + c_2|\tau|^{s-3}, \tag{9.2}$$

where $c_1, c_2 \ge 0$ are constants and $s > 3$ is fixed with $s \in (3,\infty]$ for $n = 1$, $s \in (3,\infty)$ for $n = 2$, and $s \in (3, 2n/(n-2)]$ for $n = 3,4,5$.

We set $U = L^2(\Omega)$, $Y = H_0^1(\Omega)$, $W = H^{-1}(\Omega)$, $W^* = H_0^1(\Omega)$, $C = [\beta_1, \beta_2]$,

$$\mathcal{C} = \{u \in U : u(x) \in C \ \text{ on } \ \Omega\},$$

and define

$$J(y,u) = \frac{1}{2}\int_\Omega (y(x)-y_d(x))^2dx + \frac{\lambda}{2}\int_\Omega (u(x)-u_d(x))^2dx, \tag{9.3}$$

$$E(y,u) = Ay + \varphi(y) - f - gu, \tag{9.4}$$

where $A \in \mathcal{L}(H_0^1(\Omega), H^{-1}(\Omega))$ is the elliptic operator defined by $-\Delta$; i.e.,

$$\langle Ay, v\rangle_{H^{-1}(\Omega),H_0^1(\Omega)} = \int_\Omega \nabla y \cdot \nabla v \, dx \quad \forall\ y, v \in H_0^1(\Omega).$$

Remark 9.1. *Without difficulty, we could replace $-\Delta$ with other H_0^1-elliptic operators $A \in \mathcal{L}(H_0^1, H^{-1})$.*

Then we can write (9.1) in the form

$$\underset{y \in Y, u \in U}{\text{minimize}} \quad J(y,u) \quad \text{subject to} \quad E(y,u) = 0, \quad u \in \mathcal{C}. \tag{9.5}$$

We now begin with our investigation of the optimal control problem.

Lemma 9.2. *The operator $E : Y \times U \to W$ defined in (9.4) is twice continuously differentiable with derivatives*

$$\begin{aligned} &E_y(y,u) = A + \varphi'(y) \cdot I, \\ &E_u(y,u) = -g \cdot I, \\ &E_{yu}(y,u) = 0, \quad E_{uy}(y,u) = 0, \quad E_{uu}(y,u) = 0 \\ &E_{yy}(y,u)(v_1, v_2) = \varphi''(y) v_1 v_2. \end{aligned}$$

Proof. By Proposition A.12 and (9.2), the superposition operator

$$L^s(\Omega) \ni u \mapsto \varphi(u) \in L^{s'}(\Omega), \quad \frac{1}{s} + \frac{1}{s'} = 1,$$

is twice continuously differentiable, since

$$\frac{s - 2s'}{s'} = \frac{s}{s'} - 2 = s - 3.$$

The choice of s implies the embeddings

$$H_0^1(\Omega) \hookrightarrow L^s(\Omega), \quad L^{s'}(\Omega) \hookrightarrow H^{-1}(\Omega).$$

Therefore, the operator $y \in H_0^1(\Omega) \mapsto \varphi(y) \in H^{-1}(\Omega)$ is twice continuously differentiable, too, and thus also E. The form of the derivatives is obvious; see Propositions A.11 and A.12. ◻

Lemma 9.3. *For every $u \in U$, the state equation $E(y,u) = 0$ possesses a unique solution $y = y(u) \in Y$.*

Proof. Integrating (9.2) twice, we see that there exist constants $C_i', C_i \ge 0$ with

$$|\varphi'(\tau)| \le C_1' + C_2'|\tau|^{s-2}, \quad |\varphi(\tau)| \le C_1 + C_2|\tau|^{s-1}. \tag{9.6}$$

Therefore, by Proposition A.10,

$$y \in L^t(\Omega) \mapsto \varphi(y) \in L^{\frac{t}{s-1}}(\Omega) \quad \text{is continuous for all } s-1 < t < \infty, \tag{9.7}$$

$$y \in L^t(\Omega) \mapsto \varphi'(y) \in L^{\frac{t}{s-2}}(\Omega) \quad \text{is continuous for all } s-2 < t < \infty. \tag{9.8}$$

Now, let

$$\theta(t) = \int_0^t \varphi(\tau)d\tau.$$

Then $\theta'(t) = \varphi(t)$, and from (9.6) and Proposition A.12 it follows that the mapping $y \in L^t \mapsto \theta(y) \in L^{t/s}$ is twice continuously differentiable for all $s \le t < \infty$ with first derivative $v \mapsto \varphi(y)v$ and second derivative $(v,w) \mapsto \varphi'(y)vw$. Since $H_0^1(\Omega) \hookrightarrow L^s(\Omega)$, this also holds for $y \in H_0^1(\Omega) \mapsto \theta(y) \in L^1(\Omega)$. Now consider, for fixed $u \in \mathcal{C}$, the function $e : H_0^1(\Omega) \mapsto \mathbb{R}$,

$$e(y) = \frac{1}{2}\int_\Omega \nabla y(x)\cdot\nabla y(x)dx + \int_\Omega \theta(y(x))dx - (f + gu, y)_{L^2}.$$

This function is twice continuously differentiable with

$$e'(y) = Ay + \varphi(y) - f - gu = E(y,u),$$

$$e''(y)(v,v) = \langle Av, v\rangle_{H^{-1},H_0^1} + \int_\Omega \varphi'(y(x))v(x)v(x)dx \ge \|v\|^2_{H_0^1}.$$

Therefore, by standard existence and uniqueness results for strongly convex optimization problems (see, e.g., [204, Prop. 25.22]), there exists a unique solution $y = y(u) \in H_0^1(\Omega)$ of $E(y,u) = 0$. Thus, for all u, there is a unique solution $y = y(u)$ of the state equation. ☐

Next, we discuss the existence of solutions of the optimal control problem for the cases $n = 1,2,3$. To simplify the presentation, we assume $s \in (3,4]$ in the case $n = 3$.

Lemma 9.4. *Let* $1 \le n \le 5$ *and assume* $s \in (3, 2n/(n-2))$ *in the case* $n \ge 3$. *Then the optimal control problem* (9.5) *admits a solution.*

Proof. By Lemma 9.3 there exists a (feasible) minimizing sequence (y_k, u_k) for the optimal control problem, which, due to the structure of J, is bounded in $L^2 \times L^2$. Note that in the case $\beta_1, \beta_2 \in \mathbb{R}$ the particular form of $\mathcal{C}$ even implies that $\|u_k\|_{L^\infty} \le \max\{|\beta_1|, |\beta_2|\}$, but we do not need this here. From $E(y_k,u_k) = 0$ and $(\varphi(y) - \varphi(0))y \ge 0$ we obtain

$$\begin{aligned}
\|y_k\|^2_{H_0^1} &\le \langle Ay_k, y_k\rangle_{H^{-1},H_0^1} + \int_\Omega [\varphi(y_k)(x) - \varphi(0)]y_k(x)dx \\
&= (f + gu_k - \varphi(0), y_k)_{L^2} \\
&\le \big(\|f\|_{L^2} + \|g\|_{L^\infty}\|u_k\|_{L^2} + \text{meas}(\Omega)^{1/2}|\varphi(0)|\big)\|y_k\|_{L^2} \\
&\le C\big(\|f\|_{L^2} + \|g\|_{L^\infty}\|u_k\|_{L^2} + |\varphi(0)|\big)\|y_k\|_{H_0^1}.
\end{aligned}$$

This implies that (y_k) is bounded in H_0^1. Due to the boundedness of the sequence (y_k,u_k) in $H_0^1 \times L^2$ and the weak sequential closedness of $\mathcal{C}$ we can select a subsequence such that $y_k \rightharpoonup y^*$ weakly in H_0^1 and $u_k \rightharpoonup u^* \in \mathcal{C}$ weakly in L^2. Since $H_0^1 \hookrightarrow L^t$ *compactly* for all $1 \le t < \infty$ if $n \le 2$, and all $1 \le t < 2n/(n-2)$ if $n = 3,4,5$, we see, since we required $3 < s < 2n/(n-2)$ for $n = 3,4,5$, that the embedding $H_0^1 \hookrightarrow L^s$ is compact and that φ maps L^s continuously to $L^{s'}$ with $1/s' + 1/s = 1$; i.e., $s' = s/(s-1)$, see (9.7) with $t = s$. We conclude that the weak convergence $y_k \rightharpoonup y^*$ in H_0^1 implies $y_k \to y^*$ strongly in L^s and thus $\varphi(y_k) \to \varphi(y^*)$ strongly in $L^{s'} \hookrightarrow H^{-1}$. Now

$$\begin{aligned} f + gu_k &\to f + gu^* && \text{weakly in } L^2 \hookrightarrow H^{-1}, \\ f + gu_k = Ay_k + \varphi(y_k) &\to Ay^* + \varphi(y^*) && \text{weakly in } H^{-1} \end{aligned}$$

shows $E(y^*, u^*) = 0$. Therefore, (y^*, u^*) is feasible. Furthermore, J is continuous and convex, and thus weakly lower semicontinuous. From the weak convergence $(y_k, u_k) \to (y^*, u^*)$ we thus conclude that (y^*, u^*) solves the problem. □

9.1.1 Black-Box Approach

In Lemma 9.3 it was proved that the state equation admits a unique solution $y(u)$. Therefore, we can introduce the reduced objective function

$$j(u) = J(y(u), u)$$

and consider the equivalent reduced problem

$$\underset{u \in U}{\text{minimize}} \quad j(u) \quad \text{subject to} \quad u \in \mathcal{C}. \tag{9.9}$$

From Lemma 9.2 we know that E is twice continuously differentiable. Our next aim is to apply the implicit function theorem to prove that $y(u)$ is twice continuously differentiable. To this end we observe the following.

Lemma 9.5. *For all $y \in Y$ and $u \in U$, the partial derivative*

$$E_y(y, u) = A + \varphi'(y) \cdot I \in \mathcal{L}(Y, W) = \mathcal{L}\left(H_0^1, H^{-1}\right)$$

is a homeomorphism with

$$\|E_y(y, u)^{-1}\|_{W,Y} \le 1.$$

Proof. Since φ is nondecreasing, we have $\varphi' \ge 0$ and thus for all $v \in H_0^1$

$$\langle E_y(y, u)v, v\rangle_{H^{-1}, H_0^1} = (v, v)_{H_0^1} + \int_\Omega \varphi'(y) v^2 dx \ge \|v\|_{H_0^1}^2.$$

Therefore, by the Lax–Milgram theorem, $E_y(y, u) \in \mathcal{L}(H_0^1, H^{-1}) = \mathcal{L}(Y, W)$ is a homeomorphism with $\|E_y(y, u)^{-1}\|_{W,Y} \le 1$. □

Therefore, we can apply the implicit function theorem to obtain the following lemma.

Lemma 9.6. *The mapping $u \in U \mapsto y(u) \in Y$ is twice continuously differentiable.*

Since the objective function J is quadratic, we thus have the following.

Lemma 9.7. *The reduced objective function $j : U \to \mathbb{R}$ is twice continuously differentiable.*

Finally, we establish the following structural result for the reduced gradient.

Lemma 9.8. *The reduced gradient $j'(u)$ has the form*

$$j'(u) = \lambda u + G(u), \quad G(u) = -gw(u) - \lambda u_d,$$

where $w = w(u)$ solves the adjoint equation

$$Aw + \varphi'(y)w = y_d - y(u). \tag{9.10}$$

The mapping $u \in U \mapsto G(u) \in L^{p'}(\Omega)$ is continuously differentiable, and thus locally Lipschitz continuous, for all $p' \in [2,\infty]$ if $n = 1$, $p' \in [2,\infty)$ if $n = 2$, and $p' \in [2, 2n/(n-2)]$ if $n \geq 3$. As a consequence, the mapping

$$u \in L^p(\Omega) \mapsto j'(u) \in L^r(\Omega)$$

is continuously differentiable for all $p \in [2,\infty]$ and all $r \in [1, \min\{p, p'\}]$.

Proof. Using the adjoint representation of j', we see that

$$j'(u) = J_u(y(u),u) + E_u(y(u),u)^* w(u) = \lambda(u - u_d) - gw(u),$$

where $w = w(u)$ solves the adjoint equation $E_y(y(u),u)^* w = -J_y(y(u),u)$, which has the form (9.10). Since $E_y(y(u),u)^*$ is a homeomorphism by Lemma 9.5, the adjoint state $w(u)$ is unique. Further, since E_y, $y(u)$, and J_y are continuously differentiable, we can use the implicit function theorem to prove that the mapping $u \in U \mapsto w(u) \in W^*$ is continuously differentiable, and thus, in particular, locally Lipschitz continuous.

For p' as given in the lemma, the embedding $W = H_0^1 \hookrightarrow L^{p'}$ implies that the operator $G(u) = -gw(u) - \lambda u_d$ is continuously differentiable, and thus locally Lipschitz continuous, as a mapping from U to $L^{p'}$. The lemma's last assertion follows immediately. $\square$

Our aim is to apply our class of semismooth Newton methods to compute critical points of problem (9.9), i.e., to solve the VIP

$$u \in \mathcal{C}, \quad (j'(u), v - u)_{L^2} = 0 \quad \forall\, v \in \mathcal{C}. \tag{9.11}$$

The solutions of (9.11) enjoy the following regularity property.

Lemma 9.9. *Every solution $\bar{u} \in U$ of (9.11) satisfies $\bar{u} \in L^\infty(\Omega)$ if $\beta_1, \beta_2 \in \mathbb{R}$, and $\bar{u} \in L^{p'}(\Omega)$ with p' as in Lemma 9.8, otherwise.*

Proof. For $\beta_1, \beta_2 \in \mathbb{R}$ we have $\mathcal{C} \subset L^\infty(\Omega)$ and the assertion is obvious. For $\beta_1 = -\infty$, $\beta_2 = +\infty$, it follows from (9.11) that

$$0 = j'(\bar{u}) = \lambda \bar{u} + G(\bar{u}),$$

and thus $\bar{u} = -\lambda^{-1} G(\bar{u}) \in L^{p'}(\Omega)$ by Lemma 9.8. For $\beta_1 > -\infty$, $\beta_2 = +\infty$ we conclude in the same way $\mathbf{1}_{\{\bar{u} \neq \beta_1\}} j'(\bar{u}) = 0$, and thus

$$\mathbf{1}_{\{\bar{u} \neq \beta_1\}} \bar{u} = -\lambda^{-1} \mathbf{1}_{\{\bar{u} \neq \beta_1\}} G(\bar{u}) \in L^{p'}(\Omega).$$

Furthermore,

$$\mathbf{1}_{\{\bar{u}=\beta_1\}}\bar{u} = \beta_1 \mathbf{1}_{\{\bar{u}=\beta_1\}} \in L^\infty(\Omega).$$

The case $\beta_1 = -\infty$, $\beta_2 < +\infty$ can be treated in the same way. □

With the results developed above we have everything at hand to prove the semismoothness of the superposition operator Π arising from equation reformulations

$$\Pi(u) = 0, \quad \Pi(u) \overset{\text{def}}{=} \pi(u, j'(u)) \tag{9.12}$$

of problem (9.11), where π is an MCP-function for the interval $[\beta_1, \beta_2]$. In the following, we distinguish the two variants of reformulations that were discussed in section 5.1.2.

First Reformulation

Here, we discuss reformulations based on a general MCP-function $\pi = \phi_{[\beta_1,\beta_2]}$ for the interval $C = [\beta_1, \beta_2]$.

Theorem 9.10. *The problem assumptions imply that Assumptions* 5.10 (a), (b) *(with $Z = \{0\}$) are satisfied with $F = j'$ for any $p \in [2,\infty]$; any $p' \leq p$ with $p' \in [2,\infty]$ if $n = 1$, $p' \in [2,\infty)$ if $n = 2$, and $p' \in [2, 2n/(n-2)]$ if $n \geq 3$; and any $r \in [1, p']$.*

In particular, if π satisfies Assumptions 5.10 (c), (d), *then Theorem* 5.11 *yields the $\partial^\circ\Pi$-semismoothness of the operator Π. Here, the differential $\partial^\circ\Pi(u)$ consists of all operators $M \in \mathcal{L}(L^p, L^r)$,*

$$M = d_1 I + d_2 \cdot j''(u), \quad d \in L^\infty(\Omega)^2, \quad d \in \partial\pi\big(u, j'(u)\big) \quad \text{on } \Omega. \tag{9.13}$$

Proof. The assertions follow immediately from the boundedness of Ω, Lemma 9.8, and Theorem 5.11. □

Concerning higher-order semismoothness, we have the following.

Theorem 9.11. *Suppose that the operator $y \in H_0^1(\Omega) \mapsto \varphi(y) \in H^{-1}(\Omega)$ is three times continuously differentiable. This can, e.g., be satisfied if φ has suitable properties.*

Then Assumptions 5.12 (a), (b) *with $Z = \{0\}$ and $\alpha = 1$ are satisfied by $F = j'$ for $r = 2$, any $p \in (2,\infty]$, and all $p' \leq p$ with $p' \in (2,\infty]$ if $n = 1$, $p' \in (2,\infty)$ if $n = 2$, and $p' \in (2, 2n/(n-2)]$ if $n \geq 3$.*

In particular, if π satisfies Assumptions 5.12 (c), (d), *then Theorem* 5.13 *yields the β-order $\partial^\circ\Pi$-semismoothness of the operator $\Pi(u) = \pi(u, j'(u))$, where β is given by Theorem* 3.50. *The differential $\partial^\circ\Pi(u)$ consists of all operators $M \in \mathcal{L}(L^p, L^2)$ of the form* (9.13).

Proof. If $y \in H_0^1 \mapsto \varphi(y) \in H^{-1}$ is three times continuously differentiable, then also E and, thus, by the implicit function theorem, $y(u)$ is three times continuously differentiable. Hence, $j' : L^2 \to L^2$ is twice continuously differentiable and therefore its derivative is locally Lipschitz continuous. The same then holds true for $u \in L^p \mapsto j'(u) \in L^r$.

The assertions now follow from the boundedness of Ω, Lemma 9.8, and Theorem 5.13. □

Remark 9.12. *The Hessian operator j'' can be obtained via the adjoint representation in section* A.1 *of the appendix. In section* 9.1.3 *it is described how finite element discretizations of j, j', j'', $\partial^\circ\Phi$, etc., can be computed.*

Second Reformulation

We now consider the case where

$$\Pi(u) = u - P_{[\beta_1,\beta_2]}(u - \lambda^{-1} j'(u))$$

is chosen to reformulate the problem as equation $\Pi(u) = 0$.

Theorem 9.13. *The problem assumptions imply that Assumptions* 5.14 (a), (b) (*with $Z = \{0\}$*) *are satisfied with $F = j'$ for $r = 2$ and any $p' \in (2,\infty]$ if $n = 1$, $p' \in (2,\infty)$ if $n = 2$, and $p' \in (2, 2n/(n-2)]$ if $n \ge 3$.*

In particular, Theorem 5.15 *yields the $\partial^\circ\Pi$-semismoothness of the operator Π. Here, the differential $\partial^\circ\Pi(u)$ consists of all operators $M \in \mathcal{L}(L^r, L^r)$,*

$$\begin{aligned} M &= I + \lambda^{-1} d \cdot G_u(u), \\ d \in L^\infty(\Omega), \quad & d \in \partial P_{[\beta_1,\beta_2]}\big(-\lambda^{-1} G(u)\big) \quad \text{on } \Omega. \end{aligned} \tag{9.14}$$

Proof. The assertions follow immediately from the boundedness of Ω, Lemma 9.8, and Theorem 5.15. □

A result establishing higher-order semismoothness analogous to Theorem 9.11 can also be established, but we do not formulate it here.

Remark 9.14. *Since $j''(u) = \lambda I + G_u(u)$, the adjoint representation of section* A.1 *can be used to compute $G_u(u)$.*

Regularity

For the application of semismooth Newton methods, a regularity condition such as in Assumption 3.64 (a) has to hold. For the problem under consideration, we can establish regularity by using the sufficient condition of Theorem 4.8. Since this condition was established for NCPs (but can be extended to other situations), we consider the case of the NCP, i.e., $\beta_1 = 0$, $\beta_2 = \infty$. To apply Theorem 4.8, we have to verify the conditions of Assumption 4.6.

The assumptions (a)–(d) follow immediately from Lemma 9.8 for p' as in the lemma and any $p \in [p', \infty]$. Note hereby that $G'(u) = j''(u) - \lambda I$ is self-adjoint. Assumption (e) requires that the Hessian operator $j''(\bar{u})$ is coercive on the tangent space of the strongly active constraints, which is an infinite-dimensional analogue of the strong second-order sufficient condition for optimality. The remaining assumptions (f)–(h) only concern the NCP-function and are satisfied for $\phi = \phi^{FB}$ as well as $\phi(x) = x_1 - P_{[0,\infty)}(x_1 - \lambda^{-1} x_2)$, the NCP-function used in the second reformulation.

Application of Semismooth Newton Methods

In conclusion, we have shown that problem (9.1) satisfies all assumptions that are required to prove superlinear convergence of our class of (projected) semismooth Newton methods. Here, both types of reformulations are appropriate, the one of section 5.1.1 and the semismooth reformulation of section 4.2, the latter yielding a smoothing-step-free method. Numerical results are given in section 9.1.7.

9.1.2 All-at-Once Approach

We now describe, in some less detail, how mixed semismooth Newton methods can be applied to solve the all-at-once KKT-system. The continuous invertibility of $E_y(y,u) = A+\varphi'(y)I \in \mathcal{L}(H_0^1, H^{-1})$ guarantees that Robinson's regularity condition is satisfied, so that every solution $(\bar{y},\bar{u})$ satisfies the KKT conditions (5.24)–(5.26), where $\bar{w} \in W = H_0^1(\Omega)$ is a multiplier. The Lagrange function $L: Y \times U \times W \to \mathbb{R}$ is given by

$$\begin{aligned} L(y,u,w) &= J(y,u)+\langle E(y,u),w\rangle_{H^{-1},H_0^1} \\ &= J(y,u)+\langle Aw,y\rangle_{H^{-1},H_0^1}+\langle\varphi(y),w\rangle_{H^{-1},H_0^1} \\ &\quad -(f,w)_{L^2}-(gu,w)_{L^2}. \end{aligned}$$

Now, using the results of the previous sections, we obtain the following.

Lemma 9.15. *The Lagrange function L is twice continuously differentiable with derivatives*

$$\begin{aligned} &L_y(y,u,w) = J_y(y,u)+E_y(y,u)^*w = y-y_d+Aw+\varphi'(y)w, \\ &L_u(y,u,w) = J_u(y,u)+E_u(y,u)^*w = \lambda(u-u_d)-gw, \\ &L_w(y,u,w) = E(y,u), \\ &L_{yy}(y,u,w) = (1+\varphi''(y)w)I, \\ &L_{yu}(y,u,w) = 0, \quad L_{uy}(y,u,w) = 0, \quad L_{uu}(y,u,w) = 0. \end{aligned}$$

Since $L_w = E$, we have $L_{wy} = E_y$, etc.; see Lemma 9.2 for formulas. Furthermore, L_u can be written in the form

$$L_u(y,u,w) = \lambda u + G(y,u,w), \quad G(y,u,w) = -gw-\lambda u_d.$$

The mapping $(y,u,w) \in Y \times U \times W \mapsto G(y,u,w) \in L^{p'}(\Omega)$ is continuous affine linear for all $p' \in [2,\infty]$ if $n=1$, $p' \in [2,\infty)$ if $n=2$, and $p' \in [2,2n/(n-2)]$ if $n \geq 3$. As a consequence, the mapping

$$(y,u,w) \in Y \times L^p(\Omega) \times W \mapsto L_u(y,u,w) \in L^r(\Omega)$$

is continuous affine linear for all $p \in [2,\infty]$ and all $r \in [1,\min\{p,p'\}]$.

Proof. The differentiability properties and the form of the derivatives is an immediate consequence of Lemma 9.2. The mapping properties of L_u are due to the fact that the embedding $H_0^1 \hookrightarrow L^{p'}$ is continuous. □

For KKT-triples we have the following regularity result.

Lemma 9.16. *Every KKT-triple* $(\bar{y},\bar{u},\bar{w}) \in Y \times U \times W$ *of* (9.11) *satisfies* $\bar{u} \in L^\infty(\Omega)$ *if* $\beta_1,\beta_2 \in \mathbb{R}$, *and* $\bar{u} \in L^{p'}(\Omega)$ *with* p' *as in Lemma* 9.15, *otherwise.*

Proof. The proof of Lemma 9.9 can be easily adjusted. □

From Lemma 9.15 we conclude that Assumptions 5.17 (a)–(c) are satisfied for $r = 2$, all $p \in [2,\infty]$, and all $p' \le p$ as in the lemma. Hence, using an MCP-function π that satisfies Assumption 5.17 (d), we can write the KKT conditions in the form (5.27), and Theorem 5.19 yields the semismoothness of Σ. Furthermore, Lemma 9.15 implies that Assumption 5.27 is satisfied for $p = p'$, and we thus can compute smoothing steps as described in Theorem 5.29. Therefore, if the generalized differential is regular near the KKT-triple $(\bar{y},\bar{u},\bar{w}) \in Y \times L^p(\Omega) \times W$, $p = p'$ (cf. Lemma 9.16), the semismooth Newton methods of section 3.2.3 are applicable and converge superlinearly.

In a similar way, we can deal with the second mixed reformulation, which is based on Assumption 5.20.

9.1.3 Finite Element Discretization

For the discretization of the state equation, we follow [78, Ch. IV.2.5; 79, App. 1.6.4]. Let $\Omega \subset \mathbb{R}^2$ be a bounded polygonal domain and let $\mathcal{T}^h$ be a regular triangulation of Ω:

- $\mathcal{T}^h = \{T_i^h : T_i^h \text{ is a triangle, } i = 1,\dots,m^h\}$.
- $\bigcup_{T^h \in \mathcal{T}^h} T^h = \bar{\Omega}$, $\operatorname{int} T_i^h \cap \operatorname{int} T_j^h = \emptyset$ for all $i \neq j$.
- For all $i \neq j$, $T_i^h \cap T_j^h$ is either a common edge, a common vertex, or the empty set.
- The parameter h denotes the length of the longest edge of all triangles in the triangulation.

Now, we define

$$V^h = \{v^h \in C^0(\bar{\Omega}) : v^h|_T \text{ affine linear for all } T \in \mathcal{T}^h\},$$
$$V_0^h = \{v^h \in V^h : v^h|_{\partial\Omega} = 0\}.$$

Further, denote by Σ^h the set of all vertices in the triangulation $\mathcal{T}^h$ and by

$$\Sigma_0^h = \{P \in \Sigma^h : P \notin \partial\Gamma\}$$

the set of all interior vertices of $\mathcal{T}^h$.

For every $P \in \Sigma_0^h$ there exists a unique function $\beta_P^h \in V_0^h$ with $\beta_P^h(P) = 1$ and $\beta_P^h(Q) = 0$ for all $Q \in \Sigma^h$, $Q \neq P$. The set $\beta^h = \{\beta_P^h : P \in \Sigma_0^h\}$ is a basis of V_0^h, and we can write every $v^h \in V_0^h$ uniquely in the form

$$v^h = \sum_{P \in \Sigma_0^h} v_P^h \beta_P^h \quad \text{with} \quad v_P^h = v^h(P).$$

The space $H^h \subset L^\infty(\Omega)$ is defined by

$$H^h = \{u^h \in L^\infty(\Omega) : u^h|_T \text{ constant for all } T \in \mathcal{T}\}.$$

Here, the specific values of u^h on the edges of the triangles (which are null sets) are not relevant. The set of functions $\eta^h = \{\eta_T^h : T \in \mathcal{T}^h\}$, $\eta_T^h = 1$ on T and $\eta_T^h = 0$, otherwise, forms a basis of H^h, and for all $u^h \in H^h$ there holds

$$u^h = \sum_{T \in \mathcal{T}^h} u_T^h \eta_T^h, \quad \text{where} \quad u^h|_T \equiv u_T^h.$$

For every $P \in \Sigma_0^h$, let Ω_P^h be the polygon around P whose boundary connects midpoints of edges emanating from P with midpoints of triangles containing P and this edge. By χ_P^h, we denote the characteristic function of Ω_P, being equal to one on Ω_P^h and vanishing on $\Omega \setminus \Omega_P$. Finally, we introduce the linear operator $L^h : C^0(\bar{\Omega}) \cap H_0^1(\Omega) \to L^\infty(\Omega)$,

$$L^h v = \sum_{P \in \Sigma^h} v(P) \chi_P^h.$$

Obviously, $L^h v$ is constant on $\operatorname{int} \Omega_P$ with value $v(P)$.

We choose H^h for the discrete control space and V_0^h for the discrete state space. Now, we discretize the state equation as follows:

$$(y^h, v^h)_{H_0^1} + \int_\Omega \varphi(L^h y^h)(L^h v^h)dx = (f + g u^h, v^h)_{L^2} \quad \forall\, v^h \in V_0^h. \tag{9.15}$$

It is easy to see that

$$\begin{aligned}\int_\Omega \varphi(L^h y^h)(L^h \beta_P^h)dx &= \varphi(y_P^h)(L^h \beta_P^h, L^h \beta_P^h)_{L^2} \\ &= \text{meas}(\Omega_P)\varphi(y_P^h) = \frac{1}{3}\sum_{T \ni P} \text{meas}(T)\varphi(y_P^h).\end{aligned}$$

The objective function J is discretized by

$$J^h(y^h, u^h) = \frac{1}{2}\int_\Omega (L^h y^h - y_d)^2 dx + \frac{\lambda}{2}\int_\Omega (u^h - u_d)^2 dx.$$

Remark 9.17. *For the first integral in J^h we also could have used*

$$\int_\Omega (y^h - y_d)^2 dx,$$

but in coordinate form this would result in a quadratic term of the form $\frac{1}{2}\mathbf{y}^{h^T}\hat{\mathbf{M}}^h\mathbf{y}^h$, with nondiagonal matrix $\hat{\mathbf{M}}^h$, $\hat{\mathbf{M}}_{ij}^h = (\beta_i^h, \beta_j^h)_{L^2}$, which would make the numerical computations more expensive.

The discrete feasible set is

$$\mathcal{C}^h = H^h \cap \mathcal{C}.$$

Thus, we can write down the fully discrete optimal control problem:

$$\begin{aligned} \underset{y^h \in V_0^h, u^h \in H^h}{\text{minimize}} \quad & \frac{1}{2}\int_\Omega (L^h y^h - y_d)^2 dx + \frac{\lambda}{2}\int_\Omega (u^h - u_d)^2 dx \\ \text{subject to} \quad & (y^h, v^h)_{H_0^1} + (\varphi(L^h y^h), L^h v^h)_{L^2} \\ & = (f + g u^h, v^h)_{L^2} \quad \forall\, v^h \in V_0^h,\ u^h \in \mathcal{C}^h. \end{aligned} \tag{9.16}$$

Next, we intend to write (9.16) in coordinate form. To this end, let

$$\Sigma_0^h = \left\{ P_1^h, \dots, P_{n^h}^h \right\}, \quad \beta_i^h = \beta_{P_i^h}^h, \quad \eta_l^h = \eta_{T_l^h}^h.$$

Further, we write $\mathbf{y}^h \in \mathbb{R}^{n^h}$ for the coordinates of $y^h \in V_0^h$ with respect to the basis $\beta^h = \{\beta_i^h\}$ and $\mathbf{u}^h \in \mathbb{R}^{m^h}$ for the coordinates of $u^h \in H^h$ with respect to the basis $\eta^h = \{\eta_l^h\}$. We define the matrices $\mathbf{A}^h, \mathbf{S}^h \in \mathbb{R}^{n^h \times n^h}$,

$$\mathbf{A}_{ij}^h = (\beta_i^h, \beta_j^h)_{H_0^1}, \quad \mathbf{S}_{ij}^h = (L^h \beta_i^h, L^h \beta_j^h)_{L^2} \tag{9.17}$$

(note that $\mathbf{S}^h$ is diagonal and positive definite), the vectors $\mathbf{f}^h, \varphi(\mathbf{y}^h) \in \mathbb{R}^{n^h}$,

$$\mathbf{f}_i^h = (\beta_i^h, f)_{L^2}, \quad \varphi(\mathbf{y}^h)_i = \varphi(\mathbf{y}_i^h),$$

and the matrix $\mathbf{G}^h \in \mathbb{R}^{n^h \times m^h}$,

$$\mathbf{G}_{il}^h = (\beta_i^h, g\eta_l^h)_{L^2}.$$

Then (9.15) is equivalent to the nonlinear system of equations

$$\mathbf{A}^h \mathbf{y}^h + \mathbf{S}^h \varphi(\mathbf{y}^h) = \mathbf{f}^h + \mathbf{G}^h \mathbf{u}^h. \tag{9.18}$$

Further, in coordinates we can write J^h as

$$\mathbf{J}^h(\mathbf{y}^h, \mathbf{u}^h) = \frac{1}{2}\mathbf{y}^{h\,T} \mathbf{S}^h \mathbf{y}^h - \mathbf{y}_d^{h\,T} \mathbf{S}^h \mathbf{y}^h + \frac{\lambda}{2}\mathbf{u}^{h\,T} \mathbf{M}^h \mathbf{u}^h - \lambda \mathbf{u}_d^{h\,T} \mathbf{M}^h \mathbf{u}^h + \gamma,$$

where the mass matrix $\mathbf{M}^h \in \mathbb{R}^{m^h \times m^h}$, the vectors $\mathbf{y}_d^h \in \mathbb{R}^{n^h}$, $\mathbf{u}_d^h \in \mathbb{R}^{m^h}$, and the scalar γ are defined by

$$\begin{aligned} \mathbf{M}_{kl}^h &= (\eta_k^h, \eta_l^h)_{L^2}, \quad (\mathbf{y}_d^h)_i = \frac{1}{\text{meas}(\Omega_{P_i})}\int_{\Omega_{P_i}} y_d(x)dx, \\ (\mathbf{M}^h \mathbf{u}_d^h)_l &= (\eta_l^h, u_d)_{L^2}, \quad \gamma = \frac{1}{2}\|y_d\|_{L^2}^2 + \frac{\lambda}{2}\|u_d\|_{L^2}^2. \end{aligned}$$

Finally, we note that $u^h \in \mathcal{C}^h$ if and only if its η^h-coordinates $\mathbf{u}^h$ satisfy $\mathbf{u}^h \in \boldsymbol{\mathcal{C}}^h$, where

$$\boldsymbol{\mathcal{C}}^h = \{\mathbf{u}^h \in \mathbb{R}^{m^h} : \mathbf{u}_l^h \in C,\ l = 1, \dots, m^h\}.$$

Thus, we can write down the fully discrete optimal control problem in coordinate form:

$$\begin{aligned} &\underset{\mathbf{y}^h \in \mathbb{R}^{n^h}, \mathbf{u}^h \in \mathbb{R}^{m^h}}{\text{minimize}} && \mathbf{J}^h(\mathbf{y}^h, \mathbf{u}^h) \\ &\text{subject to} && \mathbf{A}^h \mathbf{y}^h + \mathbf{S}^h \varphi(\mathbf{y}^h) = \mathbf{f}^h + \mathbf{G}^h \mathbf{u}^h, \quad \mathbf{u}^h \in \mathfrak{C}^h. \end{aligned} \tag{9.19}$$

It is advisable to consider problem (9.19) only in conjunction with the coordinate-free version (9.16), since (9.16) still contains all the information on the underlying function spaces while problem (9.19) does not. To explain this in more detail, we give a very simple example (readers familiar with discretizations of optimal control problems can skip the example).

Example 9.18. Let us consider the trivial problem

$$\underset{u \in L^2(\Omega)}{\text{minimize}} \quad j(u) \stackrel{\text{def}}{=} \frac{1}{2}\|u\|_{L^2}^2.$$

Since $j'(u) = u$, from every point $u \in L^2$ a gradient step with stepsize 1 brings us to the solution $u^* \equiv 0$. Of course, for a proper discretization of this problem, we expect a similar behavior. Discretizing $U = L^2(\Omega)$ by H^h as above, and j by $j^h(u^h) = j(u^h) = \|u^h\|_{L^2}^2/2$, we have $j^{h'}(u^h) = u^h$ and thus, after one gradient step with stepsize 1, we have found the solution. Consequently, if $\mathbf{u}^h$ are the η^h-coordinates of u^h, then the η^h-coordinates $\mathbf{j}^{h'}(\mathbf{u}^h)$ of $j^h(u^h) = u^h$ are $\mathbf{j}^{h'}(\mathbf{u}^h) = \mathbf{u}^h$, and the step $-\mathbf{j}^{h'}(\mathbf{u}^h)$ brings us from $\mathbf{u}^h$ to the solution 0.

However, the following approach yields a completely different result: In coordinate form, the discretized problem reads

$$\underset{\mathbf{u}^h \in \mathbb{R}^{m^h}}{\text{minimize}} \quad \mathbf{j}^h(\mathbf{u}^h) \quad \text{with} \quad \mathbf{j}^h(\mathbf{u}^h) = \frac{1}{2}\mathbf{u}^{h^T} \mathbf{M}^h \mathbf{u}^h.$$

Differentiating $\mathbf{j}^h(\mathbf{u}^h)$ with respect to $\mathbf{u}^h$ yields

$$\frac{d}{d\mathbf{u}^h}\mathbf{j}^h(\mathbf{u}^h) = \mathbf{M}^h \mathbf{u}^h = \mathbf{M}^h \mathbf{j}^{h'}(\mathbf{u}^h).$$

Since $\|\mathbf{M}^h\| = O(h^2)$, this Euclidean gradient is very short and a gradient step of stepsize 1 will provide almost no progress. Therefore, it is crucial to work with gradients that are represented with respect to the correct inner product, in our case the one induced by the matrix $\mathbf{M}^h$, which corresponds to the inner product of H^h, the discretization of L^2. ∎

9.1.4 Discrete Black-Box Approach

We proceed by discussing the black-box approach, applied to the discrete optimal control problem (9.16). It is straightforward to derive analogues of Lemmas 9.2–9.8 for the discrete optimal control problem. In particular, the discrete state equation (9.15) possesses a unique solution operator $u^h \in H^h \mapsto y^h(u^h) \in V_0^h$ which is twice continuously differentiable. The reduced objective function is $j^h(u^h) = J^h(y^h(u^h), u^h)$ where $y^h = y^h(u^h)$ solves (9.15), or, in coordinate form, $\mathbf{j}^h(\mathbf{u}^h) = \mathbf{J}^h(\mathbf{y}^h(\mathbf{u}^h), \mathbf{u}^h)$, where $\mathbf{y}^h = \mathbf{y}^h(\mathbf{u}^h)$ solves (9.18).

The discrete adjoint equation is given by the variational equation

$$\forall v^h \in V_0^h:$$
$$(v^h, w^h)_{H_0^1} + (\varphi'(L^h y^h)L^h v^h, L^h w^h)_{L^2} = \langle -J^h_{y^h}(y^h, u^h), v^h\rangle_{H^{-1},H_0^1}.$$

The coordinates $\mathbf{w}^h \in \mathbb{R}^{n_h}$ of the discrete adjoint state $w^h \in V_0^h$ are thus given by

$$\big(\mathbf{A}^h + \mathbf{T}^h(\mathbf{y}^h)\big)\mathbf{w}^h = -\mathbf{S}^h(\mathbf{y}^h - \mathbf{y}_d^h),$$

where

$$\mathbf{T}^h(\mathbf{y}^h) = \mathbf{S}^h \operatorname{diag}\big(\varphi'(\mathbf{y}_1^h), \dots, \varphi'(\mathbf{y}_{n^h}^h)\big).$$

The discrete reduced gradient $j^{h\prime}(u^h) \in H^h$ satisfies

$$\begin{aligned}(j^{h\prime}(u^h), z^h)_{L^2} &= (J^h_{u^h}(y^h, u^h), z^h)_{L^2} + (w^h, -gz^h)_{L^2}\\ &= (\lambda(u^h - u_d) - gw^h, z^h)_{L^2}.\end{aligned}$$

Now observe that

$$\begin{aligned}\Big(\textstyle\sum_k (\mathbf{M}^{h^{-1}}\mathbf{G}^{h^T}\mathbf{w}^h)_k \eta_k^h, \sum_l \eta_l^h \mathbf{z}_l^h\Big)_{L^2} &= \mathbf{z}^{h^T}\mathbf{M}^h\mathbf{M}^{h^{-1}}\mathbf{G}^{h^T}\mathbf{w}^h\\ = \mathbf{z}^{h^T}\mathbf{G}^{h^T}\mathbf{w}^h = (w^h, gz^h)_{L^2} &= (gw^h, z^h)_{L^2}.\end{aligned}$$

Hence, the η^h-coordinates of $j^{h\prime}(u^h)$ are

$$\mathbf{j}^{h\prime}(\mathbf{u}^h) = \lambda(\mathbf{u}^h - \mathbf{u}_d^h) - \mathbf{M}^{h^{-1}}\mathbf{G}^{h^T}\mathbf{w}^h.$$

As already illustrated in Example 9.18, the vector $\mathbf{j}^{h\prime}(\mathbf{u}^h)$ is *not* the usual gradient of $\mathbf{j}^h(\mathbf{u}^h)$ with respect to $\mathbf{u}^h$, which corresponds to the gradient representation with respect to the Euclidean inner product. In fact, we have

$$\frac{d}{d\mathbf{u}^h}\mathbf{j}^h(\mathbf{u}^h) = \lambda\mathbf{M}^h(\mathbf{u}^h - \mathbf{u}_d^h) - \mathbf{G}^{h^T}\mathbf{w}^h = \mathbf{M}^h\mathbf{j}^{h\prime}(\mathbf{u}^h). \tag{9.20}$$

Rather, $\mathbf{j}^{h\prime}(\mathbf{u}^h)$ is the gradient representation with respect to the inner product of H^h, which is represented by the matrix $\mathbf{M}^h$.

Writing down the first-order necessary conditions for the discrete reduced problem (9.16), we obtain

$$u^h \in \mathcal{C}^h, \quad (j^{h\prime}(u^h), v^h - u^h)_{L^2} \geq 0 \quad \forall\, v^h \in \mathcal{C}^h. \tag{9.21}$$

In coordinate form, this becomes

$$\mathbf{u}^h \in \mathcal{C}^h, \quad \mathbf{j}^{h\prime}(\mathbf{u}^h)^T\mathbf{M}^h(\mathbf{v}^h - \mathbf{u}^h) \geq 0 \quad \forall\, \mathbf{v}^h \in \mathcal{C}^h. \tag{9.22}$$

Since $\mathbf{M}^h$ is diagonal positive definite, we can write (9.21) equivalently as

$$\mathbf{u}_l^h - P_C(\mathbf{u}_l^h - \mathbf{j}^{h\prime}(\mathbf{u}^h)_l) = 0, \quad l = 1, \dots, m^h.$$

This is the discrete analogue of the condition

$$u - P_C(u - j'(u)) = 0,$$

which we used to express the continuous problem in the form

$$\Pi(u) \stackrel{\text{def}}{=} \pi(u, j'(u)) = 0, \tag{9.23}$$

where $\pi = \phi_{[\alpha,\beta]}$ is a continuous MCP-function for the interval $[\alpha,\beta]$. As in the function space context, we apply an MCP-function $\pi = \phi_{[\alpha,\beta]}$ to reformulate (9.22) equivalently in the form

$$\mathbf{\Pi}^h(\mathbf{u}^h) \stackrel{\text{def}}{=} \begin{pmatrix} \pi\left(\mathbf{u}_1^h, \mathbf{j}^{h'}(\mathbf{u}^h)_1\right) \\ \vdots \\ \pi\left(\mathbf{u}_{m^h}^h, \mathbf{j}^{h'}(\mathbf{u}^h)_{m^h}\right) \end{pmatrix} = 0. \tag{9.24}$$

This is the discrete version of the equation reformulation (9.12). If π is semismooth then, due to the continuous differentiability of $\mathbf{j}^{h'}$, also $\mathbf{\Pi}^h$ is semismooth and finite-dimensional semismooth Newton methods can be applied. We expect a close relationship between the resulting discrete semismooth Newton method and the semismooth Newton method for the original problem in function space. This relation is established in the following considerations:

First, we have to identify the discrete correspondent to the generalized differential $\partial^\circ\Pi(u)$ in Theorem 9.10. Let $B \in \partial^\circ\Pi(u)$. Then there exists $d \in (L^\infty)^2$ with $d(x) \in \partial\pi(u(x), j'(u)(x))$ on Ω such that $B = d_1 I + d_2 \cdot j''(u)$. Replacing u by u^h and j by j^h, a suitable discretization of B is obtained by

$$B^h = d_1^h I + d_2^h \cdot j^{h''}(u^h), \tag{9.25}$$

$$d_i^h \in H^h, \quad d^h(x) \in \partial\pi\left(u^h(x), j^{h'}(u^h)(x)\right), \quad x \in \Omega. \tag{9.26}$$

Since u^h and $j^{h'}(u^h)$ are elements of H^h, they are constant on each triangle $T_l \in \mathcal{T}^h$ with values $\mathbf{u}_l^h$ and $\mathbf{j}^{h'}(\mathbf{u}^h)_l$, respectively. Denoting by $\mathbf{d}_i^h$ the η^h-coordinates of $d_i^h \in H^h$, the functions d_i^h are constant on every triangle T_l with values $\mathbf{d}_{il}^h$. Therefore, (9.26) is equivalent to

$$(\mathbf{d}_{1l}^h, \mathbf{d}_{2l}^h) \in \partial\pi\left(\mathbf{u}_l^h, \mathbf{j}^{h'}(\mathbf{u}^h)_l\right), \quad 1 \le l \le m^h.$$

Let $\mathbf{j}^{h''}(\mathbf{u}^h) \in \mathbb{R}^{m^h \times m^h}$ denote the matrix representation of $j^{h''}(u^h)$ with respect to the H^h inner product. More precisely, $\mathbf{j}^{h''}(\mathbf{u}^h)\mathbf{z}^h$ are the η^h-coordinates of $j^{h''}(u^h)z^h$; thus, for all $z^h, \tilde{z}^h \in H^h$ and corresponding coordinate vectors $\mathbf{z}^h, \tilde{\mathbf{z}}^h$, we have

$$(z^h, j^{h''}(u^h)\tilde{z}^h)_{L^2} = \mathbf{z}^{h^T}\mathbf{M}^h\mathbf{j}^{h''}(\mathbf{u}^h)\tilde{\mathbf{z}}^h.$$

The matrix representation of B^h with respect to the H^h inner product is

$$\mathbf{B}^h = \mathbf{D}_1^h + \mathbf{D}_2^h\mathbf{j}^{h''}(\mathbf{u}^h),$$

where $\mathbf{D}_i^h = \operatorname{diag}(\mathbf{d}_i^h)$. In fact, using that $\mathbf{M}^h$ is diagonal, we obtain

$$\begin{aligned}
(\eta_k^h, B^h\eta_l^h)_{L^2} &= (\eta_k^h, d_1^h\eta_l^h)_{L^2} + (\eta_k^h, d_2^h j^{h''}(u^h)\eta_l^h)_{L^2}\\
&= (d_1^h\eta_k^h, \eta_l^h)_{L^2} + (d_2^h\eta_k^h, j^{h''}(u^h)\eta_l^h)_{L^2}\\
&= \mathbf{d}_{1k}^h(\eta_k^h, \eta_l^h)_{L^2} + \mathbf{d}_{2k}^h(\eta_k^h, j^{h''}(u^h)\eta_l^h)_{L^2}\\
&= \mathbf{d}_{1k}^h\mathbf{M}_{kl}^h + \mathbf{d}_{2k}^h\big(\mathbf{M}^h\mathbf{j}^{h''}(\mathbf{u}^h)\big)_{kl}\\
&= \mathbf{M}_{kl}^h\mathbf{d}_{1k}^h + \mathbf{M}_{kk}^h\mathbf{d}_{2k}^h\big(\mathbf{j}^{h''}(\mathbf{u}^h)\big)_{kl}\\
&= \big(\mathbf{M}^h(\mathbf{D}_1^h + \mathbf{D}_2^h\mathbf{j}^{h''}(\mathbf{u}^h))\big)_{kl}.
\end{aligned}$$

Therefore, the matrix representation of the discrete correspondent to $\partial^\circ\Pi(u)$ is $\partial^\circ\mathbf{\Pi}^h(\mathbf{u}^h)$, the set consisting of all matrices $\mathbf{B}^h \in \mathbb{R}^{m^h\times m^h}$ with

$$\mathbf{B}^h = \mathbf{D}_1^h + \mathbf{D}_2^h\mathbf{j}^{h''}(\mathbf{u}^h), \tag{9.27}$$

where $\mathbf{D}_1^h$ and $\mathbf{D}_2^h$ are diagonal matrices such that

$$\big((\mathbf{D}_1^h)_{ll}, (\mathbf{D}_2^h)_{ll}\big) \in \partial\pi\left(\mathbf{u}_l^h, \mathbf{j}^{h'}(\mathbf{u}^h)_l\right), \quad l = 1,\ldots,m^h.$$

Next, we show that there is a very close relationship between $\partial^\circ\mathbf{\Pi}^h$ and finite-dimensional subdifferentials of the function $\mathbf{\Pi}^h$. To establish this relation, let us first note that the coordinate representation $\mathbf{j}^{h''}(\mathbf{u}^h)$ of $j^{h''}(u^h)$ satisfies

$$\mathbf{j}^{h''}(\mathbf{u}^h) = \frac{d}{d\mathbf{u}^h}\mathbf{j}^{h'}(\mathbf{u}^h).$$

In fact, we have for all $z^h, \tilde{z}^h \in H^h$ and corresponding coordinate vectors $\mathbf{z}^h, \tilde{\mathbf{z}}^h$

$$\begin{aligned}
\mathbf{z}^{hT}\mathbf{M}^h\mathbf{j}^{h''}(\mathbf{u}^h)\tilde{\mathbf{z}}^h &= (z^h, j^{h''}(u^h)\tilde{z}^h)_{L^2} = \mathbf{z}^{hT}\frac{d^2}{d\mathbf{u}^{h2}}\mathbf{j}^h(\mathbf{u}^h)\tilde{\mathbf{z}}^h\\
&= \mathbf{z}^{hT}\frac{d}{d\mathbf{u}^h}(\mathbf{M}^h\mathbf{j}^{h'})(\mathbf{u}^h)\tilde{\mathbf{z}}^h = \mathbf{z}^{hT}\mathbf{M}^h\frac{d}{d\mathbf{u}^h}\mathbf{j}^{h'}(\mathbf{u}^h)\tilde{\mathbf{z}}^h,
\end{aligned}$$

where we have used (9.20). This shows that for the rows of $\partial^\circ\mathbf{\Pi}^h$ there holds

$$\partial^\circ\mathbf{\Pi}_l^h = \partial\pi \circ \left\{\frac{d}{d\mathbf{u}_l^h}\begin{pmatrix}\mathbf{u}_l^h\\ \mathbf{j}^{h'}(\mathbf{u}^h)_l\end{pmatrix}\right\}$$

in the sense of Proposition 3.8 and that, by Propositions 3.4 and 3.8, $\mathbf{\Pi}_l^h$ is $\partial^\circ\mathbf{\Pi}_l^h$-semismooth if π is semismooth. Therefore, $\mathbf{\Pi}^h$ is $\partial^\circ\mathbf{\Pi}^h$-semismooth by Proposition 3.6. If π is α-order semismooth and $\mathbf{j}^{h'}$ is differentiable with α-Hölder continuous derivative, then the above reasoning yields that $\mathbf{\Pi}^h$ is even α-order $\partial^\circ\mathbf{\Pi}^h$-semismooth.

Finally, there is also a close relationship between $\partial^\circ\mathbf{\Pi}^h$ and $\partial_C\mathbf{\Pi}^h$. In fact, by the chain rule for Clarke's generalized gradient we have

$$\partial_C\mathbf{\Pi}^h(\mathbf{u}^h) \subset \partial^\circ\mathbf{\Pi}^h(\mathbf{u}^h).$$

Under additional conditions (e.g., if π or $-\pi$ is regular), equality holds. If we do not have equality, working with the differential $\partial^\circ \mathbf{\Pi}^h$ has the advantage that $\partial\pi$ and the derivatives of its arguments can be computed independently of each other, whereas in general the calculation of $\partial_C \mathbf{\Pi}^h(\mathbf{u}^h)$ is more difficult.

We collect the obtained results in the following theorem.

Theorem 9.19. *The discretization of the equation reformulation* (9.23) *of* (9.1) *in coordinate form is given by* (9.24). *Further, the multifunction $\partial^\circ \mathbf{\Pi}^h$, where $\partial^\circ \mathbf{\Pi}^h(\mathbf{u}^h)$ consists of all $\mathbf{B}^h \in \mathbb{R}^{m^h \times m^h}$ defined in* (9.27), *is the discrete analogue of the generalized differential $\partial^\circ \Pi$. We have*

$$\partial_C \mathbf{\Pi}^h(\mathbf{u}^h) \subset \partial^\circ \mathbf{\Pi}^h(\mathbf{u}^h)$$

with equality if, e.g., π or $-\pi$ is regular.

If π is semismooth, then $\mathbf{\Pi}^h$ is $\partial^\circ \mathbf{\Pi}^h$-semismooth and also semismooth in the usual sense. Further, if π is α-order semismooth and if j^h (and thus $\mathbf{j}^h$) is twice continuously differentiable with α-Hölder continuous second derivative, then $\mathbf{\Pi}^h$ is α-order $\partial^\circ \mathbf{\Pi}^h$-semismooth and also α-order semismooth in the usual sense.

Having established the $\partial^\circ \mathbf{\Pi}^h$-semismoothness of $\mathbf{\Pi}^h$, we can use any variant of the semismooth Newton methods in sections 3.2.3–3.2.5 to solve the semismooth equation (9.24). We stress that in finite dimensions no smoothing step is required to obtain fast local convergence. However, since the finite-dimensional problem (9.24) is a discretization of the continuous problem (9.12), we should, if necessary, incorporate a discrete version of a smoothing step to ensure that the algorithm exhibits mesh-independent behavior.

The resulting instance of Algorithm 3.10 then becomes the following.

Algorithm 9.20. (inexact semismooth Newton method)

0. Choose an initial point $\mathbf{u}_0^h \in \mathbb{R}^{m_h}$ and set $k = 0$.

1. Compute the discrete state $\mathbf{y}_k^h \in \mathbb{R}^{n^h}$ by solving the discrete state equation
$$\mathbf{A}^h \mathbf{y}_k^h + \mathbf{S}^h \varphi(\mathbf{y}_k^h) = \mathbf{f}^h + \mathbf{G}^h \mathbf{u}_k^h.$$

2. Compute the discrete adjoint state $\mathbf{w}_k^h \in \mathbb{R}^{n^h}$ by solving the discrete adjoint equation
$$\left(\mathbf{A}^h + \mathbf{T}^h(\mathbf{y}_k^h)\right)\mathbf{w}_k^h = -\mathbf{S}^h(\mathbf{y}^h - \mathbf{y}_d^h).$$

3. Compute the discrete reduced gradient
$$\mathbf{j}_k^{h\prime} = \lambda(\mathbf{u}_k^h - \mathbf{u}_d^h) - \mathbf{M}^{h^{-1}} \mathbf{G}^{h^T} \mathbf{w}^h$$
and the vector $\mathbf{\Pi}_k^h \in \mathbb{R}^{n^h}$, $(\mathbf{\Pi}_k^h)_l = \pi\left((\mathbf{u}_k^h)_l, \mathbf{j}_{kl}^{h\prime}\right)$.

4. If $(\mathbf{\Pi}_k^{h^T} \mathbf{M}^h \mathbf{\Pi}_k^h)^{1/2} \le \varepsilon$, then STOP with result $\mathbf{u}^{h*} = \mathbf{u}_k^h$.

5. Compute $\mathbf{B}_k^h \in \partial^\circ \mathbf{\Pi}^h(\mathbf{u}_k^h)$ (details are given below).

6. Compute $\mathbf{s}_k^h \in \mathbb{R}^{m^h}$ by solving the semismooth Newton system (details are given below)

$$\mathbf{B}_k^h \mathbf{s}_k^h = -\mathbf{\Pi}_k^h,$$

and set $\mathbf{u}_{k+1}^{h,0} = \mathbf{u}_k^h + \mathbf{s}_k^h$.

7. Perform a smoothing step (if necessary): $\mathbf{u}_{k+1}^{h,0} \mapsto \mathbf{u}_{k+1}^h$.
8. Increment k by one and go to step 1.

Remark 9.21.

(a) *We can allow for inexactness in the matrices $\mathbf{B}_k^h$, which results in an instance of Algorithm 3.16. In fact, as was shown in Theorem 3.18, besides the uniformly bounded invertibility of the matrices $\mathbf{B}_k^h$ we only need that*

$$\inf_{\mathbf{B} \in \partial^\circ \mathbf{\Pi}^h(\mathbf{u}_k^h)} \|(\mathbf{B} - \mathbf{B}_k^h)\mathbf{s}_k^h\| = o(\|\mathbf{s}_k^h\|)$$

as $\|\mathbf{s}_k^h\| \to 0$ to achieve superlinear convergence.

(b) *We also can achieve that the iteration stays feasible with respect to a closed convex set $\mathcal{K}^h$ which contains the solution of (9.24). This can be achieved by incorporating a projection onto $\mathcal{K}^h$ in the algorithm after the smoothing step and results in an instance of Algorithm 3.62. In the following, we only consider the projection-free algorithm and the projected version with projection onto $\mathfrak{C}^h$, which is given by coordinatewise projection onto C.*

(c) *The efficiency of the algorithm crucially depends on the efficient solvability of the Newton equation in step 6. We propose an efficient method in section 9.1.5.*

(d) *We observed in Lemma 9.8 that $j'(u) = \lambda u + G(u)$, where*

$$u \in U \mapsto G(u) = -gw(u) - \lambda u_d \in L^{p'}(\Omega)$$

is locally Lipschitz continuous with $p' > 2$. We concluded that a smoothing step is given by the scaled projected gradient step

$$u \mapsto P_C(u - \lambda^{-1} j'(u)) = P_C(u_d + \lambda^{-1} gw(u)).$$

Therefore, a discrete version of the smoothing step is given by

$$\mathbf{u}^h \mapsto P_C\left(\mathbf{u}^h - \lambda^{-1}\mathbf{j}^{h'}(\mathbf{u}^h)\right) = P_C\left(\mathbf{u}_d^h + \lambda^{-1}\mathbf{M}^{h^{-1}}\mathbf{G}^{h^T}\mathbf{w}^h\right). \tag{9.28}$$

Due to the smoothing property of G we also can apply a smoothing-step-free semismooth Newton method by choosing

$$\pi(x) = x_1 - P_C(x_1 - \lambda^{-1} x_2)$$

for the reformulation, which results in

$$\Pi(u) = u - P_C\left(-\lambda^{-1} G(u)\right) = u - P_C\left(u_d + \lambda^{-1} gw(u)\right).$$

In the discrete algorithm, this corresponds to

$$\begin{aligned}\mathbf{\Pi}^h(\mathbf{u}^h) &= \mathbf{u}^h - P_C\left(\mathbf{u}^h - \lambda^{-1}\mathbf{j}^{h\prime}(\mathbf{u}^h)\right)\\ &= \mathbf{u}^h - P_C\left(\mathbf{u}_d^h + \lambda^{-1}\mathbf{M}^{h^{-1}}\mathbf{G}^{h^T}\mathbf{w}^h\right).\end{aligned} \tag{9.29}$$

In section 9.1.7, *we present numerical results for both variants, the one with general MCP-function* π *and smoothing step* (9.28), *and the smoothing-step-free algorithm with* $\mathbf{\Pi}^h$ *as defined in* (9.29).

9.1.5 Efficient Solution of the Newton System

We recall that a matrix $\mathbf{B}_k^h \in \mathbb{R}^{m^h \times m^h}$ is contained in $\partial^\circ \mathbf{\Pi}^h(\mathbf{u}_k^h)$ if and only if

$$\mathbf{B}_k^h = \mathbf{D}_{k1}^h + \mathbf{D}_{k2}^h \mathbf{j}^{h\prime\prime}(\mathbf{u}_k^h),$$

where $\mathbf{D}_{k1}^h$ and $\mathbf{D}_{k2}^h$ are diagonal matrices such that

$$\left((\mathbf{D}_{k1}^h)_{ll}, (\mathbf{D}_{k2}^h)_{ll}\right) \in \partial\pi\left((\mathbf{u}_k^h)_l, \mathbf{j}^{h\prime}(\mathbf{u}_k^h)_l\right). \tag{9.30}$$

Further, for the choices of functions π we are going to use, namely ϕ_C^{FB} and $\phi_C^{E,\sigma}: x \mapsto \phi_C^E(x_1, \sigma x_2)$, $\sigma > 0$, the computation of $\partial\pi$, and thus of the matrices $\mathbf{D}_{ki}^h$, is straightforward. Concerning the calculation of $\partial\phi_C^{E,\sigma}$, see Proposition 5.6; for the computation of $\partial\phi_C^{FB}$, we refer to [70]. In both cases, there exist constants $c_i > 0$ such that for all $x \in \mathbb{R}^2$ and all $d \in \partial\pi(x)$ holds

$$0 \le d_1, d_2 \le c_1, \quad d_1 + d_2 \ge c_2.$$

In particular, the matrices $\mathbf{D}_{ki}^h$ are positive semidefinite with uniformly bounded norms, and $\mathbf{D}_{k1}^h + \mathbf{D}_{k2}^h$ is positive definite with uniformly bounded inverse.

We observed earlier the relation

$$\mathbf{j}^{h\prime\prime}(\mathbf{u}^h) = \mathbf{M}^{h^{-1}} \frac{d^2}{d\mathbf{u}^{h^2}} \mathbf{j}^h(\mathbf{u}^h).$$

For the computation of the right-hand side we use the adjoint representation of section A.1, applied to problem (9.19). The state equation for this problem is $\mathbf{E}^h(\mathbf{y}^h, \mathbf{u}^h) = 0$ with

$$\mathbf{E}^h(\mathbf{y}^h, \mathbf{u}^h) = \mathbf{A}^h\mathbf{y}^h + \mathbf{S}^h\varphi(\mathbf{y}^h) - \mathbf{f}^h - \mathbf{G}^h\mathbf{u}^h,$$

and the Lagrange function is given by

$$\mathbf{L}^h(\mathbf{y}^h, \mathbf{u}^h) = \mathbf{J}^h(\mathbf{y}^h, \mathbf{u}^h) + \mathbf{w}^{h^T}\mathbf{E}^h(\mathbf{y}^h, \mathbf{u}^h).$$

Observe that

$$\begin{aligned}&\frac{d}{d\mathbf{y}^h}\mathbf{E}^h(\mathbf{y}^h, \mathbf{u}^h) = \mathbf{A}^h + \mathbf{T}^h(\mathbf{y}^h),\\ &\frac{d}{d\mathbf{u}^h}\mathbf{E}^h(\mathbf{y}^h, \mathbf{u}^h) = -\mathbf{G}^h,\\ \frac{d^2\mathbf{L}^h}{d(\mathbf{y}^h, \mathbf{u}^h)^2}(\mathbf{y}^h, \mathbf{u}^h, \mathbf{w}^h) &= \begin{pmatrix} \mathbf{S}^h + \mathbf{S}^h \operatorname{diag}(\varphi''(\mathbf{y}^h))\operatorname{diag}(\mathbf{w}^h) & 0 \\ 0 & \lambda\mathbf{M}^h \end{pmatrix}.\end{aligned}$$

Therefore, introducing the diagonal matrix

$$\mathbf{Z}^h(\mathbf{y}^h, \mathbf{w}^h) = \mathbf{S}^h\big(I + \mathrm{diag}(\varphi''(\mathbf{y}^h))\mathrm{diag}(\mathbf{w}^h)\big),$$

and omitting the arguments for brevity, we obtain by the adjoint formula

$$\begin{aligned}\frac{d^2}{d\mathbf{u}^{h^2}}\mathbf{j}^h(\mathbf{u}^h) &= \begin{pmatrix}\left(\frac{d\mathbf{E}^h}{d\mathbf{y}^h}\right)^{-1}\frac{d\mathbf{E}^h}{d\mathbf{u}^h}\\ -I\end{pmatrix}^T \frac{d^2\mathbf{L}^h}{d(\mathbf{y}^h,\mathbf{u}^h)^2}\begin{pmatrix}\left(\frac{d\mathbf{E}^h}{d\mathbf{y}^h}\right)^{-1}\frac{d\mathbf{E}^h}{d\mathbf{u}^h}\\ -I\end{pmatrix}\\ &= \mathbf{G}^{h^T}(\mathbf{A}^h + \mathbf{T}^h(y^h))^{-1}\mathbf{Z}^h(\mathbf{y}^h,\mathbf{w}^h)(\mathbf{A}^h + \mathbf{T}^h(\mathbf{y}^h))^{-1}\mathbf{G}^h + \lambda\mathbf{M}^h.\end{aligned}$$

The Hessian $\mathbf{j}^{h''}(\mathbf{u}^h)$ with respect to the inner product of H^h is thus given by

$$\mathbf{j}^{h''}(\mathbf{u}^h) = \mathbf{M}^{h^{-1}}\mathbf{G}^{h^T}(\mathbf{A}^h + \mathbf{T}^h(\mathbf{y}^h))^{-1}\mathbf{Z}^h(\mathbf{y}^h,\mathbf{w}^h)(\mathbf{A}^h + \mathbf{T}^h(\mathbf{y}^h))^{-1}\mathbf{G}^h + \lambda I.$$

Therefore, the matrices $\mathbf{B}^h \in \partial^\circ \mathbf{\Pi}^h(\mathbf{u}^h)$ are given by

$$\mathbf{B}^h = \mathbf{D}^h + \mathbf{D}_2^h\mathbf{M}^{h^{-1}}\mathbf{G}^{h^T}(\mathbf{A}^h + \mathbf{T}^h(\mathbf{y}^h))^{-1}\mathbf{Z}^h(\mathbf{y}^h,\mathbf{w}^h)(\mathbf{A}^h + \mathbf{T}^h(\mathbf{y}^h))^{-1}\mathbf{G}^h,$$

where $\mathbf{D}_1^h$ and $\mathbf{D}_2^h$ satisfy (9.30) and

$$\mathbf{D}^h \stackrel{\text{def}}{=} \mathbf{D}_1^h + \lambda\mathbf{D}_2^h.$$

Note that $\mathbf{D}^h$ is diagonal, positive definite, and $\mathbf{D}^h$ as well as $\mathbf{D}^{h^{-1}}$ are bounded uniformly in $\mathbf{u}^h$.

Since computing $(\mathbf{A}^h + \mathbf{T}^h(\mathbf{y}^h))^{-1}\mathbf{v}^h$ means solving the linearized state equation, it is not a priori clear that the Newton equation in step 6 of Algorithm 9.20 can be solved efficiently. It is also important to observe that the main difficulties are caused by the structure of the Hessian $\mathbf{j}^{h''}$, not so much by the additional factors $\mathbf{D}_1^h$ and $\mathbf{D}_2^h$ appearing in $\mathbf{B}^h$. In other words, it is also not straightforward how the Newton system for the unconstrained reduced optimal control problem can be solved efficiently.

However, the matrix $\mathbf{B}^h$ is a discretization of the operator

$$(d_1 + \lambda d_2)I + d_2 g\cdot(A + \varphi' I)^{-1}[(1+\varphi'' w)I](A+\varphi' I)^{-1}(gI).$$

Hence, one possibility to solve the discretized semismooth Newton system efficiently is to use the compactness of the operator

$$(A+\varphi' I)^{-1}[(1+\varphi'' w)I](A+\varphi' I)^{-1}[gI]$$

to apply multigrid methods of the second kind [91, Ch. 16]. These methods are suitable for solving problems of the form

$$u = Ku + f,$$

where $K : U \mapsto V \hookrightarrow U$ (compact embedding). The application of $(A+\varphi' I)^{-1}$ to a function, i.e., application of $(\mathbf{A}^h + \mathbf{T}^h(\mathbf{y}^h))^{-1}$ to a vector, can be done efficiently by using, once again, multigrid methods. We believe that this approach has computational potential. In our computations, however, we use a different strategy that we describe now.

To develop this approach, we consider the Newton system

$$\mathbf{B}^h\mathbf{s}^h = -\mathbf{\Pi}^h(\mathbf{u}^h) \tag{9.31}$$

and derive an equivalent system of equations that, under certain assumptions, can be solved efficiently. Here, we use the relations that we observed in section 5.2.3 between the semismooth Newton system of the reduced Newton system and the semismooth Newton system obtained for the all-at-once approach. To this end, consider the system

$$\left(\begin{array}{ccc|c} \frac{d^2}{d\mathbf{y}^{h^2}}\mathbf{L}^h & \frac{d^2}{d\mathbf{y}^h d\mathbf{u}^h}\mathbf{L}^h & \frac{d^2}{d\mathbf{y}^h d\mathbf{w}^h}\mathbf{L}^h & 0 \\ \mathbf{D}_2^h\mathbf{M}^{h^{-1}}\frac{d^2}{d\mathbf{u}^h d\mathbf{y}^h}\mathbf{L}^h & \mathbf{D}_1^h + \mathbf{D}_2^h\mathbf{M}^{h^{-1}}\frac{d^2}{d\mathbf{u}^{h^2}}\mathbf{L}^h & \mathbf{D}_2^h\mathbf{M}^{h^{-1}}\frac{d^2}{d\mathbf{u}^h d\mathbf{w}^h}\mathbf{L}^h & -\mathbf{\Pi}^h \\ \frac{d^2}{d\mathbf{w}^h d\mathbf{y}^h}\mathbf{L}^h & \frac{d^2}{d\mathbf{w}^h d\mathbf{u}^h}\mathbf{L}^h & \frac{d^2}{d\mathbf{w}^{h^2}}\mathbf{L}^h & 0 \end{array}\right).$$

Using the particular form of $\mathbf{L}^h$, this becomes

$$\left(\begin{array}{ccc|c} \mathbf{Z}^h & 0 & \mathbf{A}^h + \mathbf{T}^h & 0 \\ 0 & \mathbf{D}^h & -\mathbf{D}_2^h\mathbf{M}^{h^{-1}}\mathbf{G}^{h^T} & -\mathbf{\Pi}^h \\ \mathbf{A}^h + \mathbf{T}^h & -\mathbf{G}^h & 0 & 0 \end{array}\right).$$

Performing the transformation

$$\text{Row } 1 \to \text{Row } 1 - \mathbf{Z}^h(\mathbf{A}^h + \mathbf{T}^h)^{-1} \times \text{Row } 3$$

yields the equivalent system

$$\left(\begin{array}{ccc|c} 0 & \mathbf{Z}^h(\mathbf{A}^h + \mathbf{T}^h)^{-1}\mathbf{G}^h & \mathbf{A}^h + \mathbf{T}^h & 0 \\ 0 & \mathbf{D}^h & -\mathbf{D}_2^h\mathbf{M}^{h^{-1}}\mathbf{G}^{h^T} & -\mathbf{\Pi}^h \\ \mathbf{A}^h + \mathbf{T}^h & -\mathbf{G}^h & 0 & 0 \end{array}\right), \tag{9.32}$$

and by the transformation

$$\text{Row } 2 \to \text{Row } 2 + (\mathbf{D}_2^h\mathbf{M}^{h^{-1}}\mathbf{G}^{h^T})(\mathbf{A}^h + \mathbf{T}^h)^{-1} \times \text{Row } 1,$$

we arrive at

$$\left(\begin{array}{ccc|c} 0 & \mathbf{Z}^h(\mathbf{A}^h + \mathbf{T}^h)^{-1}\mathbf{G}^h & \mathbf{A}^h + \mathbf{T}^h & 0 \\ 0 & \mathbf{B}^h & 0 & -\mathbf{\Pi}^h \\ \mathbf{A}^h + \mathbf{T}^h & -\mathbf{G}^h & 0 & 0 \end{array}\right).$$

This shows that $\mathbf{B}^h$ appears as a Schur complement of (9.32). Hence, if we solve (9.32), we also have a solution of the Newton system (9.31).

For deriving an efficient strategy for solving (9.32), we first observe that $\mathbf{D}^h$ is diagonal and nonsingular. Further, the diagonal matrix $\mathbf{Z}^h$ is invertible if and only if

$$\varphi''(\mathbf{y}^h)_i\mathbf{w}_i^h \neq 1 \quad \forall\, l = 1,\ldots,n^h. \tag{9.33}$$

In particular, this holds true if $\varphi''(\mathbf{y}^h)_i\mathbf{w}_i^h$ is small for all i. If, e.g., the state equation is linear, then $\varphi'' \equiv 0$. Further, if $\mathbf{y}^h$ is sufficiently close to the data $\mathbf{y}_d^h$, then the right-hand side of the adjoint equation is small and thus $\mathbf{w}^h$ is small. Both cases result in a positive definite diagonal matrix $\mathbf{Z}^h$. If (9.33) happens to be violated, we can perform a small perturbation of $\mathbf{Z}^h$ (but sufficiently large to avoid numerical instabilities) to make it nonsingular.

With $\mathbf{D}^h$ and $\mathbf{Z}^h$ being invertible, we transform (9.32) according to

$$\text{Row } 3 \to -\text{Row } 3 + (\mathbf{A}^h + \mathbf{T}^h){\mathbf{Z}^h}^{-1} \times \text{Row } 1 - \mathbf{G}^h {\mathbf{D}^h}^{-1} \times \text{Row } 2,$$

and obtain

$$\left(\begin{array}{ccc|c} \mathbf{Z}^h & 0 & \mathbf{A}^h + \mathbf{T}^h & 0 \\ 0 & \mathbf{D}^h & -\mathbf{D}_2^h {\mathbf{M}^h}^{-1} {\mathbf{G}^h}^T & -\mathbf{\Pi}^h \\ 0 & 0 & \mathbf{Q}^h & \mathbf{G}^h {\mathbf{D}^h}^{-1} \mathbf{\Pi}^h \end{array}\right),$$

where

$$\mathbf{Q}^h = \mathbf{G}^h {\mathbf{D}^h}^{-1} \mathbf{D}_2^h {\mathbf{M}^h}^{-1} {\mathbf{G}^h}^T + (\mathbf{A}^h + \mathbf{T}^h){\mathbf{Z}^h}^{-1}(\mathbf{A}^h + \mathbf{T}^h).$$

The matrix ${\mathbf{D}^h}^{-1} \mathbf{D}_2^h {\mathbf{M}^h}^{-1}$ is diagonal and positive definite. Hence, $\mathbf{Q}^h$ is symmetric positive definite if $\mathbf{Z}^h$ is positive definite. Furthermore, $\mathbf{Q}^h$ can be interpreted as the discretization of the differential operator

$$\frac{d_2 g^2}{d_1 + \lambda d_2} I + (A + \varphi'(y)I)\left(\frac{1}{1 + \varphi''(y)w} I\right)(A + \varphi'(y)I),$$

which is elliptic if $(1 + \varphi''(y)w)$ is positive on Ω.

Hence, fast solvers (multigrid, preconditioned conjugate gradient, etc.) can be used to solve the system

$$\mathbf{Q}^h \mathbf{v}^h = \mathbf{G}^h {\mathbf{D}^h}^{-1} \mathbf{\Pi}^h. \tag{9.34}$$

Then, the solution $\mathbf{s}^h$ of the Newton system (9.31) is obtained as

$$\mathbf{s}^h = -\mathbf{\Pi}^h + {\mathbf{D}^h}^{-1} \mathbf{D}_2^h {\mathbf{M}^h}^{-1} {\mathbf{G}^h}^T \mathbf{v}^h.$$

9.1.6 Discrete All-at-Once Approach

The detailed considerations of the black-box approach can be carried out in a similar way for semismooth reformulations of the KKT-system of the discretized optimal control problem. We think there is no need to discuss this in detail. In the discrete all-at-once approach, $\mathbf{L}_{\mathbf{u}^h}^h = {\mathbf{M}^h}^{-1}(d/d\mathbf{u}^h)\mathbf{L}^h$ plays the role of ${\mathbf{j}^h}'$, and the resulting system to solve has the structure

$$\left(\begin{array}{ccc|c} \mathbf{Z}^h & 0 & \mathbf{A}^h + \mathbf{T}^h & -(d/d\mathbf{y}^h)\mathbf{L}^h \\ 0 & \mathbf{D}^h & -\mathbf{D}_2^h {\mathbf{M}^h}^{-1} {\mathbf{G}^h}^T & -\mathbf{\Pi}^h \\ \mathbf{A}^h + \mathbf{T}^h & -\mathbf{G}^h & 0 & -(d/d\mathbf{w}^h)\mathbf{L}^h \end{array}\right);$$

see section 9.1.5. If a globalization is used, it is important to formulate the merit function by means of the correct norms,

$$\frac{1}{2}\left[\frac{d\mathbf{L}^h}{d\mathbf{y}^h}\right]^T {\mathbf{A}^h}^{-1} \frac{d\mathbf{L}^h}{d\mathbf{y}^h} + \frac{1}{2} {\mathbf{\Pi}^h}^T \mathbf{M}^h \mathbf{\Pi}^h + \frac{1}{2}\left[\frac{d\mathbf{L}^h}{d\mathbf{y}^h}\right]^T {\mathbf{A}^h}^{-1} \frac{d\mathbf{L}^h}{d\mathbf{w}^h},$$

and to represent gradients with respect to the correct inner products.

9.1.7 Numerical Results

We now present numerical results for problem (9.1). The domain is the unit square $\Omega = (0,1) \times (0,1)$. For φ we choose $\varphi(y) = y^3$, which satisfies the growth condition with $s = 4$. The choice of the other data is oriented on [20, Ex. 5.1.1] (therein, however, the state equation is linear and corresponds to $\varphi \equiv 0$):

$$\begin{aligned} \beta_1 &= -\infty, \quad \beta_2 = 0, \\ y_d(x) &= \frac{1}{6}\sin(2\pi x_1)\sin(2\pi x_2)e^{2x_1}, \\ u_d &\equiv 0, \quad \lambda = 10^{-3}. \end{aligned} \tag{9.35}$$

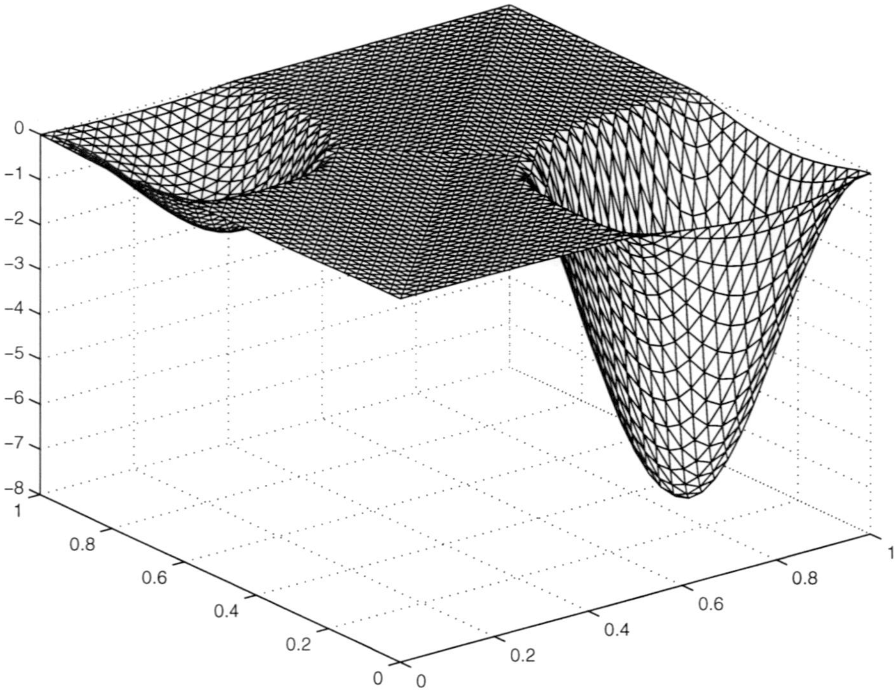

Figure 9.1 Optimal control $\bar{u}$ ($h = 1/32$).

Figure 9.1 shows the computed optimal control on $\mathcal{T}^{1/32}$ and Figure 9.2 the corresponding state. The code was implemented in MATLAB® Version 7.8.0.347 (R2009a) 64-bit (glnxa64), using sparse matrix computations. Although MATLAB is quite efficient, it usually cannot compete with Fortran or C implementations, which should be kept in mind when evaluating the runtimes given below. The computations were performed under openSUSE 11.2 Linux on an HP Compaq™ workstation with an Intel® Core™2 Duo CPU E8600 operating at 3.33 GHz.

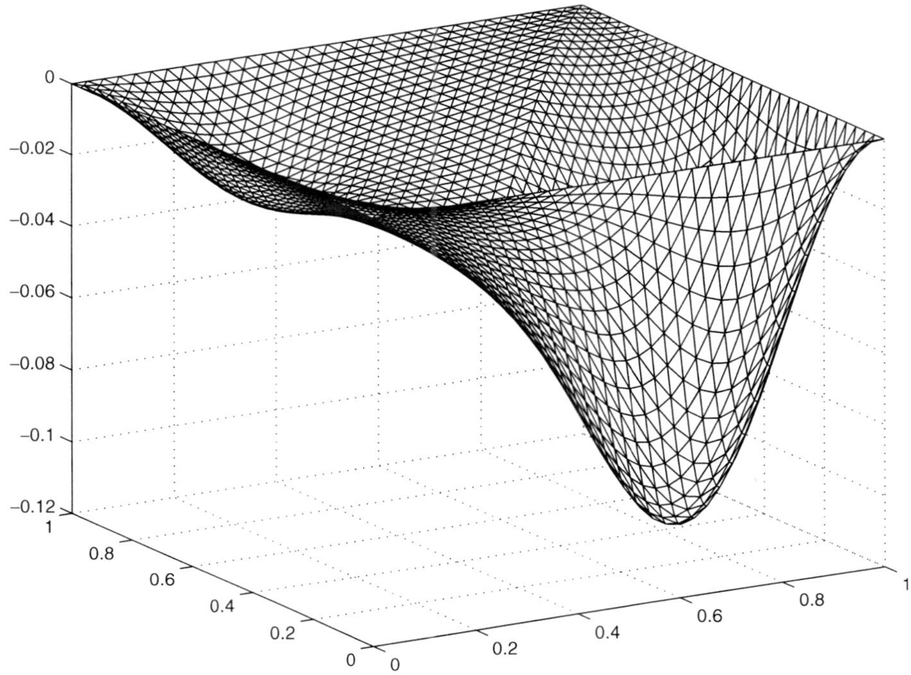

Figure 9.2 Optimal state $y(\bar{u})$ $(h = 1/32)$.

We present results for

1. Reformulations of the black-box VIP (9.11),

2. Reformulations of the all-at-once KKT-system (5.24)–(5.26),

to which we apply two variants of the semismooth Newton method,

1. Algorithm 3.10 (no constraints),

2. Algorithm 3.22 with $\mathcal{K} = \mathcal{C}$.

In both cases we consider the following choices of MCP-functions:

1. $\pi(x) = x_1 - P_{(-\infty,0]}(x_1 - \lambda^{-1}x_2)$ (smoothing-step-free algorithm).

2. $\pi(x) = -\phi^{FB}(-x)$.

We obtain eight (actually six, see below) variants of algorithms, which are denoted by A111–A222, where the three numbers express the choices for the three criteria given above. For instance, A221 stands for Algorithm 3.22, applied to the KKT-system, with $\mathcal{K} = \mathcal{C}$ and

$\pi(x) = x_1 - P_{(-\infty,0]}(x_1 - \lambda^{-1}x_2)$. Since in the class Axy2 we compute smoothing steps as described in section 4.1, and the smoothing step contains already a projection onto $\mathcal{C}$, we have A112=A122, A212=A222. We will use the names A112 and A212 in what follows.

Using Multigrid Techniques

For the efficient solution of the discrete state equation (needed in the black-box approach), and the linearized state equation (needed in the all-at-once approach), we use a conjugate gradient method that is preconditioned by one multigrid (MG) V-cycle with one red-black Gauss–Seidel iteration as presmoother and one adjoint red-black Gauss–Seidel iteration as postsmoother. Standard references on multigrid methods include [30, 91, 92, 201]. Our semismooth Newton methods with MG-preconditioned conjugate gradient solver of the Newton systems belong to the class of Newton multilevel methods [59]. For other multigrid approaches to variational inequalities we refer to [28, 114, 115, 139, 140, 141].

For the solution of the semismooth Newton system we solve the Schur complement equation (9.34) by a multigrid-preconditioned conjugate gradient method as just described.

The grid hierarchy is generated as follows: The coarsest triangulation $\mathcal{T}^1$ is shown in Figure 9.3. Given $\mathcal{T}^{2h}$, the next finer triangulation $\mathcal{T}^h$ is obtained by replacing any triangle in $\mathcal{T}^{2h}$ with four triangles, introducing the edge midpoints of the coarse triangles as new vertices; see Figure 9.4, which displays $\mathcal{T}^{1/2}$. Table 9.1 shows the resulting number

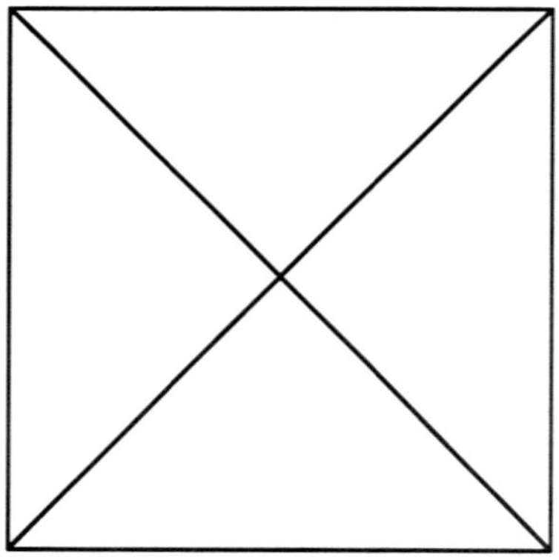

Figure 9.3 Coarsest triangulation, $\mathcal{T}^1$.

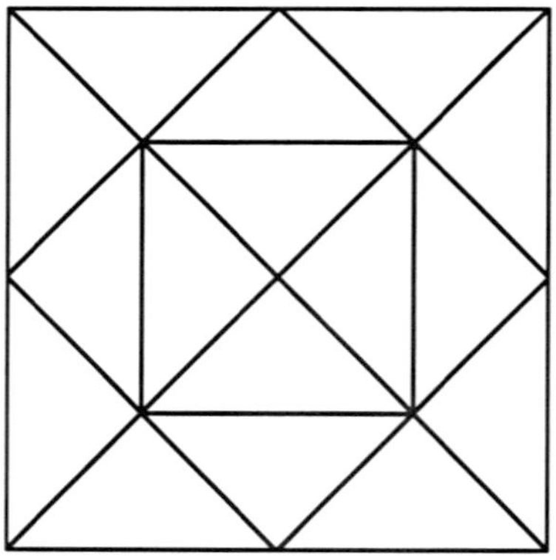

Figure 9.4 Second triangulation, $\mathcal{T}^{1/2}$.

Table 9.1 Degrees of freedom for different mesh sizes.

h	Number of interior vertices	Number of triangles
1/16	481	1024
1/32	1985	4096
1/64	8065	16384
1/128	32513	65536
1/256	130561	262144

of interior vertices and the number of triangles for each triangulation level. There is a second strategy to use the multilevel philosophy: We can perform a nested iteration over the discrete optimal control problems on the grid hierarchy. We first (approximately) solve the discrete optimal control problem on the coarsest level. We then interpolate this solution to obtain an initial point for the discrete optimal control problem on the next finer level, which we again solve approximately, and so forth. As we will see, this approach is very efficient.

Black-Box Approach

We now present numerical results for semismooth Newton methods applied to the first-order necessary conditions of the reduced problem (9.9). We thus consider the three algorithms A111, A121, and A112. The initial point is $u_0 \equiv -1$. We do not use a globalization since (as is often the case for control problems) the undamped semismooth Newton method converges without difficulties. We stress that if the nonmonotone trust-region method of section 7.4 is used, the globalization parameters can be chosen in such a way that the method essentially behaves like the pure Newton method.

Table 9.2 Iteration history of algorithm A111.

h	k	$\|u_k - \bar{u}\|_{L^2}$	$\|u_k - \bar{u}\|_{L^\infty}$	$\chi(u_k)$
$\frac{1}{16}$	0	1.623e+00	6.416e+00	4.794e−03
	1	9.454e−02	1.099e+00	1.022e−04
	2	5.958e−04	1.354e−02	5.949e−04
	3	3.611e−10	1.824e−09	3.552e−13
$\frac{1}{32}$	0	1.627e+00	6.477e+00	4.805e−03
	1	9.191e−02	1.098e+00	9.934e−05
	2	1.429e−03	5.833e−02	1.428e−03
	3	8.267e−11	4.141e−10	7.712e−14
$\frac{1}{64}$	0	1.628e+00	6.482e+00	4.807e−03
	1	9.052e−02	1.097e+00	9.769e−05
	2	1.347e−03	5.959e−02	1.346e−03
	3	6.616e−11	4.170e−10	1.254e−14
$\frac{1}{128}$	0	1.628e+00	6.487e+00	4.808e−03
	1	9.019e−02	1.098e+00	9.732e−05
	2	1.247e−03	6.325e−02	1.246e−03
	3	3.911e−08	1.001e−05	3.911e−08
	4	7.098e−11	4.621e−10	1.292e−15
$\frac{1}{256}$	0	1.628e+00	6.488e+00	4.808e−03
	1	8.988e−02	1.098e+00	9.697e−05
	2	1.309e−03	6.469e−02	1.308e−03
	3	1.735e−07	8.885e−05	1.735e−07
	4	8.936e−12	6.651e−11	1.446e−15

Table 9.3 Iteration history of algorithm A121.

h	k	$\|u_k - \bar{u}\|_{L^2}$	$\|u_k - \bar{u}\|_{L^\infty}$	$\chi(u_k)$
$\frac{1}{16}$	0	1.623e+00	6.416e+00	4.794e−03
	1	9.454e−02	1.099e+00	1.022e−04
	2	3.266e−05	1.309e−04	5.467e−08
	3	2.210e−10	1.115e−09	4.921e−14
$\frac{1}{32}$	0	1.627e+00	6.477e+00	4.805e−03
	1	9.191e−02	1.098e+00	9.934e−05
	2	5.613e−05	2.547e−04	9.082e−08
	3	5.024e−11	2.521e−10	1.086e−14
$\frac{1}{64}$	0	1.628e+00	6.482e+00	4.807e−03
	1	9.052e−02	1.097e+00	9.769e−05
	2	5.348e−05	2.404e−04	8.643e−08
	3	6.774e−11	4.115e−10	2.206e−15
$\frac{1}{128}$	0	1.628e+00	6.487e+00	4.808e−03
	1	9.019e−02	1.098e+00	9.732e−05
	2	4.679e−05	2.091e−04	7.538e−08
	3	7.116e−11	4.611e−10	8.160e−16
$\frac{1}{256}$	0	1.628e+00	6.488e+00	4.808e−03
	1	8.988e−02	1.098e+00	9.697e−05
	2	5.097e−05	2.295e−04	8.212e−08
	3	1.405e−07	6.736e−05	1.405e−10

To be independent of the choice of the MCP-function, we work with the termination condition

$$\chi(u_k) = \|u_k - P_C(u_k - j'(u_k))\|_{L^2} \le \varepsilon,$$

or, in terms of the discretized problem,

$$\left[\mathbf{u}_k^h - P_C(\mathbf{u}_k^h - \mathbf{j}_k^{h'})\right]^T \mathbf{M}^h \left[\mathbf{u}_k^h - P_C(\mathbf{u}_k^h - \mathbf{j}_k^{h'})\right] \le \varepsilon^2.$$

Except for this, the method we use agrees with Algorithm 9.20. We work with $\varepsilon = 10^{-8}$. Smaller values can be chosen as well, but it does not appear to be very reasonable to choose ε much smaller than the discretization error. The nonlinear state equation is solved by a Newton iteration, where, in each iteration, a linearized state equation has to be solved. For the computation of j' we solve the adjoint equation. All PDE solves are done by a multigrid-cg method as described above.

In our first set of tests we choose $\lambda = 0.001$ and consider problems on the triangulations $\mathcal{T}^h$ for $h = 2^{-k}$, $k = 4,5,6,7,8$. See Table 9.1 for the corresponding number of triangles and interior nodes, respectively.

The results are collected in Tables 9.2–9.4. Table 9.2 contains the results for A111, Table 9.3 the results for A121, and Table 9.4 the results for A112. Listed are the iteration k, the L^2-distance to the (discrete) solution ($\|u_k - \bar{u}\|_{L^2}$), the L^∞-distance to the (discrete) solution ($\|u_k - \bar{u}\|_{L^\infty}$), and the norm of the projected gradient ($\chi(u_k)$). For all three variants

Table 9.4 Iteration history of algorithm A112.

h	k	$\|u_k - \bar{u}\|_{L^2}$	$\|u_k - \bar{u}\|_{L^\infty}$	$\chi(u_k)$
$\frac{1}{16}$	0	1.623e+00	6.416e+00	4.794e−03
	1	6.695e−01	2.447e+00	1.630e−03
	2	8.454e−02	3.252e−01	1.463e−04
	3	9.587e−04	3.444e−03	1.601e−06
	4	1.841e−09	6.849e−09	3.216e−12
$\frac{1}{32}$	0	1.627e+00	6.477e+00	4.805e−03
	1	6.713e−01	2.466e+00	1.667e−03
	2	9.340e−02	3.645e−01	1.654e−04
	3	1.845e−03	7.274e−03	3.068e−06
	4	1.540e−07	6.725e−07	2.516e−10
$\frac{1}{64}$	0	1.628e+00	6.482e+00	4.807e−03
	1	6.718e−01	2.479e+00	1.682e−03
	2	9.462e−02	3.731e−01	1.673e−04
	3	1.737e−03	7.040e−03	2.891e−06
	4	2.760e−07	1.316e−06	4.394e−10
$\frac{1}{128}$	0	1.628e+00	6.487e+00	4.808e−03
	1	6.719e−01	2.483e+00	1.687e−03
	2	9.502e−02	3.794e−01	1.682e−04
	3	1.773e−03	7.063e−03	2.959e−06
	4	4.496e−07	1.930e−06	7.355e−10
$\frac{1}{256}$	0	1.628e+00	6.488e+00	4.808e−03
	1	6.719e−01	2.485e+00	1.688e−03
	2	9.503e−02	3.796e−01	1.682e−04
	3	1.770e−03	7.067e−03	2.954e−06
	4	6.020e−07	2.442e−06	9.997e−10

of the algorithm we observe mesh-independent convergence behavior, and superlinear rate of convergence of order >1. Only 3–4 iterations are needed until termination.

Table 9.5 shows for all three algorithms the total number of iterations (Iter.) of state equation solves (State), of linearized state equation solves (Lin. State), and of adjoint equation solves (Adj. State), and the total solution time in seconds (Time). The total number of solves of the semismooth Newton system coincides with the number of iterations, Iter. All solves of the linearized state equations are performed within the Newton method for the solution of the state equation. For algorithms A111 and A121, a total of Iter + 1 state solves and Iter + 1 adjoint state solves are required. Algorithm A112 requires in addition one state solve and one adjoint state solve per iteration for the computation of the smoothing step. We see that usually two Newton iterations are sufficient to solve the nonlinear state equation. Observe that the total computing time increases approximately linearly with the degrees of freedom. This shows that we indeed achieve multigrid efficiency. We note that algorithms A111 and A121 are superior to A112 in computing time. The main reason for this is that A112 requires the extra state equation and adjoint equation solves for the smoothing step.

Table 9.5 Performance summary for algorithms A111, A121, and A112.

Alg.	h	Iter.	State	Lin. state	Adj. state	Time
A111	1/16	3	4	10	4	0.19 s
	1/32	3	4	9	4	0.32 s
	1/64	3	4	7	4	0.82 s
	1/128	4	5	8	5	4.38 s
	1/256	4	5	8	5	23.14 s
A121	1/16	3	4	9	4	0.19 s
	1/32	3	4	8	4	0.31 s
	1/64	3	4	7	4	0.82 s
	1/128	3	4	7	4	3.66 s
	1/256	3	4	7	4	19.11 s
A112	1/16	4	9	21	9	0.45 s
	1/32	4	9	19	9	0.57 s
	1/64	4	9	19	9	1.59 s
	1/128	4	9	19	9	6.82 s
	1/256	4	9	18	9	35.04 s

Table 9.6 Performance summary for algorithms A112 and A122 without smoothing step.

h	Iter.	State	Lin. state	Adj. state	Time
Algorithm A112 without smoothing step					
1/16	7	8	16	8	0.38 s
1/32	7	8	15	8	0.63 s
1/64	7	8	15	8	1.83 s
1/128	7	8	15	8	8.29 s
1/256	7	8	14	8	42.87 s
Algorithm A122 without smoothing step					
1/16	4	5	13	5	0.26 s
1/32	4	5	11	5	0.42 s
1/64	4	5	11	5	1.17 s
1/128	4	5	11	5	5.14 s
1/256	4	5	10	5	26.38 s

In a second test we focus on the importance of the smoothing step. To this end, we have run algorithms A112 and A122 without smoothing steps (A112 is without projection whereas A122 contains a projection). The results are shown in Table 9.6. We see that A112 without smoothing steps needs an average of 7 iterations, whereas the regular algorithm A112, see Table 9.5, needs only 4 iterations on average. This shows that the smoothing step indeed has benefits, but that the algorithm still exhibits reasonable efficiency if the smoothing step is removed. If we do not perform a smoothing step, but include a projection (A122 without smoothing step), the performance of the algorithm is not affected by omitting the smoothing step, at least for the problem under consideration. We recall that the role of the smoothing

Table 9.7 Iteration history of algorithm A111 for a degenerate problem.

h	k	$\|u_k - \bar{u}\|_{L^2}$	$\|u_k - \bar{u}\|_{L^\infty}$	$\chi(u_k)$
$\frac{1}{128}$	0	1.628e+00	6.487e+00	2.053e-03
	1	9.019e-02	1.098e+00	1.006e-04
	2	1.657e-07	4.789e-07	1.536e-07
	3	8.811e-12	2.260e-11	1.624e-17

Table 9.8 Performance summary of algorithm A111 for a degenerate problem.

h	Iter.	State	Lin. state	Adj. state	Time
1/16	3	4	9	4	0.15 s
1/32	3	4	9	4	0.27 s
1/64	3	4	7	4	0.79 s
1/128	3	4	7	4	3.63 s
1/256	3	4	7	4	18.99 s

step is to avoid large discrepancies between $\|u_k - \bar{u}\|_{L^p}$ and $\|u_k - \bar{u}\|_{L^r}$, i.e., to avoid large (peak-like) deviations of u_k from $\bar{u}$ on small sets; see Example 3.57. It is intuitively clear that a projection step can help in cutting off such peaks (but there is no guarantee).

In our next test we show that lack of strict complementarity does not affect the superlinear convergence of the algorithms. Denoting by j the reduced objective function for the data (9.35) and by $\bar{u}$ the corresponding solution, we now choose $u_d = \lambda^{-1} j'(\bar{u})$. With these new data, the (new) gradient vanishes identically on Ω at $\bar{u}$ so that strict complementarity is violated. A representative run for this degenerated problem is shown in Table 9.7 (A111, $h = 1/128$). Here, $\mathbf{u}_d^h$ was obtained from the discrete solution and the discrete gradient. Similarly to the nondegenerate case, the algorithms show mesh-independent behavior; see Table 9.8. We have not included further tables for this problem since they would look essentially like those for the nondegenerate problem.

All-at-Once Approach

We now present numerical experiments for semismooth Newton methods applied to the all-at-once approach. Since the state equation is nonlinear, the advantage of this approach is that we do not have to solve the state equation in every iteration. On the other hand, the main work is solving the Newton system so that an increase of iterations in the semismooth Newton method can compensate for this saving of time.

We choose $u_0 \equiv -1$, $y_0 \equiv 0$, $w_0 \equiv 0$. Better choices for y_0 and w_0 are certainly possible. Our termination condition is

$$\chi(y_k, u_k, w_k) = \big(\|L_u(y_k, u_k, w_k) - P_C(u_k - L_u(y_k, u_k, w_k))\|_{L^2}^2 \\ + \|L_y(y_k, u_k, w_k)\|_{H^{-1}}^2 + \|E(y_k, u_k)\|_{H^{-1}}^2\big)^{1/2} \le \varepsilon$$

with $\varepsilon = 10^{-8}$. The all-at-once semismooth Newton system is solved by reducing it to the same Schur complement as was used for solving the black-box Newton equation, and by

Table 9.9 Iteration history of algorithm A212.

h	k	$\|u_k - \bar{u}\|_{L^2}$	$\|u_k - \bar{u}\|_{L^\infty}$	$\chi(y_k, u_k, w_k)$
	0	1.628e+00	6.487e+00	1.903e-01
	1	6.797e-01	2.514e+00	3.225e-01
$\frac{1}{128}$	2	1.176e-01	4.743e-01	4.007e-02
	3	3.025e-03	1.197e-02	1.068e-03
	4	1.756e-06	6.915e-06	6.767e-07
	5	3.000e-13	1.206e-12	8.823e-11
	0	1.628e+00	6.488e+00	1.903e-01
	1	6.797e-01	2.516e+00	3.225e-01
$\frac{1}{256}$	2	1.156e-01	4.645e-01	3.949e-02
	3	2.935e-03	1.171e-02	1.041e-03
	4	2.079e-06	8.583e-06	7.832e-07
	5	6.203e-14	2.888e-13	5.604e-11

Table 9.10 Performance summary for the algorithms A211, A221, and A212.

h	Algorithm					
	A211		A221		A212	
	Iter.	Time	Iter.	Time	Iter.	Time
1/16	5	0.15 s	5	0.15 s	5	0.20 s
1/32	5	0.25 s	5	0.25 s	5	0.26 s
1/64	5	0.77 s	5	0.74 s	5	0.76 s
1/128	5	3.65 s	5	3.60 s	5	3.78 s
1/256	5	19.31 s	5	19.38 s	5	20.11 s

applying MG-preconditioned cg. Only the right-hand side is different. Table 9.9 shows two representative runs of algorithm A212. Furthermore, Table 9.10 contains information on the performance of the algorithms A211, A221, and A212 for different mesh sizes.

In comparison with the black-box algorithms, we see that the all-at-once approach and the black-box approach are comparably efficient. As an advantage of the all-at-once approach we note that the smoothing step can be performed with minimum additional cost, whereas in the black-box approach it requires one additional solve of both, state and adjoint equation. We believe that the more expensive it is to solve the state equation (due to nonlinearity), the more favorable is the all-at-once approach.

Nested Iteration

Next, we present numerical results for the nested iteration approach. Here, we start on the grid $\mathcal{T}^{1/2}$, solve the problem with termination threshold $\varepsilon = 10^{-5}$, compute from its solution an initial point for the problem on the next finer grid $\mathcal{T}^{1/4}$, and so on. On the finest level we solve with termination threshold $\varepsilon = 10^{-8}$. Table 9.11 shows the number of iterations per level and the total execution time for the nested version of algorithm A111. Comparison with Table 9.11 shows that the nested version of A111 needs less than half the

Table 9.11 Performance summary for nested iteration version of algorithm A111.

h	Iter.	State	Lin. state	Adj. state	h	Iter.	State	Lin. state	Adj. state
1/2	1	2	6	2	1/32	2	3	5	3
1/4	2	3	7	3	1/64	1	2	4	2
1/8	2	3	6	3	1/128	1	2	4	2
1/16	2	3	6	3	1/256	1	2	3	2
Total Time: 8.95 s									

time to solve the problem as the unnested version (8.95 vs. 23.14 seconds). The use of nested iteration is thus very promising. Furthermore, it is very robust since, except for the coarsest problem, the Newton iteration is started with a very good initial point.

Discussion of the Results

From the presented numerical results we draw the following conclusions:

- The proposed methods allow us to use fast iterative solvers for their implementation. This leads to runtimes of optimal order in the sense that they are approximately proportional to the number of unknowns.
- The class of semismooth Newton methods performs very efficiently and exhibits mesh-independent behavior. We observe superlinear convergence as predicted by our theory.
- Both black-box and all-at-once approaches lead to efficient and robust algorithms which are comparable in runtime. If smoothing steps are used, the all-at-once approach is advantageous since it does not require additional state and adjoint state solves to compute the smoothing step.
- Lack of strict complementarity does not affect the fast convergence of the algorithms. This confirms our theory, which does not require strict complementarity.
- The choice of the MCP-function $\pi(x) = x_1 - P_C(x_1 - \lambda^{-1}x_2)$ appears to be preferable to $\pi(x) = -\phi^{FB}(-x)$ for this class of problems, at least with the black-box approach. The main reason for this is the additional cost of the smoothing step.
- The performance of the ϕ^{FB}-based algorithms, which from a theoretical point of view require a smoothing step, degrades by a certain margin if the smoothing step is turned off. This, however, is compensated for if we turn on the projection step. Our numerical experience indicates that this effect is problem dependent. It should be mentioned that so far we have never observed a severe deterioration of performance when switching off the smoothing step. But we stress that pathological situations like the one in Example 3.57 can occur, and that they result in a stagnation of convergence on fine grids (we have tried this, but do not include numerical results here).

We conclude this section by noting that many other optimal control problems can be handled in a similar way. In particular, Neumann boundary control can be used instead of distributed control. Furthermore, the control of other types of PDEs by semismooth Newton methods is possible, e.g., Neumann boundary control of the wave equation [146] and Neumann boundary control of the heat equation [31, 195]. The optimal control of the incompressible Navier–Stokes equations is considered in Chapter 10.

9.2 Obstacle Problems

In this section we study the class of obstacle problems described in section 1.6.2. In addition, this class of problems was also discussed as an example accompanying the theory in Chapter 8 on state constrained and related problems. Instead of proceeding as in Examples 8.6, 8.39, and 8.41, we use this section to further illustrate the technique of dual regularization, which, as was shown in section 8.2.4, is equivalent to the Moreau–Yosida regularization.

Obstacle problems of the following or similar type arise in many applications, e.g., potential flow of perfect fluids, lubrication, wake problems, etc.; see, e.g., [79] and the references therein. We describe the problem in terms of the obstacle problem for an elastic membrane. Our investigations can also be carried out in a similar way for elastic contact problems in solid mechanics; see, e.g., [197]. Furthermore, semismooth Newton methods were successfully applied to nonlinear contact problems and to problems with friction [94, 117, 116].

For $q \in [2,\infty)$, let $g \in W^{2,q}(\Omega)$ represent a (lower) obstacle located over the nonempty bounded open set $\Omega \subset \mathbb{R}^2$ with sufficiently smooth boundary, and denote by $y \in H_0^1(\Omega)$ the position of a membrane, and by $f \in L^q(\Omega)$ external forces. For compatibility we assume $g \leq 0$ on $\partial\Omega$, which is assumed to be sufficiently smooth. Then $y \in H_0^1(\Omega)$ solves the variational inequality

$$\begin{aligned} & y \geq g \quad \text{on } \Omega, \\ & a(y, v-y) - (f, v-y)_{L^2} \geq 0 \quad \forall\, v \in H_0^1(\Omega), \ v \geq g \ \text{ on } \Omega, \end{aligned} \tag{9.36}$$

where

$$a : H_0^1(\Omega) \times H_0^1(\Omega) \to \mathbb{R}, \quad a(y,z) = \sum_{i,j} a_{ij} \frac{\partial y}{\partial x_i} \frac{\partial z}{\partial x_j},$$

$a_{ij} = a_{ji} \in C^1(\bar{\Omega})$, and a being H_0^1-elliptic, i.e.,

$$a(y,y) \geq \nu \|y\|_{H_0^1}^2 \quad \forall\, y \in H_0^1(\Omega)$$

with a constant $\nu > 0$. The bounded bilinear form a induces a bounded linear operator $A \in \mathcal{L}(H_0^1, H^{-1})$ via $a(v,w) = \langle Av, w\rangle_{H^{-1},H_0^1}$ for all $v, w \in H_0^1(\Omega)$. The ellipticity of a and the Lax–Milgram theorem imply that $A \in \mathcal{L}(H_0^1, H^{-1})$ is a homeomorphism with $\|A^{-1}\|_{H^{-1},H_0^1} \leq \nu^{-1}$, and regularity results imply that $A^{-1} \in \mathcal{L}(L^2, H^2)$.

Introducing the closed convex set

$$\mathcal{F} = \{y \in H_0^1(\Omega) : y \geq g \ \text{ on } \Omega\}$$

and the objective function $J : H_0^1(\Omega) \mapsto \mathbb{R}$,

$$J(y) \stackrel{\text{def}}{=} \frac{1}{2} a(y,y) - (f,y)_{L^2},$$

we can write (9.36) equivalently as optimization problem

$$\text{minimize} \quad J(y) \quad \text{subject to} \quad y \in \mathcal{F}. \tag{9.37}$$

The ellipticity of a implies that J is strictly convex with $J(y) \to \infty$ as $\|y\|_{H_0^1} \to \infty$. Hence, using that $\mathcal{F}$ is a closed and convex subset of the Hilbert space $H_0^1(\Omega)$, we see that (9.37) possesses a unique solution $\bar{y} \in \mathcal{F}$ [65, Prop. II.1.2]. Further, regularity results [29, Thm. I.1] ensure that $\bar{y} \in H_0^1(\Omega) \cap W^{2,q}(\Omega)$.

9.2.1 Dual Problem

Since (9.37) is not posed in an L^p-setting, we derive an equivalent dual problem, which, as we will see, is posed in $L^2(\Omega)$. Denoting by $I_{\mathcal{F}} : H_0^1(\Omega) \to \mathbb{R} \cup \{+\infty\}$, the indicator function of $\mathcal{F}$, i.e., $I_{\mathcal{F}}(y)(x) = 0$ for $x \in \mathcal{F}$ and $I_{\mathcal{F}}(y)(x) = +\infty$ for $x \notin \mathcal{F}$, we can write (9.37) in the form

$$\inf_{y \in H_0^1(\Omega)} \; J(y) + I_{\mathcal{F}}(y). \tag{9.38}$$

The corresponding (Fenchel–Rockafellar) dual problem [65, Ch. III.4] (we choose $F = I_{\mathcal{F}}$, $G = J$, $\Lambda = I$, $u = y$, and $p^* = -u$ in the terminology of [65]) is

$$\sup_{u \in H^{-1}(\Omega)} \; -J^*(u) - I_{\mathcal{F}}^*(-u), \tag{9.39}$$

where $J^* : H^{-1}(\Omega) \to \mathbb{R} \cup \{+\infty\}$ and $I_{\mathcal{F}}^* : H^{-1}(\Omega) \to \mathbb{R} \cup \{+\infty\}$ are the conjugate functions of F and $I_{\mathcal{F}}$, respectively:

$$J^*(u) = \sup_{y \in H_0^1(\Omega)} \langle u, y \rangle_{H^{-1}, H_0^1} - J(u), \tag{9.40}$$

$$I_{\mathcal{F}}^*(u) = \sup_{y \in H_0^1(\Omega)} \langle u, y \rangle_{H^{-1}, H_0^1} - I_{\mathcal{F}}(y). \tag{9.41}$$

Let $y_0 \in H_0^1(\Omega)$ be such that $I_{\mathcal{F}}(y_0) = 0$; e.g., $y_0 = \bar{y}$. Then J is continuous at y_0 and $I_{\mathcal{F}}$ is bounded at y_0. Furthermore, since $I_{\mathcal{F}} \geq 0$, the ellipticity implies $J(y) + I_{\mathcal{F}}(y) \to \infty$ as $\|y\|_{H_0^1} \to \infty$. Therefore, [65, Thm. III.4.2] applies so that (9.38) and (9.39) possess solutions $\bar{y}$ (this we knew already) and $\bar{u}$, respectively, and for any pair of solutions holds

$$J(\bar{y}) + I_{\mathcal{F}}(\bar{y}) + J^*(\bar{u}) + I_{\mathcal{F}}^*(-\bar{u}) = 0.$$

Further, the following extremality relations hold:

$$J(\bar{y}) + J^*(\bar{u}) - \langle \bar{u}, \bar{y} \rangle_{H^{-1}, H_0^1} = 0, \tag{9.42}$$

$$I_{\mathcal{F}}(\bar{y}) + I_{\mathcal{F}}^*(-\bar{u}) + \langle \bar{u}, \bar{y} \rangle_{H^{-1}, H_0^1} = 0. \tag{9.43}$$

This implies

$$\bar{u} \in \partial J(\bar{y}), \tag{9.44}$$

$$-\bar{u} \in \partial I_{\mathcal{F}}(\bar{y}). \tag{9.45}$$

In our case J is smooth, which yields

$$\bar{u} = J'(\bar{y}) = A\bar{y} - f. \tag{9.46}$$

We know that the primal solution $\bar{y}$ is unique, and thus the dual solution $\bar{u}$ is unique too, by (9.46). Further, by regularity, $\bar{y} \in H_0^1(\Omega) \cap W^{2,q}(\Omega)$, which, via (9.46), implies $\bar{u} \in L^q(\Omega)$.

The supremum in the definition of J^*, see (9.40), is attained for $y = A^{-1}(f+u)$, with value

$$\begin{aligned} J^*(u) &= \langle u, y\rangle_{H^{-1},H_0^1} - \frac{1}{2}\langle Ay, y\rangle_{H^{-1},H_0^1} + \langle f, y\rangle_{H^{-1},H_0^1} \\ &= \frac{1}{2}\langle f+u, A^{-1}(f+u)\rangle_{H^{-1},H_0^1}. \end{aligned}$$

For $u \in L^2(\Omega)$ we can write

$$J^*(u) = \frac{1}{2}(f+u, A^{-1}(f+u))_{L^2}.$$

Further, see also [29, p. 19; 65, Ch. IV.4],

$$I_{\mathcal{F}}^*(u) = \sup_{y \in H_0^1} \langle u, y\rangle_{H^{-1},H_0^1} - I_{\mathcal{F}}(y) = \sup_{y \in \mathcal{F}} \langle u, y\rangle_{H^{-1},H_0^1}.$$

For $u \in L^2(\Omega)$ we have

$$I_{\mathcal{F}}^*(u) = \sup_{y \in \mathcal{F}} (u, y)_{L^2} = \begin{cases} (g, u)_{L^2} & \text{if } u \le 0 \text{ on } \Omega, \\ +\infty & \text{otherwise.} \end{cases}$$

Therefore, using the regularity of $\bar{y}$ and $\bar{u}$, we can write (9.39) in the form

$$\underset{u \in L^2(\Omega)}{\text{maximize}} \quad -\frac{1}{2}(f+u, A^{-1}(f+u))_{L^2} + (g, u)_{L^2} \quad \text{subject to} \quad u \ge 0, \tag{9.47}$$

and we know that $\bar{u} \in L^q(\Omega)$. We recall that from the dual solution $\bar{u}$ we can recover the primal solution $\bar{y}$ from the identity (9.46): $\bar{y} = A^{-1}(f + \bar{u})$.

In the following we prefer to write (9.47) as a minimization problem:

$$\underset{u \in L^2(\Omega)}{\text{minimize}} \quad \frac{1}{2}(f+u, A^{-1}(f+u))_{L^2} - (g, u)_{L^2} \quad \text{subject to} \quad u \ge 0. \tag{9.48}$$

Example 9.22. In the case $A = -\Delta$ the primal problem is

$$\underset{y \in H_0^1(\Omega)}{\text{minimize}} \quad \frac{1}{2}\|y\|_{H_0^1}^2 - (f, y)_{L^2} \quad \text{subject to} \quad y \ge g,$$

and the dual (minimization) problem reads

$$\underset{u \in L^2(\Omega)}{\text{minimize}} \quad \frac{1}{2}\|f+u\|_{H^{-1}}^2 - (g, u)_{L^2} \quad \text{subject to} \quad u \ge 0,$$

where $\|u\|_{H^{-1}} = \|\Delta^{-1}u\|_{H_0^1}$ is the norm dual to $\|\cdot\|_{H_0^1}$. ∎

We collect our results in the following theorem.

Theorem 9.23. *Under the problem assumptions, the obstacle problem* (9.36) *possesses a unique solution* $\bar{y} \in H_0^1(\Omega)$, *and this solution is contained in* $W^{2,q}(\Omega)$. *The dual problem* (9.39) *possesses a unique solution* $\bar{u} \in H^{-1}(\Omega)$ *as well. Primal and dual solution are linked via the equation*

$$A\bar{y} = f + \bar{u}.$$

In particular, $\bar{u} \in L^q(\Omega)$, *and the dual (minimization) problem can be written in the form* (9.48).

9.2.2 Regularized Dual Problem

Problem (9.48) is not coercive in the sense that for $\|u\|_{L^2} \to \infty$ the objective function tends to $+\infty$. Hence, we consider the regularized problem

$$\begin{aligned} &\underset{u \in L^2(\Omega)}{\text{minimize}} \quad j_\lambda(u) \overset{\text{def}}{=} \frac{1}{2}(f+u, A^{-1}(f+u))_{L^2} + \frac{\lambda}{2}\|u-u_d\|_{L^2}^2 - (g,u)_{L^2} \\ &\text{subject to} \quad u \ge 0 \text{ on } \Omega \end{aligned} \tag{9.49}$$

with $u_d \in L^{p'}(\Omega)$, $p' \in (2,\infty)$, and (small) regularization parameter $\lambda > 0$. This problem has the following properties.

Theorem 9.24. *The objective function of problem* (9.49) *is strongly convex and*

$$j_\lambda(u) \to \infty \quad \text{as } \|u\|_{L^2} \to \infty.$$

In particular, (9.49) *possesses a unique solution* $\bar{u}_\lambda \in L^2(\Omega)$, *and this solution lies in* $L^{p'}(\Omega)$. *The derivative of* j_λ *has the form*

$$j_\lambda'(u) = \lambda(u-u_d) + A^{-1}(f+u) - g \overset{\text{def}}{=} \lambda u + G(u). \tag{9.50}$$

Here, the mapping $G(u) = A^{-1}(f+u) - g - \lambda u_d$ *maps* $L^2(\Omega)$ *continuously and affine linearly into* $L^{p'}(\Omega)$.

Proof. Obviously, j_λ is a smooth quadratic function, and with $z = A^{-1}(f+u)$,

$$\begin{aligned} j_\lambda(u) &= \frac{\lambda}{2}\|u-u_d\|_{L^2}^2 + \frac{1}{2}a(z,z) - (g,u)_{L^2} \\ &\ge \frac{\lambda}{2}\|u-u_d\|_{L^2}^2 - \|g\|_{L^2}\|u\|_{L^2} \to \infty \end{aligned}$$

as $\|u\|_{L^2} \to \infty$. Therefore, since $\{u \in L^2(\Omega) : u \ge 0\}$ is closed and convex, we see that (9.49) possesses a unique solution $\bar{u}_\lambda \in L^2(\Omega)$.

Certainly, $j_\lambda'(u)$ is given by (9.50), and the fact that $A \in \mathcal{L}(H_0^1, H^{-1})$ implies that

$$G : u \in L^2(\Omega) \mapsto A^{-1}(f+u) - g - \lambda u_d \in H_0^1(\Omega) + L^{p'}(\Omega) \hookrightarrow L^{p'}(\Omega)$$

is continuous affine linear. From the optimality conditions for (9.49) we conclude

$$j'_\lambda(\bar{u}_\lambda) = 0 \quad \text{on } \{x \in \Omega : \bar{u}_\lambda(x) \neq 0\}.$$

Hence,

$$\bar{u}_\lambda = \mathbf{1}_{\{\bar{u}_\lambda \neq 0\}} \bar{u}_\lambda = -\lambda^{-1} \mathbf{1}_{\{\bar{u}_\lambda \neq 0\}} G(\bar{u}_\lambda) \in L^{p'}(\Omega). \qquad \square$$

Corollary 9.25. *Under the problem assumptions, $F = j'_\lambda$ satisfies Assumption* 3.38 (a) *for any $p \in [2,\infty)$, any $p' \in [2,\infty)$ with $p' \le p$ and $u_d \in L^{p'}(\Omega)$, any $1 \le r < p'$, and all $\alpha \in (0,1]$. Furthermore, F satisfies Assumption* 4.1 *for $r = 2$ and all $p \in (2,\infty)$ with $u_d \in L^p(\Omega)$. Finally, F also satisfies Assumptions* 4.6 (a)–(e) *for all $p \in [2,\infty)$ and all $p' \in (2,\infty)$.*

Proof. The corollary is an immediate consequence of Theorem 9.24 and the L^2-coercivity of j_λ. $\square$

Remark 9.26. *Corollary* 9.25 *provides all the assumptions that are needed to establish the semismoothness of NCP-function-based reformulations. In fact, for general NCP-functions Theorem* 3.50 *is applicable, whereas for the special choice $\pi(x) = x_1 - P_{[0,\infty)}(x_1 - \lambda^{-1}x_2)$ we can use Theorem* 4.4*. Furthermore, the sufficient condition for regularity of Theorem* 4.8 *is applicable. Hence, we can apply our class of semismooth Newton methods to solve problem* (9.49).

Next, we derive bounds for the approximation errors $\|\bar{u}_\lambda - \bar{u}\|_{H^{-1}}$ and $\|\bar{y}_\lambda - \bar{y}\|_{H_0^1}$, where $\bar{y}_\lambda = A^{-1}(f + \bar{u}_\lambda)$.

Theorem 9.27. *Let $\bar{u}$ and $\bar{u}_\lambda$ denote the solutions of* (9.48) *and* (9.49)*, respectively. Then $\bar{y} = A^{-1}(f + \bar{u})$ solves the obstacle problem* (9.36) *and with $\bar{y}_\lambda = A^{-1}(f + \bar{u}_\lambda)$ holds, as $\lambda \to 0^+$:*

$$\|\bar{u}_\lambda - \bar{u}\|_{H^{-1}} = o(\lambda^{1/2}), \tag{9.51}$$

$$\|\bar{y}_\lambda - \bar{y}\|_{H_0^1} = o(\lambda^{1/2}). \tag{9.52}$$

Proof. By Theorems 9.23 and 9.24 we know that the dual problem (9.48) and the regularized dual problem (9.49) possess unique solutions $\bar{u}, \bar{u}_\lambda \in L^p(\Omega)$. Now

$$\begin{aligned} j_\lambda(\bar{u}_\lambda) &\le j_\lambda(\bar{u}) = j(\bar{u}) + \frac{\lambda}{2}\|\bar{u} - u_d\|_{L^2}^2 \le j(\bar{u}_\lambda) + \frac{\lambda}{2}\|\bar{u} - u_d\|_{L^2}^2 \\ &= j_\lambda(\bar{u}_\lambda) + \frac{\lambda}{2}\big(\|\bar{u} - u_d\|_{L^2}^2 - \|\bar{u}_\lambda - u_d\|_{L^2}^2\big). \end{aligned}$$

This proves

$$\|\bar{u}_\lambda - u_d\|_{L^2} \le \|\bar{u} - u_d\|_{L^2}. \tag{9.53}$$

Further,

$$\begin{aligned} j(\bar{u}) &\le j(\bar{u}_\lambda) = j_\lambda(\bar{u}_\lambda) - \frac{\lambda}{2}\|\bar{u}_\lambda - u_d\|_{L^2}^2 \le j_\lambda(\bar{u}) - \frac{\lambda}{2}\|\bar{u}_\lambda - u_d\|_{L^2}^2 \\ &= j(\bar{u}) + \frac{\lambda}{2}\big(\|\bar{u} - u_d\|_{L^2}^2 - \|\bar{u}_\lambda - u_d\|_{L^2}^2\big) \le j(\bar{u}) + \frac{\lambda}{2}\|\bar{u} - u_d\|_{L^2}^2. \end{aligned} \tag{9.54}$$

Therefore,

$$0 \leq j(\bar{u}_\lambda) - j(\bar{u}) \leq \frac{\lambda}{2}\|\bar{u} - u_d\|_{L^2}^2 = O(\lambda) \quad \text{as } \lambda \to 0^+.$$

Now let $\lambda_k \to 0^+$. Since

$$\mathcal{M} = \{u \in L^2(\Omega) : u \geq 0,\ \|u - u_d\|_{L^2} \leq \|\bar{u} - u_d\|_{L^2}\}$$

is closed, convex, and bounded, there exists a subsequence and a point $\tilde{u} \in \mathcal{M}$ such that $u_{\lambda_{k'}} \to \tilde{u}$ weakly in L^2. Since j is convex and continuous, it is weakly lower semicontinuous, so that

$$j(\bar{u}) \leq j(\tilde{u}) \leq \liminf_{k' \to \infty} j(u_{\lambda_{k'}}) = \liminf_{k' \to \infty} \big[j(\bar{u}) + O(\lambda_{k'})\big] = j(\bar{u}).$$

Hence $\tilde{u}$ is a solution of (9.48) and therefore $\tilde{u} = \bar{u}$, since $\bar{u}$ is the unique solution. By a subsequence-subsequence argument we conclude that

$$\bar{u}_\lambda \to \bar{u} \quad \text{weakly in } L^2(\Omega) \quad \text{as } \lambda \to 0^+. \tag{9.55}$$

Since $u \mapsto \|u - u_d\|_{L^2}$ is convex and continuous, hence weakly lower semicontinuous, we obtain from (9.53) and (9.55)

$$\|\bar{u} - u_d\|_{L^2} \leq \liminf_{\lambda \to 0^+} \|\bar{u}_\lambda - u_d\|_{L^2},$$
$$\|\bar{u} - u_d\|_{L^2} \geq \limsup_{\lambda \to 0^+} \|\bar{u}_\lambda - u_d\|_{L^2},$$

which proves

$$\|\bar{u}_\lambda - u_d\|_{L^2} \to \|\bar{u} - u_d\|_{L^2} \quad \text{as } \lambda \to 0^+. \tag{9.56}$$

Since L^2 is a Hilbert space, (9.55) and (9.56) imply

$$\bar{u}_\lambda \to \bar{u} \ \text{ in } \ L^2 \quad \text{as } \lambda \to 0^+. \tag{9.57}$$

Hence, (9.54) implies

$$j(\bar{u}_\lambda) - j(\bar{u}) = o(\lambda).$$

Since $\bar{u}$ solves (9.48), there holds $(j'(\bar{u}), \bar{u}_\lambda - \bar{u})_{L^2} \geq 0$. Therefore,

$$\begin{aligned} j(\bar{u}_\lambda) - j(\bar{u}) &= (j'(\bar{u}), \bar{u}_\lambda - \bar{u})_{L^2} + \frac{1}{2}(\bar{u}_\lambda - \bar{u}, j''(\bar{u})(\bar{u}_\lambda - \bar{u}))_{L^2} \\ &\geq \frac{1}{2}(\bar{u}_\lambda - \bar{u}, j''(\bar{u})(\bar{u}_\lambda - \bar{u}))_{L^2} = \frac{1}{2}(\bar{u}_\lambda - \bar{u}, A^{-1}(\bar{u}_\lambda - \bar{u}))_{L^2}. \end{aligned}$$

Hence, with $v = \bar{u}_\lambda - \bar{u}$ and $w = A^{-1}v$,

$$\begin{aligned} \|v\|_{H^{-1}}^2 &= \|Aw\|_{H^{-1}}^2 \leq \|A\|_{H_0^1, H^{-1}}^2 \|w\|_{H_0^1}^2 \leq \|A\|_{H_0^1, H^{-1}}^2 \kappa^{-1} \langle Aw, w\rangle_{H^{-1}, H_0^1} \\ &\leq \kappa^{-1}\|A\|_{H_0^1, H^{-1}}^2 \langle v, w\rangle_{L^2} \leq 2\kappa^{-1}\|A\|_{H_0^1, H^{-1}}^2 (j(\bar{u}_\lambda) - j(\bar{u})) \\ &= 2\kappa^{-1}\|A\|_{H_0^1, H^{-1}}^2 o(\lambda). \end{aligned}$$

This proves (9.51). The solution of the obstacle problem is $\bar{y} = A^{-1}(f+\bar{u})$. For $\bar{y}_\lambda = A^{-1}(f+\bar{u}_\lambda)$ the following holds:

$$\begin{aligned}\|\bar{y}_\lambda - \bar{y}\|^2_{H_0^1} &= \|A^{-1}(\bar{u}_\lambda - \bar{u})\|^2_{H_0^1} = \|w\|^2_{H_0^1} \le \kappa^{-1}\langle Aw, w\rangle_{H^{-1},H_0^1} \\ &= \kappa^{-1}(\bar{u}_\lambda - \bar{u}, A^{-1}(\bar{u}_\lambda - \bar{u}))_{L^2} \le 2\kappa^{-1}(j(\bar{u}_\lambda) - j(\bar{u})) \\ &= 2\kappa^{-1}o(\lambda).\end{aligned}$$

The proof is complete. □

Remark 9.28. *The parameter λ has to be chosen sufficiently small to ensure that the error is not larger than the discretization error. Our approach will be to successively reduce λ.*

9.2.3 Discretization

We use the same finite element spaces as in section 9.1.3. A straightforward discretization yields the discrete obstacle problem (in coordinate form)

$$\underset{\mathbf{y}^h\in\mathbb{R}^{n^h}}{\text{minimize}} \quad \frac{1}{2}\mathbf{y}^{h^T}\mathbf{A}^h\mathbf{y}^h - \mathbf{f}^{h^T}\mathbf{y}^h \quad \text{subject to} \quad \mathbf{y}^h \ge \mathbf{g}^h. \tag{9.58}$$

Here, $\mathbf{g}^h \in \mathbb{R}^{n^h}$, $\mathbf{g}^h_i = g(P^h_i)$, approximates the obstacle. Furthermore, $\mathbf{f}^h_i = (\beta^h_i, f)_{L^2}$ and $\mathbf{A}^h_{ij} = (A\beta^h_i, \beta^h_j)_{H^{-1},H_0^1}$. The corresponding dual problem is

$$\begin{aligned}&\underset{\mathbf{u}^h\in\mathbb{R}^{n^h}}{\text{minimize}} \quad \frac{1}{2}(\mathbf{f}^h + \mathbf{S}^h\mathbf{u}^h)^T\mathbf{A}^{h^{-1}}(\mathbf{f}^h + \mathbf{S}^h\mathbf{u}^h) - \mathbf{g}^{h^T}\mathbf{S}^h\mathbf{u}^h \\ &\text{subject to} \quad \mathbf{u}^h \ge 0.\end{aligned} \tag{9.59}$$

Here, $\mathbf{S}^h \in \mathbb{R}^{n^h\times n^h}$ is defined as in (9.17). The discrete regularized dual problem then is given by

$$\begin{aligned}\text{minimize}_{\mathbf{u}^h\in\mathbb{R}^{n^h}} \quad & \mathbf{j}^h_\lambda(\mathbf{u}^h) \overset{\text{def}}{=} \frac{1}{2}(\mathbf{f}^h + \mathbf{S}^h\mathbf{u}^h)^T\mathbf{A}^{h^{-1}}(\mathbf{f}^h + \mathbf{S}^h\mathbf{u}^h) \\ & + \frac{\lambda}{2}(\mathbf{u}^h - \mathbf{u}^h_d)^T\mathbf{S}^h(\mathbf{u}^h - \mathbf{u}^h_d) - \mathbf{g}^{h^T}\mathbf{S}^h\mathbf{u}^h \\ \text{subject to} \quad & \mathbf{u}^h \ge 0,\end{aligned} \tag{9.60}$$

where, e.g., $[\mathbf{S}^h\mathbf{u}^h_d]_i = (L^h\beta^h_i, L^hu_d)_{L^2}$. From the solution $\bar{\mathbf{u}}^h_\lambda$ of (9.60) we compute $\mathbf{y}^h_\lambda$ via $\mathbf{A}^h\bar{\mathbf{y}}^h_\lambda = \mathbf{f}^h + \mathbf{S}^h\bar{\mathbf{u}}^h_\lambda$.

The gradient of $\mathbf{j}^{h'}_\lambda$ and the Hessian $\mathbf{j}^{h''}_\lambda$ of $\mathbf{j}^h_\lambda$ with respect to the $\mathbf{S}^h$ inner product are given by

$$\begin{aligned}\mathbf{j}^{h'}_\lambda(\mathbf{u}^h) &= \mathbf{A}^{h^{-1}}(\mathbf{f}^h + \mathbf{S}^h\mathbf{u}^h) + \lambda(\mathbf{u}^h - \mathbf{u}^h_d) - \mathbf{g}^h, \\ \mathbf{j}^{h''}_\lambda(\mathbf{u}^h) &= \mathbf{A}^{h^{-1}}\mathbf{S}^h + \lambda I.\end{aligned}$$

Choosing a Lipschitz continuous and semismooth NCP-function ϕ, we reformulate (9.60) in the form

$$\mathbf{\Phi}^h(\mathbf{u}^h) \stackrel{\text{def}}{=} \begin{pmatrix} \phi\big(\mathbf{u}_1^h, \mathbf{j}_\lambda^{h\prime}(\mathbf{u}^h)_1\big) \\ \vdots \\ \phi\big(\mathbf{u}_{n^h}^h, \mathbf{j}_\lambda^{h\prime}(\mathbf{u}^h)_{n^h}\big) \end{pmatrix} = 0. \tag{9.61}$$

This is the discrete counterpart of the semismooth reformulation in function space

$$\Phi(u) \stackrel{\text{def}}{=} \phi\big(u, j_\lambda'(u)\big) = 0.$$

As in section 9.1.4, we can argue that an appropriate discretization of $\partial^\circ\Phi$ is $\partial^\circ\mathbf{\Phi}^h(\mathbf{u}^h)$, the set of all matrices $\mathbf{B}^h \in \mathbb{R}^{n^h\times n^h}$ with

$$\mathbf{B}^h = \mathbf{D}_1^h + \mathbf{D}_2^h \mathbf{j}_\lambda^{h\prime\prime}(\mathbf{u}^h), \tag{9.62}$$

where $\mathbf{D}_1^h$ and $\mathbf{D}_2^h$ are diagonal matrices such that

$$\big((\mathbf{D}_1^h)_{ll}, (\mathbf{D}_2^h)_{ll}\big) \in \partial\phi\big(\mathbf{u}_l^h, \mathbf{j}_\lambda^{h\prime}(\mathbf{u}^h)_l\big), \quad l = 1, \ldots, n^h.$$

Again, we have the inclusion

$$\partial_C \mathbf{\Phi}^h(\mathbf{u}^h) \subset \partial^\circ \mathbf{\Phi}^h(\mathbf{u}^h)$$

with equality if ϕ or $-\phi$ is regular. With the same argumentation as in the derivation of Theorem 9.19 we can show that $\mathbf{\Phi}^h$ is $\partial^\circ\mathbf{\Phi}^h$-semismooth (and thus also semismooth in the usual sense). Semismoothness of higher order can be proved analogously. Hence, we can apply our semismooth Newton methods to solve (9.61). The details of the resulting algorithm, which are not given here, parallel Algorithm 9.20. The central task is to solve the semismooth Newton system (we suppress the subscript k)

$$[\mathbf{D}_1^h + \mathbf{D}_2^h \mathbf{j}_\lambda^{h\prime\prime}(\mathbf{u}^h)]\mathbf{s}^h = -\mathbf{\Phi}^h(\mathbf{u}^h).$$

Using the structure of $\mathbf{j}_\lambda^{h\prime\prime}$ and that $(\mathbf{D}_1^h + \lambda\mathbf{D}_2^h)$ is diagonal and positive definite for our choices of ϕ, we see that this is equivalent to $\mathbf{s}^h = \mathbf{S}^{h^{-1}}\mathbf{A}^h\mathbf{v}^h$, where $\mathbf{v}^h$ solves

$$[\mathbf{A}^h + \mathbf{S}^h(\mathbf{D}_1^h + \lambda\mathbf{D}_2^h)^{-1}\mathbf{D}_2^h]\mathbf{v}^h = -\mathbf{S}^h(\mathbf{D}_1^h + \lambda\mathbf{D}_2^h)^{-1}\mathbf{\Phi}^h(\mathbf{u}^h).$$

This can be viewed as a discretization of the PDE

$$Av + \frac{d_2}{d_1 + \lambda d_2} v = \frac{-1}{d_1 + \lambda d_2}\Phi(u).$$

Therefore, we can apply a multigrid method to compute $\mathbf{v}^h$, from which $\mathbf{s}^h$ can be obtained easily.

9.2.4 Numerical Results

We consider the following problem:

$$\begin{aligned}
\Omega &= (0,1)\times(0,1),\\
g &= -\frac{1}{4}+\frac{1}{2}\sin(\pi x_1)\sin(\pi x_2),\\
f &= -5\sin(2\pi x_1)\sin(2\pi x_2)\left(\frac{1}{2}+e^{2x_1+x_2}\right).
\end{aligned}\tag{9.63}$$

The triangulation is the same as in section 9.1.7. Again, the code was implemented in MATLAB Version 7.8.0.347 (R2009a) 64-bit (glnxa64), using sparse matrix computations, and was run under openSUSE 11.2 Linux on an HP Compaq™ workstation with an Intel Core 2 Duo CPU E8600 operating at 3.33 GHz. To obtain sufficiently accurate solutions, the regularization parameter has to be chosen appropriately. Here, we use a nested iteration approach and determine λ in dependence on the current mesh size. It is known [79, App. I.3] that, under appropriate conditions, the described finite element discretization leads to approximation errors $\|\bar{y}^h-\bar{y}\|_{H_0^1}=O(h)$. Since we have shown in Theorem 9.27 that $\|\bar{y}_\lambda-\bar{y}\|_{H_0^1}=o(\lambda^{1/2})$, we choose λ of the order h^2; more precisely, we work with

$$\lambda=\lambda_h=\frac{h^2}{10}.$$

We then solve problem (9.60) for $h=1/2$ until

$$\chi(u_k)=\|u_k-P_{[0,\infty)}(u_k-j_\lambda'(u_k))\|_{L^2}\le\varepsilon \tag{9.64}$$

with $\varepsilon=10^{-5}$ (in the corresponding discrete norms), interpolate this coarse solution to obtain an initial point on $\mathcal{T}^{1/4}$, solve this problem (now with $\lambda=\lambda_{1/4}$) until (9.64) is satisfied, interpolate again, and repeat this procedure until we have reached the finest grid on which we iterate until (9.64) holds with $\varepsilon=10^{-8}$. To further reduce the effect of regularization, we always use as u_d the interpolated solution from the next coarser grid (the same point that we use as initial point). On $\mathcal{T}^{1/2}$ we choose $u_d=u_0\equiv 0$. The obstacle is shown in Figure 9.5, the state solution $\bar{y}_\lambda$ for $\lambda=\lambda_{1/64}$ is displayed in Figure 9.6, and the dual solution $\bar{u}_\lambda$ is depicted in Figure 9.7. Note that $\{x:\bar{u}(x)\neq 0\}$ is the contact region, and that for our choice of λ the solution $\bar{u}$ is approximated up to a fraction of the discretization error by $\bar{u}_\lambda$. It can be seen that $\bar{u}$ is discontinuous at the boundary of the contact region.

In the numerical tests it turned out that it is not advantageous to let λ^{-1} become too large in the smoothing steps. Hence, we set $\gamma=\min\{10^5,\lambda^{-1}\}$ and work with smoothing steps of the form $S_k(u)=P_{[0,\infty)}(u-\gamma j_\lambda'(u))$. On the other hand, even very small λ does not cause any problems in the NCP-function $\phi(x)=x_1-P_{[0,\infty)}(x_1-\lambda^{-1}x_2)$. We consider two methods: The smoothing-step-free algorithm A111 with $\phi(x)=x_1-P_{[0,\infty)}(x_1-\lambda^{-1}x_2)$, and algorithm A112 with ϕ^{FB} and smoothing step as just described. It turns out that without globalization the projected variant A121 tends to cycle when λ becomes very small. Since incorporating a globalization requires additional evaluations of j_λ and/or its gradient, which is expensive due to the presence of A^{-1}, we do not present numerical results for a globalized version of A121.

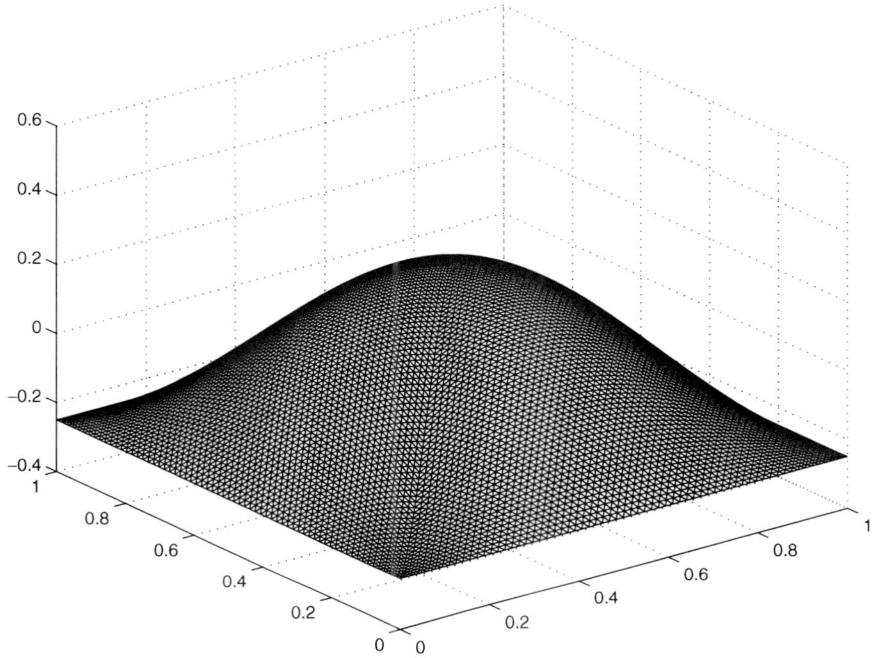

Figure 9.5 The obstacle g ($h = 1/64$).

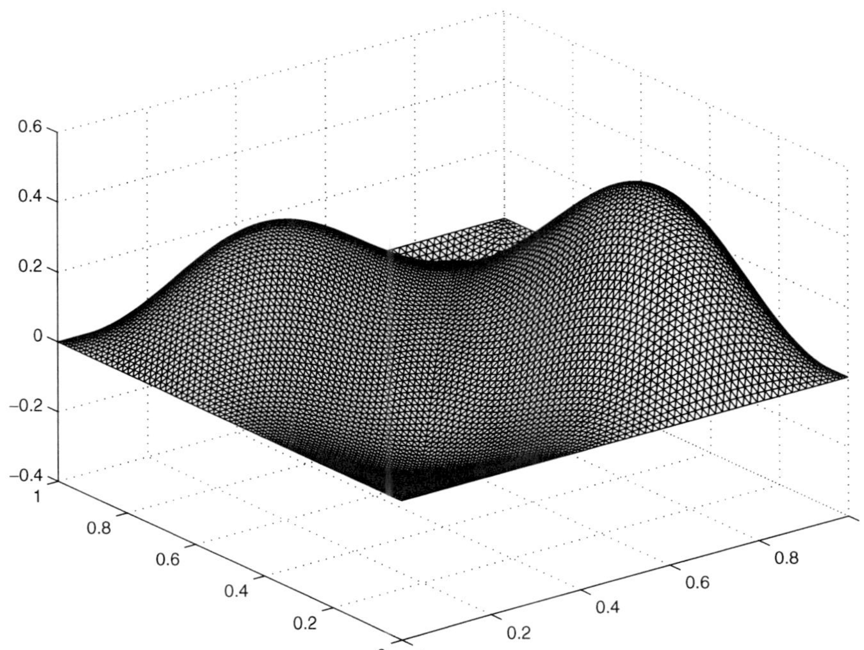

Figure 9.6 Computed state $\bar{y}_\lambda$ ($h = 1/64$).

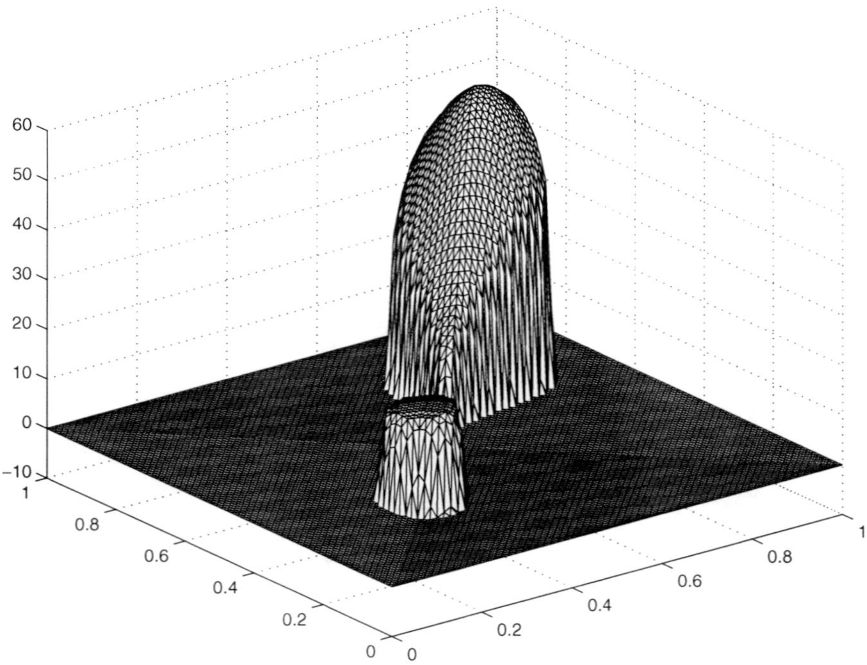

Figure 9.7 Computed dual solution $\bar{u}_\lambda$ ($h = 1/64$).

In Table 9.12 (A111) and Table 9.14 (A112) we show, for each level of the nested iteration, the value of λ, the number of iterations performed on this level (Iter.), and the number of PDE solves. Furthermore, the (discrete) distance $\|y^* - \bar{y}\|_{H_0^1}$ of the (discrete) computed solution y^* to the (discrete) solution $\bar{y}$ corresponding to $\lambda = 0$ and the (discrete) distance $\|y^* - \bar{y}_\lambda\|_{H_0^1}$ of the (discrete) computed solution y^* to the (discrete) solution $\bar{y}_\lambda$ corresponding to $\lambda = h_{\text{final}}^2/10$ are shown. The total runtime is also given. We see that on each level only a few Newton iterations are performed. In Table 9.13 the iteration history of A111 on the finest level is shown for $h_{\text{final}} = 1/256$. Obviously, the convergence is superlinear with rate >1, and we observe mesh-independent performance of the methods. Furthermore, the runtime increases approximately linearly with the number of unknowns. In conclusion, it can be seen that, similarly to the optimal control problem in section 9.1, the algorithms offer all the favorable properties that are predicted by our theory. For this application, the smoothing-step-free algorithm with the projection-based NCP-function leads to significantly shorter solution times than the algorithm with Fischer–Burmeister function and smoothing step. This is mainly caused by the additional PDE solves needed for the smoothing steps. As for the use of multigrid methods, it would be interesting to investigate if, instead of multilevel Newton methods, nonlinear multigrid methods can be used successfully.

Furthermore, we stress that many other variational inequalities can be treated in a similar way. In particular, this applies to certain kinds of the following problems: problems with constraints on the boundary, time-dependent VIPs, quasi-variational inequalities [16, 18], and VIPs of the second kind.

Table 9.12 Performance summary for nested iteration version of algorithm A111.

h	λ	Iter.	PDE solves	h	λ	Iter.	PDE solves
$h_{\text{final}} = 1/64$							
1/2	2.500e−02	1	2	1/32	9.766e−05	3	4
1/4	6.250e−03	2	3	1/64	2.441e−05	4	5
1/8	1.563e−03	2	3				
1/16	3.906e−04	3	4				
$\|y^* - \bar{y}\|_{H_0^1} = 2.375\text{e}{-03}$							
$\|y^* - \bar{y}_\lambda\|_{H_0^1} = 1.978\text{e}{-10}$							
Total Time: 0.46 s							
$h_{\text{final}} = 1/128$							
1/2	2.500e−02	1	2	1/32	9.766e−05	3	4
1/4	6.250e−03	2	3	1/64	2.441e−05	3	4
1/8	1.563e−03	2	3	1/128	6.104e−06	4	5
1/16	3.906e−04	3	4				
$\|y^* - \bar{y}\|_{H_0^1} = 8.671\text{e}{-04}$							
$\|y^* - \bar{y}_\lambda\|_{H_0^1} = 3.572\text{e}{-10}$							
Total Time: 1.20 s							
$h_{\text{final}} = 1/256$							
1/2	2.500e−02	1	2	1/32	9.766e−05	3	4
1/4	6.250e−03	2	3	1/64	2.441e−05	3	4
1/8	1.563e−03	2	3	1/128	6.104e−06	3	4
1/16	3.906e−04	3	4	1/256	1.526e−06	4	5
$\|y^* - \bar{y}\|_{H_0^1} = 3.024\text{e}{-04}$							
$\|y^* - \bar{y}_\lambda\|_{H_0^1} = 5.594\text{e}{-11}$							
Total Time: 5.39 s							

Table 9.13 Iteration history of algorithm A111 on the final level $h = h_{\text{final}} = 1/256$.

Algorithm A111			
k	$\|y_k - \bar{y}_\lambda\|_{H_0^1}$	$\|y_k - \bar{y}\|_{H_0^1}$	$\chi(u_k)$
0	1.701e-03	1.862e-03	5.812e-05
1	5.648e-04	6.199e-04	3.273e-01
2	2.682e-05	3.034e-04	1.706e-02
3	2.333e-09	3.024e-04	7.343e-07
4	5.594e-11	3.024e-04	8.139e-11

Table 9.14 Performance summary for nested iteration version of algorithm A112.

<table>
<tr><th>h</th><th>λ</th><th>Iter.</th><th>PDE solves</th><th>h</th><th>λ</th><th>Iter.</th><th>PDE solves</th></tr>
<tr><td colspan="8">$h_{\text{final}} = 1/64$</td></tr>
<tr><td>1/2</td><td>2.500e−02</td><td>4</td><td>9</td><td>1/32</td><td>9.766e−05</td><td>4</td><td>9</td></tr>
<tr><td>1/4</td><td>6.250e−03</td><td>3</td><td>7</td><td>1/64</td><td>2.441e−05</td><td>5</td><td>11</td></tr>
<tr><td>1/8</td><td>1.563e−03</td><td>4</td><td>9</td><td></td><td></td><td></td><td></td></tr>
<tr><td>1/16</td><td>3.906e−04</td><td>4</td><td>9</td><td></td><td></td><td></td><td></td></tr>
<tr><td colspan="8">$\|y^* - \bar{y}\|_{H_0^1} = 2.374\text{e}{-03}$</td></tr>
<tr><td colspan="8">$\|y^* - \bar{y}_\lambda\|_{H_0^1} = 1.632\text{e}{-07}$</td></tr>
<tr><td colspan="8">Total Time: 0.85 s</td></tr>
<tr><td colspan="8">$h_{\text{final}} = 1/128$</td></tr>
<tr><td>1/2</td><td>2.500e−02</td><td>4</td><td>9</td><td>1/32</td><td>9.766e−05</td><td>4</td><td>9</td></tr>
<tr><td>1/4</td><td>6.250e−03</td><td>3</td><td>7</td><td>1/64</td><td>2.441e−05</td><td>4</td><td>9</td></tr>
<tr><td>1/8</td><td>1.563e−03</td><td>4</td><td>9</td><td>1/128</td><td>6.104e−06</td><td>6</td><td>13</td></tr>
<tr><td>1/16</td><td>3.906e−04</td><td>4</td><td>9</td><td></td><td></td><td></td><td></td></tr>
<tr><td colspan="8">$\|y^* - \bar{y}\|_{H_0^1} = 8.670\text{e}{-04}$</td></tr>
<tr><td colspan="8">$\|y^* - \bar{y}_\lambda\|_{H_0^1} = 3.153\text{e}{-08}$</td></tr>
<tr><td colspan="8">Total Time: 2.81 s</td></tr>
<tr><td colspan="8">$h_{\text{final}} = 1/256$</td></tr>
<tr><td>1/2</td><td>2.500e−02</td><td>4</td><td>9</td><td>1/32</td><td>9.766e−05</td><td>4</td><td>9</td></tr>
<tr><td>1/4</td><td>6.250e−03</td><td>3</td><td>7</td><td>1/64</td><td>2.441e−05</td><td>4</td><td>9</td></tr>
<tr><td>1/8</td><td>1.563e−03</td><td>4</td><td>9</td><td>1/128</td><td>6.104e−06</td><td>4</td><td>9</td></tr>
<tr><td>1/16</td><td>3.906e−04</td><td>4</td><td>9</td><td>1/256</td><td>1.526e−06</td><td>5</td><td>11</td></tr>
<tr><td colspan="8">$\|y^* - \bar{y}\|_{H_0^1} = 3.024\text{e}{-04}$</td></tr>
<tr><td colspan="8">$\|y^* - \bar{y}_\lambda\|_{H_0^1} = 2.272\text{e}{-08}$</td></tr>
<tr><td colspan="8">Total Time: 12.22 s</td></tr>
</table>

9.3 L^1-optimization

Recently, semismooth Newton methods have been used to solve optimization problems involving L^1-functionals [35, 42, 82, 183]. These functionals are of importance since, in the last years, it has been observed and intensively investigated that L^1-regularization promotes sparsity. In various applications, it is desirable to compute sparse optimal controls, where sparsity means that the support of the optimal control is small. This can, e.g., be used to model actuator placement problems. Instead of considering the placement of finitely many actuators directly, a distributed actuator (expressed through a distributed control) is modeled and combined with sparsity constraints. The support and amplitude of the sparse optimal control then indicates where actuators should be placed.

The fact that L^1-functionals favor sparse solutions can be motivated as follows. If we want a regularization that penalizes deviation from zero of a variable t, we need to choose a penalty function r with $r(0) = 0$ that increases immediately if t moves away from zero. If we introduce a parameter $\alpha > 0$ and require that $r = r_\alpha$ is convex and that the level set $\{t : r_\alpha(t) \le \alpha\}$ has the form $[-1, 1]$, then the steepest slope $|r'_\alpha(0, \pm 1)|$ that we can achieve at $t = 0$ under these constraints is $|r'_\alpha(0, \pm 1)| = \alpha$ and it is achieved by, e.g., $\alpha|t|$. We could add any convex function $z(t)$ that satisfies $z(t) = 0$ on $[-1, 1]$, but the choice $r_\alpha(t) = \alpha|t|$ appears to be most natural. This shows that a function of the form $\alpha\|x\|_1$ is the right choice for a convex penalty that promotes sparsity of the vector x. Similarly, functions of the form $\alpha\|u\|_{L^1}$ are appropriate to promote sparsity (i.e., small support) of a function $u \in L^1(\Omega)$. Of course, this penalization (or regularization) has a drawback—it is nonsmooth. But, as we will see, semismooth Newton methods are nevertheless applicable.

Since the space $L^1(\Omega)$ is weaker than $L^2(\Omega)$ and also not reflexive, it is not fully appropriate to just replace the usual L^2-regularization by an L^1-regularization. Hence, we keep the L^2-regularization and add an L^1-regularization for penalizing large supports of the control u.

We look at a familiar elliptic optimal control problem, but now augmented with an L^1-regularization,

$$\min_{y \in H_0^1(\Omega), u \in L^2(\Omega)} \quad J_0(y) + \frac{\lambda}{2}\|u\|_{L^2}^2 + \alpha\|u\|_{L^1} \quad \text{subject to} \quad Ay = u,$$

with an $H_0^1(\Omega)$-elliptic operator $A \in \mathcal{L}(H_0^1, H^{-1})$ with coercivity constant $\nu > 0$. The functional $J_0 : H_0^1(\Omega) \to \mathbb{R}$ is twice continuously differentiable. We also could include control bounds, but this would make the semismooth reformulation more complicated, which we want to avoid for the moment. We set

$$J(y, u) := J_0(y) + \frac{\lambda}{2}\|u\|_{L^2}^2 + \alpha\|u\|_{L^1}$$

and define the reduced objective function

$$j(u) := J(A^{-1}u, u) = j_1(u) + \alpha r(u)$$

with

$$j_1(u) := j_0(u) + \frac{\lambda}{2}\|u\|_{L^2}^2, \quad j_0(u) := J_0(A^{-1}u), \quad r(u) := \|u\|_{L^1}.$$

We next derive optimality conditions. We do this via the directional derivative $j'(u, s)$ of j and obtain for a local solution $\bar{u} \in L^2(\Omega)$

$$\bar{u} \in L^2(\Omega), \quad j'(\bar{u}, u - \bar{u}) \ge 0 \quad \forall\, u \in L^2(\Omega). \tag{9.65}$$

To compute the directional derivative of j, we first observe that the functions j_0 and j_1 are twice continuously differentiable with

$$j_1'(u) = j_0'(u) + \lambda u = (A^{-1})^* J_0'(A^{-1}u) + \lambda u \in L^2(\Omega).$$

Further, for all $u, s \in L^2(\Omega)$, there holds

$$j'(u,s) = \lim_{t\to 0^+} \frac{1}{t}(j_1(u+ts) - j_1(u) + \alpha(r(u+ts) - r(u))) = (j_1'(u), s)_{L^2} + \alpha r'(u,s),$$

where

$$r'(u,s) = \lim_{t\to 0^+} \frac{1}{t}(r(u+ts) - r(u)) = \lim_{t\to 0^+} \int_\Omega \frac{1}{t}(|u(x)+ts(x)| - |u(x)|)dx.$$

The argument $a_t(x) := \frac{1}{t}(|u(x)+ts(x)| - |u(x)|)$ of the integral converges pointwise to $a(u,s)$, where

$$a(u,s)(x) = \begin{cases} -s(x), & u(x) < 0, \\ |s(x)|, & u(x) = 0, \\ s(x), & u(x) > 0. \end{cases}$$

Furthermore,

$$|a_t| \le |s| \quad \text{in } \Omega.$$

Hence, by the dominated convergence theorem,

$$r'(u,s)(x) = \lim_{t\to 0^+} \int_\Omega a_t(x)dx = \int_\Omega a(u,s)(x)dx.$$

Therefore, we can write (9.65) as follows:

$$\int_\Omega [j_1'(\bar{u})(x)s(x) + \alpha a(\bar{u},s)(x)]dx \ge 0 \quad \forall\, s \in L^2(\Omega). \tag{9.66}$$

Since $a(\bar{u},s)(x)$ is a superposition and thus depends only on the point values $\bar{u}(x)$ and $s(x)$, we see that for every measurable set $M \subset \Omega$, there holds

$$1_M[j_1'(\bar{u})s + \alpha a(\bar{u},s)] = 1_M[j_1'(\bar{u})(1_M s) + \alpha a(\bar{u}, 1_M s)],$$

where $1_M(x) = 1$ for $x \in M$, and $1_M(x) = 0$, otherwise. From this, it follows that

$$\int_M [j_1'(\bar{u})(x)s(x) + \alpha a(\bar{u},s)(x)]dx \ge 0$$

for all $s \in L^2(\Omega)$ and all measurable sets $M \subset \Omega$. Hence, (9.66) is equivalent to

$$j_1'(\bar{u})s + \alpha a(\bar{u},s) \ge 0 \quad \text{in } \Omega \quad \forall\, s \in L^2(\Omega).$$

Using the structure of a, we obtain

$$j_1'(\bar{u})(x) \begin{cases} = \alpha, & \bar{u}(x) < 0, \\ = -\alpha, & \bar{u}(x) > 0, \\ \in [-\alpha, \alpha], & \bar{u}(x) = 0, \end{cases} \quad \text{in } \Omega. \tag{9.67}$$

The last case might deserve an explanation: On the set $Z = \{\bar{u} = 0\}$, there holds

$$0 \le 1_Z(j_1'(\bar{u})s + \alpha a(\bar{u},s)) = 1_Z(j_1'(\bar{u})s + \alpha|s|) \quad \forall\, s \in L^2(\Omega),$$

which is equivalent to $|j_1'(\bar{u})| \le \alpha$ in Z, and this is exactly the condition stated above.

It is not difficult to see that (9.67) is equivalent to $\bar{u}$ being a solution of the nonsmooth operator equation

$$j_1'(u) - P_{[-\alpha,\alpha]}(j_1'(u) - \gamma u) = 0,$$

where the parameter $\gamma > 0$ can be chosen arbitrarily. Using that j_1' has the structure

$$j_1'(u) = j_0'(u) + \lambda u,$$

we obtain the equation

$$j_0'(u) + \lambda u - P_{[-\alpha,\alpha]}(j_0'(u) + (\lambda - \gamma)u) = 0. \tag{9.68}$$

We see that $j_0'(u) = (A^{-1})^* J_0'(A^{-1}u)$ has a smoothing property in the sense that it is continuously differentiable (and thus locally Lipschitz continuous) from $L^2(\Omega)$ to $H_0^1(\Omega)$. Therefore, choosing $p > 2$ such that $H_0^1(\Omega) \hookrightarrow L^p(\Omega)$, j_0' is continuously differentiable (and thus locally Lipschitz continuous) from $L^2(\Omega)$ to $L^p(\Omega)$. If we set $\gamma = \lambda$ to cancel the u-term inside the projection, (9.68) becomes

$$j_0'(u) + \lambda u - P_{[-\alpha,\alpha]}(j_0'(u)) = 0. \tag{9.69}$$

Since $v \mapsto P_{[-\alpha,\alpha]}(v)$ is semismooth from $L^p(\Omega)$ to $L^2(\Omega)$, we see that this is a semismooth operator equation. Hence, a semismooth Newton iteration without smoothing step can be applied to this equation.

Alternatively, we can use that $\bar{u}$ solves (9.69) to construct an L^2-L^p-smoothing step via

$$S(u) := \frac{1}{\lambda}\left[P_{[-\alpha,\alpha]}(j_0'(u)) - j_0'(u)\right].$$

Since, for an arbitrary choice of $\gamma > 0$, the operator

$$j_0'(u) + \lambda u - P_{[-\alpha,\alpha]}(j_0'(u) + (\lambda - \gamma)u)$$

is semismooth from $L^p(\Omega)$ to $L^2(\Omega)$, it is also possible to work with this operator and to combine it with the above smoothing step to construct a semismooth Newton iteration.

For further details on L^1 optimization problems, their solution by semismooth Newton methods, and numerical applications, we refer to [35, 42, 82, 183].

Chapter 10

Optimal Control of Incompressible Navier–Stokes Flow

10.1 Introduction

The Navier–Stokes equations describe viscous fluid flow and are thus of central interest for many simulations of practical importance (e.g., in aerodynamics, hydrodynamics, medicine, weather forecasting, and environmental and ocean sciences). Currently, significant efforts are made to develop and analyze optimal control techniques for the Navier–Stokes equations. In particular, control of the incompressible Navier–Stokes equations has been investigated intensively in, e.g., [1, 22, 23, 57, 74, 86, 87, 88, 89, 90, 96, 107, 110]. Our aim is to show that the developed class of semismooth Newton methods can be applied to the constrained distributed control of the incompressible Navier–Stokes equations. We follow [191, 194]. Related results can be found in, e.g., [49, 50, 51].

We consider instationary incompressible flow in two space dimensions. The set $\Omega \subset \mathbb{R}^2$ occupied by the fluid is assumed to be nonempty, open, and bounded with sufficiently smooth boundary $\partial\Omega$. By $t \in [0,T]$, $T > 0$, we denote time, and by $x = (x_1, x_2)^T$ the spatial position. For the time-space domain we introduce the notation $Q = (0,T) \times \Omega$. The state of the fluid is determined by its velocity field $y = (y_1, y_2)^T$ and its pressure P, both depending on t and x. Throughout, we work in dimensionless form.

The Navier–Stokes equations can be written in the form

$$\begin{aligned}
y_t - \nu\Delta y + (y\cdot\nabla)y + \nabla P &= Ru + f && \text{in } Q,\\
\nabla\cdot y &= 0 && \text{in } Q,\\
y &= 0 && \text{in } (0,T)\times\partial\Omega,\\
y(0,\cdot) &= y_0 && \text{in } \Omega.
\end{aligned} \tag{10.1}$$

Here, $\nu > 0$ is the kinematic viscosity, y_0 is a given initial state at time $t = 0$ satisfying $\nabla\cdot y_0 = 0$, $u(t,x)$ is the control, R is a linear operator, and $f(t,x)$ are given data. The precise functional analytic setting is given in section 10.2 below. In (10.1) the following notation is used:

$$\nabla \cdot y = (y_1)_{x_1} + (y_2)_{x_2}, \qquad \Delta y = \begin{pmatrix} \Delta y_1 \\ \Delta y_2 \end{pmatrix} = \begin{pmatrix} (y_1)_{x_1x_1} + (y_1)_{x_2x_2} \\ (y_2)_{x_1x_1} + (y_2)_{x_2x_2} \end{pmatrix},$$

$$(v \cdot \nabla) y = \begin{pmatrix} v_1(y_1)_{x_1} + v_2(y_1)_{x_2} \\ v_1(y_2)_{x_1} + v_2(y_2)_{x_2} \end{pmatrix}, \quad \nabla P = \begin{pmatrix} P_{x_1} \\ P_{x_2} \end{pmatrix}.$$

We perform time-dependent control on the right-hand side. To this end, assume a nonempty and bounded open set $\Omega_c \subset \mathbb{R}^k$ and a control operator $R \in \mathcal{L}(L^2(\Omega_c)^l, H^{-1}(\Omega)^2)$, and choose as control space $U = L^2(Q_c)^l$, $Q_c = (0,T) \times \Omega_c$.

Example 10.1. For time-dependent control of the right-hand side on a subset $\Omega_c \subset \Omega$ of the spatial domain, we can choose $R \in \mathcal{L}(L^2(\Omega_c)^2, H^{-1}(\Omega)^2)$,

$$(Rv)(x) = v(x) \quad \text{for } x \in \Omega_c, \quad (Rv)(x) = 0, \quad \text{otherwise.} \qquad \blacksquare$$

Given a closed convex feasible set $\mathcal{C} \subset U$, the control problem consists of finding a control $u \in \mathcal{C}$ which, together with the corresponding solution (y, P) of the state equation (10.1), minimizes the objective function $J(y,u)$. Specifically, we consider tracking-type objective functions of the form

$$J(y,u) = \frac{1}{2}\int_0^T \int_\Omega \|Ny - z_d\|_2^2 dx dt + \frac{\lambda}{2}\int_0^T \int_{\Omega_c} \|u - u_d\|_2^2 d\omega dt. \tag{10.2}$$

Here, $N : H_0^1(\Omega)^2 \mapsto L^2(\Omega)^m$, $m \geq 1$, is a bounded linear operator, $z_d \in L^2(Q)^m$ is a desired candidate state observation to which we would like Ny to drive by optimal control, $\lambda > 0$ is a regularization parameter, and $u_d \in L^{p'}(Q_c)^l$, $p' > 2$, are given data.

10.2 Functional Analytic Setting of the Control Problem

In our analysis we will consider weak solutions of the Navier–Stokes equations. To make this precise, we first introduce several function spaces which provide a standard framework for the analysis of the Navier–Stokes equations [76, 151, 185].

10.2.1 Function Spaces

We work in the spaces

$$\begin{aligned}
&\mathcal{V} = \{v \in C_c^\infty(\Omega)^2 : \nabla \cdot v = 0\}, \\
&H = \text{closure of } \mathcal{V} \text{ in } L^2(\Omega)^2, \\
&V = \text{closure of } \mathcal{V} \text{ in } H_0^1(\Omega)^2, \\
&L^p(X) = L^p(0,T;X), \quad W = \{v \in L^2(V) : v_t \in L^2(V^*)\}, \\
&C(X) = C(0,T;X) = \{v : [0,T] \mapsto X,\ v \text{ continuous}\},
\end{aligned}$$

with inner products and norms

$$(v,w)_H = (v,w)_{L^2(\Omega)^2} = \int_\Omega \Big(\sum_i v_i w_i\Big)\,dx,$$

$$(v,w)_V = (v,w)_{H_0^1(\Omega)^2} = \int_\Omega \Big(\sum_{i,j} [v_i]_{x_j}[w_i]_{x_j}\Big),$$

$$\|y\|_{L^p(X)} = \left(\int_0^T \|y(t)\|_X^p dt\right)^{1/p}, \quad \|y\|_{L^\infty(X)} = \operatorname*{ess\,sup}_{0<t<T} \|y(t)\|_X,$$

$$\|v\|_W = \left(\int_0^T (\|v\|_V^2 + \|v_t\|_{V^*}^2)dt\right)^{1/2}, \quad \|y\|_{C(X)} = \sup_{0\le t\le T} \|y(t)\|_X.$$

Here, the dual space V^* of V is chosen in such a way that

$$V \hookrightarrow H = H^* \hookrightarrow V^*$$

is a Gelfand triple. The following relations between the introduced spaces hold:

$$W \hookrightarrow C(H) \hookrightarrow L^\infty(H), \quad L^p(V)^* = L^q(V^*), \quad \frac{1}{p}+\frac{1}{q} = 1, \quad 1 < p,q < \infty,$$
$$L^p(V) \hookrightarrow L^q(V), \quad 1 \le q \le p \le \infty.$$

10.2.2 The Control Problem

For the state space and control space, respectively, we choose

$$Y = W \quad \text{state space}, \qquad U = L^2(Q_c)^l \quad \text{control space}.$$

The data of the control problem are

- the initial state $y_0 \in H$,
- the right-hand side data $f \in L^2(H^{-1}(\Omega)^2)$,
- the right-hand side control operator $R \in \mathcal{L}(L^2(\Omega_c)^l, H^{-1}(\Omega)^2)$ such that

  $$w \in W^{4/3} \stackrel{\text{def}}{=} \{v \in L^2(V) : v_t \in L^{4/3}(V^*)\} \mapsto R^*w \in L^{p'}(Q_c)^l$$

  is well defined and continuous with $p' > 2$,
- the objective function $J : Y \times U \to \mathbb{R}$ as defined in (10.2), with data $z_d \in L^2(Q)^m$, $u_d \in L^{p'}(Q_c)^l$, observation operator $N \in \mathcal{L}(H_0^1(\Omega), L^2(\Omega)^m)$, and regularization parameter $\lambda > 0$,
- the feasible set $\mathcal{C} \subset U$, which is nonempty, closed, and convex. In order to apply the semismooth Newton method, we will assume later in this chapter that

  $$\mathcal{C} = \{u \in U : u(t,\omega) \in C,\ (t,\omega) \in Q_c\}, \tag{10.3}$$

  where $C \subset \mathbb{R}^l$ is a closed convex set.

Remark 10.2. *For the choice of R discussed in Example 10.1 and $2 < p' < 7/2$, we can use the embedding $W^{4/3} \hookrightarrow L^{p'}(\Omega)^2$ established in Lemma 10.12 below, to see that*

$$w \in W^{4/3} \mapsto R^* w = w|_{Q_c} \in L^{p'}(Q)^2$$

is continuous.

For the weak formulation of the Navier–Stokes equations it is convenient to introduce the trilinear form

$$b : V \times V \times V \to \mathbb{R},$$
$$b(u,v,w) = \int_\Omega w^T (u \cdot \nabla) v dx = \int_\Omega w^T v_x u dx = \int_\Omega \sum_{i,j} u_i (v_j)_{x_i} w_j dx.$$

The variational form of (10.1) is obtained by applying test functions $v \in V$ to the momentum equation:

$$\frac{d}{dt}(y,v)_H + \nu(y,v)_V + b(y,y,v)$$
$$= \langle Ru + f, v\rangle_{H^{-1}(\Omega)^2, H_0^1(\Omega)^2} \quad \forall\, v \in V \qquad \text{in } (0,T), \tag{10.4}$$
$$y(0,\cdot) = y_0 \qquad \text{in } \Omega. \tag{10.5}$$

Note here that the incompressibility condition $\nabla \cdot y = 0$ is absorbed in the definition of the state space W. Further, the pressure term drops out since $\nabla \cdot v = 0$, and thus integration by parts yields

$$\langle \nabla P, v\rangle_{H^{-1}(\Omega)^2, H_0^1(\Omega)^2} = -(P, \nabla \cdot v)_{L^2(\Omega)^2} = 0.$$

Furthermore, the initial condition (10.5) makes sense for $y \in W$, since $W \hookrightarrow C(H)$. For the well-definedness of (10.4), and also for our analysis, it is important to know the following facts about the trilinear form b.

Lemma 10.3. *There exists a constant $c > 0$ such that, for all $u, v, w \in V$,*

$$b(u,v,w) = -b(u,w,v), \tag{10.6}$$
$$|b(u,v,w)| \le c\|u\|_{L^4(\Omega)^2}\|v\|_V\|w\|_{L^4(\Omega)^2}, \tag{10.7}$$
$$|b(u,v,w)| \le c\|u\|_H^{1/2}\|u\|_V^{1/2}\|v\|_V\|w\|_H^{1/2}\|w\|_V^{1/2} \le c\|u\|_V\|v\|_V\|w\|_V. \tag{10.8}$$

Proof (sketch). Equation (10.6) results from integration by parts using $\nabla \cdot u = 0$, (10.7) follows by applying Hölder's inequality (see [185, Ch. III Lem. 3.4]); and (10.8) follows from $V \hookrightarrow H$ and the estimate [185, Ch. III Lem. 3.3]

$$\|v\|_{L^4(\Omega)} \le 2^{1/4}\|v\|_{L^2(\Omega)}^{1/2}\|\nabla v\|_{L^2(\Omega)^2}^{1/2} \quad \forall\, v \in H_0^1(\Omega). \qquad \square \tag{10.9}$$

Equations (10.4) and (10.5) can be written as operator equation

$$E(y,u) = 0 \tag{10.10}$$

with $E : W \times U \to Z^*$, $Z^* \stackrel{\text{def}}{=} L^2(V^*) \times H$. For convenience, we introduce the following operators: For all $y, v, w \in V$, all $u \in L^2(\Omega_c)^l$, and all $z \in L^2(\Omega)^m$,

$$\begin{aligned}
&A \in \mathcal{L}(V, V^*), && \langle Av, w\rangle_{V^*,V} = (v, w)_V,\\
&B \in \mathcal{L}(V, \mathcal{L}(V, V^*)), && \langle B(y)v, w\rangle_{V^*,V} = b(y, v, w),\\
&R^\pi \in \mathcal{L}(L^2(\Omega_c)^l, V^*), && \langle R^\pi u, v\rangle_{V^*,V} = \langle Ru, v\rangle_{H^{-1}(\Omega)^2, H_0^1(\Omega)^2},\\
&N^\pi \in \mathcal{L}(V, L^2(\Omega)^m), && (N^\pi v, z)_{L^2(\Omega)^m} = (Nv, z)_{L^2(\Omega)^m}.
\end{aligned}$$

Further, we define $f^\pi \in L^2(V^*)$ by

$$\langle f^\pi, v\rangle_{V^*,V} = \langle f, v\rangle_{H^{-1}(\Omega)^2, H_0^1(\Omega)^2} \quad \forall\, v \in V.$$

Using these notations, the operator E assumes the form

$$E(y,u) = \begin{pmatrix} E_1(y,u) \\ E_2(y,u) \end{pmatrix} = \begin{pmatrix} y_t + \nu Ay + B(y)y - R^\pi u - f^\pi \\ y(0,\cdot) - y_0 \end{pmatrix}.$$

Thus, we can write the optimal control problem in abstract form:

$$\text{minimize} \quad J(y,u) \quad \text{subject to} \quad E(y,u) = 0 \quad \text{and} \quad u \in \mathcal{C}. \tag{10.11}$$

10.3 Analysis of the Control Problem

10.3.1 State Equation

Concerning existence and uniqueness of solutions to the state equations (10.4) and (10.5), we have the following.

Proposition 10.4. *For all $u \in U$ and $y_0 \in H$, there exists a unique $y = y(u) \in W$ such that $E(y,u) = 0$. Furthermore, with $r(u) = R^\pi u + f^\pi$,*

$$\|y\|_{C(H)} \le \|y_0\|_H + \frac{1}{\sqrt{\nu}} \|r(u)\|_{L^2(V^*)}, \tag{10.12}$$

$$\|y\|_{L^2(V)} \le \frac{1}{\sqrt{\nu}} \|y_0\|_H + \frac{1}{\nu} \|r(u)\|_{L^2(V^*)}, \tag{10.13}$$

$$\|y\|_W \le c\big(\|y_0\|_H + \|r(u)\|_{L^2(V^*)} + \|y_0\|_H^2 + \|r(u)\|_{L^2(V^*)}^2\big). \tag{10.14}$$

The constant c depends only on ν.

Proof. The existence and uniqueness is established in, e.g., [151, Thm. 3.3], together with the energy equality

$$\frac{1}{2}\|y(t)\|_H^2 + \nu \int_0^t \|y(s)\|_V^2 ds = \frac{1}{2}\|y_0\|_H^2 + \int_0^t \langle r(u)(s), y(s)\rangle_{V^*,V} ds, \tag{10.15}$$

which holds for all $t \in [0,T]$ and is obtained by choosing $v = y(t)$ as a test function in (10.4), integrating from 0 to t, and using

$$2\int_0^t \langle y_t(s), y(s)\rangle_{V^*,V} ds = \|y(t)\|_H^2 - \|y(0)\|_H^2.$$

By the Cauchy–Schwarz and Young inequalities we have

$$\int_0^t \left|\langle r(u)(s), y(s)\rangle_{V^*,V}\right| ds \le \int_0^t \|r(u)(s)\|_{V^*}\|y(s)\|_V ds$$
$$\le \frac{1}{2\nu}\int_0^t \|r(u)(s)\|_{V^*}^2 ds + \frac{\nu}{2}\int_0^t \|y(s)\|_V^2 ds.$$

Hence, (10.15) yields

$$\|y(t)\|_H^2 + \nu\int_0^t \|y(s)\|_V^2 ds \le \|y_0\|_H^2 + \frac{1}{\nu}\int_0^t \|r(u)(s)\|_{V^*}^2 ds,$$

which proves (10.12) and (10.13). The state equation (10.4) yields for all $v \in L^2(V)$, using (10.6), (10.8), and Hölder's inequality,

$$\int_0^T \left|\langle y_t, v\rangle_{V^*,V}\right| dt \le \int_0^T \left(\nu|(y,v)_V| + |b(y,y,v)| + |\langle r(u), v\rangle_{V^*,V}|\right) dt$$
$$\le \int_0^T \left(\nu\|y\|_V + c\|y\|_H\|y\|_V + \|r(u)\|_{V^*}\right)\|v\|_V dt$$
$$\le \left(\nu\|y\|_{L^2(V)} + c\|y\|_{L^\infty(H)}\|y\|_{L^2(V)} + \|r(u)\|_{L^2(V^*)}\right)\|v\|_{L^2(V)}.$$

With the Young inequality, (10.12), and (10.13), it follows that (10.14) holds. □

We know already that the state equation possesses a unique solution $y(u)$. Our aim is to show that the reduced control problem

$$\text{minimize} \quad j(u) \stackrel{\text{def}}{=} J(y(u), u) \quad \text{subject to} \quad u \in \mathcal{B} \tag{10.16}$$

can be solved by the semismooth Newton method. In particular, we must show that j is twice continuously differentiable. This will be done based on the implicit function theorem, which requires us to investigate the differentiability properties of the operator E. In this context, it is convenient to introduce the trilinear form

$$\beta : V \times V \times V \to \mathbb{R}, \quad \beta(u,v,w) = b(u,v,w) + b(v,u,w). \tag{10.17}$$

The following estimates are used several times. In their derivation, and throughout the rest of this chapter (if not stated differently), c denotes a generic constant that may differ from instance to instance.

From (10.6), (10.8), and $V \hookrightarrow H$, it follows for all $u, v, w \in V$ that

$$|\beta(u,v,w)| \le |b(u,w,v)| + |b(v,w,u)|$$
$$\le c\|u\|_H^{1/2}\|u\|_V^{1/2}\|v\|_H^{1/2}\|v\|_V^{1/2}\|w\|_V \tag{10.18}$$
$$\le c\|u\|_H^{1/2}\|u\|_V^{1/2}\|v\|_V\|w\|_V. \tag{10.19}$$

Further, (10.18) and Hölder's inequality with exponents $(\infty, 4, \infty, 4, 2)$ yield for all $u, v \in L^2(V) \cap L^\infty(H) \hookleftarrow W$, and all $w \in L^2(V)$

$$\int_0^T |\beta(u,v,w)| dt \le c\int_0^T \|u\|_H^{1/2}\|u\|_V^{1/2}\|v\|_H^{1/2}\|v\|_V^{1/2}\|w\|_V dt$$
$$\le c\|u\|_{L^\infty(H)}^{1/2}\|u\|_{L^2(V)}^{1/2}\|v\|_{L^\infty(H)}^{1/2}\|v\|_{L^2(V)}^{1/2}\|w\|_{L^2(V)}. \tag{10.20}$$

In particular, for all $u, v \in W$, and $w \in L^2(V)$,

$$\int_0^T |\beta(u,v,w)|dt \leq c\|u\|_W \|v\|_W \|w\|_{L^2(V)}. \tag{10.21}$$

Finally, (10.19) and Hölder's inequality with exponents $(\infty,4,4,2)$ give for all $u \in L^2(V) \cap L^\infty(H)$, $v \in L^4(V)$, and $w \in L^2(V)$

$$\begin{aligned}\int_0^T |\beta(u,v,w)|dt &\leq c\int_0^T \|u\|_H^{1/2}\|u\|_V^{1/2}\|v\|_V\|w\|_V dt \\ &\leq c\|u\|_{L^\infty(H)}^{1/2}\|u\|_{L^2(V)}^{1/2}\|v\|_{L^4(V)}\|w\|_{L^2(V)}.\end{aligned} \tag{10.22}$$

We now prove that the state equation is infinitely Fréchet differentiable.

Proposition 10.5. *Let $y_0 \in H$ and $(y,u) \in W \times U$. Then the operator $E : W \times U \to Z^*$ is twice continuously differentiable with Lipschitz continuous first derivative, constant second derivative, and vanishing third and higher derivatives. The derivatives are given by*

$$E_1'(y,u)(v,w) = v_t + \nu Av + B(y)v + B(v)y - R^\pi w, \tag{10.23}$$

$$E_2'(y,u)(v,w) = v(0,\cdot), \tag{10.24}$$

$$E_1''(y,u)(v,w)(\hat{v},\hat{w}) = B(\hat{v})v + B(v)\hat{v}, \tag{10.25}$$

$$E_2''(y,u)(v,w)(\hat{v},\hat{w}) = 0. \tag{10.26}$$

Proof. Since E_2 is linear and continuous, the assertions on E_2' and E_2'' are obvious. Thus, we only have to consider E_1. If E_1 is differentiable, then formal differentiation shows that E_1' has the form stated in (10.23). This operator maps $(v,w) \in W \times U$ continuously to $L^2(V^*)$. In fact, for all $z \in L^2(V)$, we obtain the following using (10.21):

$$\begin{aligned}&\int_0^T \left|\langle v_t + \nu Av + B(y)v + B(v)y - R^\pi w, z\rangle_{V^*,V}\right| dt \\ &\leq \int_0^T \Big(\|v_t\|_{V^*}\|z\|_V + \nu\|v\|_V\|z\|_V + |\beta(y,v,z)| + \|R^\pi w\|_{V^*}\|z\|_V\Big)dt \\ &\leq \Big(\|v_t\|_{L^2(V^*)} + \nu\|v\|_{L^2(V)} + c\|y\|_W\|v\|_W + \|R^\pi\|_{U,L^2(V^*)}\|w\|_U\Big)\|z\|_{L^2(V)}.\end{aligned}$$

Next, we show that E_1 is differentiable with its derivative given by (10.23). Using the linearity of A, $B(v)$, $v \mapsto B(v)$, and R^π, we obtain for all $y, v \in W$, $u, w \in U$

$$\begin{aligned}&E_1(y+v,u+w) - E_1(y,u) - (v_t + \nu Av + B(y)v + B(v)y - R^\pi w) \\ &\quad = B(y+v)(y+v) - B(y)y - B(y)v - B(v)y = B(v)v.\end{aligned}$$

For all $z \in L^2(V)$ there holds by (10.6), (10.8), and Hölder's inequality

$$\begin{aligned}&\int_0^T \left|\langle B(v)v, z\rangle_{V^*,V}\right| dt = \int_0^T |b(v,v,z)|dt \leq \int_0^T c\|v\|_V\|v\|_H\|z\|_V dt \\ &\leq c\|v\|_{L^2(V)}\|v\|_{L^\infty(H)}\|z\|_{L^2(V)} \leq c\|v\|_W^2\|z\|_{L^2(V)},\end{aligned}$$

which proves the Fréchet differentiability of E_1. Note that E_1' depends affine linearly on $(y,u) \in W \times U$. It remains to show that the mapping

$$E_1' : W \times U \mapsto \mathcal{L}(W \times U, L^2(V^*))$$

is continuous at $(0,0)$. But this follows from

$$|\langle E_1'(y,u)(v,w) - E_1'(0,0)(v,w), z\rangle_{V^*,V}| = |\beta(y,v,z)| \le c\|y\|_W \|v\|_W \|z\|_{L^2(V)}$$

for all $y, v \in W$, all $u, w \in U$, and all $z \in L^2(V)$, where we have used (10.21). As a consequence, E_1' is affine linear and continuous, and thus Lipschitz, and E_1 is twice continuously differentiable with constant second derivative as given in (10.25). Further, since E'' is constant, it follows that $E^{(k)} = 0$ for all $k \ge 3$. □

The next result concerns the linearized state equation. The proof can be obtained by standard methods; the interested reader is referred to [107, 110].

Proposition 10.6. *Let $y_0 \in H$ and $(y,u) \in W \times U$. Then the operator $E_y(y,u) \in \mathcal{L}(W, Z^*)$ is a homeomorphism, or, in more detail: For all $y \in W$, $g \in L^2(V^*)$, and $v_0 \in H$, the linearized Navier–Stokes equations*

$$\begin{aligned} v_t + \nu A v + B(y)v + B(v)y &= g \quad \text{in } L^2(V^*), \\ v(0,\cdot) &= v_0 \quad \text{in } H \end{aligned} \tag{10.27}$$

possess a unique solution $v \in W$. Furthermore, the following estimate holds:

$$\|v_t\|_{L^2(V^*)} + \|v\|_{L^2(V)} + \|v\|_{L^\infty(H)} \le c\|v\|_W \tag{10.28}$$

$$\le c(\|y\|_{L^2(V)}, \|y\|_{L^\infty(H)})(\|g\|_{L^2(V^*)} + \|v_0\|_H) \tag{10.29}$$

$$\le c(\|y\|_W)(\|g\|_{L^2(V^*)} + \|v_0\|_H), \tag{10.30}$$

where the functions $c(\cdot)$ depend locally Lipschitz on their arguments.

Proposition 10.7. *The mapping*

$$(y,u) \in W \times U \mapsto E_y(y,u)^{-1} \in \mathcal{L}(Z^*, W)$$

is Lipschitz continuous on bounded sets. More precisely, there exists a locally Lipschitz continuous function c such that, for all $(y_i, u_i) \in W \times U$, $i = 1,2$, the following holds:

$$\|E_y(y_1,u_1)^{-1} - E_y(y_2,u_2)^{-1}\|_{Z^*,W} \le c(\|y_1\|_W, \|y_2\|_W)\|y_1 - y_2\|_W.$$

Proof. Let $z = (g, v_0) \in Z^* = L^2(V^*) \times H$ be arbitrary and set, for $i = 1,2$, $v_i = E_y(y_i,u_i)^{-1} z$. Then, with $y_{12} = y_1 - y_2$, $u_{12} = u_1 - u_2$, and $v_{12} = v_1 - v_2$, we have $v_{12}(0) = 0$ and

$$\begin{aligned} 0 &= (E_1)_y(y_1,u_1)v_1 - (E_1)_y(y_2,u_2)v_2 \\ &= (v_{12})_t + \nu A v_{12} + B(y_1)v_1 + B(v_1)y_1 - B(y_2)v_2 - B(v_2)y_2 \\ &= (v_{12})_t + \nu A v_{12} + B(y_2)v_{12} + B(v_{12})y_2 + B(y_{12})v_1 + B(v_1)y_{12} \\ &= (E_1)_y(y_2,u_{12})v_{12} + B(y_{12})v_1 + B(v_1)y_{12}, \\ 0 &= (E_2)_y(y_1,u_1)v_1 - (E_2)_y(y_1,u_1)v_2 = v_{12}(0,\cdot). \end{aligned}$$

Therefore,

$$E_y(y_2,u_{12})v_{12} = \begin{pmatrix} -B(y_{12})v_1 - B(v_1)y_{12} \\ 0 \end{pmatrix},$$

and thus, by Proposition 10.6 and (10.21),

$$\begin{aligned} \|v_{12}\|_W &\le c(\|y_2\|_W)(\|B(y_{12})v_1 + B(v_1)y_{12}\|_{L^2(V^*)}) \\ &\le c(\|y_2\|_W)\|v_1\|_W\|y_{12}\|_W \\ &\le c(\|y_2\|_W)c(\|y_1\|_W)(\|g\|_{L^2(V^*)} + \|v_0\|_H)\|y_{12}\|_W \\ &\le c(\|y_1\|_W, \|y_2\|_W)\|y_{12}\|_W\|z\|_{Z^*}, \end{aligned}$$

where $c(\cdot)$ are locally Lipschitz continuous functions. □

10.3.2 Control-to-State Mapping

In this section we show that the control-to-state mapping $u \in U \mapsto y(u) \in W$ is infinitely differentiable and that $y(u)$, $y'(u)$, and $y''(u)$ are Lipschitz continuous on bounded sets.

Theorem 10.8. *The solution operator $u \in U \mapsto y(u) \in W$ of* (10.10) *is infinitely continuously differentiable. Further, there exist locally Lipschitz continuous functions $c(\cdot)$ such that for all $u, u_1, u_2, v, w \in U$ there holds*

$$\|y(u)\|_W \le c(\|y_0\|_H, \|r\|_{L^2(V^*)}), \tag{10.31}$$

$$\|y'(u)\|_W \le c(\|y_0\|_H, \|r\|_{L^2(V^*)}), \tag{10.32}$$

$$\|y_1 - y_2\|_W \le c(\|y_0\|_H, \|r_1\|_{L^2(V^*)}, \|r_2\|_{L^2(V^*)})\|u_1 - u_2\|_U, \tag{10.33}$$

$$\begin{aligned} \|(y_1' - y_2')v\|_W &\le c(\|y_0\|_H, \|r_1\|_{L^2(V^*)}, \|r_2\|_{L^2(V^*)}) \\ &\quad \cdot \|R^\pi(u_1 - u_2)\|_{L^2(V^*)}\|R^\pi v\|_{L^2(V^*)}, \end{aligned} \tag{10.34}$$

$$\begin{aligned} \|(y_1'' - y_2'')(v,w)\|_W &\le c(\|y_0\|_H, \|r_1\|_{L^2(V^*)}, \|r_2\|_{L^2(V^*)}) \\ &\quad \cdot \|R^\pi(u_1 - u_2)\|_{L^2(V^*)}\|R^\pi v\|_{L^2(V^*)}\|R^\pi w\|_{L^2(V^*)}, \end{aligned} \tag{10.35}$$

with $r = R^\pi u + f^\pi$, $r_i = R^\pi u_i + f^\pi$, $y_i = y(u_i)$, $y_i' = y'(u_i)$, and $y_i'' = y''(u_i)$.

Proof. Since E is infinitely continuously differentiable by Proposition 10.5 and the partial derivative $E_y(y(u),u) \in \mathcal{L}(W,Z^*)$ is a homeomorphism according to Proposition 10.6, the implicit function theorem yields that $u \in U \mapsto y(u) \in W$ is infinitely continuously differentiable.

The estimate (10.31) is just a restatement of (10.14) in Proposition 10.4. Using (10.31) and Proposition 10.6, we see that the derivative $u \in U \mapsto y'(u) \in \mathcal{L}(U,W)$ satisfies, setting $y = y(u)$, for all $v \in U$,

$$\begin{aligned} \|y'(u)v\|_W = \|E_y(y,u)^{-1}E_u(y,u)v\|_W &\le \|E_y(y,u)^{-1}\|_{Z^*,W}\|E_u(y,u)v\|_{Z^*} \\ &\le c(\|y\|_W)\|E_u(y,u)v\|_{Z^*} \le c(\|y_0\|_H, \|r\|_{L^2(V^*)})\|R^\pi v\|_{L^2(V^*)} \end{aligned}$$

with $c(\cdot)$ being locally Lipschitz. This proves (10.32).

Using (10.32), we obtain for all $u_1, u_2 \in U$, setting $u_{12} = u_1 - u_2$ and $u(\tau) = \tau u_1 + (1-\tau)u_2$,

$$\begin{aligned}\|y_1 - y_2\|_W &\le \int_0^1 \|y'(u(\tau))u_{12}\|_W d\tau \\ &\le \int_0^1 c\big(\|y_0\|_H, \|r(u(\tau))\|_{L^2(V^*)}\big)\|R^\pi u_{12}\|_{L^2(V^*)} d\tau \\ &\le c\big(\|y_0\|_H, \|r_1\|_{L^2(V^*)}, \|r_2\|_{L^2(V^*)}\big)\|R^\pi (u_1 - u_2)\|_{L^2(V^*)}\end{aligned}$$

with a locally Lipschitz function c. Therefore, (10.33) is shown.

From Proposition 10.7, (10.31), and (10.33), we obtain, for all $v \in U$,

$$\begin{aligned}\|(y_1' - y_2')v\|_W &= \|E_y(y_1,u_1)^{-1}E_u(y_1,u_1)v - E_y(y_2,u_2)^{-1}E_u(y_2,u_2)v\|_W \\ &\le c(\|y_1\|_W, \|y_2\|_W)\|y_1 - y_2\|_W \|R^\pi v\|_{L^2(V^*)} \\ &\le c(\|y_0\|_H, \|r_1\|_{L^2(V^*)}, \|r_2\|_{L^2(V^*)})\|R^\pi (u_1 - u_2)\|_{L^2(V^*)}\|R^\pi v\|_{L^2(V^*)}\end{aligned}$$

with $c(\cdot)$ being locally Lipschitz continuous. This establishes (10.34).

Finally, differentiating the equation $E(y(u),u) = 0$ twice yields, for all $u, v, w \in U$, with $y = y(u)$,

$$\begin{aligned}&E_y(y,u)y''(u)(v,w) + E_{yy}(y,u)(y'(u)v, y'(u)w) \\ &+ E_{yu}(y,u)(y'(u)v, w) + E_{uy}(y,u)(v, y'(u)w) + E_{uu}(y,u)(v,w) = 0.\end{aligned}$$

Now, we use that $E_u v = (-R^\pi v, 0)^T$ is constant to conclude that

$$\begin{aligned}y''(u)(v,w) &= -E_y(y,u)^{-1}E_{yy}(y,u)(y'(u)v, y'(u)w) \\ &= -E_y(y,u)^{-1}\big(B(y'(u)v)y'(u)w + B(y'(u)w)y'(u)v\big).\end{aligned}$$

From this, Proposition 10.7, (10.33), and (10.34), we see that (10.35) holds true. □

10.3.3 Adjoint Equation

Next, given a control $u \in U$ and a state $y \in W$, we analyze the adjoint equation

$$E_y(y,u)^* \begin{pmatrix} w \\ h \end{pmatrix} = g, \tag{10.36}$$

which can be used for the representation of the gradient $j'(u)$. In fact (see section A.1 of the appendix), we have with $y = y(u)$

$$j'(u) = J_u(y,u) + E_u(y,u)^* \begin{pmatrix} w \\ h \end{pmatrix}, \quad \text{where} \quad E_y(y,u)^* \begin{pmatrix} w \\ h \end{pmatrix} = -J_y(y,u).$$

Proposition 10.9. (a) *For every $u \in U$ and $y \in W$, the adjoint equation* (10.36) *possesses a unique solution $(w,h) \in Z = L^2(V) \times H$ for all $g \in W^*$. Moreover,*

$$\|w\|_{L^2(V)} + \|h\|_H \le c\|(w,h)\|_Z \le c(\|y\|_W)\|g\|_{W^*}, \tag{10.37}$$

where $c(\cdot)$ is locally Lipschitz.

(b) *Assume now that $g \in L^{4/3}(V^*) \cap W^*$. Then the adjoint equation can be written in the form*

$$-\frac{d}{dt}(w,v)_H + \nu(w,v)_V + \beta(y,v,w) = \langle g,v\rangle_{V^*,V} \quad \forall\, v \in V \quad \text{on } (0,T), \tag{10.38}$$

$$w(T,\cdot) = 0 \qquad \text{on } \Omega, \tag{10.39}$$

$$h - w(0,\cdot) = 0 \qquad \text{on } \Omega. \tag{10.40}$$

Furthermore, $w_t \in L^{4/3}(V^*) \cap W^*$, $w \in C(V^*)$, and

$$\|w_t\|_{W^*} \le c(\|y\|_W)\|g\|_{W^*}, \tag{10.41}$$

$$\|w_t\|_{L^{4/3}(V^*)} \le c(\|y\|_W)\|g\|_{W^*} + \|g\|_{L^{4/3}(V^*)} \tag{10.42}$$

with $c(\cdot)$ being locally Lipschitz continuous.

Proof. (a) From Proposition 10.6 we know that $E_y(y,u) \in \mathcal{L}(W,Z^*)$ is a homeomorphism and thus also $E_y(y,u)^* \in \mathcal{L}(Z,W^*)$ is a homeomorphism. Hence, the adjoint equation possesses a unique solution $(w,h) \in Z = L^2(V) \times H$ that depends linearly and continuously on $g \in W^*$. More precisely, Proposition 10.6 yields

$$\begin{aligned}\|w\|_{L^2(V^*)} + \|h\|_H &\le c\|(w,h)\|_Z = c\|(E_y(y,u)^*)^{-1}g\|_Z\\ &\le c\|(E_y(y,u)^*)^{-1}\|_{W^*,Z}\|g\|_{W^*}\\ &= c\|E_y(y,u)^{-1}\|_{Z^*,W}\|g\|_{W^*} \le c(\|y\|_W)\|g\|_{W^*},\end{aligned}$$

where $c(\cdot)$ depends locally Lipschitz on $\|y\|_W$.

(b) For the rest of the proof we assume $g \in W^* \cap L^{4/3}(V^*)$. We proceed by showing that the adjoint equation coincides with (10.38). Using the trilinear form β defined in (10.17), the adjoint state $(w,h) \in L^2(V) \times H$ satisfies for all $v \in W$

$$\int_0^T \big(\langle v_t, w\rangle_{V^*,V} + \nu(v,w)_V + \beta(y,v,w) - \langle g,v\rangle_{V^*,V}\big)dt + (v(0),h)_H = 0. \tag{10.43}$$

In particular, we obtain for $v \in W$ replaced by φv with $\varphi \in C_c^\infty(0,T)$ and $v \in V$

$$-\frac{d}{dt}(w,v)_H + \nu(w,v)_V + \beta(y,v,w) = \langle g,v\rangle_{V^*,V} \quad \forall\, v \in V \quad \text{on } (0,T),$$

in the sense of distributions, which is (10.38). As a result of (10.22), we have that $z \in L^4(V) \mapsto \beta(y,z,w)$ is linear and continuous and therefore an element of $L^4(V)^* = L^{4/3}(V^*)$. For $v \in V$ this implies $\beta(y,v,w) \in L^{4/3}(0,T)$. Further, $\langle g,v\rangle_{V^*,V} \in L^{4/3}(0,T)$ and $(w,v)_V \in L^2(0,T)$, hence

$$\frac{d}{dt}(w,v)_H = \nu(w,v)_V + \beta(y,v,w) - \langle g,v\rangle_{V^*,V} \in L^{4/3}(0,T).$$

This shows that $(w,v)_H \in H^{1,4/3}(0,T)$. For all $v \in V$ and all $\varphi \in C^\infty([0,T])$ there holds $\varphi v \in W$. We choose these particular test functions in (10.43) and integrate by parts (which is allowed since $C^\infty([0,T]) \hookrightarrow H^{1,4}(0,T))$. This gives

$$0 = \int_0^T \Big((v,w)_H \varphi' + \big(\nu(v,w)_V + \beta(y,v,w) - \langle g,v\rangle_{V^*,V}\big)\varphi \Big) dt + (v,h)_H \varphi(0)$$

$$= \int_0^T \big(-\frac{d}{dt}(w,v)_H + \nu(w,v)_V + \beta(y,v,w) - \langle g,v\rangle_{V^*,V}\big)\varphi dt$$
$$+ (v,h-w(0))_H\varphi(0) + (v,w(T))_H\varphi(T).$$

The integral vanishes, since (10.38) was already shown to hold. Considering all $\varphi \in C^\infty([0,T])$ with $\varphi(0)=0$ proves (10.39), whereas (10.40) follows by considering all $\varphi \in C^\infty([0,T])$ with $\varphi(T)=0$.

Finally, we solve (10.38) for w_t and apply (10.21) to derive, for all $z \in W$,

$$\langle w_t, z\rangle_{W^*,W} \le \int_0^T \big(\nu|(w,z)_V| + |\beta(y,z,w)|\big)dt + |\langle g,z\rangle_{W^*,W}|$$
$$\le \nu\|w\|_{L^2(V)}\|z\|_{L^2(V)} + c\|y\|_W\|w\|_{L^2(V)}\|z\|_W + \|g\|_{W^*}\|z\|_W.$$

Further, for all $z \in L^4(V)$,

$$\int_0^T \langle w_t, z\rangle_{V^*,V} dt \le \int_0^T \big|\nu(w,z)_V + \beta(y,z,w) - \langle g,z\rangle_{V^*,V}\big| dt$$
$$\le \big(\nu\|w\|_{L^{4/3}(V)} + c\|y\|_W\|w\|_{L^2(V)} + \|g\|_{L^{4/3}(V^*)}\big)\|z\|_{L^4(V)},$$

where we have used Hölder's inequality and (10.22). Application of (10.37) completes the proof of (10.41) and (10.42). The assertion $w \in C(V^*)$ follows from the embedding $\{w \in L^2(V) : w_t \in L^{4/3}(V^*)\} \hookrightarrow C(V^*)$. □

Our next aim is to estimate the distance of two adjoint states (w_i, h_i), $i = 1,2$, that correspond to different states y_i and right-hand sides g_i.

Proposition 10.10. *For given $y_i \in W$ and $g_i \in W^* \cap L^{4/3}(V^*)$, $i = 1,2$, let $(w_i,h_i) \in L^2(V) \times H$ denote the corresponding solutions of the adjoint equation* (10.36) *with state y_i and right-hand side g_i. Then $w_i \in L^2(V) \cap C(V^*)$, $(w_i)_t \in W^* \cap L^{4/3}(V^*)$, $h_i = w_i(0)$, and*

$$\begin{aligned}&\|w_1 - w_2\|_{L^2(V)} + \|(w_1-w_2)_t\|_{L^{4/3}(V^*)} + \|h_1 - h_2\|_H \\ &\quad \le c(\|y_1\|_W, \|y_2\|_W)\big(\|g_1 - g_2\|_{W^*} + \|g_1\|_{W^*}\|y_1 - y_2\|_W\big) \\ &\quad + \|g_1 - g_2\|_{L^{4/3}(V^*)},\end{aligned} \tag{10.44}$$

where $c(\cdot)$ is locally Lipschitz continuous.

Proof. The existence and regularity results are those stated in Proposition 10.9. Introducing the differences $w_{12} = w_1 - w_2$, $h_{12} = h_1 - h_2$, $y_{12} = y_1 - y_2$, and $g_{12} = g_1 - g_2$, we have $w_{12}(T) = 0$ and $h_{12} = w_{12}(0)$ on Ω and, on $(0,T)$,

$$-\frac{d}{dt}(w_{12}, v)_H + \nu(w_{12}, v)_V + \beta(y_1, v, w_1) - \beta(y_2, v, w_2) = \langle g_{12}, v\rangle_{V^*,V}.$$

Rearranging terms yields

$$-\frac{d}{dt}(w_{12},v)_H+\nu(w_{12},v)_V+\beta(y_2,v,w_{12})=\langle g_{12},v\rangle_{V^*,V}-\beta(y_{12},v,w_1).$$

Therefore, (w_{12},h_{12}) is the solution of the adjoint equation for the state y_2 and the right-hand side

$$g=g_{12}-\ell,\quad \ell:v\mapsto\beta(y_{12},v,w_1).$$

From (10.21), (10.22) we know that $\ell\in W^*\cap L^{4/3}(V^*)$ and

$$\|\ell\|_{W^*}+\|\ell\|_{L^{4/3}(V^*)}\le c\|y_{12}\|_W\|w_1\|_{L^2(V)}.$$

Therefore, by Proposition 10.9,

$$\begin{aligned}
\|w_{12}\|_{L^2(V)}+\|(w_{12})_t\|_{L^{4/3}(V^*)}+\|h_{12}\|_H &\le c(\|y_2\|_W)\|g\|_{W^*}+\|g\|_{L^{4/3}(V^*)}\\
&\le c(\|y_2\|_W)\big(\|g_{12}\|_{W^*}+c\|w_1\|_{L^2(V)}\|y_{12}\|_W\big)+\|g_{12}\|_{L^{4/3}(V^*)}\\
&\quad+c\|w_1\|_{L^2(V)}\|y_{12}\|_W\\
&\le c(\|y_2\|_W)\big(\|g_{12}\|_{W^*}+\|w_1\|_{L^2(V)}\|y_{12}\|_W\big)+\|g_{12}\|_{L^{4/3}(V^*)}\\
&\le c(\|y_2\|_W)\big(\|g_{12}\|_{W^*}+c(\|y_1\|_W)\|g_1\|_{W^*}\|y_{12}\|_W\big)+\|g_{12}\|_{L^{4/3}(V^*)}\\
&\le c(\|y_1\|_W,\|y_2\|_W)\big(\|g_{12}\|_{W^*}+\|g_1\|_{W^*}\|y_{12}\|_W\big)+\|g_{12}\|_{L^{4/3}(V^*)},
\end{aligned}$$

where $c(\cdot)$ is locally Lipschitz. The proof is complete. □

10.3.4 Properties of the Reduced Objective Function

We will now show that the reduced objective function j meets all requirements that are needed to apply semismooth Newton methods for the solution of the control problem (10.16). We have, since J is quadratic,

$$\begin{aligned}
J_u(y,u)&=\lambda(u-u_d), & J_y(y,u)&=N^{\pi *}(N^\pi y-z_d),\\
J_{uu}(y,u)&=\lambda I, & J_{uy}(y,u)&=0,\\
J_{yu}(y,u)&=0, & J_{yy}(y,u)&=N^{\pi *}N^\pi.
\end{aligned}$$

Since $u\in U\mapsto y(u)\in W$ is infinitely differentiable and y, y', and y'' are Lipschitz continuous on bounded sets, see Theorem 10.8, we obtain that $j(u)=J(y(u),u)$ is infinitely differentiable with j, j', and j'' being Lipschitz continuous on bounded sets.

Further, using the adjoint representation of the gradient, and the fact that $E_u v=(-R^\pi v,0)^T$, we have, with $y=y(u)$,

$$j'(u)=J_u(y,u)-R^{\pi *}w=\lambda(u-u_d)-R^*w, \tag{10.45}$$

where w solves the adjoint equation (10.38), (10.39) with right-hand side

$$g=-J_y(y,u)=-N^{\pi *}(N^\pi y-z_d)\in L^2(V^*)\hookrightarrow W^*\cap L^{4/3}(V^*). \tag{10.46}$$

Therefore, we have the following.

Theorem 10.11. *The reduced objective function $j : U = L^2(Q_c)^l \to \mathbb{R}$ is infinitely differentiable with j, j', and j'' being Lipschitz continuous on bounded sets. The reduced gradient has the form*

$$
\begin{aligned}
j'(u) &= \lambda u + G(u), \\
G(u) &= -R^* w - \lambda u_d,
\end{aligned}
$$

where w is the adjoint state. In particular, the operator G maps $L^2(Q_c)^l$ Lipschitz continuously on bounded sets to $L^{p'}(Q_c)^l$. Further, $G : L^2(Q_c)^l \to L^2(Q_c)^l$ is continuously differentiable with $G'(u) = G'(u)^*$ being bounded on bounded sets in $\mathcal{L}(L^2(Q_c)^l, L^{p'}(Q_c)^l)$.

Proof. The properties of j follow from Theorem 10.8 and (10.45). The Lipschitz continuity assertion on G follows from (10.44), (10.33), and (10.46). Further, $G(u) = j'(u) - \lambda u$ is, considered as a mapping $L^2(Q_c)^l \to L^2(Q_c)^l$, continuously differentiable with derivative $G'(u) = j''(u) - \lambda I$. In particular, we see that G' is self-adjoint. Now consider $G'(u)$ for all $u \in B_\rho = \rho B_{L^2(Q_c)^l}$. On this set G maps Lipschitz continuously into $L^{p'}(Q_c)^l$. Denoting the Lipschitz rank by L_ρ, we now prove $\|G'(u)\|_{L^2(Q_c)^l, L^{p'}(Q_c)^l} \le L_\rho$ for all $u \in B_\rho$. In fact, for all $u \in B_\rho$ and all $v \in L^2(Q_c)^l$ we have $u + tv \in B_\rho$ for $t > 0$ small enough and thus

$$
\|G'(u)v\|_{L^{p'}(Q_c)^l} = \lim_{t \to 0^+} t^{-1} \|G(u+tv) - G(u)\|_{L^{p'}(Q_c)^l} \le L_\rho \|v\|_{L^2(Q_c)^l}. \qquad \square
$$

For illustration, we consider the case where $\Omega_c \subset \Omega$, $l = 2$, and

$$
(Rv)(x) = v(x) \quad \text{for } x \in \Omega_c, \quad (Rv)(x) = 0, \quad \text{otherwise.}
$$

We need the following embedding.

Lemma 10.12. *For all $1 \le p < 7/2$ and all $v \in L^2(V)$ with $v_t \in L^{4/3}(V^*)$ there holds*

$$
\|v\|_{L^p(Q)^2} \le c\big(\|v_t\|_{L^{4/3}(V^*)} + \|v\|_{L^2(V)}\big).
$$

Proof. In [10] it is proved that for all $1 \le q < 8$ there holds

$$
W^{4/3} = \{v \in L^2(V) : v_t \in L^{4/3}(V^*)\} \hookrightarrow L^q(H)
$$

(the embedding is even compact). We proceed by showing that for all $p \in [1, 7/2)$ there exists $q \in [1, 8)$ such that $L^q(H) \cap L^2(V) \hookrightarrow L^p(Q)^2$. Due to the boundedness of Q it suffices to consider all $p \in [2, 7/2)$. Recall that $V \hookrightarrow L^s(\Omega)^2$ for all $s \in [1, \infty)$. Now let $r = 4$, $r' = 4/3$,

$$
\theta = 1 - \frac{3}{2p} \in \left[\frac{1}{4}, \frac{4}{7}\right) \quad \text{and} \quad s = \frac{6}{7 - 2p} \in [2, \infty).
$$

Then there holds

$$
\begin{aligned}
&\frac{\theta}{2} + \frac{1-\theta}{s} = \frac{1}{p}, && \frac{1}{r} + \frac{1}{r'} = 1, \\
&q = \theta p r = 4p - 6 \in [2, 8), && (1-\theta) p r' = 2.
\end{aligned}
$$

Thus, we can apply the interpolation inequality and Hölder's inequality to conclude

$$\begin{aligned}
\|v\|^p_{L^p(Q)^2} &= \int_0^T \|v\|^p_{L^p(\Omega)^2} dt \le c\int_0^T \|v\|^{\theta p}_{L^2(\Omega)^2}\|v\|^{(1-\theta)p}_{L^s(\Omega)^2} dt \\
&\le c\left(\int_0^T \|v\|^{\theta pr}_{L^2(\Omega)^2} dt\right)^{1/r}\left(\int_0^T c\|v\|^{(1-\theta)pr'}_{L^s(\Omega)^2} dt\right)^{1/r'} \\
&= c\|v\|^{\theta p}_{L^q(H)^2}\|v\|^{(1-\theta)p}_{L^2(L^s(\Omega)^2)} \\
&\le c\big(\|v_t\|_{L^{4/3}(V^*)} + \|v_t\|_{L^2(V)}\big)^{\theta p}\|v\|^{(1-\theta)p}_{L^2(V)} \\
&\le c\big(\|v_t\|_{L^{4/3}(V^*)} + \|v_t\|_{L^2(V)}\big)^{p}. \qquad \square
\end{aligned}$$

For $2 < p' < 7/2$ we thus have that

$$w \in W^{4/3} \hookrightarrow L^{p'}(Q)^2 \mapsto R^*w = w|_{Q_c} \in L^{p'}(Q_c)^2$$

is continuous, and thus Theorem 10.11 is applicable.

10.4 Application of Semismooth Newton Methods

We now consider the reduced problem (10.16) with feasible set of the form (10.3), and reformulate its first-order necessary optimality conditions in the form of the nonsmooth operator equation

$$\Psi(u) = 0,$$

$$\Psi(u)(t,\omega) = u(t,\omega) - P_C\big(u(t,\omega) - \lambda^{-1}j'(u)(t,\omega)\big), \quad (t,\omega) \in Q_c.$$

Let us assume that P_C is semismooth. Then, for $r = 2$ and any p' as specified, Theorem 10.11 shows that Assumption 5.14 is satisfied by $F = j'$. Therefore, Theorem 5.15 is applicable and yields the $\partial_C^\circ\Psi$-semismoothness of $\Psi : L^2(Q_c)^l \to L^2(Q_c)^l$.

If we prefer to work with a reformulation by means of a different Lipschitz continuous and semismooth function $\pi : \mathbb{R}^l \times \mathbb{R}^l \to \mathbb{R}^l$,

$$\pi(a,b) = 0 \quad \Longleftrightarrow \quad a - P_C(a-b) = 0,$$

in the form

$$\Pi(u) \stackrel{\text{def}}{=} \pi\big(u, j'(u)\big) = 0, \tag{10.47}$$

we can use Theorem 5.11 to establish the semismoothness of the resulting operator as a mapping $L^p(Q_c)^l \to L^2(Q_c)^l$ for any $p \ge p'$. A smoothing step is then provided by $S(u) = P_C(u - \lambda^{-1}j'(u))$. An example for π would be $\pi(a,b) = a - P_C(a - \sigma b)$ with fixed $\sigma > 0$. Therefore, our class of semismooth Newton methods is applicable to both reformulations.

We also can apply the sufficient condition for regularity of Theorem 4.8. Since this condition was established in the framework of NCPs, we consider now the case $U = L^2(Q_c)$ and $C = [0,\infty)$. Then, we immediately see that Theorem 10.11 provides everything to verify Assumption 4.6, provided that $j''(\bar{u})$ is coercive on the tangent space of the strongly active constraints as assumed in (e), and that the used NCP-function $\pi = \phi$ satisfies (f)–(h). The coercivity condition can be interpreted as a strong second-order sufficient condition for optimality; see [62, 195].

10.5 Numerical Results

For our numerical tests, we consider a lid driven cavity flow problem, where the full right-hand side can be controlled, i.e., $\Omega_c = \Omega$, $U = L^2(Q_T)^2$, $Ru = u$. We use a velocity tracking objective functional, i.e., $Ny = y$ with target velocity y_d.

Two problem settings will be considered: One with pointwise bound constraints on the control, $u_1 \in [\alpha,\beta]$ and $u_2 \in [\alpha,\beta]$ a.e. on Q_T with $\alpha,\beta \in \mathbb{R}$, $\alpha < \beta$; this corresponds to $C = [\alpha,\beta]^2$.

The second setting is with pointwise constraints on the norm of the control, i.e., $u_1(t,x)^2 + u_2(t,x)^2 \le r^2$ a.e. on Q_T with $r > 0$; this corresponds to $C = \{(a,b)^T \in \mathbb{R}^2 : a^2 + b^2 \le r^2\}$.

The discretization we use parallels the one described in [19]. We sketch it for the case of a general polyhedral domain and Dirichlet boundary conditions, which includes our particular cavity flow problem as a special case. For discretization in space we use a triangulation $\mathcal{T}^h$ of the polygonal domain Ω with h denoting the maximal triangle diameter. In our test problem, we have $\Omega = (0,1)^2$. We construct $\mathcal{T}^{h/2}$ from $\mathcal{T}^h$ by subdividing every triangle into four congruent subtriangles by connecting midpoints of the edges. Denote by L^h the space of continuous, piecewise linear finite elements over the triangulation $\mathcal{T}^h$ and set $V^h = (L^{h/2})^2$. The space discretization uses the discrete pressure space $P^h = \{p^h \in L^h : p^h(x_0^h) = 0\}$, where x_0^h is a fixed node of $\mathcal{T}^h$, and the discrete spatial velocity space

$$V_b^h = \{v^h \in V^h : v^h(x^h) = b(x^h) \text{ in all boundary nodes } x^h \text{ of } \mathcal{T}^{h/2}\}.$$

Here, b are the Dirichlet data for y on $\partial\Omega$. For notational convenience, we assume that b does not depend on time. Furthermore, let $V_0^h = \{v^h \in V^h : v^h|_{\partial\Omega} = 0\}$. In our concrete test problem, we have $\Omega = (0,1)^2$, $b = (0,0)^T$ on $\{0,1\} \times [0,1] \cup [0,1] \times \{0\}$, and $b = (1,0)^T$ on $(0,1) \times \{1\}$. For the control, the spatial space $X^h = V^h$, equipped with the mass lumped L^2 inner product, is used. Time discretization is done by finite differences on an equidistant grid of size $\Delta t = T/n_T$ on I. On the control space $U^h = (X^h)^{n_T}$ we work with the inner product $(u^h, v^h)_{U^h} = \Delta t \sum_{n=1}^{n_T} (u^{h,n}, v^{h,n})_{X^h}$. The time stepping scheme is fully implicit in the linear terms and fully explicit in the nonlinear term: Denoting by $y^{h,n} \in V_b^h$, $p^{h,n} \in P^h$, and $u^{h,n} \in X^h$ the approximations to $y(t_n)$, $p(t_n)$, and $u(t_n)$ at time $t_n = n\Delta t$, and setting $(\cdot,\cdot) = (\cdot,\cdot)_{L^2(\Omega)^2}$, the discrete state equation becomes

$$\begin{aligned}
&y^{h,0} = y_0^h = \text{approximation of } y_0 \text{ in } V_b^h,\\
&\text{for } n = 1,\dots,n_T:\\
&\frac{1}{\Delta t}(y^{h,n} - y^{h,n-1}, v^h) + \nu(\nabla y^{h,n}, \nabla v^h)_{L^2(\Omega)^{2\times 2}}\\
&\quad + ((y^{h,n-1}\cdot\nabla)y^{h,n-1}, v^h) - (p^{h,n}, \nabla\cdot v^h)_{L^2(\Omega)} = (f + u^{h,n}, v^h) \quad \forall\, v^h \in V_0^h,\\
&(\nabla\cdot y^{h,n}, q^h)_{L^2(\Omega)} = 0 \quad \forall\, q^h \in P^h.
\end{aligned}$$

A generalized Stokes problem needs to be solved in each time step, for which efficient methods are available; see, e.g., [27]. The discretized admissible set is $\mathcal{C}^h = U^h \cap \mathcal{C}$, and the discretized objective function is chosen as

$$J^h(y^h, u^h) = \frac{\Delta t}{2}\sum_{n=1}^{n_T} \|y^{h,n} - y_d^{h,n}\|_{L^2(\Omega)^2}^2 + \frac{\lambda}{2}\|u^h - u_d^h\|_{U^h}^2,$$

where $y_d^h \in (V^h)^{n_T}$ and $u_d^h \in U^h$ approximate y_d and u_d, respectively. Now, using standard nodal basis functions, we obtain unique representations $y^h = b^h + B_y^h \mathbf{y}^h$, where $b^h \in V_b^h$ is fixed, $p^h = B_p^h \mathbf{p}^h$, and $u^h = B_u^h \mathbf{u}^h$ by the coordinate vectors $\mathbf{y}^h \in \mathbb{R}^{n_y}$, $\mathbf{p}^h \in \mathbb{R}^{n_p}$, and $\mathbf{u}^h \in \mathbb{R}^{n_u}$. With $\mathbf{y}_1^h, \mathbf{y}_2^h \in \mathbb{R}^{n_y/2}$ and $\mathbf{u}_1^h, \mathbf{u}_2^h \in \mathbb{R}^{\frac{n_u}{2}}$ we denote the coordinate vectors corresponding to y_1, y_2, u_1, and u_2, respectively. The discrete state equation can then be written as

$$\mathbf{E}^h(\mathbf{y}^h, \mathbf{p}^h, \mathbf{u}^h) = 0, \quad \mathbf{E}^h : \mathbb{R}^{n_y} \times \mathbb{R}^{n_p} \times \mathbb{R}^{n_u} \to \mathbb{R}^{n_y} \times \mathbb{R}^{n_p}. \tag{10.48}$$

We set $\mathbf{J}^h(\mathbf{y}^h, \mathbf{u}^h) = J^h(b^h + B_y^h \mathbf{y}^h, B_u^h \mathbf{u}^h)$ and $\mathbf{j}^h(\mathbf{u}^h) = \mathbf{J}^h(\mathbf{y}^h(\mathbf{u}^h), \mathbf{u}^h)$, where $\mathbf{y}^h(\mathbf{u}^h)$ and $\mathbf{p}^h(\mathbf{u}^h)$ denote the solution of (10.48) corresponding to $\mathbf{u}^h$.

10.5.1 The Pointwise Bound-Constrained Problem

The discretized reduced optimal control problem with bound-constrained controls then reads

$$\min_{\mathbf{u}^h} \ \mathbf{j}^h(\mathbf{u}^h) \quad \text{subject to} \quad \alpha \le \mathbf{u}^h \le \beta \ \text{(componentwise)}. \tag{10.49}$$

The first-order optimality conditions for a solution $\bar{\mathbf{u}}^h$ of (10.49) can be written as

$$\bar{\mathbf{u}}^h \in [\alpha, \beta]^{n_u}, \quad \nabla \mathbf{j}^h(\bar{\mathbf{u}}^h)^T(\mathbf{u}^h - \bar{\mathbf{u}}^h) \ge 0 \quad \forall\, \mathbf{u}^h \in [\alpha, \beta]^{n_u}, \tag{10.50}$$

where $\nabla \mathbf{j}^h$ is the Euclidean gradient. For proper scaling, we need to transform the gradient to the U^h inner product, which is represented by the diagonal lumped mass matrix $\mathbf{M}^h$. We obtain $\nabla \mathbf{j}^h(\bar{\mathbf{u}}^h)^T(\mathbf{u}^h - \bar{\mathbf{u}}^h) = \mathbf{g}^h(\bar{\mathbf{u}}^h)^T \mathbf{M}^h(\mathbf{u}^h - \bar{\mathbf{u}}^h)$ with $\mathbf{g}^h(\mathbf{u}^h) = (\mathbf{M}^h)^{-1} \nabla \mathbf{j}^h(\mathbf{u}^h)$. The discrete version of (10.47), which is equivalent to (10.50), is

$$\mathbf{\Pi}^h(\mathbf{u}^h) \stackrel{\text{def}}{=} \Big[\pi(\mathbf{u}_i^h, \mathbf{g}^h(\mathbf{u}^h)_i)\Big]_{1 \le i \le n_u} = 0.$$

Here, we use the MCP-function $\pi : \mathbb{R}^2 \to \mathbb{R}$, $\pi(a,b) = a - P_{[\alpha,\beta]}(a-b)$. If required, smoothing steps can be computed via

$$\mathbf{S}^h(\mathbf{u}^h) = \Big[P_{[\alpha,\beta]}(\mathbf{u}_i^h - \lambda^{-1}\mathbf{g}^h(\mathbf{u}^h)_i)\Big]_{1 \le i \le n_u}.$$

Denoting by $\mathbf{u}^h$ the current iterate, the discrete semismooth Newton system reads

$$[\mathbf{D}_1^h + \mathbf{D}_2^h(\mathbf{M}^h)^{-1}\nabla^2 \mathbf{j}^h(\mathbf{u}^h)]\mathbf{s}^h = -\mathbf{\Pi}^h(\mathbf{u}^h),$$

where $\nabla^2 \mathbf{j}^h$ is the Euclidean Hessian of $\mathbf{j}^h$, and $\mathbf{D}_1^h$ and $\mathbf{D}_2^h$ are diagonal matrices satisfying $(\mathbf{D}_1^h, \mathbf{D}_2^h)_{ii} \in \partial\pi(\mathbf{u}_i^h, \mathbf{g}^h(\mathbf{u}^h)_i)$. In particular, we may choose $(\mathbf{D}_1^h, \mathbf{D}_2^h)_{ii} = (0,1)$ on $I \stackrel{\text{def}}{=} \{i : \mathbf{u}_i^h - \mathbf{g}^h(\mathbf{u}^h)_i \in [\alpha,\beta]\}$, and $(\mathbf{D}_1^h, \mathbf{D}_2^h)_{ii} = (1,0)$ on $A \stackrel{\text{def}}{=} \{i : \mathbf{u}_i^h - \mathbf{g}^h(\mathbf{u}^h)_i \notin [\alpha,\beta]\}$. This shows that the semismooth Newton system can be reduced to a symmetric linear system of the form

$$\nabla^2 \mathbf{j}^h(\mathbf{u}^h)_{II}\mathbf{s}_I^h = -\mathbf{M}_{II}^h \mathbf{\Pi}^h(\mathbf{u}^h)_I - \nabla^2 \mathbf{j}^h(\mathbf{u}^h)_{IA}\mathbf{s}_A^h, \tag{10.51}$$

where $\mathbf{s}_A^h = -\mathbf{\Pi}^h(\mathbf{u}^h)_A$.

Since the direct computation and storage of the matrix $\nabla^2 \mathbf{j}^h(\mathbf{u}^h)_{II}$ is too expensive, we solve (10.51) by a truncated conjugate gradient iteration in the subspace of U^h corresponding to the components in I. This means that CG is preconditioned by $(\mathbf{M}^h_{II})^{-1}$. Gradient computation is done by the adjoint method (in Euclidean space): Given the current state $\mathbf{y}^h = \mathbf{y}^h(\mathbf{u}^h)$, $\mathbf{p}^h = \mathbf{p}^h(\mathbf{u}^h)$, solve the adjoint equation

$$\mathbf{E}^h_{(\mathbf{y}^h,\mathbf{p}^h)}(\mathbf{y}^h,\mathbf{p}^h,\mathbf{u}^h)^T\mathbf{w}^h = -\begin{pmatrix}\nabla_{\mathbf{y}^h}\mathbf{J}^h(\mathbf{y}^h,\mathbf{u}^h)\\ 0\end{pmatrix}$$

by reverse time stepping, and compute

$$\nabla \mathbf{j}^h(\mathbf{u}^h) = \nabla_{\mathbf{u}^h}\mathbf{J}^h(\mathbf{y}^h,\mathbf{u}^h) + \mathbf{E}^h_{\mathbf{u}^h}(\mathbf{y}^h,\mathbf{p}^h,\mathbf{u}^h)^T\mathbf{w}^h.$$

For the Hessian-vector products required in the CG method, we use the following adjoint-based Hessian representation (see section A.1.2 in the appendix):

$$\nabla^2 \mathbf{j}^h(\mathbf{u}^h) = \left(-[\mathbf{E}^h_{\mathbf{u}^h}]^T[\mathbf{E}^h_{(\mathbf{y}^h,\mathbf{p}^h)}]^{-T}, \mathbf{I}\right)\nabla^2_{(\mathbf{y}^h,\mathbf{p}^h,\mathbf{u}^h)}\mathbf{L}^h\begin{pmatrix}-[\mathbf{E}^h_{(\mathbf{y}^h,\mathbf{p}^h)}]^{-1}\mathbf{E}^h_{\mathbf{u}^h}\\ \mathbf{I}\end{pmatrix}, \tag{10.52}$$

where we have suppressed the arguments, $\mathbf{I}$ is the identity, and $\mathbf{L}^h$ denotes the Lagrange function

$$\mathbf{L}^h(\mathbf{y}^h,\mathbf{p}^h,\mathbf{u}^h,\mathbf{w}^h) \stackrel{\text{def}}{=} \mathbf{J}^h(\mathbf{y}^h,\mathbf{u}^h) + (\mathbf{w}^h)^T\mathbf{E}^h(\mathbf{y}^h,\mathbf{p}^h,\mathbf{u}^h).$$

Computing a product $\nabla^2 \mathbf{j}^h(\mathbf{u}^h)\mathbf{v}^h$ via (10.52) requires one linearized state solve, multiplication with $\nabla^2_{(\mathbf{y}^h,\mathbf{p}^h,\mathbf{u}^h)}\mathbf{L}^h$, and one adjoint solve.

As already mentioned, in our test problem we consider flow in a cavity, with $\Omega = (0,1)^2$ and boundary conditions $y = (1,0)$ on $(0,1)\times\{1\}$, $y = 0$ on the rest of $\partial\Omega$. We work with a MATLAB implementation of the Navier–Stokes solver and of the discrete semismooth Newton method described above. The triangulation is uniform with $2\cdot 128^2 = 32768$ pressure triangles, $129^2 = 16641$ pressure nodes, $2\cdot 256^2 = 131072$ velocity triangles, and $257^2 = 66049$ velocity nodes. The number of time steps is $n_T = 500$ with final time $T = 0.5$, and the kinematic viscosity is $\nu = 10^{-3}$. As initial velocity y_0 we choose the stationary zero control Navier–Stokes flow. The target flow y_d is the stationary zero control Stokes flow (Figure 10.1). For the bounds on the control we use $\alpha = -0.5$, $\beta = 0.5$. Since the unconstrained optimal control ranges approximately from -2.6 to 4.5, the bounds restrict the control significantly. In the regularization we choose $\lambda = 0.01$ and $u_d = 0$. The initial point for the Newton iteration is $u_0 = 0$ (zero control). The saddle point systems arising from the generalized Stokes problems in each time step are solved by a preconditioned CG method applied to the pressure Schur complement. For solves involving the discrete operators corresponding to $w \mapsto \frac{1}{\Delta t}w - \nu\Delta w$, a preconditioned CG method is used, where a preconditioner is obtained by incomplete Cholesky factorizations, which are computed with `icfs` [148]. A preconditioner for the Schur complement is constructed based on these incomplete factorizations.

The conjugate gradient method for the solution of (10.51) is terminated if the initial residual is reduced by a factor ≤ 0.01. As this is often observed for semismooth Newton methods in optimal control, it turns out that a globalization is not required. However, it turns out that, for this problem, global convergence of the undamped semismooth Newton iteration and the choice $\pi(a,b) = a - P_{[\alpha,\beta]}(a - \lambda^{-1}b)$ instead of $\pi(a,b) = a - P_{[\alpha,\beta]}(a - b)$

Table 10.1 Iteration history (bound-constrained flow control problem).

k	CG	$\|\Pi(u_k)\|_U$	$j(u_k)$
0	–	3.606e−02	8.880e−03
1	7	2.856e−01	2.007e−03
2	7	1.284e−01	2.869e−03
3	7	3.420e−02	3.017e−03
4	7	3.453e−03	3.025e−03
5	7	4.670e−05	3.025e−03
6	8	2.238e−08	3.025e−03
7	8	5.218e−12	3.025e−03

is not achieved. The scaling by $\lambda^{-1} = 100$ appears to be too aggressive far away from the solution. For deciding if a smoothing step is required, the L^2-norm of the semismooth Newton step is compared with its L^p-norm (we use $p = 7/2$). This is reasonable since in a neighborhood of the solution $\bar{\mathbf{u}}^h$, the step $\mathbf{s}^h$ is a good approximation to the vector $\mathbf{u}^h - \bar{\mathbf{u}}^h$, where $\mathbf{u}^h$ is the current iterate. Since the Lipschitz constant of the smoothing step involves the factor λ^{-1}, we perform a smoothing step only if the L^p-norm of the step exceeds the L^2-norm of the step by a factor of λ^{-1}. This does not occur in our computations, and thus no smoothing steps arise in the iteration.

Table 10.1 shows the iteration history of the method: Iteration k, CG iterations per Newton step (CG), residual $\|\Pi(u_k)\|_U$, and objective function value $j(u_k)$. Here, the CG iterations required for computing $\mathbf{s}^h_{l-1}$, and thus $\mathbf{u}^h_l$, are listed in the row corresponding to $k = l$ (*not* $k = l-1$). This explains the "–" in the row corresponding to $k = 0$. We observe fast local convergence, as predicted by the theory, and a very moderate number of CG iterations per Newton step (7–8). For pictures of the flow and control, we refer to our second numerical experiment with pointwise ball constraints; see section 10.5.2. The results were obtained with MATLAB 7.9.0.529 (R2009b) 64-bit (glnxa64) on a 2.7 GHz Quad-Core AMD Opteron™ 8384 Processor with 64 GB memory. A comparison with the results in [109] shows that in terms of CG iterations and convergence speed the method performs comparably to the second-order methods for the unconstrained flow control problem investigated in [109]. Therefore, with the proposed method, the pointwise constrained problem can be solved with about the same effort as the unconstrained problem, which makes the algorithm very attractive.

10.5.2 The Pointwise Ball-Constrained Problem

We now consider the case of pointwise ball constraints

$$u(t,x) \in C := \{(a,b)^T : a^2 + b^2 \le r^2\} \quad \text{a.e. on } Q_T.$$

The discretized reduced flow control problem is then given by

$$\min_{\mathbf{u}^h} \ \mathbf{j}^h(\mathbf{u}^h) \qquad \text{subject to} \quad \begin{pmatrix} [\mathbf{u}^h_1]_j \\ [\mathbf{u}^h_2]_j \end{pmatrix} \in C \quad \left(1 \le j \le \frac{n_u}{2}\right), \tag{10.53}$$

where we have used the splitting of $\mathbf{u}^h$ into $\mathbf{u}^h_1$ and $\mathbf{u}^h_2$.

The first-order optimality conditions for a solution $\bar{\mathbf{u}}^h$ of (10.53) are

$$\begin{aligned}&\begin{pmatrix}[\bar{\mathbf{u}}_1^h]_j\\ [\bar{\mathbf{u}}_2^h]_j\end{pmatrix} \in C \quad \left(1 \le j \le \frac{n_u}{2}\right),\\ &\nabla \mathbf{j}^h(\bar{\mathbf{u}}^h)^T(\mathbf{u}^h - \bar{\mathbf{u}}^h) \ge 0 \quad \forall\, \mathbf{u}^h \in \mathbb{R}^{n_u},\ \begin{pmatrix}[\mathbf{u}_1^h]_j\\ [\mathbf{u}_2^h]_j\end{pmatrix} \in C \quad \left(1 \le j \le \frac{n_u}{2}\right).\end{aligned} \tag{10.54}$$

We again use the U^h inner product representation of the gradient,

$$\mathbf{g}^h(\mathbf{u}^h) = (\mathbf{M}^h)^{-1}\nabla \mathbf{j}^h(\mathbf{u}^h).$$

The optimality conditions can be rewritten as

$$\mathbf{\Pi}^h(\mathbf{u}^h) \stackrel{\text{def}}{=} \left[\pi([\mathbf{u}_1^h]_j, [\mathbf{u}_2^h]_j, \mathbf{g}_1^h(\mathbf{u}^h)_j, \mathbf{g}_2^h(\mathbf{u}^h)_j)\right]_{1\le j\le \frac{n_u}{2}} = 0.$$

Here, $\mathbf{g}_1^h(\mathbf{u}^h)$ and $\mathbf{g}_1^h(\mathbf{u}^h)$ are the subvectors of $\mathbf{g}^h(\mathbf{u}^h)$ corresponding to $\nabla_{\mathbf{u}_1^h}\mathbf{j}^h(\mathbf{u}^h)$ and $\nabla_{\mathbf{u}_2^h}\mathbf{j}^h(\mathbf{u}^h)$, respectively, and $\pi : \mathbb{R}^4 \to \mathbb{R}^2$ is defined by

$$\pi(a_1, a_2, b_1, b_2) = \begin{pmatrix}a_1\\ a_2\end{pmatrix} - P_C\left(\begin{pmatrix}a_1\\ a_2\end{pmatrix} - \begin{pmatrix}b_1\\ b_2\end{pmatrix}\right).$$

Smoothing steps can be computed via

$$\mathbf{S}^h(\mathbf{u}^h) = \left[P_C\left(\begin{pmatrix}[\mathbf{u}_1^h]_j\\ [\mathbf{u}_2^h]_j\end{pmatrix} - \lambda^{-1}\begin{pmatrix}\mathbf{g}_1^h(\mathbf{u}^h)_j\\ \mathbf{g}_2^h(\mathbf{u}^h)_j\end{pmatrix}\right)\right]_{1\le j\le \frac{n_u}{2}}.$$

There holds for $z \in \mathbb{R}^2$, $\|z\|_2 < r$, that $P_C(\hat{z}) = \hat{z}$ in a neighborhood of z and thus

$$P_C'(z) = I \quad \forall\, z \in \mathbb{R}^2,\ \|z\|_2 < r.$$

For $z \in \mathbb{R}^2$, $\|z\|_2 > r$, there holds $P_C(\hat{z}) = r\frac{\hat{z}}{\|\hat{z}\|_2}$ in a neighborhood of z and thus

$$P_C'(z) = \frac{r}{\|z\|_2}\left(I - \frac{z}{\|z\|_2}\frac{z^T}{\|z\|_2}\right).$$

Hence, Clarke's generalized Jacobian $\partial P_C(z) \subset \mathbb{R}^{2\times 2}$ is given by

$$\partial P_C(z) = \begin{cases}\{I\}, & \|z\|_2 < r,\\ \left\{\frac{r}{\|z\|_2}(I - vv^T)\right\}, & \|z\|_2 > r,\\ \{I - tvv^T : 0 \le t \le 1\}, & \|z\|_2 = r,\end{cases}$$

where $v = v(z) = \frac{z}{\|z\|_2}$.

Let the variables be ordered according to

$$\mathbf{u}^h = ([\mathbf{u}_1^h]_1, [\mathbf{u}_2^h]_1, [\mathbf{u}_1^h]_2, [\mathbf{u}_2^h]_2, \ldots, [\mathbf{u}_1^h]_{\frac{n_u}{2}}, [\mathbf{u}_2^h]_{\frac{n_u}{2}})^T,$$

which means that $[\mathbf{u}_1^h]_j = [\mathbf{u}^h]_{2j-1}$ and $[\mathbf{u}_2^h]_j = [\mathbf{u}^h]_{2j}$.

The semismooth Newton system then assumes the form

$$\left(\mathbf{I}^h - \mathbf{B}^h\left[\mathbf{I}^h - (\mathbf{M}^h)^{-1}\nabla^2\mathbf{j}^h(\mathbf{u}^h)\right]\right)\mathbf{s}^h = -\boldsymbol{\Pi}^h(\mathbf{u}^h), \tag{10.55}$$

where $\mathbf{B}^h \in \mathbb{R}^{n_u\times n_u}$ is a 2×2 block matrix $\mathbf{B}^h = \mathrm{blockdiag}(\mathbf{B}^h_{11},\dots,\mathbf{B}^h_{\frac{n_u}{2},\frac{n_u}{2}})$ with

$$\mathbf{B}^h_{jj} \in \partial P_C\left(\begin{pmatrix}[\mathbf{u}^h_1]_j\\ [\mathbf{u}^h_2]_j\end{pmatrix} - \begin{pmatrix}\mathbf{g}^h_1(\mathbf{u}^h)_j\\ \mathbf{g}^h_2(\mathbf{u}^h)_j\end{pmatrix}\right).$$

Let $K = \{j : \|([\mathbf{u}^h_1]_j,[\mathbf{u}^h_2]_j)^T\|_2 \le r\}$ and $L = \{j : \|([\mathbf{u}^h_1]_j,[\mathbf{u}^h_2]_j)^T\|_2 > r\}$. For $j \in K$ we can choose $\mathbf{B}^h_{jj} = I$. In the case $\|([\mathbf{u}^h_1]_j,[\mathbf{u}^h_2]_j)^T\|_2 = r$, this corresponds to the choice $t = 0$. For $j \in L$, there holds $\mathbf{B}^h_{jj} = \frac{r}{\|\mathbf{z}^h_j\|_2}(I - \mathbf{v}^h_j(\mathbf{v}^h_j)^T)$, with

$$\mathbf{z}^h_j = \begin{pmatrix}[\mathbf{u}^h_1]_j\\ [\mathbf{u}^h_2]_j\end{pmatrix} - \begin{pmatrix}\mathbf{g}^h_1(\mathbf{u}^h)_j\\ \mathbf{g}^h_2(\mathbf{u}^h)_j\end{pmatrix}, \quad \mathbf{v}^h_j = \frac{\mathbf{z}^h_j}{\|\mathbf{z}^h_j\|_2}.$$

Our next aim is to reduce the semismooth Newton equation to a symmetric system. Choosing $\mathbf{T}^h = \mathrm{blockdiag}(\mathbf{T}^h_{11},\dots,\mathbf{T}^h_{\frac{n_u}{2},\frac{n_u}{2}})$ with $\mathbf{T}^h_{jj} = I$ for $j \in K$ and $\mathbf{T}^h_{jj} = (\mathbf{w}^h_j,\mathbf{v}^h_j)$ for $j \in L$, where $\|\mathbf{w}^h\|_2 = 1$, $(\mathbf{w}^h_j)^T\mathbf{v}^h = 0$, the matrices $\mathbf{T}^h_{jj}$ are orthogonal; i.e., $(\mathbf{T}^h_{jj})^T\mathbf{T}^h_{jj} = I$. Furthermore,

$$(\mathbf{T}^h)^T\mathbf{B}^h\mathbf{T}^h = \mathrm{blockdiag}(\mathbf{W}^h_{11},\dots,\mathbf{W}^h_{\frac{n_u}{2},\frac{n_u}{2}}) =: \mathbf{W}^h,$$

where

$$\mathbf{W}^h_{jj} = I, \quad j \in K, \quad \mathbf{W}^h_{jj} = \frac{r}{\|\mathbf{z}^h_j\|_2}\begin{pmatrix}1 & 0\\ 0 & 0\end{pmatrix}, \quad j \in L.$$

We have

$$\begin{aligned}
&\left(\mathbf{I}^h - \mathbf{B}^h\left[\mathbf{I}^h - (\mathbf{M}^h)^{-1}\nabla^2\mathbf{j}^h(\mathbf{u}^h)\right]\right)\\
&= \left(\mathbf{I}^h - \mathbf{T}^h\mathbf{W}^h(\mathbf{T}^h)^T\left[\mathbf{I}^h - (\mathbf{M}^h)^{-1}\nabla^2\mathbf{j}^h(\mathbf{u}^h)\right]\right)\\
&= \mathbf{T}^h\left(\mathbf{I}^h - \mathbf{W}^h(\mathbf{T}^h)^T\left[\mathbf{T}^h - (\mathbf{M}^h)^{-1}\nabla^2\mathbf{j}^h(\mathbf{u}^h)\right]\mathbf{T}^h\right)(\mathbf{T}^h)^T\\
&= \mathbf{T}^h(\mathbf{M}^h)^{-1}\left(\mathbf{M}^h - \mathbf{W}^h\left[\mathbf{M}^h - (\mathbf{T}^h)^T\nabla^2\mathbf{j}^h(\mathbf{u}^h)\mathbf{T}^h\right]\right)(\mathbf{T}^h)^T.
\end{aligned}$$

Here we used that $\mathbf{M}^h = \mathrm{blockdiag}(\kappa_1 I,\dots,\kappa_{\frac{n_u}{2}} I)$ commutes with block diagonal matrices. Introducing $\mathbf{d}^h = (\mathbf{T}^h)^T\mathbf{s}^h$, we obtain that (10.55) is equivalent to

$$\left(\mathbf{M}^h - \mathbf{W}^h\left[\mathbf{M}^h - (\mathbf{T}^h)^T\nabla^2\mathbf{j}^h(\mathbf{u}^h)\mathbf{T}^h\right]\right)\mathbf{d}^h = -\mathbf{M}^h(\mathbf{T}^h)^T\boldsymbol{\Pi}^h(\mathbf{u}^h).$$

Since the second row of $\mathbf{W}^h_{jj}$ is zero for all $j \in L$, we obtain

$$[\mathbf{d}^h]_{2j} = -[(\mathbf{T}^h)^T\boldsymbol{\Pi}^h(\mathbf{u}^h)]_{2j} \quad \forall\, j \in L.$$

Now, setting $I = \{i : 1 \le i \le n_u,\ i/2 \notin L\}$, the diagonal matrix $(\mathbf{W}^h)_{II}$ is invertible and the vector $\mathbf{d}_I^h$ solves a symmetric linear system with the coefficient matrix

$$(\mathbf{W}_{II}^h)^{-1}\mathbf{M}_{II}^h - \mathbf{M}_{II}^h + \left[(\mathbf{T}^h)^T\nabla^2\mathbf{j}^h(\mathbf{u}^h)\mathbf{T}^h\right]_{II}.$$

We apply a CG method to this reduced system with preconditioner $\mathbf{W}_{II}^h(\mathbf{M}_{II}^h)^{-1}$.

Except for the ball constraint, where we choose $r = 1$, all data and parameter settings are the same as in the box-constrained case; see page 266. The target flow (zero control Stokes flow) is depicted in Figure 10.1. The optimal control at time $t = 0.1$ is displayed in Figure 10.2. Snapshots of the optimally controlled flow are shown in Figure 10.3. The iteration history in Table 10.2 shows that also for this problem the semismooth Newton method is locally fast convergent. The number of CG iterations is partially higher than in the bound constrained case (21, 21, and 15 for the 2nd, 3rd, and 4th Newton system), but the remaining 6 inexact solves need only 7–8 CG iterations each.

Table 10.2 Iteration history (ball-constrained flow control problem).

k	CG	$\|\Pi(u_k)\|_U$	$j(u_k)$
0	–	3.606e−02	8.880e−03
1	7	1.953e−01	2.007e−03
2	21	5.259e−02	2.255e−03
3	21	7.215e−03	2.267e−03
4	15	2.019e−04	2.270e−03
5	8	3.146e−05	2.270e−03
6	8	1.782e−06	2.270e−03
7	7	1.751e−07	2.270e−03
8	8	1.155e−09	2.270e−03
9	8	1.400e−12	2.270e−03

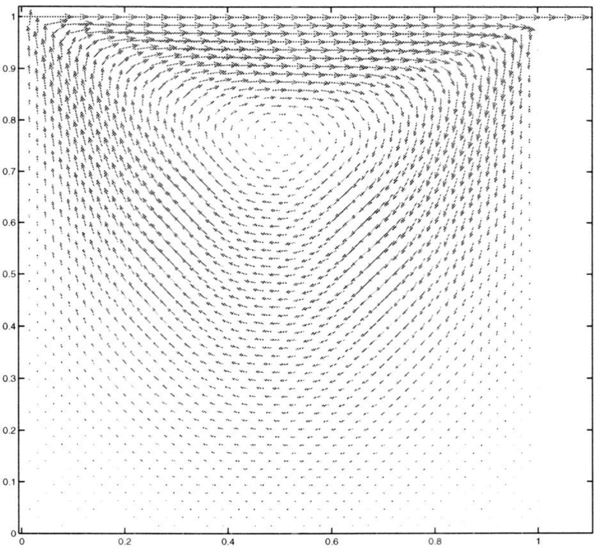

Figure 10.1 Target flow y_d (Stokes flow). The arrows are scaled by 0.1.

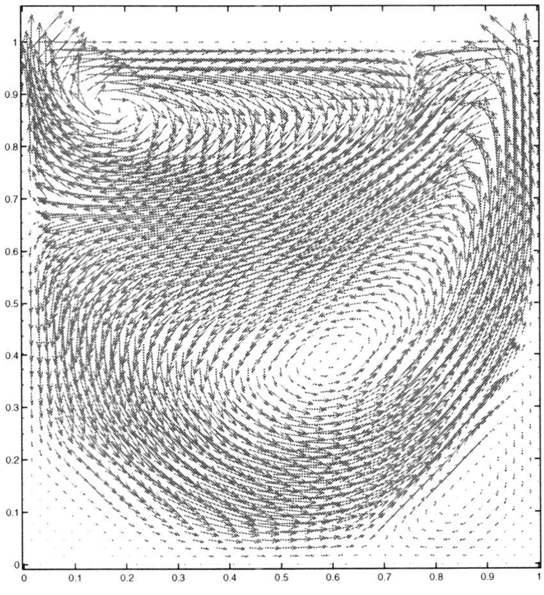

Figure 10.2 Optimal control $\bar{u}$ at $t = 0.1$. The arrows are scaled by 0.1.

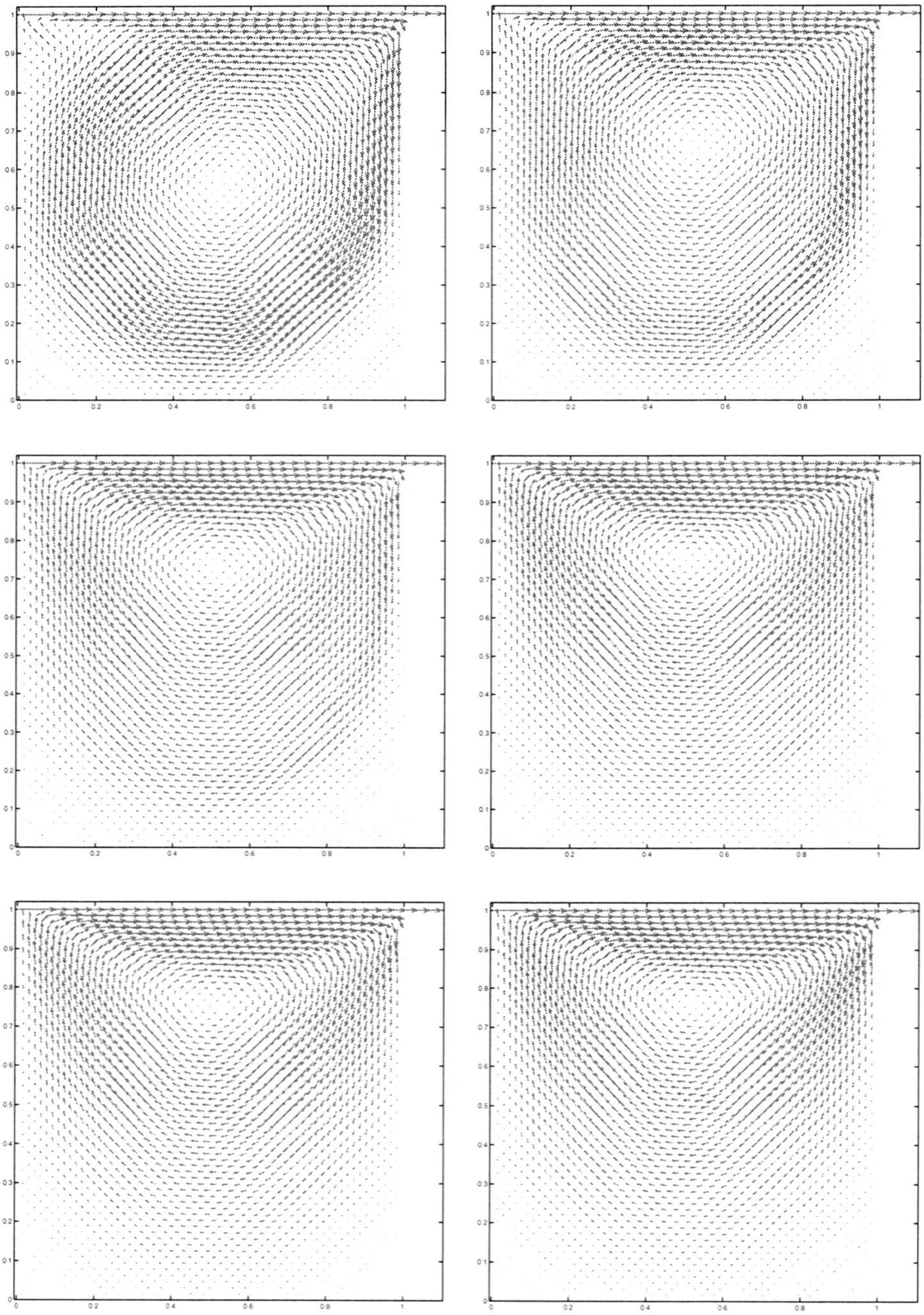

Figure 10.3 Optimally controlled flow (ball-constrained problem). The arrows are scaled by 0.1.

Chapter 11

Optimal Control of Compressible Navier–Stokes Flow

11.1 Introduction

In this chapter we show an application of semismooth Newton methods to a boundary control problem governed by the time-dependent compressible Navier–Stokes equations. The underlying Navier–Stokes solver was developed by Scott Collis [47] and the adjoint code for the computation of the reduced gradient was obtained in joint work with Scott Collis, Matthias Heinkenschloss, Kaveh Ghayour, and Stefan Ulbrich. The goal was to investigate a vortex control problem for the unsteady, compressible Navier–Stokes equations that is suitable for providing insights towards more advanced applications such as aeroacoustic noise control. A particularly interesting application is the control of noise arising from blade-vortex interaction (BVI), which can occur in machines with rotors, such as helicopters and turbines. Here, vortices shed by a preceding blade hit a subsequent blade which results in a high-amplitude, impulsive noise. For more details we refer to [43, 44, 45, 46] and the references therein.

11.2 The Flow Control Problem

In the following, we will not consider noise control. Rather, we content ourselves with solving a model problem to investigate the viability of our approach for controlling the compressible Navier–Stokes equations. This model consists in two counter-rotating viscous vortices above an infinite wall which, due to the self-induced velocity field, propagate downward and interact with the wall. As a control mechanism we use suction and blowing on part of the wall; i.e., we control the normal velocity of the fluid on this part of the wall. As the computational domain we use a rectangle:

$$\Omega = (-L_1, L_1) \times (0, L_2).$$

The wall is located at $x_2 \equiv 0$, whereas the left, right, and upper part of the boundary are "transparent" in the sense that we pose nonreflecting boundary conditions there. The domain Ω is occupied by a compressible fluid whose state is described by $y = (\rho, v_1, v_2, \theta)$ with density $\rho(t,x)$, velocities $v_i(t,x)$, $i = 1,2$, and temperature $\theta(t,x)$. Here, $t \in I \overset{\text{def}}{=} (0,T)$

is the time and $x = (x_1, x_2)$ denotes the spatial location. The state satisfies the

Compressible Navier–Stokes (CNS) Equations:

$$\frac{\partial}{\partial t}F^0(y) + \sum_{i=1}^{2} \frac{\partial}{\partial x_i} F^i(y) = \sum_{i=1}^{2} \frac{\partial}{\partial x_i} G^i(y, \nabla y) \quad \text{on } I \times \Omega,$$
$$y(0,\cdot) = y_0 \quad \text{on } \Omega.$$

Here, we have written CNS in conservative form. Boundary conditions are specified below. We have used the following notation:

$$F^0(y) = \begin{pmatrix} \rho \\ \rho v_1 \\ \rho v_2 \\ \rho E \end{pmatrix}, \quad F^1(y) = \begin{pmatrix} \rho v_1 \\ \rho v_1^2 + p \\ \rho v_1 v_2 \\ (\rho E + p) v_1 \end{pmatrix}, \quad F^2(y) = \begin{pmatrix} \rho v_2 \\ \rho v_1 v_2 \\ \rho v_2^2 + p \\ (\rho E + p) v_2 \end{pmatrix},$$

$$G^i(y, \nabla y) = \frac{1}{\text{Re}} \begin{pmatrix} 0 \\ \tau_{1i} \\ \tau_{2i} \\ \tau_{1i} v_1 + \tau_{2i} v_2 + \dfrac{\kappa}{(\gamma - 1)\text{M}^2 \text{Pr}} \theta_{x_i} \end{pmatrix}.$$

The pressure p, the total energy per unit mass E, and the stress tensor τ are given by

$$p = \frac{\rho\theta}{\gamma \text{M}^2}, \quad E = \frac{\theta}{\gamma(\gamma - 1)\text{M}^2} + \frac{1}{2}(v_1^2 + v_2^2),$$
$$\tau_{ii} = 2\mu (v_i)_{x_i} + \lambda(\nabla \cdot v), \quad \tau_{12} = \tau_{21} = \mu((v_1)_{x_2} + (v_2)_{x_1}).$$

Here μ and λ are the first and second coefficient of viscosity, respectively, κ is the thermal conductivity, M is the reference Mach number, Pr is the reference Prandtl number, and Re is the reference Reynolds number. The boundary conditions on the wall are

$$\partial\theta/\partial n = 0, \quad v_1 = 0, \quad v_2 = u \quad \text{on } \Sigma_c = I \times (-L_1, L_1) \times \{0\},$$

and on the rest of the boundary we pose nonreflecting boundary conditions that are derived from inviscid characteristic boundary conditions.

At the initial time $t = 0$ two counter-rotating viscous vortices are located in the center of Ω. Without control ($v_2 = u \equiv 0$), the vortices move downward and interact with the wall, which causes them to bounce back; see Figure 11.1. Our aim is to perform control by suction and blowing on the wall in such a way that the terminal kinetic energy is minimized. To this end, we choose the objective function

$$J(y, u) = \int_\Omega \left[\frac{\rho}{2}(v_1^2 + v_2^2)\right]_{t=T} dx + \frac{\alpha}{2}\|u\|^2_{H^1(\Sigma_c)}.$$

The first term is the kinetic energy at the final time $t = T$, whereas the second term is an H^1-regularization with respect to (t, x_1). Here, we write $\alpha > 0$ for the regularization parameter to avoid confusion with the second coefficient of viscosity. As control space, we choose $U = H^1(I, H_0^1(-L_1, L_1))$. We stress that the mathematical existence and uniqueness theory for the compressible Navier–Stokes equations, see [112, 152, 155] for state of the art references, seems not yet to be complete enough to admit a rigorous control theory.

Therefore, our choice of the control space is guided more by formal and heuristic arguments than by rigorous control theory. If the H^1-regularization is omitted or replaced by an L^2-regularization, the control exhibits increasingly heavy oscillations in time and space during the course of optimization, which indicates that the problem is ill-posed without a sufficiently strong regularization.

In the following, we want to solve the described optimal control problem, with the control subject to pointwise bound constraints. We apply our inexact semismooth Newton methods and use BFGS-updates [55, 56] to approximate the Hessian of the reduced objective function. The restriction of the control by pointwise bound constraints has the realistic interpretation that it is technically only possible to inject or draw off fluid with a certain maximum speed. We arrive at the following optimal flow control problem:

$$\begin{aligned} &\text{minimize} && J(y,u) \stackrel{\text{def}}{=} \int_\Omega \left[\frac{\rho}{2}(v_1^2+v_2^2)\right]_{t=T} dx + \frac{\alpha}{2}\|u\|^2_{H^1(\Sigma_c)} \\ &\text{subject to} && y \text{ solves CNS for the boundary conditions associated with } u, \\ &&& u_{\min} \le u \le u_{\max}. \end{aligned} \tag{11.1}$$

11.3 Adjoint-Based Gradient Computation

The computations we present in the following use results and software developed jointly with Scott Collis, Kaveh Ghayour, Matthias Heinkenschloss, and Stefan Ulbrich [44, 45, 46], in particular

1. A Navier–Stokes solver, written in Fortran90 by Scott Collis [47], was adjusted to the requirements of optimal control. For space discretization finite differences are used which are sixth order accurate in the interior of the domain. The time discretization is done by an explicit Runge–Kutta method. The code was parallelized on the basis of OpenMP.

2. Two different variants of adjoint-based gradient computation were considered:

(a) The first approach derives the adjoint Navier–Stokes equations including adjoint wall boundary conditions [45]. The derivation of adjoint boundary conditions for the nonreflecting boundary conditions turns out to be a delicate matter and will not be discussed here. Hence, in this approach we have used the (appropriately augmented) adjoint boundary conditions of the Euler equation. The gradient calculation then requires the solution of the Navier–Stokes equations, followed by the solution of the adjoint Navier–Stokes equations backward in time. Since the discretized adjoint equation is usually not the exact adjoint of the discrete state equation, this approach, which usually is called *optimize, then discretize* (OD), only yields inexact discrete gradients in general.

(b) In a second approach we have investigated the adjoint-based computation of gradients by applying the reverse mode of automatic differentiation (AD). For this, we used the AD-software Tangent Linear and Adjoint Model Compiler (TAMC) [75], a source-to-source compiler, which translates Fortran90 routines to their corresponding adjoint Fortran90 routines. This approach yields exact (up to roundoff errors) discrete gradients and is termed *discretize, then optimize* (DO).

For the computational results shown below, the DO method described in (b) was used. This approach has the advantage of providing exact discrete gradients, which is very favorable when doing optimization. In fact, descent methods based on inexact gradients require a control mechanism over the amount of inexactness, which is not a trivial task in OD-based approaches. Secondly, the use of exact gradients is very helpful in verifying the correctness of the adjoint code, since potential errors can usually be found immediately by comparing directional derivatives with the corresponding finite difference quotients.

When working with the OD approach, which has the advantage that the source code of the CNS-solver is not required, the discretization of the state equation, adjoint equation, and objective function have to be compatible (in a sense not discussed here; see, e.g., [44, 93]) to obtain gradients that are good approximations (i) of the infinite-dimensional gradients, and (ii) of the exact discrete gradients. Here, requirement (ii) is important for a successful solution of the discrete control problem, whereas (i) crucially influences the quality of the computed discrete optimal control, measured in terms of the infinite-dimensional control problem. This second issue also applies to the DO approach, but for DO it is only important to use compatible discretizations for the state equation and objective function. With respect to this interesting topic, we have used [93] as a guideline, to which we refer for further reference.

For this book, the computations were newly run, but since the code is quite complex, the implementation was not modified. In particular, the original projected line search–based globalization strategy was used.

11.4 Semismooth BFGS-Newton Method

The implementation of the semismooth Newton method uses BFGS-approximations of the Hessian matrix. The resulting semismooth Newton systems have a similar structure as those arising in the step computation of the limited-memory BFGS method L-BFGS-B by Byrd, Lu, Nocedal, and Zhu [32, 206]. Our implementation uses a similar globalization as L-BFGS-B and is described below.

11.4.1 Quasi-Newton BFGS-Approximations

In this section, we focus on the use of BFGS-approximations in semismooth Newton methods for the discretized control problem. We stress, however, that convergence results for quasi-Newton methods in infinite-dimensional Hilbert spaces are available [83, 134, 179]. Using a similar notation as in Chapter 9, the semismooth Newton system for the discrete control problem assumes the form (written in coordinates in the discrete L^2-space)

$$[\mathbf{D}_1^h]_k + [\mathbf{D}_2^h]_k \mathbf{H}_k^h \mathbf{s}_k^h = -\mathbf{\Phi}^h(\mathbf{u}_k^h)$$

with $\mathbf{H}_k^h = \mathbf{j}^{h''}(\mathbf{u}_k^h)$ and diagonal matrices $[\mathbf{D}_i^h]_k$, $|([\mathbf{D}_1^h]_k + [\mathbf{D}_2^h]_k)_{jj}| \geq \kappa$.

For the approximation of the Hessian $\mathbf{H}_k^h$ we work with

Limited-Memory BFGS-Matrices ($l \approx 10$):

$$\mathbf{B}_k^h = \mathbf{B}_0^h - \mathbf{W}_k^h \mathbf{Z}_k^h \mathbf{W}_k^{h^T} \in \mathbb{R}^{n^h \times n^h}, \quad \mathbf{W}_k^h \in \mathbb{R}^{n^h \times 2l}, \quad \mathbf{Z}_k^h \in \mathbb{R}^{2l \times 2l},$$

where we have used the compact representation of [33], to which we refer for details. The matrix $\mathbf{B}_0^h$ is the initial BFGS-matrix and should be chosen such that (a) the product $(\mathbf{B}_0^h)^{-1}\mathbf{v}^h$ can be computed in a reasonably efficient way, since this is needed in the BFGS-updates, and (b) the inner product induced by $\mathbf{B}_0^h$ approximates the original infinite-dimensional inner product on U sufficiently well. In the case of our flow control problem, we have $U = H^1(I, H_0^1(-L_1, L_1))$, and use a finite difference approximation of the underlying Laplace operator to obtain $\mathbf{B}_0^h$. Compared with the state and adjoint solves, the solution of the 2-D Helmholtz equation required to compute $(\mathbf{B}_0^h)^{-1}\mathbf{v}^h$ is negligible. The inverse of $\mathbf{M}_k^h = [\mathbf{D}_1^h]_k + [\mathbf{D}_2^h]_k \mathbf{B}_k^h$ can be computed by the Sherman–Morrison–Woodbury formula:

$$(\mathbf{M}_k^h)^{-1} = \mathbf{C}_k^h + \mathbf{C}_k^h [\mathbf{D}_2^h]_k \mathbf{W}_k^h \left(I - \mathbf{Z}_k^h {\mathbf{W}_k^h}^T \mathbf{C}_k^h [\mathbf{D}_2^h]_k \mathbf{W}_k^h\right)^{-1} \mathbf{Z}_k^h {\mathbf{W}_k^h}^T \mathbf{C}_k^h,$$

where $\mathbf{C}_k^h = ([\mathbf{D}_1^h]_k + [\mathbf{D}_2^h]_k \mathbf{B}_0^h)^{-1}$.

11.4.2 The Algorithm

We now give a sketch of the algorithm.

1. The Hessian matrix of the discrete objective function is approximated by limited-memory BFGS-matrices. Here, we choose $\mathbf{B}_0^h$ such that it represents a finite difference approximation of the inner product on U.
2. The globalization is similar as in the well-accepted L-BFGS-B method of Byrd, Lu, Nocedal, and Zhu [32, 206]:

   i. At the current point $\mathbf{u}_k^h \in \mathcal{B}^h$, the objective function $\mathbf{j}^h$ is approximated by a quadratic model $\mathbf{q}_k^h$.

   ii. Starting from $\mathbf{u}_k^h$, a generalized Cauchy point $\mathbf{u}_k^{h,c} \in \mathcal{B}^h$ is computed by an Armijo-type linesearch for $\mathbf{q}_k^h$ along the projected gradient path

   $$P_{\mathcal{B}^h}(\mathbf{u}_k^h - t\mathbf{j}_k^{h'}), \quad t \geq 0.$$

   iii. The semismooth Newton method is used to compute a Newton point $\mathbf{u}_k^{h,n}$.

   iv. By approximate minimization of $\mathbf{q}_k^h$ along the projected path

   $$P_{\mathcal{B}^h}(\mathbf{u}_k^{h,c} + t(\mathbf{u}_k^{h,n} - \mathbf{u}_k^{h,c})), \quad t \in [0,1],$$

   the point $\mathbf{u}_k^{h,q}$ is computed.

   v. The new iterate $\mathbf{u}_{k+1}^h$ is obtained by approximate minimization of $\mathbf{j}_k^h$ on the line segment $[\mathbf{u}_k^h, \mathbf{u}_k^{h,q}]$, using the algorithm by Moré–Thuente [161].

11.5 Numerical Results

We now present numerical results for the described semismooth BFGS-Newton method when applied to the flow control problem (11.1). Here are the main facts about the problem and the implementation:

- The space discretization is done by a high-order finite difference method on an 128×96 Cartesian mesh.

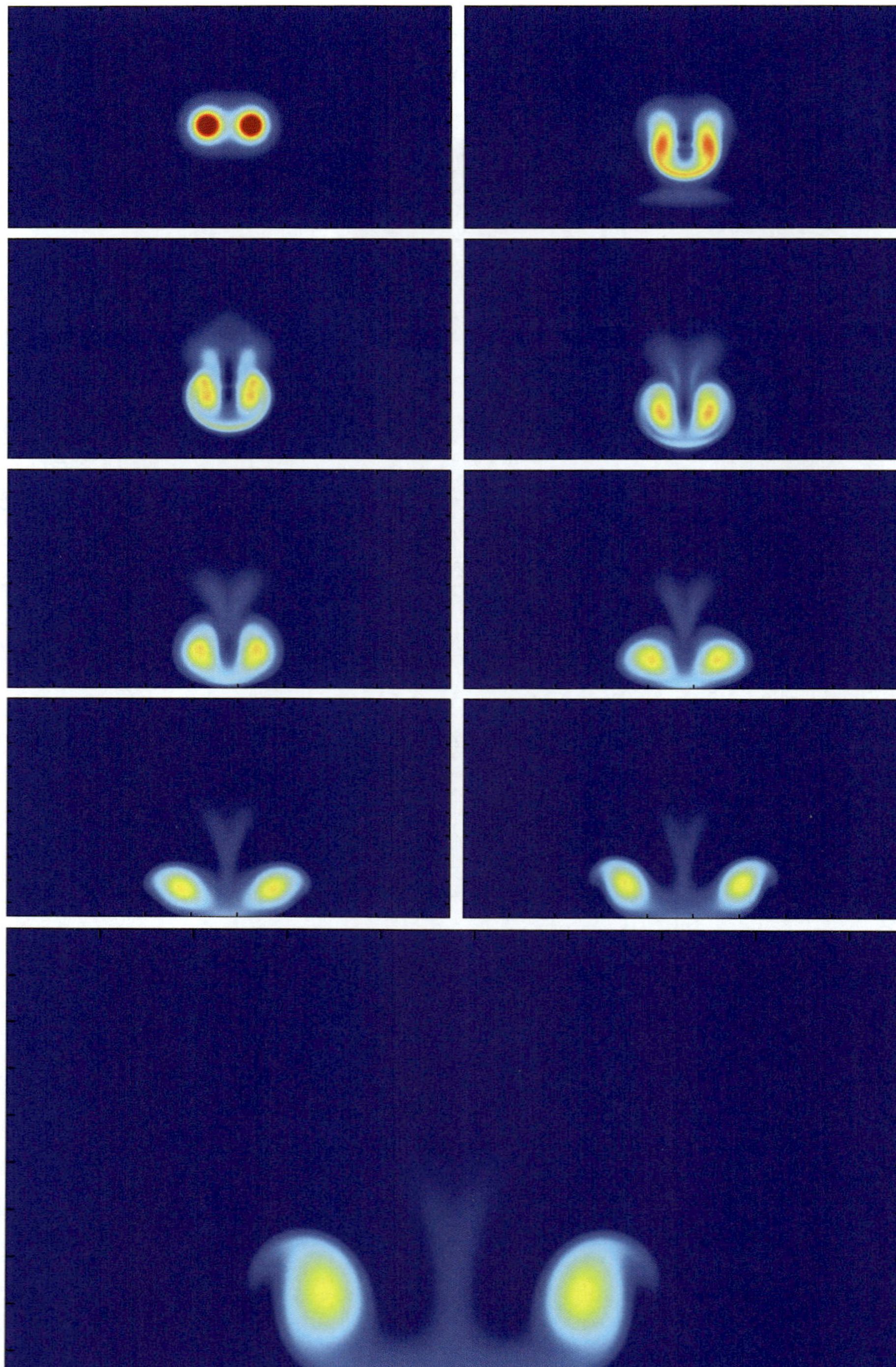

Figure 11.1 Snapshots of the state (density ρ) corresponding to no control (left to right, top to bottom), $t = 0, 3, \ldots, 24$.

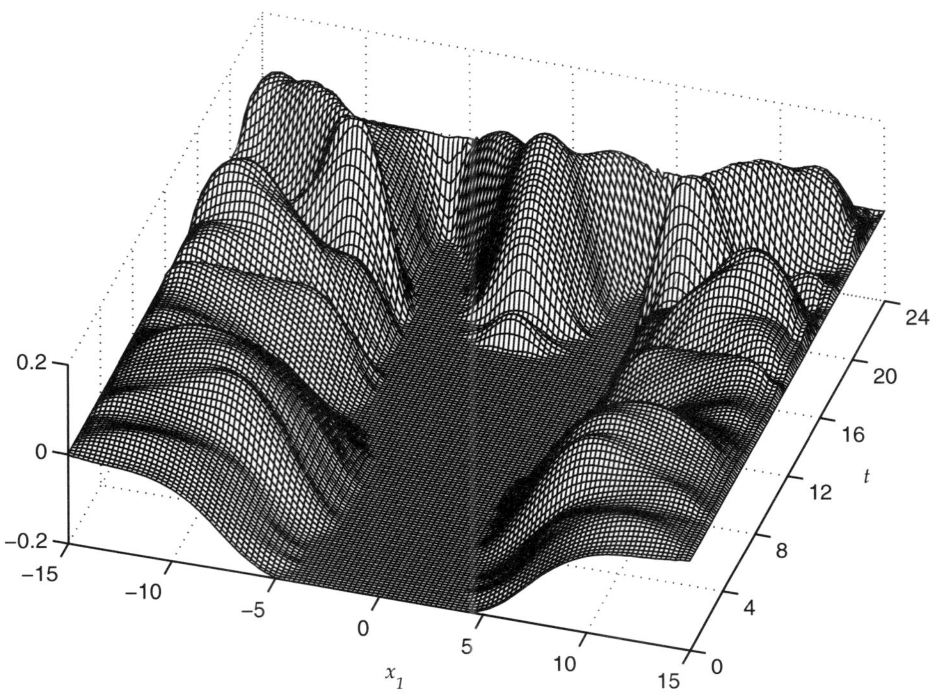

Figure 11.2 Computed optimal control $\bar{u}$.

- For the time discretization the standard 4-stage Runge–Kutta method is used, with 600 time steps and $T = 24$. This allows parallelization within each time step.
- We compute exact discrete gradients by solving the adjoint of the discrete state equation, which is obtained by the reverse mode of automatic differentiation using TAMC [75].
- As the optimization method, we use the semismooth BFGS-Newton method described above.
- Parameters: $\mathrm{Re} = 50$, $\mathrm{Pr} = 1$, $\mathrm{M} = 0.5$, $\gamma = 1.4$; regularization parameter $\alpha = 0.005$; bounds $u_{\min} = -0.2$, $u_{\max} = 0.2$.
- As the NCP-function we use a variant of the penalized Fischer–Burmeister function [36].
- The resulting problem has over 75,000 control variables and over 29,000,000 state variables, and thus is very large scale.
- The computations were performed on a 32 Core Linux machine with 2.7 GHz Quad-Core AMD Opteron 8384 Processors and 64 GB memory. We used four processors.

Figure 11.1 displays the state (the density ρ is shown) of the uncontrolled system ($v_2|_{\Sigma_c} = u \equiv 0$). We see that the vortices hit the wall and bounce back. The terminal state, at which we evaluate the kinetic energy, is shown in the last, magnified picture. The resulting terminal kinetic energy in the no-control case is

$$\text{No control } (v_2|_{\Sigma_c} = u \equiv 0): \quad E_{\text{kin}}|_{t=T} = J(y(0),0) = 7.9.$$

Figure 11.3 on the next page shows the state (represented by the density ρ) when optimal control is applied. The optimal control was obtained by 100 iterations of the BFGS-Newton method. The resulting terminal kinetic energy in the optimal control case and the objective function value ($E_{\text{kin}}|_{t=T}$ + regularization), respectively, are

$$\text{Optimal control } (v_2|_{\Sigma_c} = \bar{u}): \quad E_{\text{kin}}|_{t=T} = 0.059, \quad J(\bar{u}, y(\bar{u})) = 0.085,$$

where $\bar{u}$ denotes the computed optimal control, which is displayed in Figure 11.2 on the previous page. It can be seen in Figure 11.1 that the lower bound becomes active. In fact, the upper bound also is active at a few points, but this is not apparent from the picture. By applying optimal control the vortices are successfully absorbed. If we had displayed the kinetic energy instead of the density, the vortices would be almost invisible at the terminal time in the optimal control case, since the optimal control reduces the terminal kinetic energy to less than one hundredth of its value without control. In comparison with our computational experience for the unconstrained control problem, the semismooth Newton method performs comparably efficiently. This shows the efficiency of semismooth Newton methods for the solution of very large scale problems.

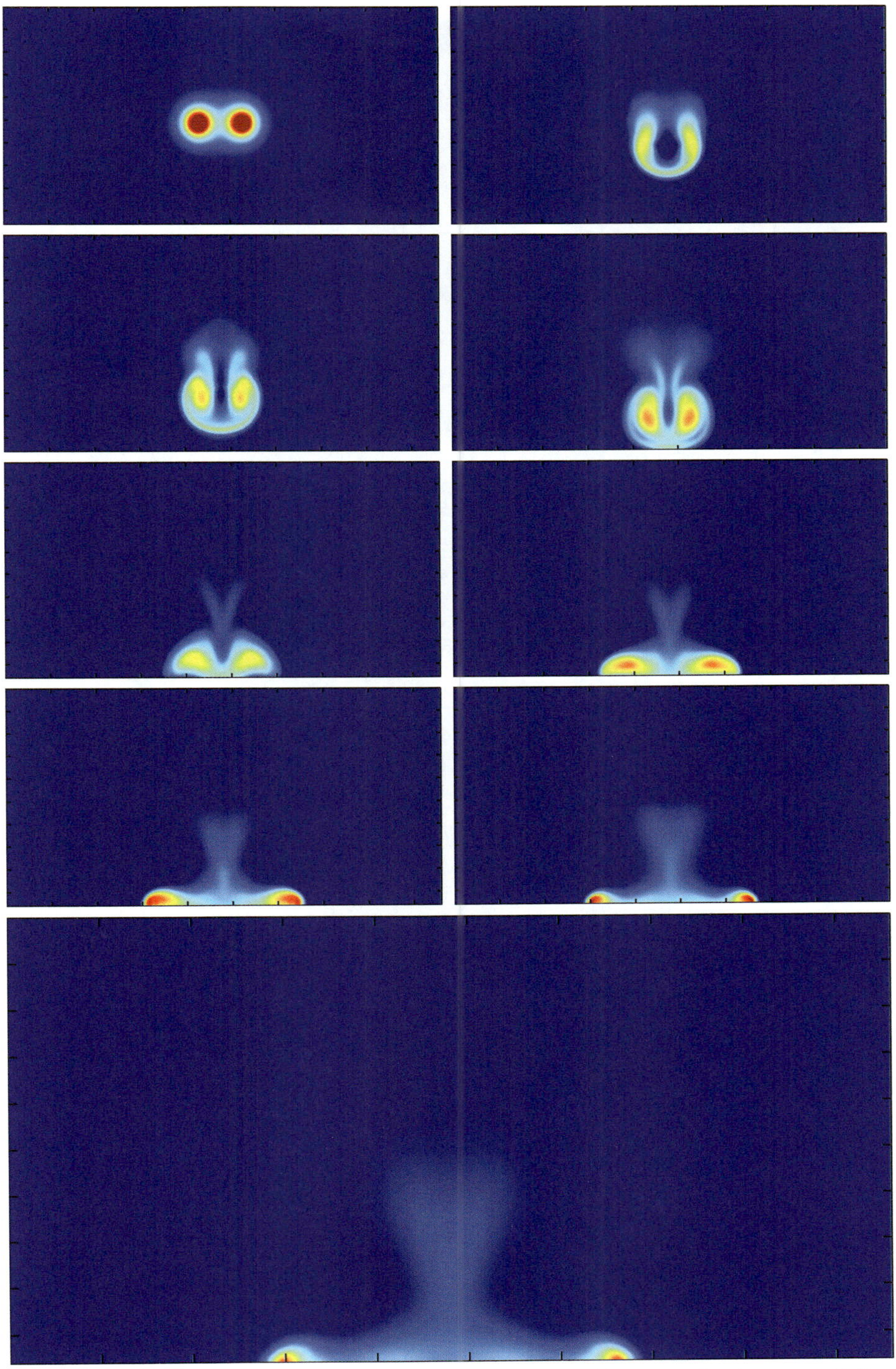

Figure 11.3 Snapshots of the state (density ρ) corresponding to optimal control (left to right, top to bottom), $t = 0, 3, \ldots, 24$.

Appendix

A.1 Adjoint Approach for Optimal Control Problems

In this appendix we describe the adjoint approach for the computation of the gradient and Hessian of the reduced objective function. We consider the abstract optimal control problem

$$\underset{y\in Y, u\in U}{\text{minimize}} \quad J(y,u) \quad \text{subject to} \quad E(y,u)=0, \quad u\in U_{ad} \tag{A.1}$$

with feasible set $U_{ad} \subset U$, objective function

$$J : Y \times U \to \mathbb{R},$$

and state equation operator

$$E : Y \times U \to W.$$

The control space U, the state space Y, and the image space W are Banach spaces. We assume the existence of a neighborhood V of U_{ad} such that, for all $u \in V$, the state equation $E(y,u) = 0$ possesses a unique solution $y = y(u)$. Then the optimal control problem (A.1) is equivalent to the reduced optimal control problem

$$\text{minimize} \quad j(u) \quad \text{subject to} \quad u \in U_{ad}, \tag{A.2}$$

where $j : U \supset V \to \mathbb{R}$, $j(u) = J(y(u),u)$ is the reduced objective function.

A.1.1 Adjoint Representation of the Reduced Gradient

We now describe the adjoint approach for the computation of $j'(u)$. To this end, we assume that J and E are Fréchet differentiable near $(y(u),u)$ and that $u \mapsto y(u)$ is Fréchet differentiable near u. According to the implicit function theorem, the latter holds, e.g., if E is continuously differentiable near $(y(u),u)$ and if the partial derivative $E_y(y(u),u)$ is continuously invertible. Under the given hypotheses the function j is differentiable near u.

We introduce a Lagrange multiplier $w \in W^*$ for the state equation in (A.1) and define the Lagrange function $L : Y \times V \times W^* \to \mathbb{R}$,

$$L(y,u,w) = J(y,u) + \langle w, E(y,u)\rangle_{W^*,W}.$$

Since $E(y(u),u) = 0$ on V, we have

$$L(y(u),u,w) = J(y(u),u) = j(u) \quad \forall\, u \in V,\ w \in W^*.$$

Hence,

$$j'(u) = y_u(u)^* L_y(y(u),u,w) + L_u(y(u),u,w) \quad \forall\, u \in V,\ w \in W^*. \tag{A.3}$$

The idea now is to choose $w \in W^*$ such that

$$L_y(y(u),u,w) = 0.$$

This equation is called the *adjoint equation* and its solution $w = w(u) \in W^*$ is the *adjoint state*. Thus, written in detail, the adjoint state $w = w(u)$ is the solution of the adjoint equation

$$J_y(y(u),u) + E_y(y(u),u)^* w = 0.$$

If we assume that $E_y(y(u),u)$ is continuously invertible, the adjoint state w is uniquely determined. For $w = w(u)$ we obtain

$$\begin{aligned} j'(u) &= y_u(u)^* L_y(y(u),u,w(u)) + L_u(y(u),u,w(u)) \\ &= L_u(y(u),u,w(u)) = J_u(y(u),u) + E_u(y(u),u)^* w(u). \end{aligned}$$

The identity

$$j'(u) = J_u(y(u),u) + E_u(y(u),u)^* w(u)$$

is called the *adjoint representation* of the reduced gradient $j'(u)$.

Therefore, the derivative $j'(u)$ can be computed as follows:

1. Compute the state $y = y(u) \in Y$ by solving the state equation
$$E(y,u) = 0.$$
2. Compute the adjoint state $w = w(u) \in W^*$ by solving the adjoint equation
$$E_y(y,u)^* w = -J_y(y,u).$$
3. Compute $j'(u) = J_u(y,u) + E_u(y,u)^* w$.

Remark A.1. *If the state equation is an initial value problem, then the adjoint equation is reverse in time. For the derivation of adjoint equations for various types of optimal control problems governed by PDEs, we refer to Lions* [150].

Remark A.2. *As described in the above derivation, there holds, see* (A.3),

$$\langle j'(u), s\rangle_{U^*,U} = \langle L_y(y(u),u,w), y_u(u)s\rangle_{Y^*,Y} + \langle L_u(y(u),u,w), s\rangle_{U^*,U}$$

for all $u \in V$, $w \in W^*$, and $s \in U$. The purpose of the adjoint state $w = w(u)$ is to achieve $\langle L_y(y(u),u,w(u)), y_u(u)s\rangle_{Y^*,Y} = 0$ for all $s \in U$. Therefore, it is actually not necessary that $L_y(y(u),u,w(u)) = 0$, but only that $\langle L_y(y(u),u,w(u)), v\rangle_{Y^*,Y} = 0$ for all v in the range $y_u(u)U$ of $y_u(u)$. This could be used to relax the requirements on the adjoint state in cases where $y_u(u)$ is not onto, which might be helpful in settings that are beyond the framework of the implicit function theorem.

A.1.2 Adjoint Representation of the Reduced Hessian

The adjoint approach can be continued to obtain adjoint formulas for the Hessian operator $j''(u)$. To this end, we assume that J and E are twice differentiable near $(y(u),u)$ and that $u \mapsto y(u)$ is twice differentiable near u. By (A.3), we have, for all $w \in W^*$ and all $v_1, v_2 \in U$, writing $y = y(u)$,

$$\begin{aligned} j''(u)(v_1,v_2) = {} & L_{yu}(y,u,w)(y_u(u)v_1,v_2) + L_{yy}(y,u,w)(y_u(u)v_1,y_u(u)v_2) \\ & + \langle L_y(y,u,w), y_{uu}(u)(v_1,v_2)\rangle_{Y^*,Y} + L_{uy}(y,u,w)(v_1,y_u(u)v_2) \\ & + L_{uu}(y,u,w)(v_1,v_2). \end{aligned}$$

If we choose $w = w(u)$, then $L_y(y(u),u,w) = 0$, and thus

$$j''(u) = T(u)^* L''_{(y,u)}(y(u),u,w(u))T(u), \tag{A.4}$$

where $L''_{(y,u)}$ denotes the second partial derivative with respect to (y,u), and

$$T(u) = \begin{pmatrix} y_u(u) \\ I_U \end{pmatrix} = \begin{pmatrix} -E_y(y(u),u)^{-1}E_u(y(u),u) \\ I_U \end{pmatrix}.$$

Here, in the second expression for $T(u)$ we assume that $E_y(y(u),u)$ is continuously invertible and use that, since $E(y(\cdot),\cdot) \equiv 0$, there holds

$$E_y(y(u),u)y_u(u) + E_u(y(u),u) = 0.$$

Remark A.3. *It is interesting to note that in the case where $E_y(y(u),u)$ is continuously invertible, the mapping $T(u)$ is a continuous linear homeomorphism from U to the null space of $E'(y(u),u)$. In fact, it is obvious that $E'(y(u),u)T(u) = 0$. Conversely, if $E_y(y(u),u)h + E_u(y(u),u)v = 0$, then*

$$h = -E_y(y(u),u)^{-1}E_u(y(u),u)v,$$

and thus

$$\begin{pmatrix} h \\ v \end{pmatrix} = T(u)v.$$

Therefore, $j''(u)$ is the restriction of the Hessian $L''_{(y,u)}(y(u),u,w(u))$ of the Lagrangian to the null space of $E'(y(u),u)$, parameterized by $v \in U \mapsto T(u)v$.

Usually, the formula (A.4) is not used to compute the complete Hessian operator. Rather, it is used to compute directional derivatives $j''(u)v$ of j'. Here is the required procedure:

1. Compute the state $y = y(u) \in Y$ by solving the state equation

$$E(y,u) = 0.$$

2. Compute the adjoint state $w = w(u) \in W^*$ by solving the adjoint equation

$$E_y(y,u)^* w = -J_y(y,u).$$

3. Compute $z = z(u) \in Y$ as the solution of the linearized state equation

$$E_y(y,u)z = -E_u(y,u)v.$$

4. Compute $h = h(u) \in W^*$ by solving the adjoint system

$$E_y(y,u)^*h = -L_{yy}(y,u,w)z - L_{yu}(y,u,w)v.$$

5. Set $j''(u)v := E_u(y,u)^*h + L_{uy}(y,u,w)z + L_{uu}(y,u,w)v$.

A.2 Several Inequalities

For convenience, we recall several well-known inequalities, which are frequently used throughout this work.

Lemma A.4 (Hölder's inequality). *Let $p_i \in [1,\infty]$, $i = 1,\dots,n$, and $p \in [1,\infty]$ satisfy*

$$\frac{1}{p_1} + \cdots + \frac{1}{p_n} = \frac{1}{p}.$$

Then, for all $f_i \in L^{p_i}(\Omega)$ there holds $f = f_1 f_2 \cdots f_n \in L^p(\Omega)$ and

$$\|f\|_{L^p} \le \|f_1\|_{L^{p_1}} \cdots \|f_n\|_{L^{p_n}}.$$

The following estimate is frequently used in Chapter 3. It follows immediately from Hölder's inequality.

Lemma A.5. *Let Ω be bounded, $1 \le p \le q \le \infty$, and*

$$\begin{aligned} c_{p,q}(\Omega) &\stackrel{\text{def}}{=} meas(\Omega)^{\frac{q-p}{pq}} && \text{if } p < q < \infty,\\ c_{p,\infty}(\Omega) &\stackrel{\text{def}}{=} meas(\Omega)^{1/p} && \text{if } p < \infty,\\ c_{p,q}(\Omega) &\stackrel{\text{def}}{=} 1 && \text{if } p = q. \end{aligned}$$

Then for all $v \in L^q(\Omega)$ there holds

$$\|v\|_{L^p} \le c_{p,q}(\Omega)\|v\|_{L^q}.$$

Lemma A.6 (Young's inequality). *Assume $a,b \ge 0$, $\eta > 0$, and $p,q \in (1,\infty)$ with $1/p + 1/q = 1$. Then, setting $0^p = 0$, there holds*

$$ab \le \frac{\eta}{p}a^p + \frac{\eta^{-q/p}}{q}b^q.$$

A.3 Elementary Properties of Multifunctions

A multifunction $\Gamma : X \supset V \rightrightarrows Y$ between Banach spaces X and Y assigns to every $x \in V$ a subset $\Gamma(x) \subset Y$ of Y, which can be empty. Γ is called closed-valued (compact-valued, nonempty-valued, etc.) if for all $x \in V$ the image set $\Gamma(x)$ is closed (compact, nonempty, etc.).

Definition A.7. [40, 177] *A multifunction $\Gamma : V \rightrightarrows \mathbb{R}^l$ defined on $V \subset \mathbb{R}^k$ is* upper semicontinuous *at $x \in V$ if for all $\varepsilon > 0$ there exists $\delta > 0$ such that*

$$\Gamma(x') \subset \{z+h : z \in \Gamma(x),\ \|h\| < \varepsilon\} \quad \forall\, x' \in V,\ \|x'-x\| < \delta.$$

Definition A.8. [40, 177] *A multifunction $\Gamma : V \rightrightarrows \mathbb{R}^l$ defined on the measurable set $V \subset \mathbb{R}^k$ is called* measurable [177, p. 160] *if it is closed-valued and if for all closed (or open, or compact* [177, Prop. 1A]*) sets $C \subset \mathbb{R}^l$ the preimage*

$$\Gamma^{-1}(C) = \{x \in V : \Gamma(x) \cap C \neq \emptyset\}$$

is measurable.

The following theorem is important.

Theorem A.9 (Measurable Selection). [40, Thm. 3.1.1] *Let $\Gamma : V \subset \mathbb{R}^k \rightrightarrows \mathbb{R}^l$ be measurable and nonempty-valued. Then there exists a measurable function $\gamma : V \to \mathbb{R}^l$ such that $\gamma(x) \in \Gamma(x)$ for all $x \in V$.*

Further results on set-valued analysis can be found in [15, 40, 177].

A.4 Nemytskii Operators

In this appendix we establish several results on superposition (or Nemytskii) operators involving differentiable outer functions. These results are used in the proof of the continuous differentiability of the merit function $u \mapsto \|\Phi(u)\|_{L^2}^2/2$ in Chapter 7 as well as in the analysis of the nonlinear elliptic control problem in section 9.1. Concerning Nemytskii operators, we also refer to [11, 12, 13].

Proposition A.10. *Let $\Omega \subset \mathbb{R}^n$ be measurable with finite measure and $1 \le p,q < \infty$. Let $f : \mathbb{R}^m \to \mathbb{R}$ be continuous and consider $F(u)(x) = f(u(x))$ for $u \in L^p(\Omega)^m$. Assume that*

$$|f(u)| \le c_1 + c_2\|u\|_2^{p/q} \quad \forall\, u \in \mathbb{R}^m \tag{A.5}$$

with constants $c_i \ge 0$. Then $F : L^p(\Omega)^m \to L^q(\Omega)$ is continuous and bounded with

$$\|F(u)\|_{L^q} \le C_1 + C_2\|u\|_{[L^p]^m}^{p/q},$$

with constants $C_i \ge 0$.

Proof. See [204, Prop. 26.6]. □

Proposition A.11. *Let $\Omega \subset \mathbb{R}^n$ be measurable with finite measure and $1 \le q < p < \infty$. Let $f : \mathbb{R}^m \to \mathbb{R}$ be continuously differentiable and consider $F(u)(x) = f(u(x))$ for $u \in L^p(\Omega)^m$. Assume that*

$$\|f'(u)\|_2 \le c_1' + c_2'\|u\|_2^{\frac{p-q}{q}} \quad \forall\, u \in \mathbb{R}^m \tag{A.6}$$

with constants $c_i' \ge 0$. Then $F : L^p(\Omega)^m \to L^q(\Omega)$ is continuously Fréchet differentiable with $F'(u)v = f'(u)v$.

Proof. We have

$$\begin{aligned} |f(u)| &\le |f(0)| + \int_0^1 |f'(tu)u|dt \le |f(0)| + \|u\|_2 \int_0^1 (c_1' + c_2'\|tu\|_2^{\frac{p-q}{q}})dt \\ &\le |f(0)| + c_1'\|u\|_2 + \frac{c_2'q}{p}\|u\|_2^{\frac{p}{q}} \le c_1 + c_2\|u\|_2^{\frac{p}{q}} \end{aligned}$$

with constants $c_i \ge 0$. Hence, by Proposition A.10, $F : L^p \to L^q$ is continuous. Further, with $r = pq/(p-q)$ there holds

$$\frac{p}{r} = \frac{p-q}{q},$$

so that $u \in L^p(\Omega)^m \to f_{u_i}(u) \in L^r(\Omega)$ is continuous by Proposition A.10. Hence,

$$\|f'(u)v\|_{L^q} \le C\|f'(u)\|_{[L^r]^m}\|v\|_{[L^p]^m},$$

showing that $M(u) : v \in [L^p]^m \mapsto f'(u)v \in L^q$ satisfies $M(u) \in \mathcal{L}([L^p]^m, L^q)$. The estimate

$$\|f'(u_1)v - f'(u_2)v\|_{L^q} \le C\|f'(u_1) - f'(u_2)\|_{[L^r]^m}\|v\|_{[L^p]^m}$$

proves that $M : [L^p]^m \to \mathcal{L}([L^p]^m, L^q)$ is continuous. Further,

$$\begin{aligned} &\|F(u+v) - F(u) - M(u)v\|_{L^q} = \|f(u+v) - f(u) - f'(u)v\|_{L^q} \\ &= \left\| \int_0^1 [f'(u+tv) - f'(u)]v dt \right\|_{L^q} \le \int_0^1 \|[f'(u+tv) - f'(u)]v\|_{L^q} dt \\ &\le \int_0^1 \|f'(u+tv) - f'(u)\|_{[L^r]^m}\|v\|_{[L^p]^m} dt \\ &= o(\|v\|_{[L^p]^m}) \quad \text{as } \|v\|_{[L^p]^m} \to 0, \end{aligned}$$

so that F is continuously Fréchet differentiable with $F' = M$. ☐

Proposition A.12. *Let $\Omega \subset \mathbb{R}^n$ be measurable with finite measure and $1 \le p, q < \infty$, $p > 2q$. Let $f : \mathbb{R} \to \mathbb{R}$ be twice continuously differentiable and consider $F(u)(x) = f(u(x))$ for $u \in L^p(\Omega)$. Assume that*

$$|f''(u)| \le c_1'' + c_2''|u|^{\frac{p-2q}{q}} \tag{A.7}$$

with constants $c_i'' \ge 0$. Then $F : L^p(\Omega) \to L^q(\Omega)$ is twice continuously Fréchet differentiable with

$$F'(u)v = f'(u)v, \quad F''(u)(v,w) = f''(u)vw. \tag{A.8}$$

Proof. As in the proof of Proposition A.11 we obtain constants $c_i' \geq 0$ with

$$|f'(u)| \leq c_1' + c_2'|u|^{\frac{p-q}{q}}.$$

Hence, by Proposition A.11, $F : L^p \to L^q$ is continuously differentiable with derivative $F'(u)v = f'(u)v$. Now consider $g(u) = f'(u)$. From (A.7) and Proposition A.11 we obtain that for $r = pq/(p-q) > q$ the operator

$$G : L^p(\Omega) \to L^r(\Omega), \quad G(u) = g(u(x)) = f'(u(x)),$$

is continuously differentiable with derivative $G'(u)v = g'(u)v = f''(u)v$. Now, define the operator $b(u; v, w) = f''(u)vw$. Then

$$\|b(u;v,w)\|_{L^q} \leq \|f''(u)v\|_{L^r}\|w\|_{L^p} \leq \|G'(u)\|_{L^p,L^r}\|v\|_{L^p}\|w\|_{L^p}.$$

Therefore, $b(u;\cdot,\cdot)$ is a continuous bilinear operator $L^p \times L^p \to L^q$ that depends continuously on $u \in L^p$. Further,

$$\begin{aligned}
\|F'(u+w)v - F'(u)v - b(u;v,w)\|_{L^q} &= \|f'(u+w)v - f'(u)v - f''(u)vw\|_{L^q} \\
&\leq \|f'(u+w) - f'(u) - f''(u)w\|_{L^r}\|v\|_{L^p} \\
&= \|G(u+w) - G(u) - G'(u)w\|_{L^r}\|v\|_{L^p} \\
&= o(\|w\|_{L^p})\|v\|_{L^p} \quad \text{as } \|v\|_{L^p}, \|w\|_{L^p} \to 0.
\end{aligned}$$

This proves that $F : L^p \to L^q$ is twice continuously differentiable with derivatives as in (A.8). □

Proposition A.13. *Let $\Omega \subset \mathbb{R}^n$ be a bounded domain and consider a k times continuously differentiable function $f : \mathbb{R} \to \mathbb{R}$. Then the operator*

$$F : C(\bar{\Omega}) \to C(\bar{\Omega}), \quad F(u)(x) = f(u(x))$$

is k times continuously F-differentiable with rth derivative

$$F^{(r)}(u)(h_1, \ldots, h_r) = f^{(r)}(u)h_1 \cdots h_r.$$

Proof. Consider first $k = 1$. Let $u \in C(\bar{\Omega})$, set $C = \|u\|_{C(\bar{\Omega})} + 1$, and observe that the continuous function $f'(s)$ is uniformly continuous on the compact set $[-C, C]$. Hence, for every fixed $\varepsilon \in (0,1)$, there exists $\delta \in (0,1)$ with $|f'(s_1) - f'(s_2)| < \varepsilon$ for all $s_1, s_2 \in [-C, C]$, $|s_1 - s_2| < \delta$. Now consider $h \in C(\bar{\Omega})$, $\|h\|_{C(\bar{\Omega})} < \delta$. Then, for all $x \in \bar{\Omega}$,

$$\begin{aligned}
&|f(u(x)+h(x)) - f(u(x)) - f'(u(x))h(x)| \\
&\quad \leq \int_0^1 |[f'(u(x)+th(x)) - f'(u(x))]h(x)|\,dt \leq \int_0^1 \varepsilon|h(x)|\,dt \leq \varepsilon\|h\|_{C(\bar{\Omega})}.
\end{aligned}$$

Here, we used that $u(x) + th(x), u(x) \in [-C, C]$, and $|th(x)| < \delta$ for all $t \in [0,1]$ and all $x \in \bar{\Omega}$. Taking the maximum over all x shows that $C(\bar{\Omega}) \ni u \mapsto f(u) \in C(\bar{\Omega})$ is F-differentiable. The derivative at u is given by $h \mapsto f'(u)h$. The continuity of the derivative follows from $|[f'(u(x)+h(x))-f'(u(x))]h_1(x)| < \varepsilon |h_1(x)|$ for all $x \in \bar{\Omega}$ and $h, h_1 \in C(\bar{\Omega})$ with $\|h\|_{C(\bar{\Omega})} < \delta$.

Now consider $k \geq 2$ and assume that the result holds for $1,\ldots,k-1$. We then have that $C(\bar{\Omega}) \ni u \mapsto f^{(k-1)}(u) \in C(\bar{\Omega})$ is continuously F-differentiable with derivative $h \mapsto f^{(k)}(u)h$. Thus, with $u, h_i \in C(\bar{\Omega})$, $1 \leq i \leq k$,

$$\begin{aligned}
&\|f^{(k-1)}(u+h_k)h_1\cdots h_{k-1} - f^{(k-1)}(u)h_1\cdots h_{k-1} - f^{(k)}(u)h_1\cdots h_k\|_{C(\bar{\Omega})} \\
&\quad \leq \|f^{(k-1)}(u+h_k) - f^{(k-1)}(u) - f^{(k)}(u)h_k\|_{C(\bar{\Omega})}\|h_1\|_{C(\bar{\Omega})}\cdots\|h_{k-1}\|_{C(\bar{\Omega})} \\
&\quad = o(\|h_k\|_{C(\bar{\Omega})})\|h_1\|_{C(\bar{\Omega})}\cdots\|h_{k-1}\|_{C(\bar{\Omega})},
\end{aligned}$$

which shows that $u \mapsto f^{(k-1)}(u)h_1\cdots h_{k-1}$ is F-differentiable with derivative as asserted. Also,

$$\begin{aligned}
&\|(f^{(k)}(u+h) - f^{(k)}(u))h_1\cdots h_k\|_{C(\bar{\Omega})} \\
&\quad \leq \|f^{(k)}(u+h) - f^{(k)}(u)\|_{C(\bar{\Omega})}\|h_1\|_{C(\bar{\Omega})}\cdots\|h_k\|_{C(\bar{\Omega})}
\end{aligned}$$

and $\|f^{(k)}(u+h) - f^{(k)}(u)\|_{C(\bar{\Omega})} \to 0$ as $\|h\|_{C(\bar{\Omega})} \to 0$ show that the kth derivative is continuous. $\square$

Bibliography

[1] F. ABERGEL AND R. TEMAM, *On some control problems in fluid mechanics*, Theoret. Comput. Fluid Dyn., 1 (1990), pp. 303–326.

[2] R. A. ADAMS AND J. J. F. FOURNIER, *Sobolev Spaces*, 2nd ed., Pure and Applied Mathematics (Amsterdam) 140, Elsevier/Academic Press, Amsterdam, 2003.

[3] E. L. ALLGOWER, K. BÖHMER, F. A. POTRA, AND W. C. RHEINBOLDT, *A mesh-independence principle for operator equations and their discretizations*, SIAM J. Numer. Anal., 23 (1986), pp. 160–169.

[4] W. ALT, *The Lagrange-Newton method for infinite-dimensional optimization problems*, Numer. Funct. Anal. Optim., 11 (1990), pp. 201–224.

[5] W. ALT, *Parametric optimization with applications to optimal control and sequential quadratic programming*, Bayreuth. Math. Schr., 35 (1991), pp. 1–37.

[6] W. ALT, *Sequential quadratic programming in Banach spaces*, in Advances in Optimization, W. Oettli and D. Pallaschke, eds., Springer, Berlin, 1992, pp. 281–301.

[7] W. ALT, *Discretization and mesh-independence of Newton's method for generalized equations*, in Mathematical Programming with Data Perturbations, A. V. Fiacco, ed., Marcel Dekker, New York, 1998, pp. 1–30.

[8] W. ALT AND K. MALANOWSKI, *The Lagrange-Newton method for nonlinear optimal control problems*, Comput. Optim. Appl., 2 (1993), pp. 77–100.

[9] W. ALT, R. SONTAG, AND F. TRÖLTZSCH, *An SQP method for optimal control of weakly singular Hammerstein integral equations*, Appl. Math. Optim., 33 (1996), pp. 227–252.

[10] H. AMANN, *Compact embeddings of vector-valued Sobolev and Besov spaces*, Glas. Mat. Ser. III, 35 (2000), pp. 161–177.

[11] J. APPELL, *Upper estimates for superposition operators and some applications*, Ann. Acad. Sci. Fenn. Ser. A I Math., 8 (1983), pp. 149–159.

[12] J. APPELL, *The superposition operator in function spaces—a survey*, Expo. Math., 6 (1988), pp. 209–270.

[13] J. APPELL AND P. P. ZABREJKO, *Nonlinear Superposition Operators*, Cambridge University Press, Cambridge, 1990.

[14] N. ARADA, E. CASAS, AND F. TRÖLTZSCH, *Error estimates for the numerical approximation of a semilinear elliptic control problem*, Comput. Optim. Appl., 23 (2002), pp. 201–229.

[15] J.-P. AUBIN AND H. FRANKOWSKA, *Set-valued Analysis*, Birkhäuser Boston, Boston, MA, 1990.

[16] C. BAIOCCHI AND A. CAPELO, *Variational and Quasivariational Inequalities*, John Wiley, New York, 1984.

[17] A. BEN-TAL AND M. ZIBULEVSKY, *Penalty/barrier multiplier methods for convex programming problems*, SIAM J. Optim., 7 (1997), pp. 347–366.

[18] A. BENSOUSSAN AND J.-L. LIONS, *Impulse Control and Quasivariational Inequalities*, Gauthier–Villars, Montrouge, 1984.

[19] M. BERGGREN, *Numerical solution of a flow-control problem: Vorticity reduction by dynamic boundary action*, SIAM J. Sci. Comput., 19 (1998), pp. 829–860.

[20] M. BERGOUNIOUX, M. HADDOU, M. HINTERMÜLLER, AND K. KUNISCH, *A comparison of a Moreau–Yosida-based active set strategy and interior point methods for constrained optimal control problems*, SIAM J. Optim., 11 (2000), pp. 495–521.

[21] M. BERGOUNIOUX, K. ITO, AND K. KUNISCH, *Primal-dual strategy for constrained optimal control problems*, SIAM J. Control Optim., 37 (1999), pp. 1176–1194.

[22] T. BEWLEY, R. TEMAM, AND M. ZIANE, *Existence and uniqueness of optimal control to the Navier-Stokes equations*, C. R. Acad. Sci. Paris Sér. I Math., 330 (2000), pp. 1007–1011.

[23] T. R. BEWLEY, R. TEMAM, AND M. ZIANE, *A general framework for robust control in fluid mechanics*, Phys. D, 138 (2000), pp. 360–392.

[24] S. C. BILLUPS, *Algorithms for Complementarity Problems and Generalized Equations*, Ph.D. thesis, University of Wisconsin, Madison, WI, 1995.

[25] J. F. BONNANS AND C. POLA, *A trust region interior point algorithm for linearly constrained optimization*, SIAM J. Optim., 7 (1997), pp. 717–731.

[26] J. M. BORWEIN AND Q. J. ZHU, *A survey of subdifferential calculus with applications*, Nonlinear Anal., 38 (1999), pp. 687–773.

[27] J. H. BRAMBLE AND J. E. PASCIAK, *Iterative techniques for time dependent Stokes problems*, Comput. Math. Appl., 33 (1997), pp. 13–30.

[28] A. BRANDT AND C. W. CRYER, *Multigrid algorithms for the solution of linear complementarity problems arising from free boundary problems*, SIAM J. Sci. Statist. Comput., 4 (1983), pp. 655–684.

[29] H. BRÉZIS, *Problèmes unilatéraux*, J. Math. Pures Appl., 51 (1972), pp. 1–168.

[30] W. L. BRIGGS, V. E. HENSON, AND S. F. MCCORMICK, *A Multigrid Tutorial*, 2nd ed., SIAM, Philadelphia, PA, 2000.

[31] J. BURGER AND M. POGU, *Functional and numerical solution of a control problem originating from heat transfer*, J. Optim. Theory Appl., 68 (1991), pp. 49–73.

[32] R. H. BYRD, P. LU, J. NOCEDAL, AND C. ZHU, *A limited memory algorithm for bound constrained optimization*, SIAM J. Sci. Comput., 16 (1995), pp. 1190–1208.

[33] R. H. BYRD, J. NOCEDAL, AND R. B. SCHNABEL, *Representations of quasi-Newton matrices and their use in limited memory methods*, Math. Program., 63 (1994), pp. 129–156.

[34] P. H. CALAMAI AND J. J. MORÉ, *Projected gradient methods for linearly constrained problems*, Math. Program., 39 (1987), pp. 93–116.

[35] E. CASAS, R. HERZOG, AND G. WACHSMUTH, *Analysis of an Elliptic Control Problem with Non-differentiable Cost Functional*, Technical report, Faculty of Mathematics, TU Chemnitz, Germany, 2010.

[36] B. CHEN, X. CHEN, AND C. KANZOW, *A penalized Fischer-Burmeister NCP-function*, Math. Program., 88 (2000), pp. 211–216.

[37] B. CHEN AND N. XIU, *A global linear and local quadratic noninterior continuation method for nonlinear complementarity problems based on Chen–Mangasarian smoothing functions*, SIAM J. Optim., 9 (1999), pp. 605–623.

[38] X. CHEN, Z. NASHED, AND L. QI, *Smoothing methods and semismooth methods for nondifferentiable operator equations*, SIAM J. Numer. Anal., 38 (2000), pp. 1200–1216.

[39] X. CHEN, L. QI, AND D. SUN, *Global and superlinear convergence of the smoothing Newton method and its application to general box constrained variational inequalities*, Math. Comp., 67 (1998), pp. 519–540.

[40] F. H. CLARKE, *Optimization and Nonsmooth Analysis*, John Wiley, New York, 1983.

[41] F. H. CLARKE, YU. S. LEDYAEV, R. J. STERN, AND P. R. WOLENSKI, *Nonsmooth Analysis and Control Theory*, Springer-Verlag, New York, 1998.

[42] C. CLASON, B. JIN, AND K. KUNISCH, *A semismooth Newton method for L^1 data fitting with automatic choice of regularization parameters and noise calibration*, SIAM J. Imaging Sci., 3 (2010), pp. 199–231.

[43] S. S. COLLIS, K. GHAYOUR, AND M. HEINKENSCHLOSS, *Optimal transpiration boundary control for aeroacoustics*, AIAA Journal, 41 (2003), pp. 1257–1270.

[44] S. S. COLLIS, K. GHAYOUR, M. HEINKENSCHLOSS, M. ULBRICH, AND S. ULBRICH, *Towards adjoint-based methods for aeroacoustic control*, in 39th Aerospace Science Meeting & Exhibit, Reno, NV, AIAA Paper 2001–0821, 2001.

[45] S. S. Collis, K. Ghayour, M. Heinkenschloss, M. Ulbrich, and S. Ulbrich, *Numerical solution of optimal control problems governed by the compressible Navier-Stokes equations*, in Optimal Control of Complex Structures (Oberwolfach, 2000), K.-H. Hoffmann, I. Lasieka, G. Leugering, J. Sprekels, and F. Tröltzsch, eds., Internat. Ser. Numer. Math. 139, Birkhäuser, Basel, 2002, pp. 43–55.

[46] S. S. Collis, K. Ghayour, M. Heinkenschloss, M. Ulbrich, and S. Ulbrich, *Optimal control of unsteady compressible viscous flows*, Internat. J. Numer. Methods Fluids, 40 (2002), pp. 1401–1429.

[47] S. S. Collis and S. K. Lele, *A Computational Investigation of Receptivity in High-speed Flow Near a Swept Leading-edge*, Technical report TF-71, Department of Mechanical Engineering, Stanford University, Stanford, CA, 1996.

[48] B. D. Craven and B. M. Glover, *An approach to vector subdifferentials*, Optimization, 38 (1996), pp. 237–251.

[49] J. C. De los Reyes, *A primal-dual active set method for bilaterally control constrained optimal control of the Navier-Stokes equations*, Numer. Funct. Anal. Optim., 25 (2005), pp. 657–683.

[50] J. C. de los Reyes and K. Kunisch, *A semi-smooth Newton method for control constrained boundary optimal control of the Navier-Stokes equations*, Nonlinear Anal., 62 (2005), pp. 1289–1316.

[51] J. C. de los Reyes and K. Kunisch, *A semi-smooth Newton method for regularized state-constrained optimal control of the Navier-Stokes equations*, Computing, 78 (2006), pp. 287–309.

[52] T. De Luca, F. Facchinei, and C. Kanzow, *A semismooth equation approach to the solution of nonlinear complementarity problems*, Math. Program., 75 (1996), pp. 407–439.

[53] T. De Luca, F. Facchinei, and C. Kanzow, *A theoretical and numerical comparison of some semismooth algorithms for complementarity problems*, Comput. Optim. Appl., 16 (2000), pp. 173–205.

[54] J. E. Dennis, Jr., and J. J. Moré, *A characterization of superlinear convergence and its application to quasi-Newton methods*, Math. Comp., 28 (1974), pp. 549–560.

[55] J. E. Dennis, Jr., and J. J. Moré, *Quasi-Newton methods, motivation and theory*, SIAM Rev., 19 (1977), pp. 46–89.

[56] J. E. Dennis, Jr., and R. B. Schnabel, *Numerical Methods for Unconstrained Optimization and Nonlinear Equations*, Prentice–Hall Inc., Englewood Cliffs, NJ, 1983.

[57] M. Desai and K. Ito, *Optimal controls of Navier–Stokes equations*, SIAM J. Control Optim., 32 (1994), pp. 1428–1446.

[58] P. DEUFLHARD AND F. A. POTRA, *Asymptotic mesh independence of Newton–Galerkin methods via a refined Mysovskii theorem*, SIAM J. Numer. Anal., 29 (1992), pp. 1395–1412.

[59] P. DEUFLHARD AND M. WEISER, *Local inexact Newton multilevel FEM for nonlinear elliptic problems*, in Computational Science for the 21st Century, M.-O. Bristeau, G. Etgen, W. Fitzgibbon, J.-L. Lions, J. Periaux, and M. Wheeler, eds., Wiley, New York, 1997, pp. 129–138.

[60] S. P. DIRKSE AND M. C. FERRIS, *The PATH solver: A non-monotone stabilization scheme for mixed complementarity problems*, Optim. Methods Softw., 5 (1995), pp. 123–156.

[61] A. L. DONTCHEV, W. W. HAGER, AND V. M. VELIOV, *Uniform convergence and mesh independence of Newton's method for discretized variational problems*, SIAM J. Control Optim., 39 (2000), pp. 961–980.

[62] J. C. DUNN AND T. TIAN, *Variants of the Kuhn–Tucker sufficient conditions in cones of nonnegative functions*, SIAM J. Control Optim., 30 (1992), pp. 1361–1384.

[63] G. DUVAUT AND J.-L. LIONS, *Inequalities in Mechanics and Physics*, Grundlehren der Mathematischen Wissenschaften 219, C. W. John, trans., Springer-Verlag, Berlin, 1976.

[64] B. C. EAVES, *On the basic theorem of complementarity*, Math. Program., 1 (1971), pp. 68–75.

[65] I. EKELAND AND R. TEMAM, *Convex Analysis and Variational Problems*, North–Holland, Amsterdam, 1976.

[66] F. FACCHINEI, A. FISCHER, AND C. KANZOW, *Regularity properties of a semismooth reformulation of variational inequalities*, SIAM J. Optim., 8 (1998), pp. 850–869.

[67] F. FACCHINEI, H. JIANG, AND L. QI, *A smoothing method for mathematical programs with equilibrium constraints*, Math. Program., 85 (1999), pp. 107–134.

[68] F. FACCHINEI AND C. KANZOW, *A nonsmooth inexact Newton method for the solution of large-scale nonlinear complementarity problems*, Math. Program., 76 (1997), pp. 493–512.

[69] F. FACCHINEI AND J. SOARES, *A new merit function for nonlinear complementarity problems and a related algorithm*, SIAM J. Optim., 7 (1997), pp. 225–247.

[70] M. C. FERRIS, C. KANZOW, AND T. S. MUNSON, *Feasible descent algorithms for mixed complementarity problems*, Math. Program., 86 (1999), pp. 475–497.

[71] A. FISCHER, *A special Newton-type optimization method*, Optimization, 24 (1992), pp. 269–284.

[72] A. FISCHER, *Solution of monotone complementarity problems with locally lipschitzian functions*, Math. Program., 76 (1997), pp. 513–532.

[73] M. FUKUSHIMA AND J.-S. PANG, *Some feasibility issues in mathematical programs with equilibrium constraints*, SIAM J. Optim., 8 (1998), pp. 673–681.

[74] A. V. FURSIKOV, *Optimal Control of Distributed Systems: Theory and Applications*, American Mathematical Society, Providence, RI, 2000.

[75] R. GIERING AND T. KAMINSKI, *Recipes for adjoint code construction*, ACM Trans. Math. Software, 24 (1998), pp. 437–474.

[76] V. GIRAULT AND P.-A. RAVIART, *Finite Element Methods for Navier-Stokes Equations*, Springer-Verlag, Berlin, 1986.

[77] B. M. GLOVER AND D. RALPH, *First order approximations to nonsmooth mappings with application to metric regularity*, Numer. Funct. Anal. Optim., 15 (1994), pp. 599–620.

[78] R. GLOWINSKI, *Numerical Methods for Nonlinear Variational Problems*, Springer-Verlag, New York, 1984.

[79] R. GLOWINSKI, J.-L. LIONS, AND R. TRÉMOLIÈRES, *Numerical Analysis of Variational Inequalities*, North–Holland, Amsterdam, 1981.

[80] M. S. GOWDA, *Inverse and implicit function theorems for H-differentiable and semismooth functions*, Optim. Methods Softw., 19 (2004), pp. 443–461.

[81] M. S. GOWDA AND G. RAVINDRAN, *Algebraic univalence theorems for nonsmooth functions*, J. Math. Anal. Appl., 252 (2000), pp. 917–935.

[82] R. GRIESSE AND D. A. LORENZ, *A semismooth Newton method for Tikhonov functionals with sparsity constraints*, Inverse Problems, 24 (2008), article 035007.

[83] A. GRIEWANK, *The local convergence of Broyden-like methods on Lipschitzian problems in Hilbert spaces*, SIAM J. Numer. Anal., 24 (1987), pp. 684–705.

[84] L. GRIPPO, F. LAMPARIELLO, AND S. LUCIDI, *A nonmonotone line search technique for Newton's method*, SIAM J. Numer. Anal., 23 (1986), pp. 707–716.

[85] W. A. GRUVER AND E. SACHS, *Algorithmic Methods in Optimal Control*, Pitman, Boston, 1981.

[86] M. D. GUNZBURGER, L. HOU, AND T. P. SVOBODNY, *Analysis and finite element approximation of optimal control problems for the stationary Navier-Stokes equations with distributed and Neumann controls*, Math. Comp., 57 (1991), pp. 123–151.

[87] M. D. GUNZBURGER, L. S. HOU, AND T. P. SVOBODNY, *Analysis and finite element approximation of optimal control problems for the stationary Navier-Stokes equations with Dirichlet controls*, RAIRO Modél. Math. Anal. Numér., 25 (1991), pp. 711–748.

[88] M. D. GUNZBURGER AND S. MANSERVISI, *The velocity tracking problem for Navier–Stokes flows with bounded distributed controls*, SIAM J. Control Optim., 37 (1999), pp. 1913–1945.

[89] M. D. GUNZBURGER AND S. MANSERVISI, *Analysis and approximation of the velocity tracking problem for Navier–Stokes flows with distributed control*, SIAM J. Numer. Anal., 37 (2000), pp. 1481–1512.

[90] M. D. GUNZBURGER AND S. MANSERVISI, *The velocity tracking problem for Navier–Stokes flows with boundary control*, SIAM J. Control Optim., 39 (2000), pp. 594–634.

[91] W. HACKBUSCH, *Multigrid Methods and Applications*, Springer-Verlag, Berlin, 1985.

[92] W. HACKBUSCH AND U. TROTTENBERG, EDS., *Multigrid Methods*, Springer-Verlag, Berlin, 1982.

[93] W. W. HAGER, *Runge-Kutta methods in optimal control and the transformed adjoint system*, Numer. Math., 87 (2000), pp. 247–282.

[94] S. HARTMANN, S. BRUNSSEN, E. RAMM, AND B. WOHLMUTH, *Unilateral non-linear dynamic contact of thin-walled structures using a primal-dual active set strategy*, Internat. J. Numer. Methods Engrg., 70 (2007), pp. 883–912.

[95] M. HEINKENSCHLOSS, *Mesh independence for nonlinear least squares problems with norm constraints*, SIAM J. Optim., 3 (1993), pp. 81–117.

[96] M. HEINKENSCHLOSS, *Formulation and analysis of a sequential quadratic programming method for the optimal Dirichlet boundary control of Navier-Stokes flow*, in Optimal Control: Theory, Algorithms and Applications, W. W. Hager and P. M. Pardalos, eds., Kluwer Academic Publishers, Dordrecht, NL, 1998, pp. 178–203.

[97] M. HEINKENSCHLOSS, *A trust region method for norm constrained problems*, SIAM J. Numer. Anal., 35 (1998), pp. 1594–1620.

[98] M. HEINKENSCHLOSS AND F TRÖLTZSCH, *Analysis of the Lagrange-SQP-Newton method for the control of a phase field equation*, Control Cybernet., 28 (1999), pp. 177–211.

[99] M. HEINKENSCHLOSS, M. ULBRICH, AND S. ULBRICH, *Superlinear and quadratic convergence of affine-scaling interior-point Newton methods for problems with simple bounds without strict complementarity assumption*, Math. Program., 86 (1999), pp. 615–635.

[100] M. HINTERMÜLLER, *Mesh independence and fast local convergence of a primal-dual active-set method for mixed control-state constrained elliptic control problems*, ANZIAM J., 49 (2007), pp. 1–38.

[101] M. HINTERMÜLLER AND M. HINZE, *Moreau–Yosida regularization in state constrained elliptic control problems: Error estimates and parameter adjustment*, SIAM J. Numer. Anal., 47 (2009), pp. 1666–1683.

[102] M. HINTERMÜLLER, K. ITO, AND K. KUNISCH, *The primal-dual active set strategy as a semismooth Newton method*, SIAM J. Optim., 13 (2003), pp. 865–888.

[103] M. HINTERMÜLLER AND K. KUNISCH, *Feasible and noninterior path-following in constrained minimization with low multiplier regularity*, SIAM J. Control Optim., 45 (2006), pp. 1198–1221.

[104] M. Hintermüller and K. Kunisch, *Path-following methods for a class of constrained minimization problems in function space*, SIAM J. Optim., 17 (2006), pp. 159–187.

[105] M. Hintermüller, F. Tröltzsch, and I. Yousept, *Mesh-independence of semismooth Newton methods for Lavrentiev-regularized state constrained nonlinear optimal control problems*, Numer. Math., 108 (2008), pp. 571–603.

[106] M. Hintermüller and M. Ulbrich, *A mesh-independence result for semismooth Newton methods*, Math. Program., 101 (2004), pp. 151–184.

[107] M. Hinze, *Optimal and Instantaneous Control of the Instationary Navier–Stokes Equations*, Habilitationsschrift, Fachbereich Mathematik, Technische Universität Berlin, 2000.

[108] M. Hinze, *A variational discretization concept in control constrained optimization: The linear-quadratic case*, Comput. Optim. Appl., 30 (2005), pp. 45–61.

[109] M. Hinze and K. Kunisch, *Second Order Methods for Optimal Control of Time-dependent Fluid Flow*, Bericht Nr. 165 des Spezialforschungsbereichs F003 Optimierung und Kontrolle, Karl-Franzens Universität Graz, Austria, 1999.

[110] M. Hinze and K. Kunisch, *Second order methods for optimal control of time-dependent fluid flow*, SIAM J. Control Optim., 40 (2001), pp. 925–946.

[111] M. Hinze, R. Pinnau, M. Ulbrich, and S. Ulbrich, *Optimization with PDE Constraints*, Mathematical Modelling: Theory and Applications 23, Springer-Verlag, Dordrecht, NL, 2008.

[112] D. Hoff, *Discontinuous solutions of the Navier–Stokes equations for multidimensional flows of heat-conducting fluids*, Arch. Ration. Mech. Anal., 139 (1997), pp. 303–354.

[113] R. H. W. Hoppe, *Multigrid algorithms for variational inequalities*, SIAM J. Numer. Anal., 24 (1987), pp. 1046–1065.

[114] R. H. W. Hoppe, *Une méthode multigrille pour la solution des problèmes d'obstacle*, RAIRO Modél. Math. Anal. Numér., 24 (1990), pp. 711–735.

[115] R. H. W. Hoppe and R. Kornhuber, *Adaptive multilevel methods for obstacle problems*, SIAM J. Numer. Anal., 31 (1994), pp. 301–323.

[116] S. Hüeber, G. Stadler, and B. I. Wohlmuth, *A primal-dual active set algorithm for three-dimensional contact problems with Coulomb friction*, SIAM J. Sci. Comput., 30 (2008), pp. 572–596.

[117] S. Hüeber and B. I. Wohlmuth, *A primal-dual active set strategy for non-linear multibody contact problems*, Comput. Methods Appl. Mech. Engrg., 194 (2005), pp. 3147–3166.

[118] A. D. Ioffe, *Nonsmooth analysis: Differential calculus of nondifferentiable mappings*, Trans. Amer. Math. Soc., 266 (1981), pp. 1–56.

[119] K. ITO AND K. KUNISCH, *Augmented Lagrangian methods for nonsmooth, convex optimization in Hilbert spaces*, Nonlinear Anal., 41 (2000), pp. 591–616.

[120] K. ITO AND K. KUNISCH, *Optimal control of elliptic variational inequalities*, Appl. Math. Optim., 41 (2000), pp. 343–364.

[121] K. ITO AND K. KUNISCH, *Semi-smooth Newton methods for state-constrained optimal control problems*, Systems Control Lett., 50 (2003), pp. 221–228.

[122] K. ITO AND K. KUNISCH, *Semi-smooth Newton methods for variational inequalities of the first kind*, M2AN Math. Model. Numer. Anal., 37 (2003), pp. 41–62.

[123] V. JEYAKUMAR, *Simple Characterizations of Superlinear Convergence for Semismooth Equations via Approximate Jacobians*, Applied Mathematics Research Report AMR98/28, University of New South Wales, Sydney, Australia, 1998.

[124] V. JEYAKUMAR, *Solving B-differentiable Equations*, Applied Mathematics Research Report AMR98/27, University of New South Wales, Sydney, Australia, 1998.

[125] V. JEYAKUMAR AND D. T. LUC, *Approximate Jacobian matrices for nonsmooth continuous maps and C^1-optimization*, SIAM J. Control Optim., 36 (1998), pp. 1815–1832.

[126] H. JIANG, M. FUKUSHIMA, L. QI, AND D. SUN, *A trust region method for solving generalized complementarity problems*, SIAM J. Optim., 8 (1998), pp. 140–157.

[127] H. JIANG AND L. QI, *A new nonsmooth equations approach to nonlinear complementarity problems*, SIAM J. Control Optim., 35 (1997), pp. 178–193.

[128] H. JIANG AND D. RALPH, *Smooth SQP methods for mathematical programs with nonlinear complementarity constraints*, SIAM J. Optim., 10 (2000), pp. 779–808.

[129] L. V. KANTOROVICH AND G. P. AKILOV, *Functional Analysis*, 2nd ed., H. L. Silcock, trans., Pergamon Press, Oxford, 1982.

[130] C. KANZOW AND A. KLUG, *An interior-point affine-scaling trust-region method for semismooth equations with box constraints*, Comput. Optim. Appl., 37 (2007), pp. 329–353.

[131] C. KANZOW AND S. PETRA, *Projected filter trust region methods for a semismooth least squares formulation of mixed complementarity problems*, Optim. Methods Softw., 22 (2007), pp. 713–735.

[132] C. KANZOW AND H. PIEPER, *Jacobian smoothing methods for nonlinear complementarity problems*, SIAM J. Optim., 9 (1999), pp. 342–373.

[133] C. KANZOW AND M. ZUPKE, *Inexact trust-region methods for nonlinear complementarity problems*, in Reformulation: Nonsmooth, Piecewise Smooth, Semismooth and Smoothing Methods (Lausanne, 1997), M. Fukushima and L. Qi, eds., Kluwer Academic Publishers, Dordrecht, NL, 1999, pp. 211–233.

[134] C. T. KELLEY AND E. W. SACHS, *A new proof of superlinear convergence for Broyden's method in Hilbert space*, SIAM J. Optim., 1 (1991), pp. 146–150.

[135] C. T. KELLEY AND E. W. SACHS, *Multilevel algorithms for constrained compact fixed point problems*, SIAM J. Sci. Comput., 15 (1994), pp. 645–667.

[136] C. T. KELLEY AND E. W. SACHS, *A trust region method for parabolic boundary control problems*, SIAM J. Optim., 9 (1999), pp. 1064–1081.

[137] N. KIKUCHI AND J. T. ODEN, *Contact Problems in Elasticity: A Study of Variational Inequalities and Finite Element Methods*, SIAM, Philadelphia, PA, 1988.

[138] D. KINDERLEHRER AND G. STAMPACCHIA, *An Introduction to Variational Inequalities and Their Applications*, Academic Press, New York, 1980. Republished as Classics in Applied Mathematics 31, SIAM, Philadelphia, 2000.

[139] R. KORNHUBER, *Monotone multigrid methods for elliptic variational inequalities. I*, Numer. Math., 69 (1994), pp. 167–184.

[140] R. KORNHUBER, *Monotone multigrid methods for elliptic variational inequalities. II*, Numer. Math., 72 (1996), pp. 481–499.

[141] R. KORNHUBER, *Adaptive Monotone Multigrid Methods for Nonlinear Variational Problems*, Teubner, Stuttgart, 1997.

[142] K. KRUMBIEGEL AND A. RÖSCH, *A virtual control concept for state constrained optimal control problems*, Comput. Optim. Appl., 43 (2009), pp. 213–233.

[143] B. KUMMER, *Newton's method for nondifferentiable functions*, in Advances in Mathematical Optimization, J. Guddat et al., eds., Akademie-Verlag, Berlin, 1988, pp. 114–125.

[144] B. KUMMER, *Newton's method based on generalized derivatives for nonsmooth functions: convergence analysis*, in Advances in Optimization, W. Oettli and D. Pallaschke, eds., Springer, Berlin, 1992, pp. 171–194.

[145] K. KUNISCH AND A. RÖSCH, *Primal-dual active set strategy for a general class of constrained optimal control problems*, SIAM J. Optim., 13 (2002), pp. 321–334.

[146] I. LASIECKA AND R. TRIGGIANI, *Regularity theory of hyperbolic equations with nonhomogeneous Neumann boundary conditions*. II. *General boundary data*, J. Differential Equations, 94 (1991), pp. 112–164.

[147] M. LAUMEN, *Newton's mesh independence principle for a class of optimal shape design problems*, SIAM J. Control Optim., 37 (1999), pp. 1070–1088.

[148] C.-J. LIN AND J. J. MORÉ, *Incomplete Cholesky factorizations with limited memory*, SIAM J. Sci. Comput., 21 (1999), pp. 24–45.

[149] C.-J. LIN AND J. J. MORÉ, *Newton's method for large bound-constrained optimization problems*, SIAM J. Optim., 9 (1999), pp. 1100–1127.

[150] J.-L. LIONS, *Optimal Control of Systems Governed by Partial Differential Equations*, S. K. Mitter, trans., Springer-Verlag, New York, 1971.

[151] P.-L. LIONS, *Mathematical Topics in Fluid Mechanics*, Vol. 1, Clarendon Press, New York, 1996.

[152] P.-L. LIONS, *Mathematical Topics in Fluid Mechanics*, Vol. 2, Clarendon Press, New York, 1998.

[153] Z.-Q. LUO, J.-S. PANG, AND D. RALPH, *Mathematical Programs with Equilibrium Constraints*, Cambridge University Press, Cambridge, 1996.

[154] O. L. MANGASARIAN, *Equivalence of the complementarity problem to a system of nonlinear equations*, SIAM J. Appl. Math., 31 (1976), pp. 89–92.

[155] A. MATSUMURA AND T. NISHIDA, *The initial value problem for the equations of motion of viscous and heat-conductive gases*, J. Math. Kyoto Univ., 20 (1980), pp. 67–104.

[156] H. MAURER AND J. ZOWE, *First and second-order necessary and sufficient optimality conditions for infinite-dimensional programming problems*, Math. Program., 16 (1979), pp. 98–110.

[157] G. P. MCCORMICK AND K. RITTER, *Methods of conjugate directions versus quasi-Newton methods*, Math. Program., 3 (1972), pp. 101–116.

[158] C. MEYER, U. PRÜFERT, AND F. TRÖLTZSCH, *On two numerical methods for state-constrained elliptic control problems*, Optim. Methods Softw., 22 (2007), pp. 871–899.

[159] C. MEYER, A. RÖSCH, AND F. TRÖLTZSCH, *Optimal control of PDEs with regularized pointwise state constraints*, Comput. Optim. Appl., 33 (2006), pp. 209–228.

[160] R. MIFFLIN, *Semismooth and semiconvex functions in constrained optimization*, SIAM J. Control Optim., 15 (1977), pp. 959–972.

[161] J. J. MORÉ AND D. J. THUENTE, *Line search algorithms with guaranteed sufficient decrease*, ACM Trans. Math. Software, 20 (1994), pp. 286–307.

[162] T. S. MUNSON, *Algorithms and Environments for Complementarity*, Ph.D. thesis, University of Wisconsin, Madison, WI, 2000.

[163] T. S. MUNSON, F. FACCHINEI, M. C. FERRIS, A. FISCHER, AND C. KANZOW, *The semismooth algorithm for large scale complementarity problems*, INFORMS J. Comput., 13 (2001), pp. 294–311.

[164] L. NIRENBERG, *On elliptic partial differential equations*, Ann. Scuola Norm. Sup. Pisa (3), 13 (1959), pp. 115–162.

[165] P. D. PANAGIOTOPOULOS, *Inequality Problems in Mechanics and Applications: Convex and Nonconvex Energy Functions*, Birkhäuser Boston, Boston, MA, 1985.

[166] J.-S. PANG AND L. QI, *Nonsmooth equations: Motivation and algorithms*, SIAM J. Optim., 3 (1993), pp. 443–465.

[167] H. QI, L. QI, AND D. SUN, *Solving Karush–Kuhn–Tucker systems via the trust region and the conjugate gradient methods*, SIAM J. Optim., 14 (2003), pp. 439–463.

[168] L. QI, *Convergence analysis of some algorithms for solving nonsmooth equations*, Math. Oper. Res., 18 (1993), pp. 227–244.

[169] L. QI, *C-differential Operators, C-differentiability and Generalized Newton Methods*, Research Report AMR96/5, University of New South Wales, Sydney, Australia, 1996.

[170] L. QI AND J. SUN, *A nonsmooth version of Newton's method*, Math. Program., 58 (1993), pp. 353–367.

[171] D. RALPH, *Rank-*1 *Support Functionals and the Rank-*1 *Generalized Jacobian, Piecewise Linear Homeomorphisms*, Ph.D. thesis, University of Wisconsin, Madison, WI, 1990.

[172] D. RALPH, *Global convergence of damped Newton's method for nonsmooth equations via the path search*, Math. Oper. Res., 19 (1994), pp. 352–389.

[173] K. RITTER, *A quasi-Newton method for unconstrained minimization problems*, in Nonlinear Programming 2 (Proc. Special Interest Group Math. Programming Sympos., Univ. Wisconsin, Madison, WI, 1974), O. L. Mangasarian, R. R. Meyer, and S. M. Robinson, eds., Academic Press, New York, 1975, pp. 55–100.

[174] S. M. ROBINSON, *Stability theory for systems of inequalities, part* II: *Differentiable nonlinear systems*, SIAM J. Numer. Anal., 13 (1976), pp. 497–513.

[175] S. M. ROBINSON, *Normal maps induced by linear transformations*, Math. Oper. Res., 17 (1992), pp. 691–714.

[176] S. M. ROBINSON, *Newton's method for a class of nonsmooth functions*, Set-Valued Anal., 2 (1994), pp. 291–305.

[177] R. T. ROCKAFELLAR, *Integral functionals, normal integrands and measurable selections*, in Nonlinear Operators and the Calculus of Variations, Lecture Notes in Math 543, J. P. Gossez, E. J. Lami Dozo, J. Mawhin, and L. Waelbroeck, eds., Springer, Berlin, 1976, pp. 157–207.

[178] R. T. ROCKAFELLAR AND R. J.-B. WETS, *Variational Analysis*, Springer-Verlag, Berlin, 1998.

[179] E. W. SACHS, *Broyden's method in Hilbert space*, Math. Program., 35 (1986), pp. 71–82.

[180] A. SCHIELA, *A simplified approach to semismooth Newton methods in function space*, SIAM J. Optim., 19 (2008), pp. 1417–1432.

[181] S. SCHOLTES, *Introduction to Piecewise Differentiable Equations*, Habilitationsschrift, Institut für Statistik und Mathematische Wirtschaftstheorie, Universität Karlsruhe, Karlsruhe, Germany, 1994.

[182] A. SHAPIRO, *On concepts of directional differentiability*, J. Optim. Theory Appl., 66 (1990), pp. 477–487.

[183] GEORG STADLER, *Elliptic optimal control problems with L^1-control cost and applications for the placement of control devices*, Comput. Optim. Appl., 44 (2009), pp. 159–181.

[184] S. STEFFENSEN AND M. ULBRICH, *A new relaxation scheme for mathematical programs with equilibrium constraints*, SIAM J. Optim., 20 (2010), pp. 2504–2539.

[185] R. TEMAM, *Navier–Stokes Equations: Theory and Numerical Analysis*, 3rd ed., North–Holland, Amsterdam, 1984.

[186] L. THIBAULT, *On generalized differentials and subdifferentials of Lipschitz vector-valued functions*, Nonlinear Anal., 6 (1982), pp. 1037–1053.

[187] PH. L. TOINT, *Global convergence of a class of trust-region methods for nonconvex minimization in Hilbert space*, IMA J. Numer. Anal., 8 (1988), pp. 231–252.

[188] PH. L. TOINT, *Non-monotone trust-region algorithms for nonlinear optimization subject to convex constraints*, Math. Program., 77 (1997), pp. 69–94.

[189] F. TRÖLTZSCH, *An SQP method for the optimal control of a nonlinear heat equation*, Control Cybernet., 23 (1994), pp. 267–288.

[190] M. ULBRICH, *Nonmonotone trust-region methods for bound-constrained semismooth equations with applications to nonlinear mixed complementarity problems*, SIAM J. Optim., 11 (2001), pp. 889–917.

[191] M. ULBRICH, *Nonsmooth Newton-like Methods for Variational Inequalities and Constrained Optimization Problems in Function Spaces*, Habilitationsschrift, Fakultät für Mathematik, Technische Universität München, München, Germany, 2001.

[192] M. ULBRICH, *On a nonsmooth Newton method for nonlinear complementarity problems in function space with applications to optimal control*, in Complementarity: Applications, Algorithms and Extensions, M. C. Ferris, O. L. Mangasarian, and J.-S. Pang, eds., Kluwer Academic Publishers, Dordrecht, NL, 2001, pp. 341–360.

[193] M. ULBRICH, *Semismooth Newton methods for operator equations in function spaces*, SIAM J. Optim., 13 (2002), pp. 805–841.

[194] M. ULBRICH, *Constrained optimal control of Navier-Stokes flow by semismooth Newton methods*, Systems Control Lett., 48 (2003), pp. 297–311.

[195] M. ULBRICH AND S. ULBRICH, *Superlinear convergence of affine-scaling interior-point Newton methods for infinite-dimensional nonlinear problems with pointwise bounds*, SIAM J. Control Optim., 38 (2000), pp. 1938–1984.

[196] M. ULBRICH AND S. ULBRICH, *Non-monotone trust region methods for nonlinear equality constrained optimization without a penalty function*, Math. Program., 95 (2003), pp. 103–135.

[197] M. ULBRICH AND S. ULBRICH, *A Multigrid Semismooth Newton Method for Contact Problems in Linear Elasticity*, preprint, Fakultät für Mathematik, Technische Universität München, 2010.

[198] M. ULBRICH, S. ULBRICH, AND M. HEINKENSCHLOSS, *Global convergence of trust-region interior-point algorithms for infinite-dimensional nonconvex minimization subject to pointwise bounds*, SIAM J. Control Optim., 37 (1999), pp. 731–764.

[199] S. VEELKEN, *A New Relaxation Scheme for Mathematical Programs with Equilibrium Constraints: Theory and Numerical Experience*, Dissertation, Fakultät für Mathematik, Technische Universität München, München, Germany, 2009.

[200] S. VOLKWEIN, *Mesh-independence for an augmented Lagrangian-SQP method in Hilbert spaces*, SIAM J. Control Optim., 38 (2000), pp. 767–785.

[201] P. WESSELING, *An Introduction to Multigrid Methods*, John Wiley, Chichester, UK, 1992.

[202] H. XU, *Set-valued approximations and Newton's methods*, Math. Program., 84 (1999), pp. 401–420.

[203] Y. YANG, D. LI, AND S. ZHOU, *A trust region method for a semismooth reformulation to variational inequality problems*, Optim. Methods Softw., 14 (2000), pp. 139–157.

[204] E. ZEIDLER, *Nonlinear Functional Analysis and Its Applications. II/B*, Springer-Verlag, New York, 1990.

[205] H. ZHANG AND W. W. HAGER, *A nonmonotone line search technique and its application to unconstrained optimization*, SIAM J. Optim., 14 (2004), pp. 1043–1056.

[206] C. ZHU, R. H. BYRD, P. LU, AND J. NOCEDAL, *Algorithm 778: L-BFGS-B: Fortran subroutines for large-scale bound-constrained optimization*, ACM Trans. Math. Software, 23 (1997), pp. 550–560.

[207] W. P. ZIEMER, *Weakly Differentiable Functions: Sobolev Spaces and Functions of Bounded Variation*, Springer-Verlag, Berlin, 1989.

[208] J. ZOWE AND S. KURCYUSZ, *Regularity and stability for the mathematical programming problem in Banach spaces*, Appl. Math. Optim., 5 (1979), pp. 49–62.

Index